Chemical Oceanography

Element Fluxes in the Sea

Chemical Oceanography: Element Fluxes in the Sea focuses on the use of chemical distributions to understand mechanisms of physical, chemical, biological, and geological processes in the ocean. After an introduction describing observed chemical concentrations, chapters focus on using chemical tracers to determine fluxes on a variety of timescales. Long-term chemical cycles are dominated by exchanges between seawater and land, sediments, and underwater volcanoes. Biological and ocean mixing processes dominate internal chemical cycles that respond to changes on hundred- to thousand-year timescales. Stable and radioactive isotopes trace the fluxes of nutrients and carbon to quantify the rates and mechanisms of chemical cycles. Anthropogenic influences – which have grown to be of the same magnitude as some natural cycles – are a specific focus throughout the book. Discussion boxes and quantitative problems help instructors to deepen student learning. Appendices enhance the book's utility as a reference text for students and researchers.

Steven R. Emerson has been a professor of Oceanography at the University of Washington for about 40 years. He taught Chemical Oceanography for most of this period while being the major advisor to 12 Ph.D. students and an equal number of post-docs. His research focuses on fluxes at the air–sea interface and the sediment–ocean interface. He is a fellow of the American Geophysical Union and the Geochemical Society.

Roberta C. Hamme is an associate professor in the School of Earth and Ocean Sciences at the University of Victoria, and holds a Canada Research Chair in Ocean Carbon Dynamics. She has taught upper-level undergraduate Chemical Oceanography since 2007. Her research focuses on understanding and quantifying the natural mechanisms that transport carbon from the surface ocean to the deep. Her main tools are measurements of dissolved gases, both bioactive gases like oxygen and inert gases such as neon, argon, and krypton.

"*Chemical Oceanography: Element Fluxes in the Sea* is completely updated from the previous version. The new version cites up-to-date, peer-reviewed literature, and includes compelling figures, discussion boxes, and problems at the end of each chapter. In addition, the links to MATLAB® and Python® toolboxes are a great resource. In my opinion, this is the best chemical oceanography textbook currently available for both undergraduate and graduate-level courses."

Annie Bourbonnais, University of South Carolina

"The choice of contents for a chemical oceanography textbook is, to some extent, a Rorschach test of the authors' view of the field. In their focus on element fluxes affected by life in the oceans, Emerson and Hamme seek to navigate the narrow channel between attention to detail – a hallmark of quantitative ocean science – and the desire to share the fabulous panorama that is the field as a whole."

Andrew Dickson, Scripps Institution of Oceanography

"In their outstanding and exceptionally well-structured textbook, Emerson and Hamme transform the way we think about chemical oceanography. While the distribution of chemical tracers in the ocean still provides the foundation of their textbook, they organize it around the biogeochemical transformations that govern these distributions. A particular focus is the question of how fast these processes operate and how we can measure these rates. To this end, they introduce many modern techniques involving various isotope systems and transient tracers in a way no other textbook has achieved so far. This is a must-read for any student, postdoc, and researcher in the field, especially in these rapidly changing times."

Nicolas Gruber, ETH Zürich

"A readable, comprehensive, authoritative account, by two distinguished chemical oceanographers, of what we know about chemical processes in the oceans, and how we have learned it. The book features deep descriptions of the oceanic cycles of oxygen, nitrogen, and especially carbon. *Chemical Oceanography* will be valuable to Earth scientists as a guide to topics in chemical oceanography, to specialists as a source of detailed information, and to students as a textbook chock full of stimulating problems and provocative topics for discussion."

Michael Bender, Princeton University

"This new book is a comprehensive and modern treatment of a broad range of marine chemistry topics. The thoughtful, well-written text and clear illustrations are a valuable resource for professors, and provide a strong foundation in the subject for advanced undergraduate and graduate students."

Abigail Renegar, Nova Southeastern University

Chemical Oceanography

Element Fluxes in the Sea

STEVEN R. EMERSON
School of Oceanography
University of Washington, USA

ROBERTA C. HAMME
School of Earth and Ocean Sciences
University of Victoria, Canada

With figures produced by Michael Peterson,
Bainbridge Island, USA

Shaftesbury Road, Cambridge CB2 8EA, United Kingdom

One Liberty Plaza, 20th Floor, New York, NY 10006, USA

477 Williamstown Road, Port Melbourne, VIC 3207, Australia

314–321, 3rd Floor, Plot 3, Splendor Forum, Jasola District Centre, New Delhi – 110025, India

103 Penang Road, #05–06/07, Visioncrest Commercial, Singapore 238467

Cambridge University Press is part of Cambridge University Press & Assessment, a department of the University of Cambridge.

We share the University's mission to contribute to society through the pursuit of education, learning and research at the highest international levels of excellence.

www.cambridge.org
Information on this title: www.cambridge.org/highereducation/9781107179899

DOI: 10.1017/9781316841174

First published 2022

A catalogue record for this publication is available from the British Library

Library of Congress Cataloging-in-Publication data
Names: Emerson, Steven (Steven R.), author. | Hamme, Roberta C., author.
Title: Chemical oceanography : element fluxes in the sea / Steven R. Emerson, Roberta C. Hamme.
Description: Cambridge, United Kingdom ; New York, NY : Cambridge University Press, 2022. | Includes bibliographical references and index.
Identifiers: LCCN 2021029696 | ISBN 9781107179899 (hardcover)
Subjects: LCSH: Chemical oceanography.
Classification: LCC GC111.2 .E64 2022 | DDC 551.46/6–dc23
LC record available at https://lccn.loc.gov/2021029696

ISBN 978-1-107-17989-9 Hardback

Additional resources for this publication at www.cambridge.org/emerson-hamme

Contents

Preface *page* xi
Acknowledgments xiii

1 Oceanography Background: Dissolved Chemicals, Circulation, and Biology in the Sea 1
1.1 The Chemical Perspective 1
1.2 Constituents of Seawater 3
1.2.1 Water in Seawater 3
1.2.2 Ions in Seawater 7
1.2.3 Salinity and Density of Seawater 10
1.2.4 Element Classification 17
1.2.5 Anthropogenic Influences 22
Discussion Box 1.1 24
1.3 Ocean Circulation 25
1.3.1 Seasonality 25
1.3.2 Wind-Driven Circulation 25
1.3.3 Interior Circulation 30
Discussion Box 1.2 34
1.4 Ocean Biology 35
1.4.1 Types of Plankton 35
1.4.2 Marine Metabolism: Estimates of Fluxes 37
Discussion Box 1.3 39
References 40
Problems for Chapter 1 41

2 Geochemical Mass Balance: Chemical Flow across the Ocean's Boundaries 44
2.1 Major Ion Mass Balance: Weathering and Authigenic Mineral Formation 45
2.1.1 The Source: Weathering and River Fluxes to the Ocean 45
2.1.2 Residence Times of Seawater Constituents 47
Discussion Box 2.1 50
2.1.3 Mackenzie and Garrels Mass Balance 50
2.1.4 The Sinks: Hydrothermal Circulation 56
Discussion Box 2.2 65
2.1.5 The Sinks: Reverse Weathering 65
2.2 Gases in the Ocean–Atmosphere System 68
2.2.1 Air–Sea Chemical Equilibrium: Henry's Law Solubility 69
Discussion Box 2.3 73
2.2.2 Gas Sources and Sinks in the Ocean–Atmosphere System 74

Appendix 2A.1 A Brief Review of Rocks and Minerals 76
Appendix 2A.2 The Meaning of Residence Time 78
Appendix 2A.3 The Kinetics of Air–Sea Gas Exchange 80
2A.3.1 The Gas Exchange Flux Equations 80
2A.3.2 Measurements of Gas Exchange Rates in the Ocean 84
2A.3.3 Gas Transfer Due to Bubbles 87
References 90
Problems for Chapter 2 94

3 Life in the Surface Ocean: Biological Production and Export 96
3.1 The Chemistry of Life 97
3.1.1 Redox Processes 97
3.1.2 The Main Elements of Organic Matter: C, H, O, N, P 99
Discussion Box 3.1 106
3.1.3 Trace Elements in Organic Matter: Fe, Zn, Mn, Ni, Cu, Co, Cd 107
Discussion Box 3.2 112
3.2 The Flux of Biologically Produced Elements from the Surface Ocean: The Ocean's Biological Pump 113
3.2.1 A Simplified Whole-Ocean Model of the Biological Pump 114
Discussion Box 3.3 118
3.2.2 Particle Fluxes and Thorium Isotope Tracers 119
3.2.3 Upper Ocean Metabolite Mass Balance: O_2, NO_3^-, DIC, DIC+δ^{13}C-DIC 121
3.2.4 O_2/Ar and O_2/N_2 Tracers 126
3.2.5 Comparing Different Methods for Determining ANCP 128
3.3 Global Distributions of Organic Carbon Export 129
3.3.1 Comparing Measured ANCP with Model Predictions 129
3.3.2 The Anthropogenic Influence: Evidence for Changes in Biological Fluxes 133
Appendix 3A.1 Measurement of Net and Gross Biological Production 134
3A.1.1 Net Primary Production Rates 134
3A.1.2 Gross Primary Production Rates 135
References 136
Problems for Chapter 3 141

4 Life in the Deep Ocean: Biological Respiration 144
4.1 Respiration below the Euphotic Zone 144
4.1.1 Oxygen Concentrations and Apparent Oxygen Utilization (AOU) 147
4.1.2 Nutrient Concentrations and Preformed Nutrients 149
4.1.3 Nitrogen and Phosphorus Cycles 152
Discussion Box 4.1 155
4.2 Respiration Rates 156
4.2.1 Oxygen Utilization Rates (OUR) 156
4.2.2 Interaction of Respiration Rate and Age 159

4.2.3 Relationship between OUR and the Biological Pump 160
4.2.4 Respiration of Particulate and Dissolved Organic Carbon (POC & DOC) 162
4.2.5 Benthic Respiration 162
Discussion Box 4.2 165
4.3 Respiration in the Absence of Oxygen 166
4.4 Anthropogenic Influences 170
References 172
Problems for Chapter 4 174

5 Marine Carbonate Chemistry 177
5.1 Acids and Bases 178
5.1.1 The Chemical Equilibrium Constant 178
5.1.2 Hydrogen Ion Exchange 180
5.1.3 Acids and Bases in Seawater 182
5.1.4 The Alkalinity of Seawater 187
Discussion Box 5.1 191
5.2 Calculating Carbonate Equilibria and pH 191
5.3 Processes that Control Alkalinity and DIC of Seawater 194
5.3.1 Terrestrial Weathering and River Inflow 194
5.3.2 Alkalinity and DIC Changes within the Ocean 195
Discussion Box 5.2 204
5.4 Mechanisms of Calcium Carbonate Dissolution and Burial 205
5.4.1 Thermodynamic Equilibrium 206
5.4.2 The Kinetics of $CaCO_3$ Dissolution 211
5.5 Anthropogenic Influences 214
Discussion Box 5.3 218
Appendix 5A.1 Carbonate System Equilibrium Equations in Seawater 220
References 221
Problems for Chapter 5 223

6 Stable Isotope Tracers 226
6.1 Isotopes in the Environment 227
6.2 Analytical Methods and Terminology 229
6.3 Equilibrium Isotope Fractionation 231
6.3.1 Oxygen Isotopes, $\delta^{18}O$, in $CaCO_3$, a Tracer for Temperature Change 232
Discussion Box 6.1 236
6.3.2 Boron Isotopes, $\delta^{11}B$, a Tracer for pH 241
6.4 Kinetic Isotope Fractionation 244
Discussion Box 6.2 245
6.4.1 $\delta^{13}C$-DIC, a Tracer of Biological Processes 246
6.4.2 Triple Isotopes of Oxygen, a Tracer for Ocean Photosynthesis 249
6.4.3 $\delta^{18}O$ in Molecular Oxygen, a Tracer for Respiration 251

6.4.4 $\delta^{15}N$, a Tracer for the Marine Nitrogen Cycle 255
6.5 Rayleigh Fractionation 259
Discussion Box 6.3 263
6.6 Anthropogenic Influences 263
Appendix 6A.1 Relating the Stable Isotope Terms K, α, ε, and δ 266
Appendix 6A.2 Derivation of the Rayleigh Fractionation Equation 267
References 268
Problems for Chapter 6 271

7 Radioisotope Tracers 274
7.1 Radioactive Decay Mechanisms and Equations 275
7.2 Atmospheric Spallation 278
7.2.1 Natural Carbon-14 278
7.2.2 Beryllium-7 285
Discussion Box 7.1 287
7.3 Uranium and Thorium Decay Series 288
7.3.1 Secular Equilibrium 289
7.3.2 The Geochemistry of Decay Series Isotopes in the Ocean 291
7.3.3 Uranium–Thorium Activities and Ocean Particle Dynamics 294
7.3.4 ^{222}Rn-^{226}Ra Disequilibrium in Surface Waters: A Tracer of Air–Sea Gas Exchange 301
7.3.5 Excess ^{210}Pb and ^{234}Th: Tracers of Coastal Sediment Accumulation 303
7.4 Anthropogenic Influences 304
Discussion Box 7.2 308
References 309
Problems for Chapter 7 311

8 The Role of the Ocean in the Global Carbon Cycle 313
8.1 Carbon Reservoirs and Fluxes 314
Discussion Box 8.1 317
8.2 Natural Ocean Processes Controlling Atmospheric CO_2 317
8.2.1 The Solubility Pump 321
8.2.2 The Biological Pump 321
8.3 Past Changes in Atmospheric CO_2 324
8.3.1 Three-Box Ocean and Atmosphere Model 326
8.3.2 Carbonate Compensation 327
8.4 Anthropogenic Influences 329
8.4.1 Atmospheric CO_2 Observations 330
8.4.2 Anthropogenic CO_2 in the Ocean (the Revelle Factor) 331
8.4.3 The Residence Time of Dissolved Carbon with Respect to Air–Sea Gas Exchange 335
Discussion Box 8.2 336
8.4.4 Measuring Ocean Anthropogenic CO_2 Uptake 338
Discussion Box 8.3 345

8.4.5 The Oxygen Cycle: Carbon's Mirror Image 345
8.4.6 Future Challenges of the Anthropogenic Influence 350
Appendix 8A.1 The Solution to the Three-Box Ocean and Atmosphere Model 351
References 353
Problems for Chapter 8 356

Appendix A **Critical Quantities for the Ocean–Atmosphere System** 358

Appendix B **Fundamental Constants and Unit Conversions** 359

Appendix C **Vapor Pressure of Water** 361

Appendix D **Atmospheric Mole Fractions, Molar Volumes, Saturation Concentrations, and Henry's Law Constants for Gases** 362

Appendix E **Viscosity, Diffusion Coefficients, and Schmidt Numbers** 369

Appendix F **Equilibrium Constants of the Carbonate and Borate Buffer Systems** 375

Appendix G **Apparent Solubility Products of Calcite and Aragonite** 379
Appendix References 381

Index 384

The plate section can be found between pp. 250 and 251.

Preface

The field of Chemical Oceanography is evolving from surveys of chemical distributions toward a focus on deriving element fluxes in the sea. By evaluating fluxes, ocean scientists glean an understanding of mechanisms that underlie circulation, biological processes, air–sea exchange, and interactions between seawater and solids on land and in the ocean basins. Because of the imprint each process leaves on the chemistry of the sea, chemical oceanography in many ways unites the various disciplines of oceanography. As anthropogenic influences on element fluxes grow stronger so must oceanographers' understanding of processes that control them, so that humanity's impact can be predicted and informed decisions can be made about managing the ocean environment to maintain a livable planet.

This book, *Chemical Oceanography: Element Fluxes in the Sea*, is both a text for teaching the subject at the level of senior undergraduates and graduate students and an aid to chemical oceanography researchers. We were inspired to write this successor to *Chemical Oceanography and the Marine Carbon Cycle* by Emerson and Hedges (2008, Cambridge University Press) to better align with the development of our teaching strategies. This new book is more than a second edition to the earlier work. In particular, we have incorporated new materials for teaching. These include Discussion Boxes throughout each chapter, designed to facilitate in-class, small-group student discussions of the material and Problems intended to solidify key concepts using quantitative approaches. These have grown out of our courses for graduate students at University of Washington and primarily undergraduates at University of Victoria. We have also developed an extensive appendix section to provide a go-to reference for key constants.

For this new text, we have reordered and streamlined topics following the way we have found most natural to teach them and to bring greater focus to the interactions between processes and the element fluxes they control. We begin the book with a background on chemical concentrations in the ocean and introductory physical and biological aspects of oceanography that influence chemical distributions (Chapter 1). This chapter sets up some of the key questions we return to throughout the book. After this introduction, long-term geological processes that control the concentrations of major ions and gases are discussed in Chapter 2. The next two chapters (3 and 4) deal with shorter-term fluxes controlled by biological processes in the upper ocean and thermocline – the impact of life on the ocean and how it is quantified. This leads naturally to Chapter 5 on the carbonate system, which combines equilibrium chemical concepts with the impact of life and circulation on ocean carbon. This chapter provides the background necessary for a detailed discussion of the global carbon cycle and the fate of fossil fuel CO_2, with which we conclude the text (Chapter 8). Chapters 6 and 7, on the way from the carbonate system to the carbon cycle, demonstrate applications of stable and radioactive isotopes as tracers of chemical fluxes. This new text contains fewer chapters than the earlier Emerson and Hedges (2008). We have incorporated parts of the earlier book into the new text, and the publisher has agreed to make three of the earlier chapters that are not in the new book freely available online.

As an aid to teaching and research we provide the following material online at the Cambridge University Press website (www.cambridge.org/emerson-hamme): (a) all figures as they appear here, (b) computer code (MATLAB and Python) for determining constants presented in the appendices, and (c) pdf copies of chapters from Emerson and Hedges (2008) that do not appear in this book (Chapters 3, 8, and 9: Thermodynamics Background, Marine Organic Geochemistry, and Molecular Diffusion and Reaction Rates, respectively).

Acknowledgments

We are both to some extent products of the University of Washington (UW) School of Oceanography in Seattle, either as a professor (Emerson) or as a graduate student (Hamme). We would like to acknowledge the influence of UW colleagues – professors, post-docs, technicians, and students – during the time we were there. Because we were Ph.D. thesis advisor and advisee, we knew our talents well enough to feel secure in undertaking the daunting task of writing a textbook together. We each appreciate the support and encouragement we received from the other when the time invested in the book seemed to encroach too greatly on other responsibilities, and the end seemed far away.

Special acknowledgment goes to the late UW Professor John Hedges, who was a co-author of the predecessor to this book (*Chemical Oceanography and the Marine Carbon Cycle*, 2008, by Emerson and Hedges). Many concepts and descriptions that were part of the earlier work were from John, and they remain. This book could not have been completed without the efforts of Michael Peterson, whose hard work produced beautiful figures in record time, and whose knowledge of chemical oceanography and skill in drafting the figures greatly improved the clarity of our explanations.

We are grateful to the students of EOS 312/504 at University of Victoria who experienced early versions of most of the discussion boxes and quantitative problems in this book. Their reactions to these were invaluable. Several colleagues across North America either "test drove" the text in their own courses on this subject or helped to edit some of the chapters. This feedback has been important to the final product and is greatly appreciated. Parts of this book were written on short sabbaticals by Emerson to the writing-friendly environments of the Whiteley Center at Friday Harbor Laboratory of the University of Washington, and Clare Hall College of Cambridge University in Cambridge, England. We would like to thank the editors at Cambridge University Press for their skill in presenting the book. Matt Lloyd, in particular, helped us to solve problems, from the concept at the beginning to the final product.

Finally, we would like to thank our families for their generosity in giving us the time we devoted to creating this textbook, which was sometimes stolen from our personal lives. Steve Emerson would like to thank his wife Julie for her support and love during this project. Roberta Hamme thanks her husband Jody for both his love and his suggestions on physical oceanography, as well as their children, Cordelia and Felix, for enlivening every day.

1 Oceanography Background: Dissolved Chemicals, Circulation, and Biology in the Sea

1.1 The Chemical Perspective 1
1.2 Constituents of Seawater 3
1.2.1 Water in Seawater 3
1.2.2 Ions in Seawater 7
1.2.3 Salinity and Density of Seawater 10
1.2.4 Element Classification 17
1.2.5 Anthropogenic Influences 22
Discussion Box 1.1 24
1.3 Ocean Circulation 25
1.3.1 Seasonality 25
1.3.2 Wind-Driven Circulation 25
1.3.3 Interior Circulation 30
Discussion Box 1.2 34
1.4 Ocean Biology 35
1.4.1 Types of Plankton 35
1.4.2 Marine Metabolism: Estimates of Fluxes 37
Discussion Box 1.3 39
References 40
Problems for Chapter 1 41

1.1 The Chemical Perspective

This book describes a chemical perspective on the science of oceanography. The goal is to understand the mechanisms that control the distributions of chemical compounds in the sea. The "chemical perspective" uses measured chemical distributions to infer oceanic biological, physical, chemical, and geological processes. This method has enormous information potential because of the variety of chemical compounds and the diversity of their chemical behaviors and distributions. It is complicated by the requirement that one must understand something about the reactions and timescales that control chemical distributions. Chemical concentrations in the sea "remember" the mechanisms that shape them over their whole oceanic lifetime. The timescales of important mechanisms range from seconds or less for very rapid photochemical reactions to more than 100 million years for the mineral forming reactions that control relatively unreactive elements in seawater.

The great range in timescales is associated with an equally large range in space scales: from chemical fluxes associated with individual microorganisms to globally distributed processes like river inflow and hydrothermal circulation.

Studies of chemical oceanography have evolved from those focused on discovering what is in seawater and the physical chemical interactions among constituents to those that seek to identify the rates and mechanisms responsible for distributions. While there is still important research that might be labeled "pure marine chemistry" (meaning studies of element speciation, thermodynamics, and kinetics in seawater), much of the field has turned to the chemical perspective described here, resulting in a fascinating array of new research frontiers.

We mention just a few of the exciting areas that are at present mature. The ocean, through its role in the global carbon cycle, is one of the key controls of the fate of anthropogenic CO_2 added to the atmosphere, which has ramifications for global climate and ocean acidification, today and into the future. The importance of dissolved metals in limiting marine biological production has been demonstrated by large-scale iron addition experiments in regions where the surface ocean is rich in phosphorus and nitrate year-round. Investigations of isotopic and chemical tracers in ocean sediments and polar ice cores provide analytical constraints for understanding how the ocean influenced climate during past glacial ages. Chemical alterations associated with hydrothermal processes at mid-ocean ridges are redox reactions that may have led to the origin of life on Earth. These are just a few examples that are relevant to oceanography and the global environment in a field that is continuously developing new research avenues.

Because the science of chemical oceanography is focused on distributions of chemical constituents in the sea, its evolution has been controlled to a large extent by analytical developments. We do not wish to dwell on analytical methods; however, discoveries of new processes often follow the development of better techniques to make accurate measurements. Probably the most recent example has been the evolution of a variety of mass spectrometers capable of precisely determining extremely low concentrations of metals, isotopes, individual organic compounds, and atmospheric gases on small samples. There have been a host of other analytical breakthroughs that are too numerous to mention but that have had a great influence on our ability to interpret the ocean's secrets.

The evolution of analytical methods has been accompanied by increased sophistication and organization in sampling the ocean. International programs that coordinate scientists from the world community have been organized to measure physical and chemical properties of seawater on a much greater scale than could be achieved by individual nations. Research vessels from many countries are used to determine large-scale sections of temperature, salinity, and chemical concentrations in all the oceans every 5 to 10 years. These programs began with the Geochemical Ocean Sections Study (GEOSECS) in the 1970s, and then evolved to the World Ocean Circulation Experiment (WOCE) and Joint Global Ocean Flux Study (JGOFS), both in the 1990s. Successors of these ship-based sampling programs have been CLIVAR and GO-SHIP for monitoring standard physical and chemical properties and GEOTRACES for the study of metal concentrations and their isotopic ratios utilizing new methods of collecting contaminant-free samples. Remote sensing of the ocean from satellites and in situ chemical analyses from autonomous platforms like profiling floats and gliders are being used to determine seasonal changes of a limited number of physical and chemical properties that can be measured in situ in almost any area of the ocean. These autonomous platforms provide the temporal and spatial detail necessary to identify changes in ocean chemistry that result from the growing anthropogenic intrusion into the ocean.

Chemical Oceanography is the most interdisciplinary of the ocean sciences because biological and geological processes change concentrations and ocean circulation distributes them. We begin our book with background information (this chapter) about the chemical constituents of seawater, the basics of ocean circulation, and marine biological processes through which we can begin to highlight some of the important processes and questions to be considered. Subsequent chapters will deal mainly with determining chemical fluxes and the mechanisms that control concentration distributions.

Our discussion of chemical oceanography and the mechanisms controlling chemical fluxes is dominated by natural processes, but the era in which it was possible to consider humanity's influence only a small factor because of the vast nature of the oceans is over. Every subject in the book except for the most long-term processes in ocean–terrestrial exchange (Chapter 2) has a section about Anthropogenic Influences. Most of humanity's effects stem from mining the earth to extract materials used to produce energy and technological devices, and from testing nuclear weapons. At present, the first of these has the greatest influence, and humanity's intrusion into the carbon and oxygen cycles can readily be observed in Chapters 3 through 8. At the time this book is being written, there is a growing presence in the ocean of waste from synthetic products of all types that will also be part of man's influence into the future.

1.2 Constituents of Seawater

1.2.1 Water in Seawater

Water accounts for approximately 96.5 percent of seawater by weight. The innate characteristics of water affect almost all the properties of the ocean (e.g., density, salinity, gas solubility, and heat capacity) and the processes (e.g., circulation, heat exchange, chemical reactions, and biochemical transformations) that occur within it. Water is so much a part of our world and daily lives that it is easy to overlook how unusual this substance is in its physical and chemical properties. A major source of this uniqueness originates at the molecular level where oxygen has a much greater affinity (higher electronegativity) for shared electrons than hydrogen. The molecule H_2O exists because two of the four electron orbitals in the outer shell of oxygen share an electron with hydrogen atoms in covalent bonds. The other two outer-shell oxygen electron orbitals are not bonded with another atom (Fig. 1.1). Because of the strong attraction of the oxygen atom for the shared electrons in the binding electron orbitals, they are slightly positively charged compared to the negative charge of the nonbinding orbitals creating a dipole moment in the water molecule.

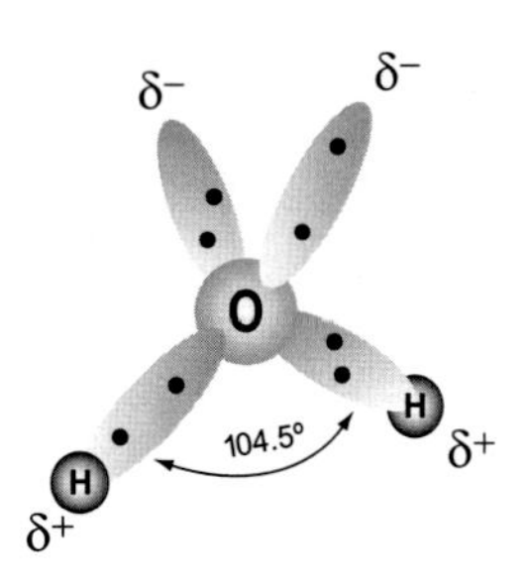

Figure 1.1 The three-dimensional structure of the water molecule. The sphere in the middle represents the oxygen atom, and the darker circles are hydrogen atoms. Ovals represent electron orbitals of the outer shell electrons in a nearly tetrahedral symmetry around the oxygen atom. Lowercase deltas, δs, indicate net charge. From Emerson and Hedges (2008).

The four electron pairs surrounding the oxygen atom tend to arrange themselves as far from each other as possible in order to minimize repulsions between these clouds of negative charge. This would ordinarily result in a tetrahedral geometry in which the angle between electron pairs is 109.5°. In the water molecule, the two nonbinding pairs are closer to the oxygen atom because there are no hydrogen atoms to attract them away from the central oxygen atom. The nonbinding orbitals thus exert a stronger repulsion against the two covalent bonding pairs, creating a distorted tetrahedral arrangement of hydrogen atoms around the oxygen center in which the angle between the two binding electron pairs is 104.5° instead of 109.5° (Fig. 1.1).

This unequal distribution of electronic charge within the H_2O molecule has enormous ramifications for how these molecules interact and react. As a result of their polarity, H_2O molecules orient in solution to minimize the proximity of like partial charges by (on average) turning hydrogens toward oxygens of neighboring molecules. Adjacent H_2O molecules undergo an additional form of intermolecular interaction known as *hydrogen bonding*. Hydrogen bonding can be thought of as an intermolecular attraction between the hydrogen atom in one water molecule and the nonbonding orbitals of the oxygen atom in a different water molecule. Hydrogen bonds are weaker than covalent bonds but are still the strongest type of intermolecular attraction. The resulting intermolecular interaction has a characteristic distance of 1.8 angstroms (1 Å = 10^{-10} cm) between the oxygen of one water molecule and the hydrogen of an adjacent water molecule and exhibits a strength of ~19 kJ mol^{-1}(1 kcal = 4.184 kJ). In contrast, the covalent H–O bond distance is approximately 1.0 Å and has a strength of 463 kJ mol^{-1}. Because the joined H_2O molecules must be aligned along the axes of their orbitals, hydrogen bonding causes H_2O molecules to exist in strongly ordered assemblies.

Pervasive hydrogen bonding is evident in all physical properties that reflect the strengths of association among H_2O molecules in both liquid water and ice. For example, liquid water has the highest heat capacity, heat of vaporization, surface tension, and dielectric constant (ion-insulating capability) of all substances. In addition, the amount of heat necessary to transform solid to liquid water is greater than for any other substance except ammonia. One gram of water absorbs over 3000 J of energy as it is heated from

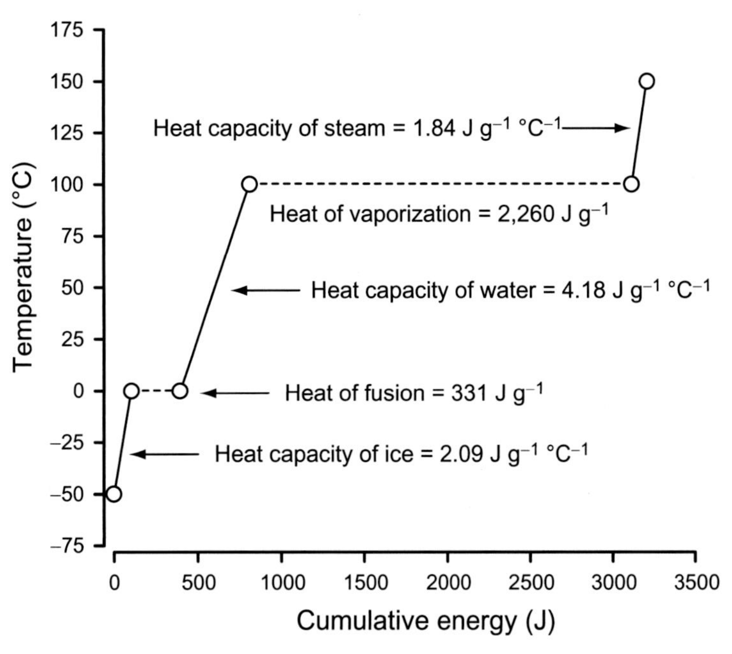

Figure 1.2 The cumulative heat energy in joules (J, x-axis) needed to heat 1 g water from –50 to +150°C. Modified from Libes (1992).

ice at –50°C to steam at +150°C (Fig. 1.2). The same amount of heat would raise the temperature of 10 g of granite (heat capacity ≈ 0.8 J g^{-1}°C^{-1}) to over 380°C. This difference occurs largely because water undergoes two (versus none for granite) phase transitions between –50°C and +150°C and has five times the heat capacity of granite. The large heat of fusion/vaporization and large heat capacity explains why the sea is able to moderate temperature changes on adjacent continental margins, transport huge amounts of heat energy from low to high latitudes in the form of ocean currents and water vapor, and absorb 93 percent of the excess heat attributed to global warming (IPCC, 2013).

The best-defined structural properties of water are at the thermal extremes for vapor and ice. Water vapor consists primarily of individual H_2O molecules with negligible intermolecular attractions holding individual H_2O molecules together. Hence water vapor has essentially no structure. At the other temperature extreme, ice has a remarkably ordered structure. The structure of Ice-I, the stable form at atmospheric pressure, orients individual molecules to maximize hydrogen bonding. In Ice-I (Fig. 1.3), all hydrogen atoms are located along the axes between the oxygen atoms of H_2O molecules that have been stretched into perfect tetrahedra. Full hydrogen bonding can be accomplished only by increasing the spacing of the water structural units in Ice-I, resulting in a solid of roughly half the density (0.92 g cm^{-3}) than would be exhibited for the closest packed structure (1.7 g cm^{-3}).

The specific structure of liquid water is poorly defined, but can be thought of as a slush of ice-like clumps floating in a pool of relatively unassociated H_2O molecules. This type of mixture helps explain many of the maxima and minima in such physical properties as density and viscosity that are often observed when liquid water is cooled or pressurized.

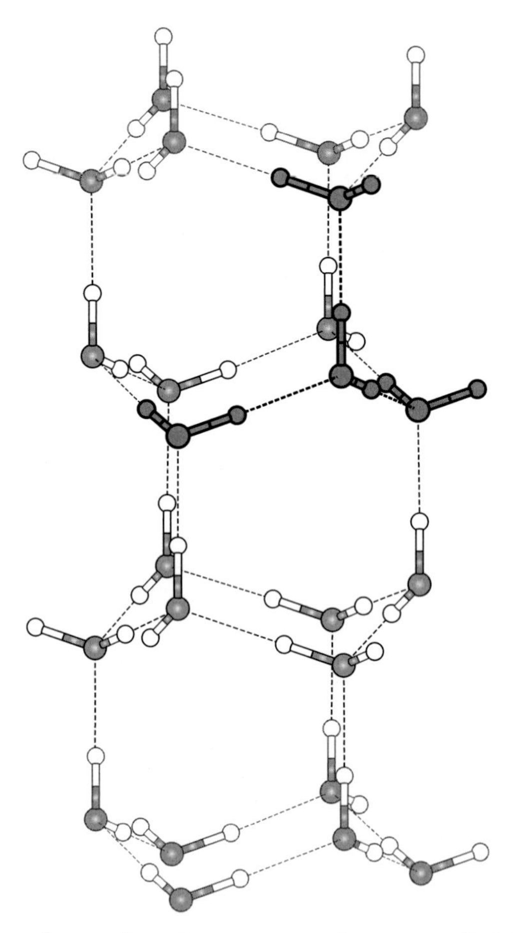

Figure 1.3 The structure of Ice-1, the predominant form of ice at one atmosphere pressure. Shaded spheres indicate oxygen atoms and open spheres hydrogen atoms. The darker shading represents four closest neighbor water molecules. Dashed lines indicate hydrogen bonding. From Emerson and Hedges (2008).

The best known of these trends is that freshwater has its greatest density near 4°C (Fig. 1.4). This phenomenon can be thought of as the net result of opposing density responses by associated and relatively free H_2O molecules to changing temperature. Lowering temperature toward the freezing point increases density by allowing slower moving, free H_2O molecules to pack closer together, but simultaneously decreases density through formation of a greater number of less dense, ice-like clusters. These two opposing effects balance each other near 4°C, below which the formation of ice-like structures predominates. Independent energy-related evidence for a high degree of association among H_2O molecules in the liquid state is that the amount of heat energy given off when liquid water freezes into ice (the heat of fusion) is only ~15 percent of the total that would theoretically be released by forming all the component hydrogen bonds from initially unassociated H_2O molecules. It is as if most liquid H_2O molecules were already in the frozen configuration. This high degree of structure in liquid water is not observed by most

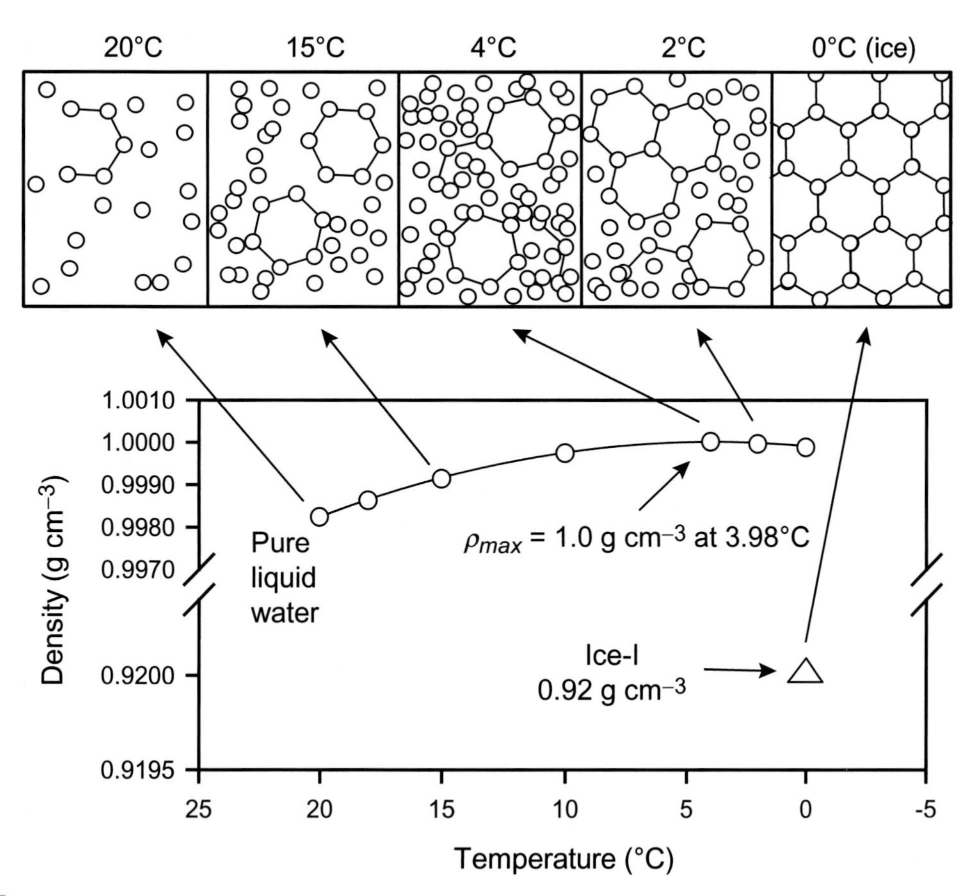

Figure 1.4 The density of liquid freshwater as a function of temperature showing that the maximum density is at 3.98°C. The schematic boxes above the density versus temperature plot indicate the relative proportions of free molecules (unconnected circles) and ordered molecules (circles connected by lines). Modified from Libes (1992).

spectrographic-based measurements, which are too slow to resolve H_2O molecules as they rapidly interconvert between different association patterns.

1.2.2 Ions in Seawater

Given that H_2O molecules have strong dipoles, it is not surprising that they interact even more strongly with charged, dissolved ions than with each other. One of the most evident indications of this strong association is *electrostriction*, the decrease in water volume when charged ions are added. This effect is illustrated in Table 1.1 for the addition of NaCl to water at concentrations similar to those in seawater. In this case, the total volume of the water–ion system shrinks by about 0.5% due to the addition of ~3 wt% NaCl. This contraction results because H_2O molecules immediately surrounding the added ions are more closely packed than in the bulk liquid due to strong ion–dipole interactions (Fig. 1.5). The H_2O molecules of hydration pack with the negative (O) portions of the molecules

Table 1.1 Effect of ions on the density and structure of liquid water

Ingredient	Mass (g)	Density (g cm^{-3})	Volume (cm^3)
H_2O (25°C)	970.78	0.993	973.5
NaCl	29.22	2.165	13.5
Simple sum	1 000.00		987.0
Actual value	1 000.00		982.0
Difference	0		5.0

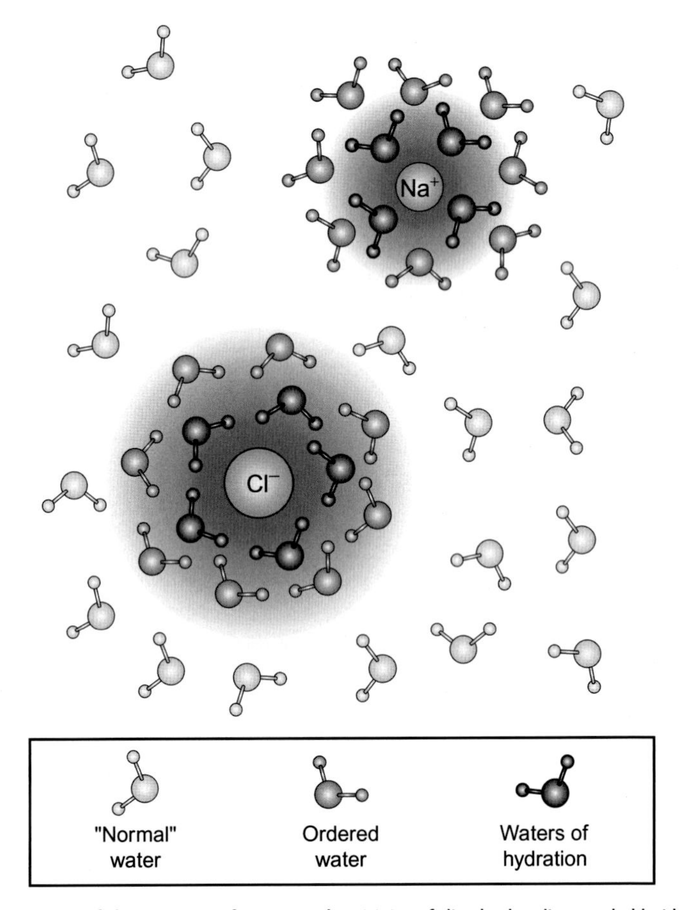

Figure 1.5 Schematic representation of the structure of water in the vicinity of dissolved sodium and chloride ions. Waters of hydration are closest to the ions and indicated by the darkest water molecules. Layers of ordered water molecules outside the waters of hydration are indicated by shaded molecules. From Emerson and Hedges (2008).

facing toward dissolved cations and the positive (H) regions facing toward dissolved anions. This reorientation capability partially nullifies the electrostatic charge of the hydrated ions and contributes toward the great ability of water to insulate, and hence dissolve, ions. Another common effect of dissolving salt in water is an increase in the

Table 1.2 Concentration units in aqueous solution and the atmosphere. Units indicated as "seawater units" are preferred in oceanography. Equivalents, eq, is equal to moles × absolute value of the charge of the species.

Name	Basis	Dimensions	Symbol	Definition
Concentrations in aqueous solution				
Molal	Mass	mol kg^{-1}	*m*	Moles per kilogram of solvent
Molar	Volume	mol L^{-1}	M	Moles per liter of solution
Normal	Volume	eq L^{-1}	N	Equivalents per liter of solution
Weight ratio	Mass	g kg^{-1}		Mass of solute per mass solution
Volume ratio	Volume	ml L^{-1}		Volume solute per volume of solution
Seawater units	Mass	mol kg^{-1}		Moles per kilogram of solution
Seawater units	Mass	eq kg^{-1}		Equivalents per kilogram of solution
Concentrations in the atmosphere				
Mole fraction	Moles	mol mol^{-1}	*X*	Moles gas per moles of dry air (= volume fraction, e.g., ppmv., for ideal gas)
Fugacity	Pressure	Pa, atm	*f*	Effective gas pressure of a single gas (= partial pressure, *p*, for ideal gas)
Partial pressure	Pressure	Pa, atm	*p*	Gas pressure of a single gas

Exponential terminology used in chemical oceanography

Name (symbol)	peta (P)	tera (T)	giga (G)	mega (M)	kilo (k)	milli (m)	micro (μ)	nano (n)	pico (p)	femto (f)	atto (a)
Unit multiplier	10^{15}	10^{12}	10^{9}	10^{6}	10^{3}	10^{-3}	10^{-6}	10^{-9}	10^{-12}	10^{-15}	10^{-18}

viscosity of the solution, because the highly structured ion–dipole assemblies act as "lumps" that inhibit flow. The magnitudes of such electrostriction and viscosity effects increase with the charge density (charge/volume) and number of each type of dissolved ion. These effects are also a function of the temperature of the system. Because liquid water at higher temperatures has less structure, there is more potential for net structural increases due to the presence of ions as temperature increases. At low temperatures, large monovalent ions with low charge densities (e.g., K^+ and Cl^-) can act as "hydrogen bond structure breakers," because the electrostatic interaction of the ions with immediately adjacent H_2O molecules is greater than the strength of the hydrogen-bonded structure of pure water. This effect results in seawater having its greatest density at the freezing point (–1.9°C).

Chemical concentrations in the ocean and atmosphere have been presented over the years in a variety of units, some of which originated in the field of chemistry and others in the field of geology (Table 1.2). Molality, molarity, normality, and volume ratio all have a long history of use in classical chemistry because of their convenience for laboratory preparations. Mass ratios (grams of solute / grams of solution) have been the preferred units in the geological literature. The modern practice in chemical oceanography, labeled "seawater units" in Table 1.2, is to present concentrations in units of moles or equivalents per kilogram of seawater. Moles and equivalents are more meaningful than mass units because reaction stoichiometry in chemical equations is presented on an atomic or molecular basis. An equivalent is equal to the number of moles times the absolute charge of the species and is thus essentially a measure of charge concentration (the difference between molarity and normality in chemistry units). The chemistry convention for equivalents is followed for seawater concentrations for all species except alkalinity, which is often defined as a measure of charge balance in solution, but in some marine literature has units of mol kg^{-1}. This exception is made to be consistent with units when alkalinity and dissolved inorganic carbon (DIC) are added and subtracted. This issue will come up again in Chapter 5 where the carbonate system is described. Mass is used in the denominator of seawater units because it is conservative at all depths of the ocean, whereas volume changes as water moves vertically because it is slightly compressible. Before launching into a detailed discussion of individual constituents, we would like to introduce the total quantity of dissolved material in seawater, salinity, and the processes that determine it.

1.2.3 Salinity and Density of Seawater

Salinity is a measure of the total mass in grams of solids dissolved in a kilogram of seawater, a mass ratio. It is composed almost entirely of elements that do not measurably change because of biological or chemical reactions in the water column. This means that while at the surface evaporation may concentrate the salt ions and precipitation (rain or snow) may dilute them. Once water moves away from the surface, its salt content does not change unless mixed with a different water mass. Thus, the concentrations of individual chemical species can be compared to salinity in order to determine their reactivity in the sea; conservative (unreactive) elements are ones that have constant or nearly constant ratios to salinity everywhere in the ocean. Relatively small changes in salinity are important in determining the density of seawater and thermohaline circulation. Salinity is also used to trace the mixing of different water masses, since salinity values are set at the ocean's surface, and can be traced for great distances within the ocean interior.

For all these reasons it is essential to have a means of rapidly and accurately measuring seawater salinity. The obvious method would be to dry seawater and weigh the leftover residue. This approach does not work very well because high temperatures (≈500°C) are required to drive off the tightly bound water in salts like magnesium chloride and sodium sulfate. At these temperatures, some of the salts of the halides (bromide and iodide) are volatile and are lost; magnesium and calcium carbonates react to form oxides releasing CO_2; and some of the hydrated calcium and magnesium chlorides decompose, giving off HCl gas. The end result of weighing the dried salts is that you come up "light," because some of the volatile elements are gone. While there were schemes created to overcome such problems, for many years the preferred method for determining salinity was titration of the chloride ion using silver nitrate,

$$Ag^+ + Cl^- \rightleftarrows AgCl(s), \tag{1.1}$$

which is quantitative. The chloride concentration, $[Cl^-]$, is related to salinity via a constant

$$S(\text{ppt}) = 1.80655 \times [Cl^-](\text{ppt}), \tag{1.2}$$

where ppt indicates parts per thousand ($g_{solute}\ kg_{seawater}^{-1}$). The titration method is time consuming and not accurate enough to determine small changes in the deep ocean, so the determination of salinity by titration gave way to more automated methods.

Salinity is at present determined by measuring the conductance of seawater using a salinometer. The modern definition of salinity uses the *Practical Salinity Scale* (PSS-78), which replaces the chlorinity–salinity relationship with a definition based on a conductivity ratio (Lewis, 1980). A seawater sample with a salinity on the practical salinity scale (S_P) of 35 ($S_P = 35$) has a conductivity equal to a KCl solution containing a mass of 32.4356 g KCl in 1 kg of solution at 15°C and 1 atm pressure. No units are necessary on the Practical Salinity Scale; however, in practice, one often sees the abbreviation "psu," meaning practical salinity units, or the designation PSS-78, meaning that the number is expressed on the 1978 practical salinity scale. Salinometers using this method are capable of an extremely high precision of 0.001 out of a typical salinity of 35.000. This is about 1 part in 35 000 – much better than most chemical titrations, which, at best, achieve routine accuracy of 1 part in 2000.

In the first decade of the twenty-first century, the international community has suggested improvements on PSS-78, so the definition of salinity is currently (at the time of writing this book) undergoing another change. First, the salinity of the standard of Atlantic surface seawater used to define PSS-78 (Standard Seawater with $S_P = 35$) now has a more accurate, independently determined composition. Second, we now know there are small but significant geographic variations of dissolved inorganic matter in seawater that are not detected by conductivity changes (Pilson, 2013; Talley et al., 2011). The new salinity scale recommended in 2010 by the IOC (Intergovernmental Oceanographic Commission), SCOR (Scientific Committee on Ocean Research), and IAPSO (International Association of the Physical Sciences of the Oceans) is called TEOS-10. The new salinity scale suggested for TEOS-10, the absolute salinity (S_A), is related to the practical salinity scale by

$$S_A = S_R + \delta S_A = \left(35.16504\ \mathrm{g\ kg^{-1}}/35\right) S_P + \delta S_A, \tag{1.3}$$

where S_R is the reference salinity, which is currently the most accurate estimate of the absolute salinity of reference Atlantic surface water, and δS_A is a geographically dependent anomaly added to correct for dissolved substances that influence density but not conductivity. The mean absolute value of the δS_A correction is 0.0107 and it ranges from zero in the surface Atlantic to 0.025 g/kg in the deep North Pacific. A geographic lookup table based on archived measurements is used to estimate the anomaly (McDougall et al., 2010). (Find out more on the TEOS-10 website; www.teos-10.org.) The authors of TEOS-10 recommend that, in order to maintain a consistent global salinity data set, observations should continue to be made based on conductivity and PSS-78 and reported in national archives as salinity on the PSS-78 scale. Throughout this book we refer to salinities that are on the Practical Salinity Scale (PSS-78) and densities that are calculated using PSS-78; however, for the most accurate calculations of density where salinity changes are small, the TEOS-10 scale should be used.

The distributions of temperature, salinity, and density in surface waters of the ocean are presented in Fig. 1.6 and sections of salinity and temperature throughout the three main ocean basins in Fig. 1.7. Salinity distributions are controlled by evaporation and

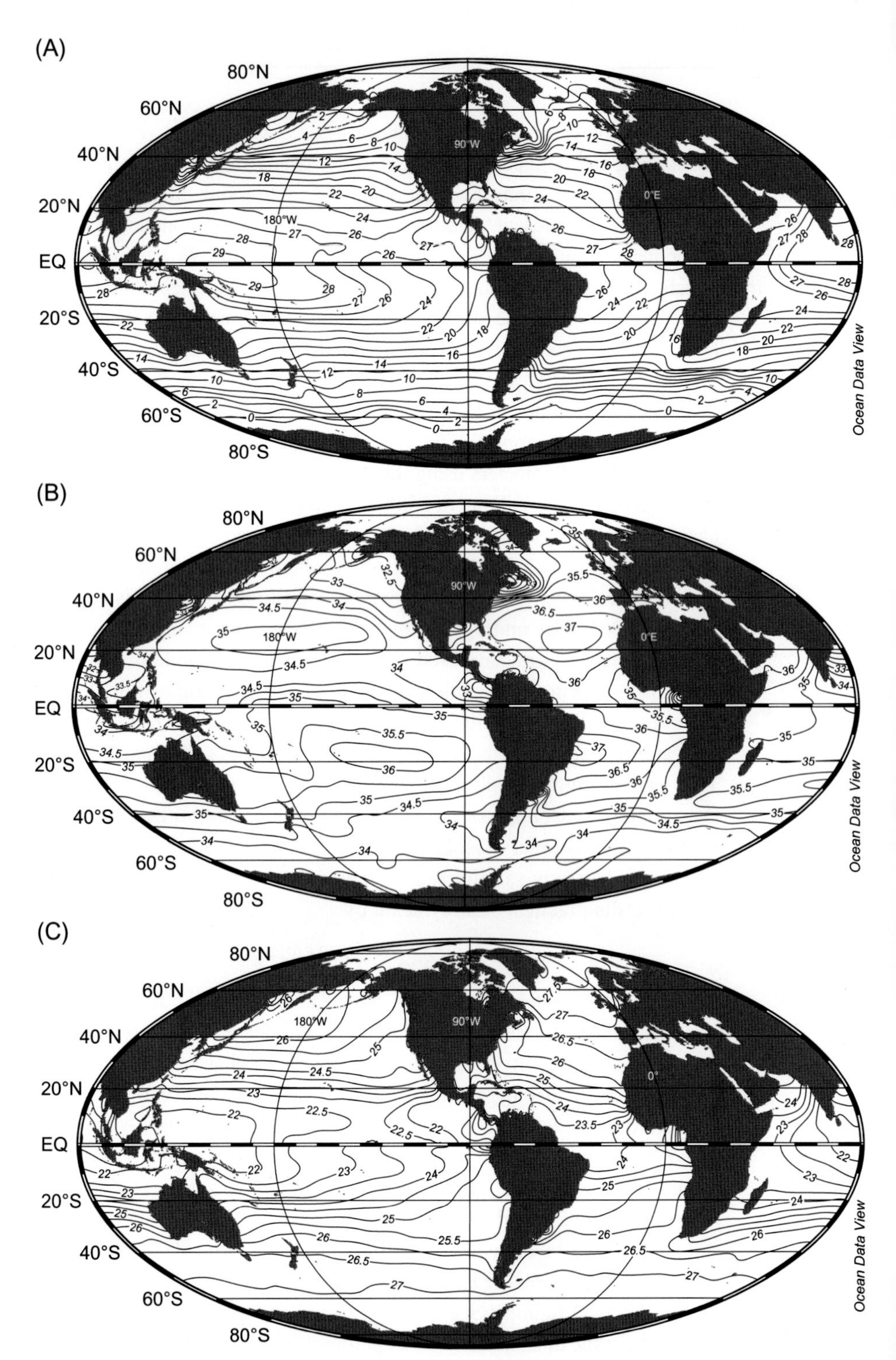

Figure 1.6 Surface ocean (A) temperature (°C), (B) salinity (PSS-78), and (C) density (σ_θ) during the winter months. Values for January, February, and March are used for the Northern Hemisphere and July, August, and September for the Southern Hemisphere, so that contours illustrate where the densest surface waters occur. The map is made from 1.0 degree-gridded data from the World Ocean Atlas 2013 (WOA13). This and other figures of ocean distributions are plotted using Ocean Data View (Schlitzer, 2021; https://odv.awi.de).

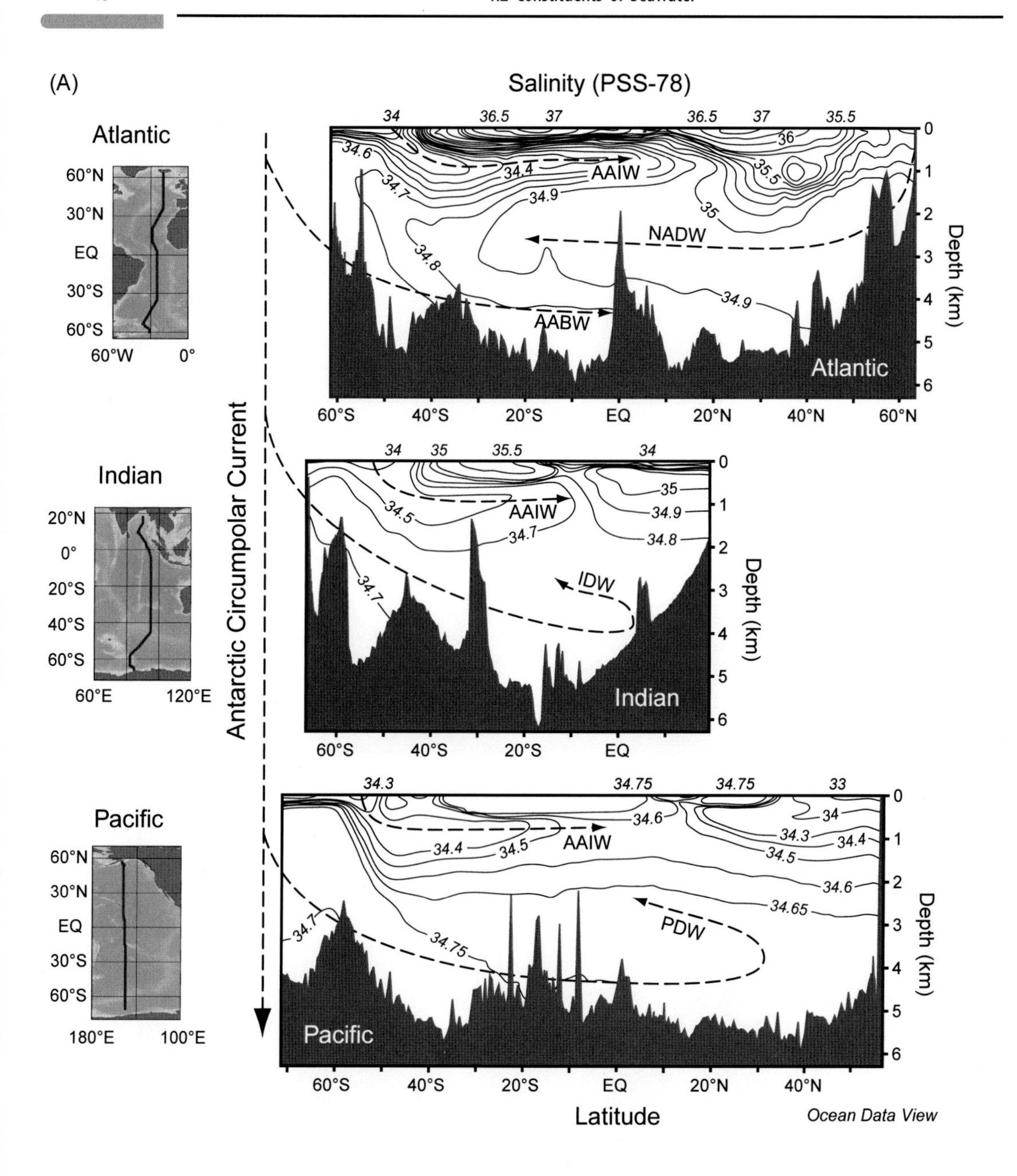

Figure 1.7 Cross sections of (A) salinity (PSS-78) and (B) temperature (°C) in the Atlantic, Indian, and Pacific Oceans. Dashed lines indicate the circulation path of Antarctic Intermediate Water (AAIW), North Atlantic Deep Water (NADW), Antarctic Bottom Water (AABW), Indian Deep Water (IDW), and Pacific Deep Water (PDW). Data are from Climate and Ocean - Variability, Predictability and Change (CLIVAR) Repeat sections P16N & P16S for the Pacific, A16N & A16S for the Atlantic, and I09N and I08S for the Indian Ocean. (Data are compiled on the Ocean Carbon Data System, OCADS website: www.ncei.noaa.gov/products/ocean-carbon-data-system/.)

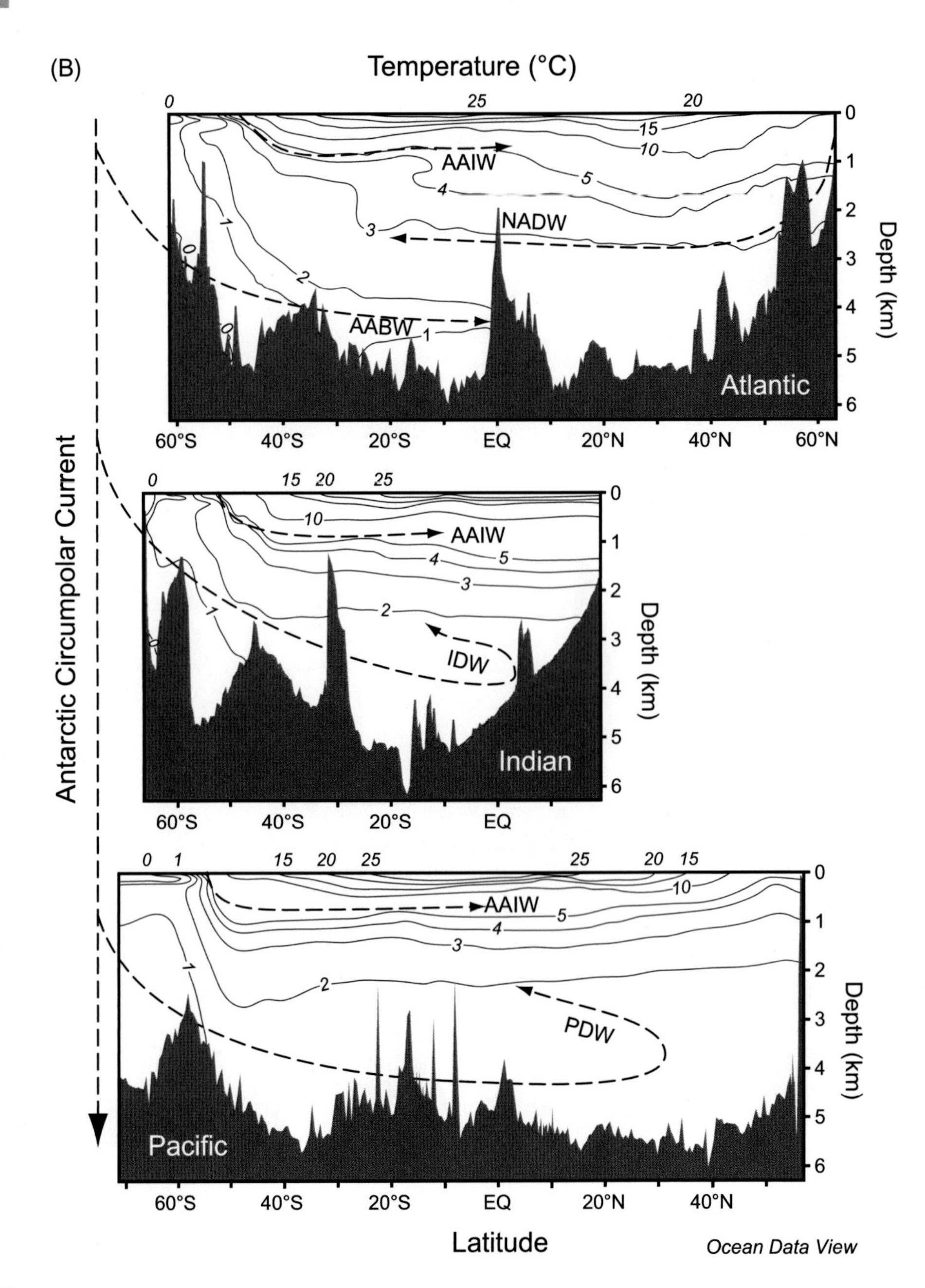

Figure 1.7 (*cont.*)

precipitation, because most major seawater constituents undergo little chemical reaction on the timescale of ocean circulation. This balance is reflected in low salinities in equatorial regions that result from extensive rain due to rising atmospheric circulation (atmospheric lows) and high salinities in the hot, dry subtropical gyres that flank the Equator to the north and south (20–35 degrees latitude) where the atmospheric circulation cells descend (atmospheric highs).

Salinity and temperature are the primary factors that determine the density of seawater. The densities of most surface seawaters range from 1024–1028 kg m^{-3}, and it is possible to evaluate density to about ±0.001 of these units. In order to avoid writing numbers with many significant figures, density is usually presented as the Greek letter sigma, σ, which has the following definition:

$$\sigma = (\rho/\rho_0 - 1) \times 1000, \tag{1.4}$$

where ρ is the density of the sample (kg m^{-3}) and ρ_0 is the maximum density of water at 3.9°C (999.974 kg m^{-3}). (Note that the numerical value of this expression is only slightly different from the shortened form of the definition, $\sigma = \rho - 1000$.) Density is calculated from temperature, salinity, and pressure (because of the compressibility of water) using the international equation of state of seawater (Millero et al., 1980). The expression above represents the in situ density of a seawater sample determined from the measured temperature, salinity, and pressure. Since water largely acquires its temperature and salinity while at the ocean surface, it is convenient to know the temperature and density a water mass would have if brought to the surface at 1 atmosphere pressure. Potential temperature, θ, is in situ temperature corrected for heat added due to compression of water at depth by hydrostatic pressure. This is the temperature a water sample would have if it were raised to the surface with no exchange of heat with the surroundings, i.e., if it changed pressure adiabatically. Density determined from potential temperature and salinity is potential density, which is indicated by sigma with a subscript θ, σ_θ. Calculation of σ_θ accounts for both the direct compression of water by pressure and the heating caused by the compression.

While all references to density in this book use σ_θ, there is another level of correction for density called *neutral density*. The reason for this further refinement is to accurately follow the most likely flow of water and its dissolved constituents as it moves from the surface ocean into the interior. For the most part, water parcels mix and advect adiabatically (i.e., without external input of heat or salt) along surfaces of constant density, which are only approximated by constant potential density. The complication is that seawater compressibility is a function of temperature, such that cold water is more compressible than warm water. Thus, if two water parcels with the same surface density but different temperature and salinity were submerged to the same pressure, the colder parcel would be denser. Neutral density surfaces are calculated as a function of latitude, longitude, and pressure (Jackett and McDougall, 1997) by taking into consideration the compressibility of water at different temperatures and salinities.

Returning to the large-scale salinity and temperature distributions in the surface ocean (Fig. 1.6), one observes that surface temperature differences in the sea are large, ranging between 30°C in equatorial surface waters to –2°C in waters that are in contact with ice. By comparison, salinity is remarkably constant with S_P = 33.0–37.0, necessitating very accurate measurements in order to distinguish differences. The average temperature and salinity of the sea are 3.50°C and S_P = 34.72, and 75 percent of all seawater is within ±4°C and ±0.3 salinity units of these values.

Note that the surface North Atlantic is nearly two salinity units saltier than the North Pacific. At first this seems counterintuitive, because more large rivers drain into the Atlantic. The reason for this difference has to do with the relative rates of evaporation in the high latitudes of the two oceans. The surface North Atlantic is on average warmer (~10.0°C) than the North Pacific (~6.7°C). Warmer water leads to warmer air, which can hold more water vapor. The warmer air thus increases evaporation and consequently salinity. The temperature difference is due to the warm Gulf Stream waters that flow north along the east coast of North America having a greater impact at high latitudes than its Pacific counterpart, the Kuroshio current. The resulting salinity difference has very important consequences for the nature of global thermohaline circulation. Because salt content influences the density of seawater, the higher salt content of North Atlantic surface waters gives them greater densities at any given temperature than North Pacific waters. This is the main reason for massive downwelling, all the way to the ocean bottom, in the North Atlantic where the water is cold and salty, but no deep-water formation in the North Pacific. The major deep-water masses of the ocean are identified by their temperature and salinity (Table 1.3) and depicted by the streamlines in Fig. 1.7. Notice that there is no North Pacific Deep Water in Table 1.3, and no water mass moving down from the surface North Pacific in Fig. 1.7.

Our explanation for the surface salinity differences between the Atlantic and Pacific so far does not provide the whole story, because it overlooks the need to budget atmospheric water transport on a global basis. In fact, the only way to cause a net salinity change in an ocean due to evaporation is via net transport of water vapor to another region on a timescale that is short with respect to the residence time (decades to centuries) of the surface water in question. Simply removing the vapor from an ocean to the atmosphere or to an adjacent landmass is insufficient if that same water rapidly returns to the source ocean. To create a salinity difference between oceans, water must be removed across a continental divide so that it either precipitates directly on another ocean or into the drainage basin of a river discharging into another ocean.

This budgetary constraint makes global salinity patterns the net result of local evaporation, wind patterns, and continental placement and topography. An ideal "vapor export window" from an ocean would be through a region where initially dry prevailing winds blow continuously over warm ocean surface waters and then across a low continental divide. Inspection of the North Atlantic Ocean shows such a window at about 20°N, where

Table 1.3 Mean potential temperature, salinity, and flow rate of major deep ocean water masses that originate near the surface. (Talley et al., 2011)

Water mass	Potential temperature (°C)	Salinity (PSS-78)	Flow estimate (Sverdrups[d])
AABW[a]	–0.7	34.65	~10
NADW[b]	2.3	34.9	15–25
AAIW[c]	3.5	34.1	4–6

[a] AABW is Antarctic Bottom Water
[b] NADW is North Atlantic Deep Water
[c] AAIW is Antarctic Intermediate Water
[d] The Sverdrup unit is 10^6 m^3 s^{-1}

the northeasterly trade winds blow westward across the Sahara, subtropical Atlantic, and then over the relatively low continental divide of Central America. The surface Atlantic Ocean expresses its highest salinity (~37.5) at this latitude, and high rainfall over western Panama and Costa Rica indicates substantial vapor export to the subtropical Pacific. In contrast, the expansive subtropical Pacific Ocean has few upwind deserts and a trade wind window that is effectively blocked by Southeast Asia. Thus, the percentage of net water loss by atmospheric circulation is much less in the larger Pacific.

This global perspective suggests that the North Atlantic Ocean is now saltier than the North Pacific as a result of the present distribution of ocean and atmosphere currents and continents over the surface of the Earth. Other distributions of ice, deserts, or continental topography, as occurred in the past, would produce very different water balances and global current systems.

Cross sections of the potential temperature and salinity of the Atlantic and Pacific Oceans (Fig. 1.7) demonstrate how water masses can be identified with distinct salinities and temperatures that are determined at the surface ocean in the area of their formation (Table 1.3). The deepest waters, Antarctic Bottom Water (AABW) and North Atlantic Deep Water (NADW), are formed at the surface in polar regions. AABW is denser because it is formed under the ice in the Weddell and Ross Seas near Antarctica and, thus, is extremely cold. NADW is not particularly cold, but is highly saline because of its source waters from the Gulf Stream and high evaporation rates in the North Atlantic. Antarctic Intermediate Water (AAIW) is both warmer and less saline than either of the deeper-water masses and thus spreads out in the ocean at shallower depths of about 1000 meters.

1.2.4 Element Classification

The shape of the dissolved concentration-depth profile of a chemical species provides major clues to which processes and reactions control it. Most chemical measurements in seawater are made on unfiltered samples, but experiments designed to operationally define "dissolved" and "particulate" phases using 0.45 μm pore-size filters indicate that, especially below the surface 100 m, nearly all the total mass of any element is in the dissolved phase. The reactivity of a chemical species determines the shape of its concentration-depth profile and its relative concentration in the major ocean basins. *Conservative* elements have a constant concentration:salinity ratio everywhere in the ocean. This means they are unreactive on the timescale of ocean circulation. Elements that are *biologically active* are relatively low in concentration in the surface euphotic zone where they are removed from solution by photosynthesis. When the biological material dies, sinks, and degrades in deeper water, concentrations are elevated. Deep waters become progressively more enriched in biologically active elements as they flow from the Atlantic to the Indian and Pacific Oceans, which is sometimes called "conveyor belt" circulation. *Adsorbed* elements are dominated by another type of reaction removing dissolved species from seawater, adsorption or scavenging of elements onto particles as they sink. An adsorbed species is relatively high in concentration in surface waters and decreases with depth, because it is removed from the dissolved phase by adsorption onto particles that originate in surface waters. In some cases, the surface ocean is enriched because of input of these elements by windblown dust. The concentration of adsorbed species decreases in deep waters along the conveyor belt circulation from the Atlantic to the Indian and Pacific Oceans as the dissolved concentration is continually removed by adsorption to particles sinking to the sea floor. An element controlled by adsorption has concentration distributions that are the mirror image of a biologically active element. Finally, *anthropogenically altered* elements are those whose distribution is significantly different now than in the past due to human activities. These are usually marked by upper water column enrichment, because these waters have been most recently in contact with the surface, where much of the input occurs.

Profiles of elements with concentration distributions that follow the conservative, biologically active, adsorbed, and anthropogenically altered categories are presented in Fig. 1.8. Magnesium (Fig. 1.8A) is conservative on the timescale of ocean circulation and has a constant Mg:salinity ratio. The small changes above 1000 meters in the profile are due to identical variations in salinity at these locations. Dissolved inorganic phosphorus (DIP, Fig. 1.8B) has a concentration distribution dominated by biological activity, where surface waters are depleted, and deep-water concentrations increase from the Atlantic to the Pacific Ocean as these waters are continuously exposed to sinking organic matter particles that are respired. Aluminum concentrations (Fig. 1.8C) in the ocean are typical for an adsorbed element. Dust input increases the concentration at the surface. Increasing exposure to sinking particles removes the concentration from seawater, lowering the concentration from the Atlantic to Pacific – precisely the opposite of that for biologically active species. Finally, the lead profile (Fig. 1.8D) is almost entirely anthropogenically altered except for values in the deepest waters. Because Pb was added to the ocean by fallout of pollution from the atmosphere, concentrations are higher in surface waters. Our ability to measure Pb in the ocean came years after the pollution of the atmosphere with tetraethyl lead in gasoline began, so it is difficult to know background Pb concentrations before the anthropogenic alteration. An approximation is that the Pb profile in the South Pacific represents "uncontaminated" background levels, and that the values in the Atlantic are altered because this is where much of the pollution Pb entered the ocean.

Conservative Elements. To within a few percentage points, conservative ions in seawater have constant concentration:salinity ratios. That is, their concentrations are not greatly affected by processes other than precipitation and evaporation– the same processes that control salinity in the ocean. Elements of high concentration tend to be conservative because they are relatively unreactive; however, conservative elements are present in all concentration ranges because some of them are both low in crustal abundance and relatively unreactive.

We define the major ions in seawater (Table 1.4) as those with concentrations greater than 10 μmol kg^{-1}. Most of the major ions are conservative (exceptions are Sr^{2+}, HCO_3^-, and CO_3^{2-}), and these ions make up more than 99.4 percent of the mass of dissolved solids in seawater. Ions Na^+ and Cl^- account for 86 percent and Na^+, Cl^-, SO_4^{2-}, and Mg^{2+} make up 97 percent. Elements with concentration profiles indicative of conservative behavior are identified by a bold border in the periodic table of Fig. 1.9. Conservative elements with concentrations less than 10 μmol kg^{-1} are found in rows 5 and 6 of the periodic table where elements with lower crustal abundances occur.

There are of course some caveats to our classification of conservative ions. Ca^{2+}, Mg^{2+}, and Sr^{2+} are not strictly conservative, as small changes on the order of 1 percent in

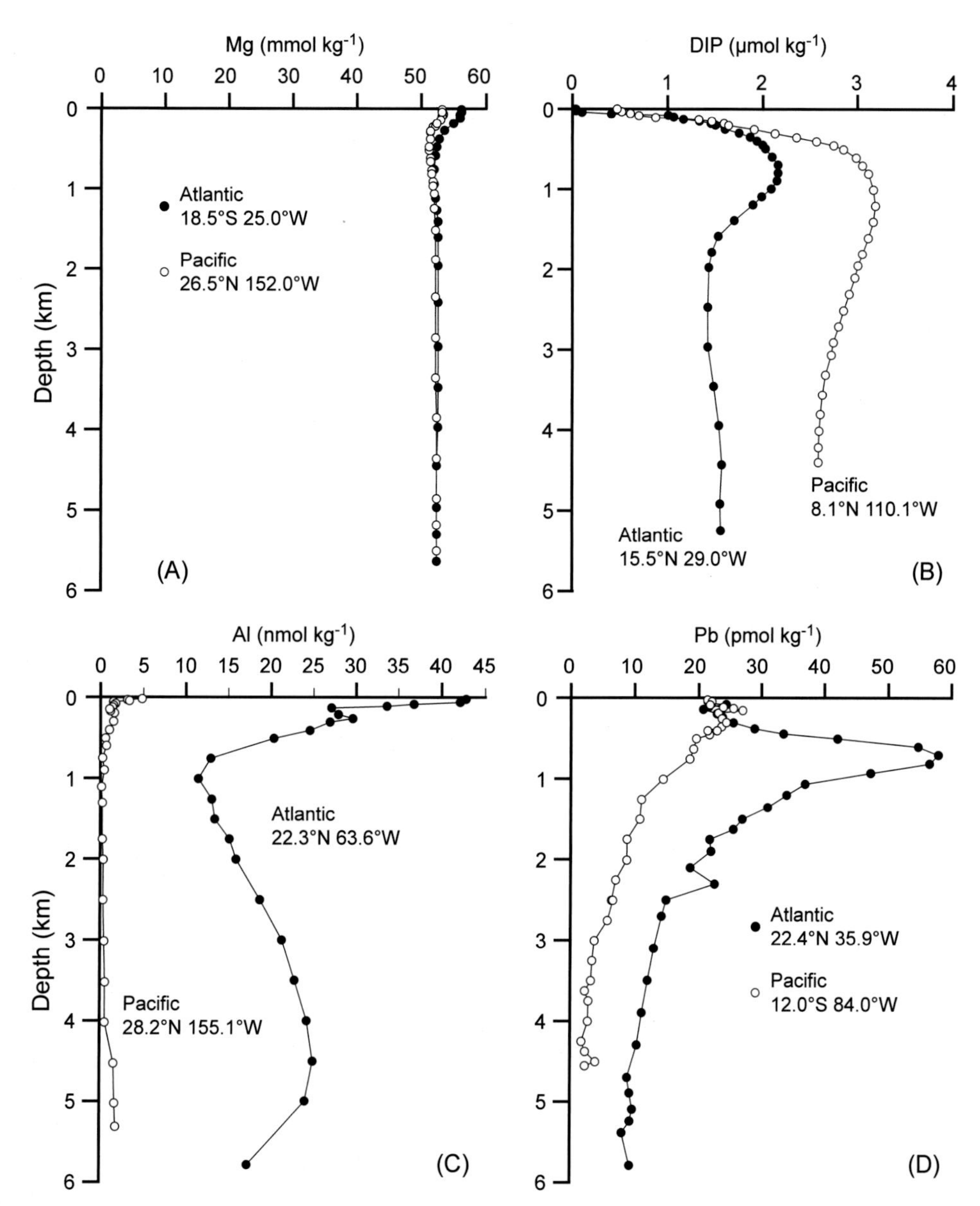

Figure 1.8 Depth profiles of element concentrations in the Pacific (open circles) and Atlantic (filled circles) Oceans. (A) Magnesium (mmol kg^{-1}) values at different depths were calculated using salinity measurements and the Mg:S relationship of Tsunogai et al. (1968). (B) Dissolved inorganic phosphorus (DIP, µmol kg^{-1}) data are from the CLIVAR and Carbon Hydrographic Data Office (CCHDO) (https://cchdo.ucsd.edu/): Pacific cruise: PR16_33RBGP604–1, station 28 and Atlantic cruise: PR_33RO200306_02, station 101. (C) Aluminum (nmol kg^{-1}). Atlantic data are from GEOTRACES Intermediate Data Product 2017, Cruise GA02, Station 27, and Pacific data are from Orians and Bruland (1986). (D) Lead (pmol kg^{-1}). Atlantic data are from GEOTRACES Intermediate Data Product 2017 cruise GA03, and Pacific data are from cruise GP16, Station 7.

Table 1.4 Major ions[a] in seawater at $S_P = 35.000$. Major ions are defined here as those charged constituents of seawater with concentrations greater than 10 μmol kg^{-1}, excluding the nutrients nitrate and silicic acid which vary in concentration.

Cations			Anions		
Species	mmol kg^{-1}	meq kg^{-1}	Species	mmol kg^{-1}	meq kg^{-1}
Na^+	469.06	469.06	Cl^-	545.86	545.86
Mg^{2+}	52.82	105.64	SO_4^{2-}	28.24	56.48
Ca^{2+}	10.28	20.56	HCO_3^-	1.80	1.80
K^+	10.21	10.21	Br^-	0.84	0.84
Sr^{2+}	0.09	0.18	CO_3^{2-}	0.25	0.51
Li^+	0.02	0.02	$B(OH)_4^-$	0.11	0.11
			F^-	0.07	0.07
ΣCations	542.48	605.67	ΣAnions	577.17	605.67

[a] The concentration cutoff for the definition of major ions traditionally consists of elements with concentrations greater than 1 mg kg^{-1}. The concentration of Li^+ is below this threshold, but it is included here to demonstrate the charge balance.

their concentration:salinity ratios have been identified. This deviation from conservative behavior was discovered by making very accurate titration and mass spectrometric measurements. If more precise methods are developed for other elements classified today as conservative, changes with respect to salinity will undoubtedly be found for some.

Gases dissolved in seawater are either chemically inert (the noble gases in the last column of Fig. 1.9) or biologically active (e.g., O_2, CO_2, and to a small extent N_2). Conservative behavior in the case of gases means that the concentrations in seawater are at or near equilibrium with their respective atmospheric gas concentrations rather than having a constant ratio with salinity. Equilibrium, or gas saturation, is primarily a function of temperature (cold water can hold more gas at equilibrium), but salinity also plays a role (saltier water can hold less gas at equilibrium). Dissolved gases at the ocean's surface are at or near saturation with respect to their atmospheric partial pressures and, if they are unreactive, maintain this concentration as they are subducted into the ocean's interior. Thus, conservative gases have concentrations that vary with temperature. Since temperature decreases with depth in the ocean, inert gas concentrations increase with depth. Atmospheric pressures and example concentrations at saturation equilibrium are presented in Table 1.5. Much more about the solubility of gases in the ocean and the processes of atmosphere–ocean exchange is presented in Chapter 2.

Biologically Active and Adsorbed Elements. Elements that are not categorized in Fig. 1.9 as conservative or "no data" are mostly some mixture of biologically active and adsorption controlled. A few of these elements have concentration distributions that are primarily controlled by biological processes: the macronutrients (concentrations in micro- to millimole kg^{-1} range: P, Si, NO_3^-, HCO_3^-), micronutrients (concentrations in the nanomole kg^{-1} range: e.g., Cd and Zn), oxygen that is consumed during respiration, and constituents of shells made by some plankton (Si). A few of the unlabeled elements appear to be mostly

Table 1.5 The major gases of the atmosphere excluding water vapor, which has a concentration of a few percent at saturation in the atmosphere and varies with temperature. Seawater equilibrium concentrations were calculated from the Henry's law coefficients at 20°C and $S_P = 35$.

Gas	Atmospheric mole fraction	Seawater equilibrium concentration (μmol kg^{-1})
N_2	7.8084×10^{-1}	4.20×10^{2}
O_2	2.0946×10^{-1}	2.26×10^{2}
Ar	9.34×10^{-3}	1.11×10^{1}
CO_2	4.10×10^{-4}	1.33×10^{1}
Ne	18.2×10^{-6}	6.8×10^{-3}
He	5.24×10^{-6}	1.7×10^{-3}
Kr	1.14×10^{-6}	2.4×10^{-3}
Xe	0.87×10^{-7}	3.4×10^{-4}

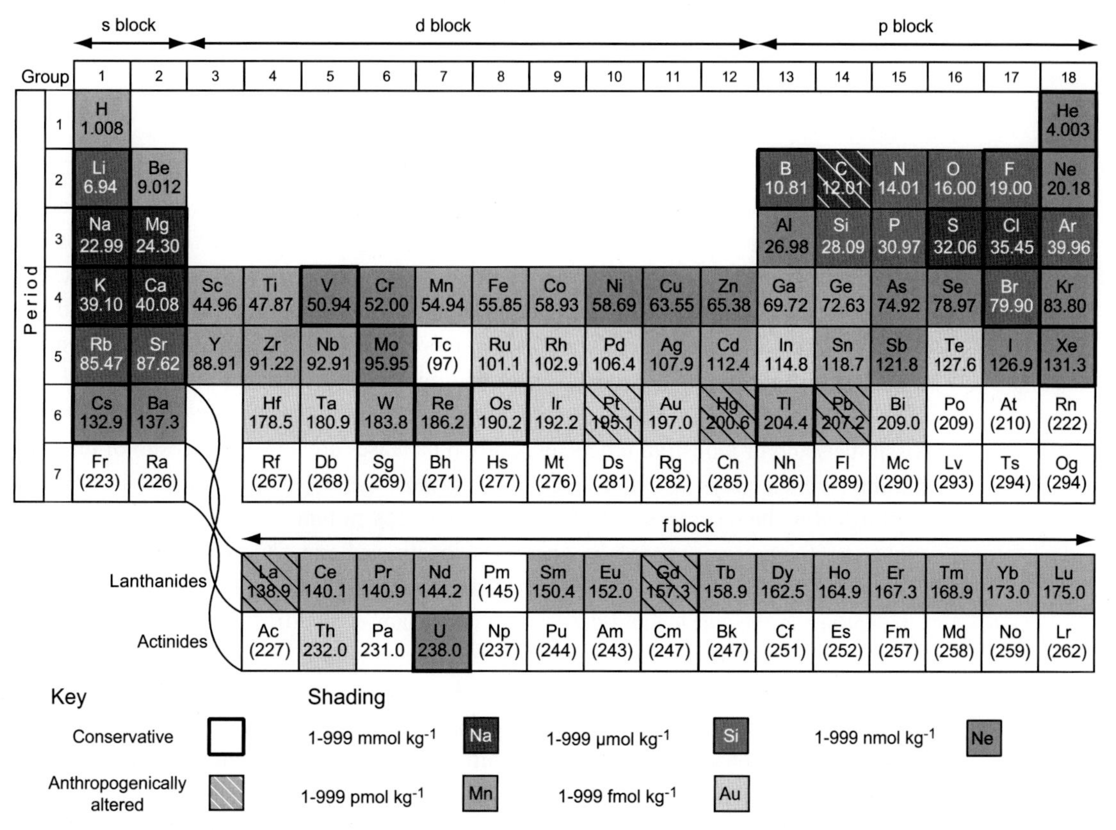

Figure 1.9 Periodic table of the elements indicating their atomic mass and seawater dissolved concentrations. Values in parentheses are the atomic mass of the longest-lived radioisotope of that element. Shading identifies the concentration range of the element dissolved in seawater. Elements with a thick border are conservative; elements that are hatched indicate measurable anthropogenic alteration (Hatje et al., 2018); and elements with no shading have uncertain or lower than femtomolar seawater concentrations. A black and white version of this figure will appear in some formats. For the color version, refer to the plate section.

controlled by adsorption, particularly Al and Th. The rest are some combination of both of these processes.

A summary of the classification of the elements in seawater is presented in Fig. 1.10, where the range of observed concentrations is plotted against the mean concentrations of elements in the ocean. First, note that this is a log–log plot and the concentration range is over 15 orders of magnitude! This is a tribute to the patience and exacting analytical expertise of chemical oceanographers. The mean concentrations (the x-axis) are dominated by the concentrations in the deep oceans – primarily the Pacific because of its size. If the range of concentrations is very small and the mean and range are the same, the element has a conservative behavior – indeed, elements with the largest concentrations, including those in Table 1.3 and a few trace metals like Mo, Re, and W, fall on the line. If the range in concentrations falls below the mean, it indicates that the element is biologically active because surface waters are depleted, as in Fig. 1.8B, e.g., the nutrients P, NO_3^-, Si, Cd, Ni, and Zn. Oxygen also falls below the line because it is biologically active, but in this case it is the deeper waters that are depleted. If the range of an element's concentration is greater than the mean, it indicates that the element concentration is influenced by either adsorption (as in Al, Fig. 1.8C, and Mn) or anthropogenic activities (as in Pb, Fig. 1.8D, and Hg). Concentrations of these elements in surface waters are greater than in deep waters.

1.2.5 Anthropogenic Influences

The concentrations of many nutrient elements and trace metals in nearshore waters are measurably altered by fossil fuel burning, industrial processes, and fertilization. However, we include only those that are clearly elevated globally in open-ocean waters in Anthropogenic Influences. Three elements in this category are highly altered – lead (Pb), mercury (Hg), and carbon (C) – and three more are suspected to be anthropogenically influenced by technological processes – gadolinium (Gd), platinum (Pt), and Lanthanum (La) (Hatje et al., 2018). These elements are delivered to the ocean as a result of the use of fossil fuels and/or industrial processes. Carbon increases are mostly the result of fossil fuel burning, which plays a major role in the carbon cycle (Chapter 8) and is now well established as being responsible for global increases in temperature. The lion's share of global heat increases caused by increases in atmospheric CO_2 ends up in the ocean because of the high heat capacity of water, and causes easily measurable changes in temperature in the surface and deep ocean (Abraham et al., 2013; IPCC, 2013).

Lead is elevated in ocean surface waters because tetraethyl lead was added to gasoline as an anti-knock compound in the 1960s and 1970s. This resulted in large anthropogenic Pb contamination of atmospheric dust, particularly downwind of industrialized nations. The contrast in concentrations between the North Atlantic and North Pacific (Fig. 1.8D) is an indication of how dramatically anthropogenic fluxes have altered lead concentrations. Mercury from fossil fuel sources also enters the ocean via particles, and it is estimated that anthropogenic inputs have tripled the preanthropogenic background level of the Hg in global ocean surface waters (Lamborg et al., 2014). The elements Gd, La, and Pt are used extensively in industrial processes: Gd is commonly used in magnetic resonance imaging, La is derived from the production of catalysts and for petroleum refining, and Pt is another common industrial catalyst.

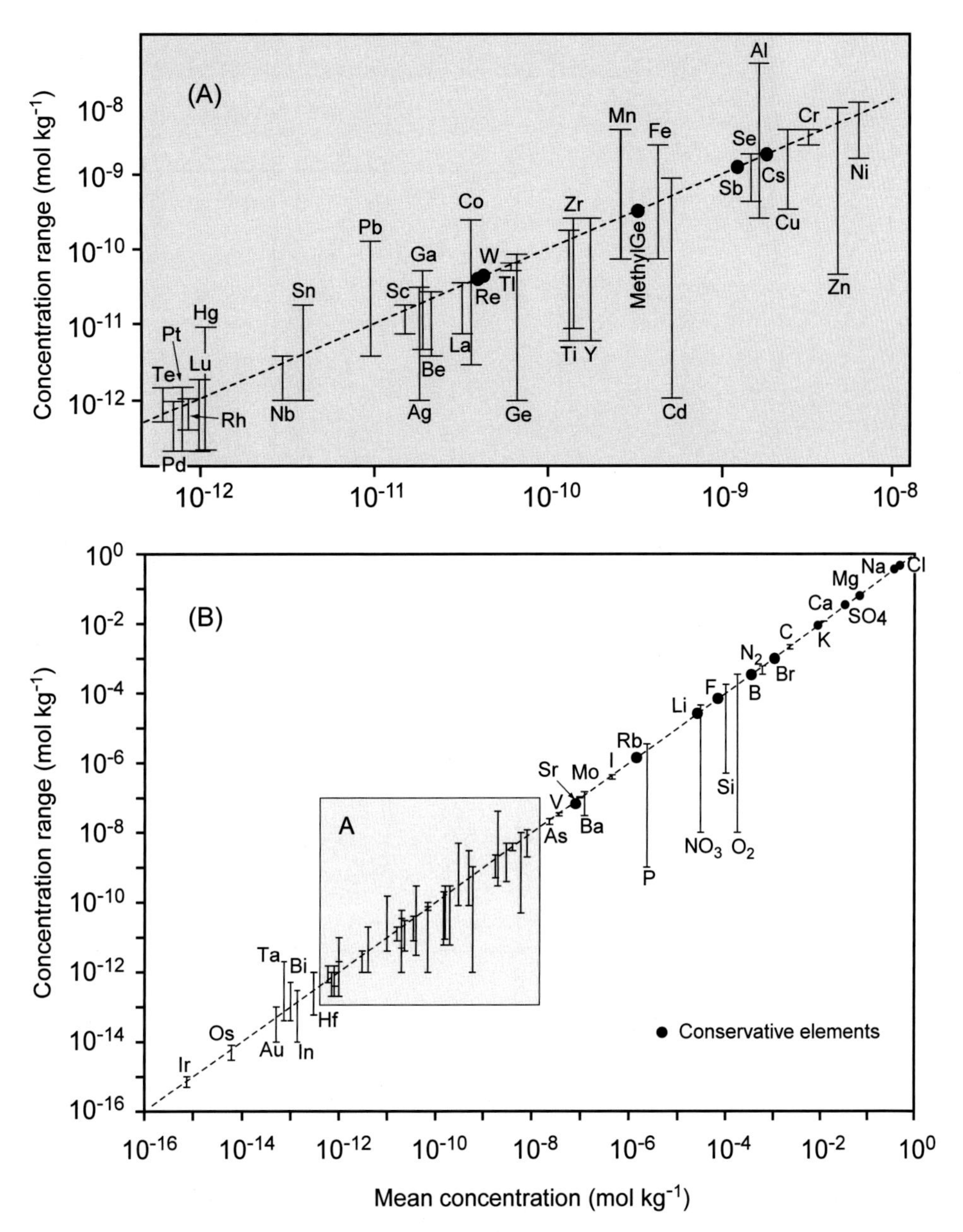

Figure 1.10 A log–log plot of element mean concentration (x-axis) against the range of observed values in the ocean (y-axis). Elements that fall on the line without a significant range in values are conservative and are indicated by a filled circle. If the element is strongly influenced by biological processes, then the range of values falls below the 45-degree line. If adsorption or anthropogenic processes are important, the bulk of the range should be above the line. A range both above and below indicates that a combination of both processes influences the concentration distribution. Data are from Bruland et al. (2014).

Discussion Box 1.1

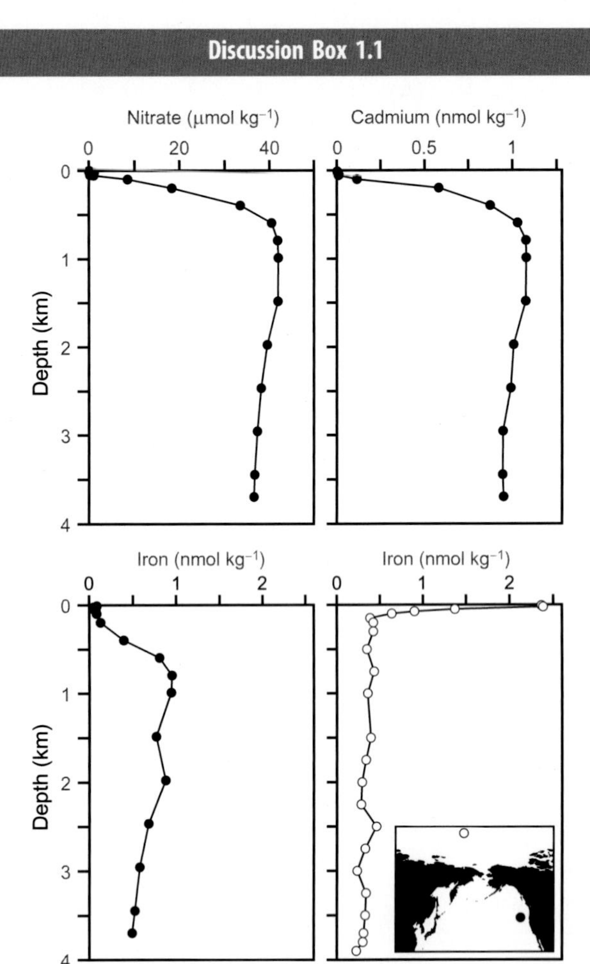

This figure shows depth profiles of nitrate, cadmium, and iron from 43°N, 133°E in the North Pacific (filled circles) as well as a depth profile of iron from 88°N, 170°W in the Arctic (open circles). Data are from the GEOTRACES Intermediate Data Product 2017.

- How would you classify nitrate and cadmium based on these profiles?
- What does the comparison between these two profiles say about the processes that control these compounds?
- How would you classify iron based on the profiles in each of these locations separately?
- What processes could be happening at each of these locations to create such different distributions?
- Speculate about what processes supply iron to the ocean. How might these be different for other elements?

1.3 Ocean Circulation

Chemical and biological reactions in the ocean combine with ocean circulation to create the distribution of observed concentrations. In order to understand the processes controlling the chemical distributions and their rates, one must know something about how the ocean circulates. The following brief overview describes the main seasonal, wind-driven, and interior circulations.

1.3.1 Seasonality

Solar heating is concentrated near the ocean surface, acting to stratify the surface by creating layers of warm water over cold. Heating decreases exponentially with depth, and even in the clearest waters, sunlight decreases to 1 percent of its surface levels by ~100 m. Turbulence induced by wind stress and convection induced by nighttime/wintertime cooling compete with solar warming stratification to create a mixed layer that is 10–100 m deep over most of the ocean. Deeper mixed layers (hundreds of meters) occur in the winter in some high-latitude regions with very strong heat loss to the atmosphere. Beneath the mixed layer is the *pycnocline* (density gradient), where waters of different densities overlie each other and are isolated from the surface mixed layer for timescales greater than 1 year. The pycnocline occupies the depth range from the winter mixed layer to 1000–1500 m.

In most of the ocean, the mixed layer shoals (becomes shallower) in the spring and summer, when sunlight is most intense, and deepens in the fall and winter, when cooling and stronger winds dominate (Fig. 1.11). These dynamics have important implications for biological and chemical distributions. A shallower mixed layer provides greater light levels to freely drifting phytoplankton by concentrating them in the most sunlit waters, leading to the potential for higher photosynthetic rates. However, a shallow mixed layer and strongly stratified pycnocline reduce mixing between the surface layer and waters beneath. Since nutrients needed for biological activity have concentrations that increase with depth, summertime reduction in mixing also decreases the rate at which biologically active elements are supplied to the surface waters. In winter, deepening of the mixed layer brings higher concentrations of biologically active elements to the surface as it entrains deeper waters but reduces exposure to light of freely drifting phytoplankton.

1.3.2 Wind-Driven Circulation

Horizontal circulation in the near-surface ocean is driven by the friction of wind on the atmosphere–ocean interface, while in deeper waters it is mostly density driven. Unequal heating of the Earth's surface creates winds in the atmosphere that impart frictional energy to ocean surface waters. The mean global atmospheric wind pattern consists of the trade winds immediately north and south of the Equator blowing from the east to west, the westerlies in mid-latitudes blowing west to east, and the polar easterlies at high latitudes blowing east to

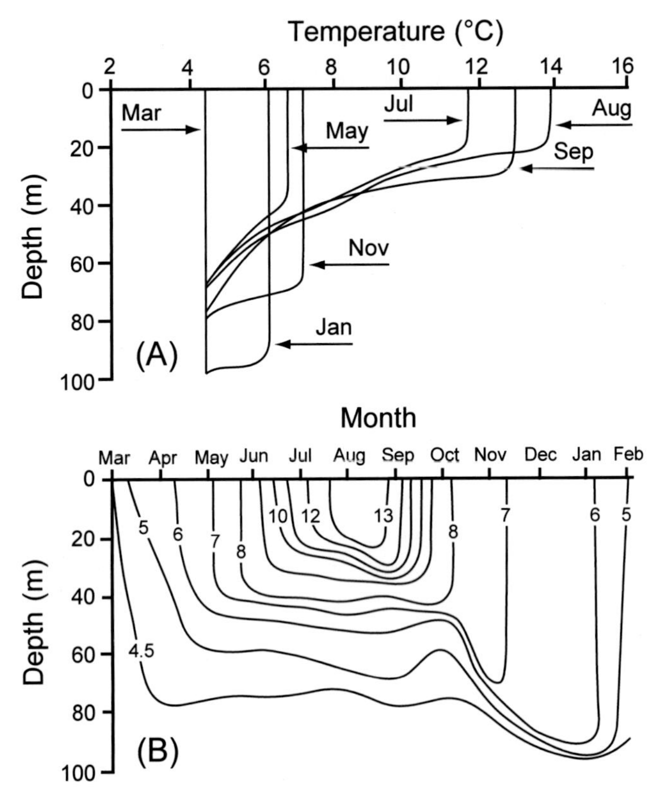

Figure 1.11 Two representations of seasonal changes in mixed layer depth in the North Pacific Ocean at 50°W, 145°N. (A) Temperature versus depth for different months. (B) Temperature contours on a depth versus time plot. Replotted from Talley et al. (2014).

west (Fig. 1.12). Surface atmospheric pressure lows occur near the Equator and at about 60° latitude due to the rising flow of air, while atmospheric highs are created by downward flow at about 30° in mid-latitudes and at the poles. The lows are accompanied by higher-than-average precipitation as rising air is cooled and decreases in its moisture-carrying capacity. Highs are characteristically dry because water-poor cool air sinks, is warmed, and increases its capacity to carry water vapor. These atmospheric circulation patterns contribute to explaining the overall surface salinity distribution presented earlier in Fig. 1.6, with higher salinity at 30° latitude and lower salinity near the Equator.

Friction of the wind on the ocean surface moves the upper 10–100 m in a mean direction that is 90° to the right of the wind in the Northern Hemisphere and 90° to the left of the wind in the Southern Hemisphere. Oceanographers call this flow *Ekman transport.* The *Coriolis force* is what causes the Ekman transport to move in a different direction than the wind. It is an apparent force, rather than a true force, used to describe motion relative to the accelerating reference frame of a rotating Earth. To understand the reason for this deviation from the direction of the wind forcing, imagine that a hypothetical particle is accelerated northward in the ocean surface somewhere near the Equator. The particle has special properties that keep it at the same depth in the surface waters and prevent friction in

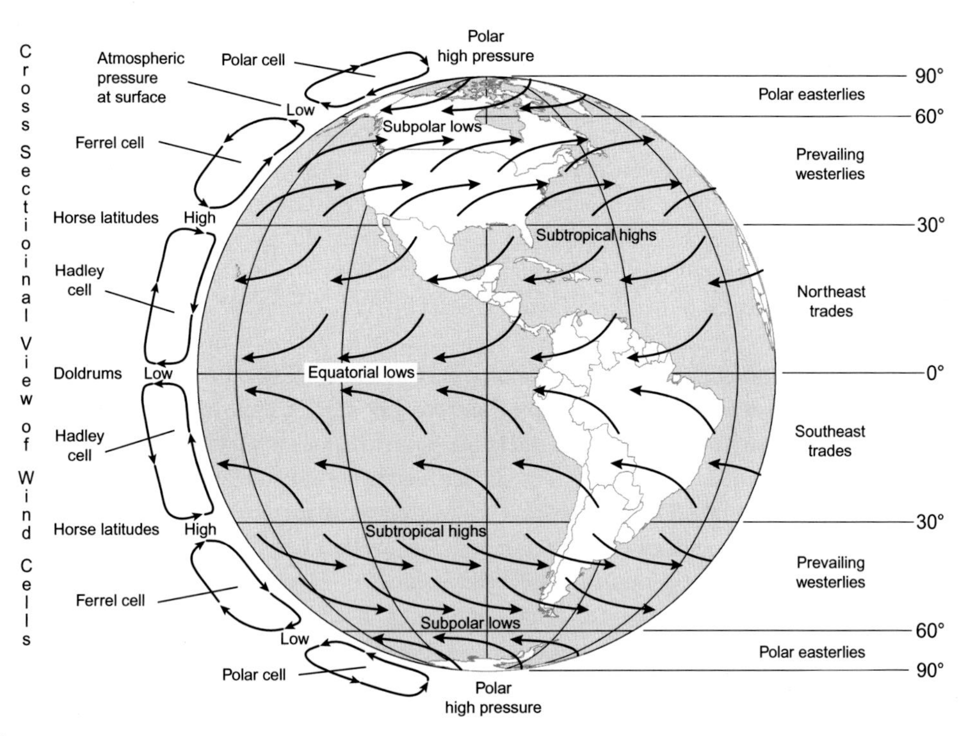

Figure 1.12 The distribution of wind direction on the Earth's surface. Redrafted from Pinet (1994).

the direction it is moving, allowing it to maintain the speed it had at the beginning of its journey. When the particle leaves the Equator, it has both the northward component of velocity and an eastward component imparted by the rotation of the Earth. Because the Earth is a sphere, the eastward velocity on its surface is greater at the Equator (455 km s^{-1}) than that at 30°N or S (402 km s^{-1}) which is in turn faster than that at 45°N or S (326 km s^{-1}). As the particle travels northward, the surface of the Earth beneath it has a slower eastward velocity. From the perspective of the Earth's surface, the particle appears to move to the right (east). What is really happening is that the particle travels in a straight line northward with respect to a rotating cylinder with a diameter of the Earth's Equator, but the surface of the Earth lags behind at higher latitudes because its diameter decreases poleward. Similarly, a particle in motion toward the south from the Equator would veer to the east. An eastward tendency for a particle in the Northern Hemisphere is movement to the right of the direction in which the particle is moving. In the Southern Hemisphere, it is to the left. The Coriolis force affects particles traveling horizontally in all directions, not just those moving north and south, as in this example.

Horizontal movement of water at the surface induces vertical motion as well. For example, the northwest-flowing trade winds in the Southern Hemisphere and southwest-flowing trade winds in the Northern Hemisphere converge at the Equator. Since the resulting mean Ekman transport is 90° to the right of the wind in the north and 90° to the left in the south, the surface waters in this region flow away from the Equator. This creates a divergence (Fig. 1.13A) in flow that is "filled in" by upwelling of water from below (*Ekman suction*).

Conversely, in the subtropics ($\approx 30°$ latitude) surface water flow converges from the north and south, causing waters to "pile up," creating a location of general downwelling (*Ekman pumping*). A similar effect is caused by the flow of winds along the coasts. For example, the flow of winds from the south (a southerly wind) along the Pacific coast of South America (Southern Hemisphere) creates a surface water flow to the left (west), which draws water away from the continent (Fig. 1.13B). Surface water transport away from the land is compensated by upwelling of deeper water along the coast. The same mechanisms create downwelling near this coast in response to a flow of wind from the north (Fig. 1.13C). Vertical movement of water caused by Ekman transport resulting from prevailing winds has important consequences for chemical and biological oceanography, because in regions of upwelling (near the Equator and on some continental margins) waters rich in biologically active elements (nutrients) are brought from below into the sunlight, resulting in high productivity and important fisheries. In locations of downwelling (i.e., the subtropical gyres), nutrient concentrations in surface waters are extremely low.

While Ekman transport is concentrated in the upper 10–100 m of the ocean, the consequences of upwelling and downwelling set up large-scale currents that are felt much deeper. Accumulation of water in areas of Ekman convergence and depletion of water in Ekman divergences cause differences in water height of a meter or so over ocean basin scales. These changes in sea level are observable by modern satellites. The differences in sea level creates pressure gradients that are felt throughout the water column. The flow of water down the hills in the center of the central gyres is turned by the Coriolis force, creating large-scale gyre transport. An example is the North Atlantic subtropical convergence zone (Fig. 1.14). As water flows downhill from the center of the gyre, it is forced to the right by the Coriolis force, creating a large-scale anticyclonic gyre (clockwise flow in the Northern Hemisphere), which has the Gulf Stream as its western, northward-flowing limb. In the area of Ekman divergence between 50°N and 70°N in the subarctic oceans, flow of water into the trough is forced to the right, causing a cyclonic gyre (counterclockwise flow in the Northern Hemisphere). Theoretically the gyre-scale transport extends to the entire depth of the ocean, but the flow substantially weakens with depth because of ocean stratification.

The strength of this gyre circulation can be predicted via the *Sverdrup balance*, which uses the concept of potential vorticity. Potential vorticity is analogous to angular momentum in that it takes into account the height of the water column and its local spin (vorticity). In its simplest form, the potential vorticity, Q, is equal to the Coriolis parameter, f, divided by the height of the water column, H:

$$Q = f/H. \tag{1.5}$$

In order to conserve potential vorticity, the Coriolis parameter and height of the water mass must vary in concert. The Coriolis parameter is a function of latitude:

$$f = 2\,\Omega \sin\varphi, \tag{1.6}$$

where Ω is the Earth's rotation rate (7.29×10^{-5} radians s^{-1}), and φ is latitude. So f varies from zero at the Equator ($\varphi = 0°$ rad) to 1.46×10^{-4} rad s^{-1} at the North Pole ($\varphi = 90°$ or 1.57 rad).

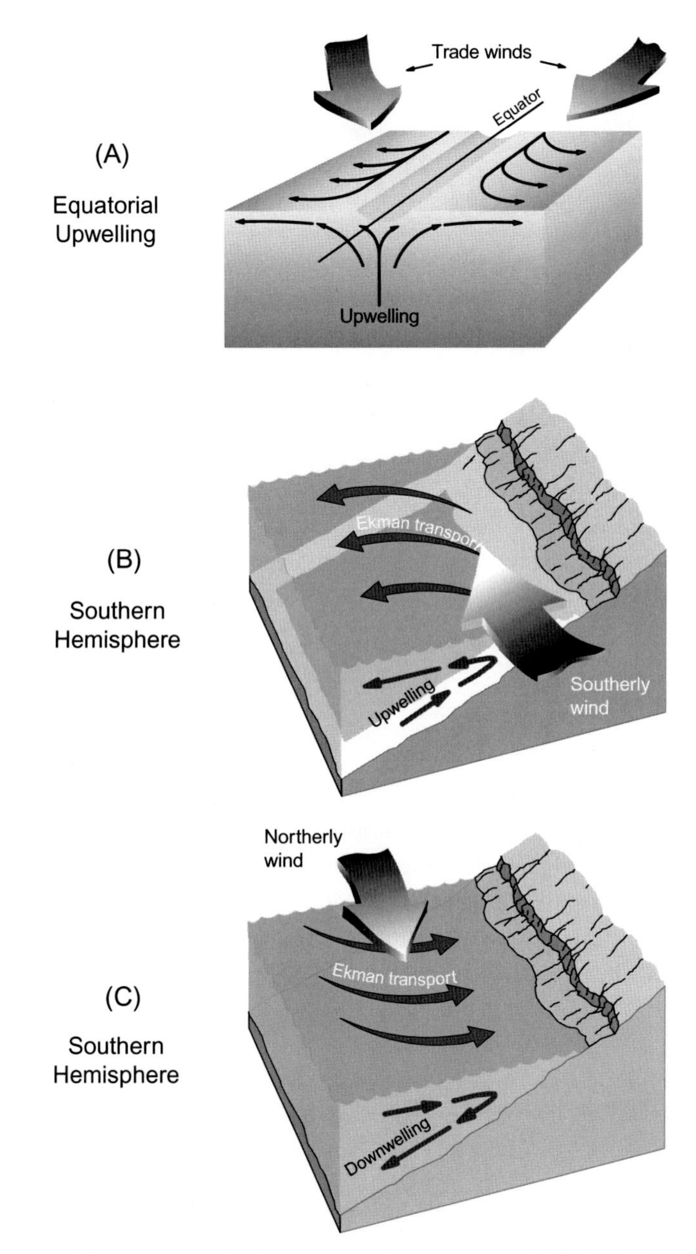

Figure 1.13 Schematic diagrams of Ekman transport in the upper ocean in response to wind forcing: (A) at the Equator, and (B) & (C) near the coastline in the Southern Hemisphere. Redrafted from Thurman (1994).

Ekman-driven downwelling in the subtropical gyres tends to "squash" the height of the water between the ocean surface and the level of no vertical motion (and broaden the diameter to conserve mass), so H goes down, which means the water mass must move toward the Equator to reduce f in order to maintain constant potential vorticity (Eq. 1.5 and

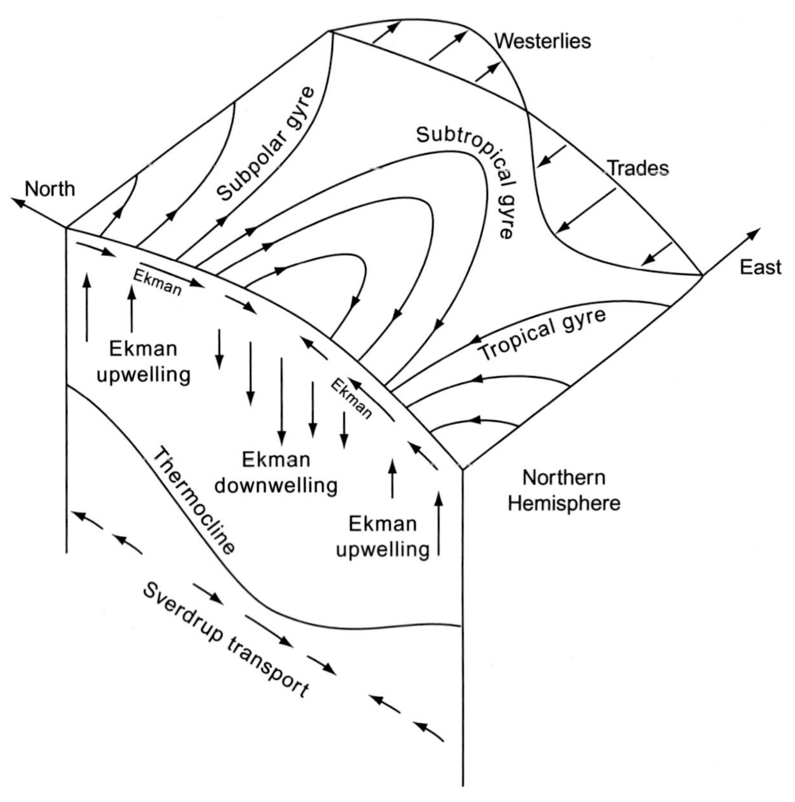

Figure 1.14 Schematic illustration of the geostrophic transport in the subtropical Northern Hemisphere. Westerly and trade winds (direction shown in the upper right) force Ekman transport, resulting in higher sea level with downwelling in the subtropics forcing the thermocline deeper, and lower sea level with upwelling in the subarctic oceans bringing the thermocline shallower. Curved lines on the surface ocean in the diagram show the resulting geostrophic transport. Redrawn from Talley et al. (2011).

Fig. 1.14). Ekman-driven upwelling in the subarctic gyre stretches the water column (and thins the diameter to conserve mass) so it must move away from the Equator by the same reasoning. These subsurface flows to and away from the Equator are called *Sverdrup transport* (Sverdrup 1947). To maintain water balance in the subtropical and subarctic regions, the return flow of water to these locations is by western boundary currents where friction (not accounted for in Eq. 1.5) can exert a torque on the water driving it in the appropriate direction – requiring that the subtropical gyres be anticyclonic (clockwise in the Northern Hemisphere) and the subarctic gyres be cyclonic (anti-clockwise in the Northern Hemisphere).

1.3.3 Interior Circulation

Thermocline Ventilation. Subsurface water was once at the surface somewhere and has generally flowed to its present location along constant density surfaces. While at the

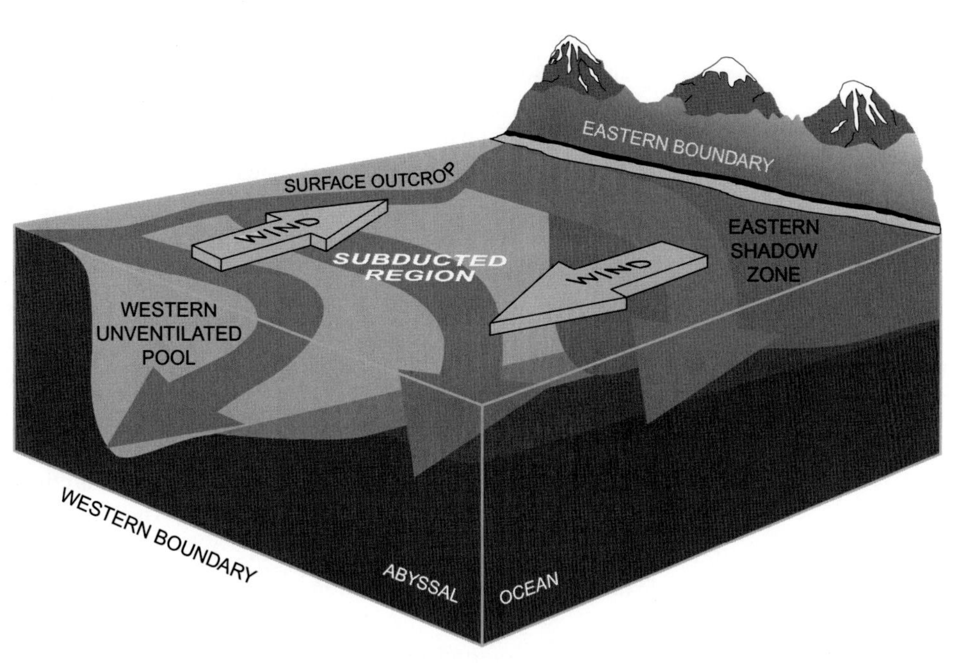

Figure 1.15 Schematic representation of ventilation of the thermocline in the Northern Hemisphere. Denser waters outcrop in the northern region near the subtropical-subarctic boundary and are subducted under the warmer subtropical surface waters into the permanent thermocline, following the southward gyre circulation. Regions that remain poorly ventilated are the eastern shadow zone and western unventilated pool. Return flow is in the western boundary current, via the Gulf Stream in the N. Atlantic and Kuroshio current in the N. Pacific. Redrafted from Talley et al. (2011).

surface in the mixed layer, waters of a particular density experience gas exchange with the atmosphere and nutrient consumption through photosynthesis. In the subtropical gyre, as those waters move equatorward with the wind-driven geostrophic circulation, they slide under the less dense water they encounter at the surface. This process is called *subduction* or *ventilation* (Fig. 1.15). Wintertime conditions create the densest surface waters and deepest mixed layers, so winter is when waters typically outcrop, exchange gases, and then subduct into the interior. This process of ventilation affects all but the warmest (least dense) surface waters, creating a structure within the ocean like a set of nested bowls, where each bowl is a different water density, spinning with the wind-driven gyre circulation. Wintertime densities near the subtropical-subarctic boundaries are particularly important "source waters" that ventilate the main thermocline. The densest waters to outcrop in the North Pacific ($\sigma_\theta = 26.6 - 26.8$) originate in the region off Japan and the Sea of Okhotsk. All denser waters in the North Pacific come from the south with intermediate and deep-water flow. In the Northwest Atlantic, the 18°C water forms a particularly large and important water mass that ventilates the thermocline. The timescale of this ocean circulation (thermocline ventilation) has been determined by measuring man-made gas tracers like tritium (^3H), chlorofluorocarbons (CFCs), and sulfur hexafluoride (SF_6) and found to be on the order of several decades (see Chapter 4).

Thermohaline Circulation. Below 1000–1500 m, temperature gradients are small and the wind-driven transport is weak. In this region, the large-scale transport is caused by thermohaline circulation. The overall water balance of the ocean below 1500 m consists of sinking of water in the polar regions (with salty Mediterranean water mixed in from the side in the Atlantic), which is balanced by mixing, upwelling, and return flow from ocean depths to the surface. What sounds like a vertical balance is in reality a complex layered structure of water masses that can be traced in three dimensions throughout the ocean by salinity, temperature, oxygen, and nutrient differences. The cross sections of salinity in Fig. 1.7A indicate deep-water masses with the temperature and salinity properties in Table 1.3. Southward-flowing North Atlantic Deep Water (NADW) in the Atlantic is bounded above and below by northward-flowing southern-source waters. Antarctic Bottom Water (AABW) flows beneath the NADW and Antarctic Intermediate Water (AAIW) flows on top. Water masses that reach the Southern Ocean are mixed in the Antarctic Circumpolar Water (ACW) that flows around Antarctica, creating one of the most vertically homogeneous regions in the world. The deep ocean circulation is illustrated by the Antarctic-centric schematic in Fig. 1.16. Waters around Antarctica play a critical role in the thermohaline circulation because deep water from all oceans is mixed in this region.

A deep-water mass, referred to as both Common Water and AABW (the latter in Fig. 1.16) enters the Indian and Pacific Oceans. In the Indian Ocean, the water flows through the Crozet Basin south of Madagascar into the Arabian Sea. Common Water enters the Pacific south of New Zealand and flows northward along the western boundary of the basin at about 4 km into the Northeast Pacific. Carbon-14 dates of the dissolved inorganic carbon of deep waters reveal that the "oldest" water in the deep sea resides in the Northeast Pacific where the ^{14}C age is >2000 years. The ^{14}C "age" is really a combination of isolation of the water from the ocean surface and mixing of different water masses. We discuss the interpretation of ^{14}C ages in more detail in Chapter 7; for now, suffice it to say that the circulation time of the deep ocean waters is on the order of 1000 years.

The original theories of deep ocean circulation assumed that the return flow from deep-water formation was uniform global ocean upwelling through the thermocline. This rising flow created the concave upward temperature and salinity profiles of the deep ocean below 1000 m. More modern concepts and tracer interpretation suggest that bottom water flowing north in the Pacific Ocean at ~4000 m depth returns southward as the Pacific Deep Water (PDW) between 2000 and 3000 m depth. Some of this water is mixed into the thermocline, while the rest rejoins the Antarctic Circumpolar Water (Fig. 1.16).

The deep-water flow in the ocean is often depicted as a "conveyor belt" in which water that originates at the surface in the North Atlantic Ocean (NADW) flows south through the Atlantic, and then north through the Indian and Pacific Oceans before it upwells and returns. The analogy is that this deep water accumulates biologically active elements from respiration and dissolution of sinking organic matter, opal, and calcite as it flows, just as a conveyor belt accumulates the rain of dust as it proceeds through a

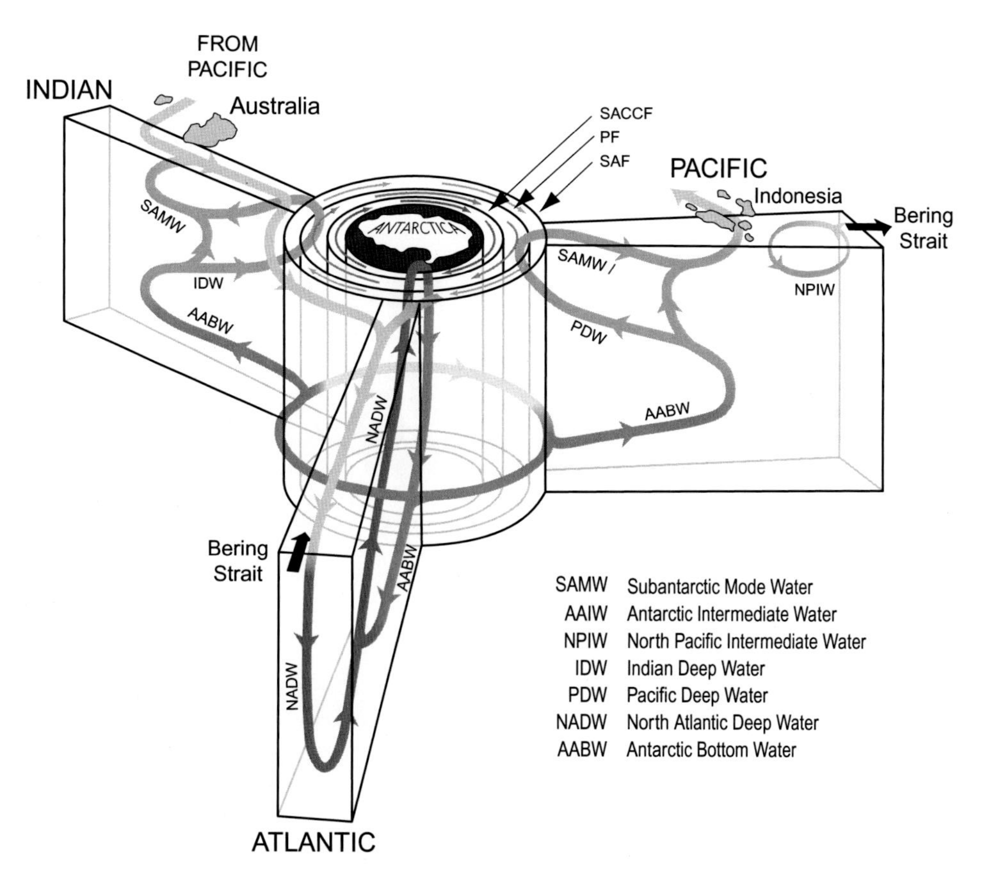

Figure 1.16 Antarctic-centric schematic of the overturning circulation indicting the major deep water flows. From bottom to top: the Antarctic Bottom Water (AABW); North Atlantic Deep Water (NADW), Pacific and Indian Deep Water (PDW, IDW), Subantarctic Mode Water and Antarctic Intermediate Water (SAMW/AAIW), and the surface return flows. Solid circles surrounding Antarctica indicate fronts in the Antarctic Circumpolar Current. They are the Subantarctic Front (SAC), Polar Front (PF), and the Southern Antarctic Circumpolar Current Front (SACCF). Redrafted from Talley et al. (2011). A black and white version of this figure will appear in some formats. For the color version, refer to the plate section.

factory (Broecker and Peng, 1982). Thus, the chemical properties of NADW become progressively more dissimilar from those of surface waters as it moves at depth away from its origin.

It is important to recognize that the conveyor belt analogy ignores the fact that deep waters also have a surface origin in Antarctica (AABW & AAIW). This is indicated clearly by the Antarctic-centric schematic of the overturning circulation in Fig. 1.16. The reason AABW is often not depicted as part of the conveyor belt circulation is that these waters have higher nutrients and lower oxygen and ^{14}C levels than most surface waters or NADW. Antarctic Bottom Water (AABW) is formed from water that upwells to the surface in the Antarctic, but it doesn't reside there long enough for biological processes and gas exchange to "reset" nutrient or ^{14}C concentrations. As depicted in Fig. 1.16, the circumpolar Southern Ocean acts more as a mixing region for all the water masses that flow there than as a location of renewal of surface-water properties to nutrient-poor and oxygen-rich values. Deep waters that enter the Indian and Pacific Oceans are a combination of waters from all the three ocean basins, while North Atlantic Deep Water has its source exclusively in the North Atlantic.

Discussion Box 1.2

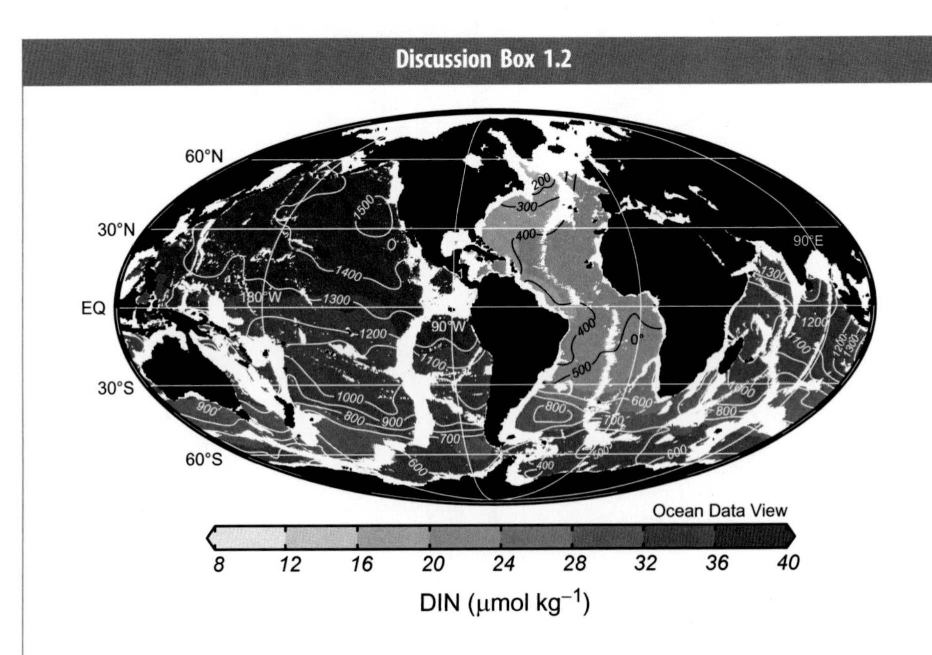

We have already seen in Fig. 1.8B that dissolved phosphate concentrations are considerably higher in the deep Pacific than in the deep Atlantic. Such a pattern is evident in all biologically active elements. The figure here (redrafted from Key et al., 2004) shows a map of dissolved inorganic nitrogen (DIN, primarily nitrate) concentrations and ages at a depth of 3000 m in the ocean. The white areas indicate regions where the ocean is shallower than 3000 m. Nitrate concentrations are shown by the shading (in units of μmol kg^{-1}). The ages of the water in different locations (time since the water was last at the surface, derived from ^{14}C measurements) are shown by the contours (in units of years). Note that even the youngest water in the deep North Atlantic appears to be 200 years old because of mixing with older waters. A black and white version of this figure will appear in some formats. For the color version, refer to the plate section.

- How would you describe the pattern in nitrate concentration?
- Based on the ^{14}C ages, how would you draw the path of the ocean's deep circulation? How does the circulation into the deep Indian Ocean relate to that into the deep Pacific Ocean?
- What processes explain the similarity between the nitrate and ^{14}C age distributions in the deep ocean? How would you expect the deep ocean distribution of other biologically active compounds to look?

1.4 Ocean Biology

Nearly all chemical reactions in the sea take place via biological metabolism or are catalyzed by biologically produced enzymes. Overwhelmingly, the most important of these processes are photosynthesis and respiration. Photosynthesis uses the Sun's energy to create ordered organic compounds that consist of roughly 65 percent proteins, 20 percent lipids, and 15 percent carbohydrates. Most of the organic matter produced during marine photosynthesis is in the form of very small (<100 μm) particles, but some is in the form of dissolved organic matter (DOM). Some of the microscopic plants and animals create shells in the form of inorganic minerals that consist primarily of calcite and aragonite ($CaCO_3$) and opal (SiO_2). About 10–30 percent of the organic matter and most of the mineral armor produced in the euphotic zone sinks or otherwise escapes by water transport and is degraded or dissolved at depth. A small fraction (0.1–1.0 percent) of the particulate organic material produced at the surface sinks all the way to the bottom and accumulates in marine sediments, while a greater fraction of the minerals escapes dissolution and is buried in the sediments.

In order to understand the role of plants and animals in distributing chemical compounds in the sea, one must understand something about the biological agents that perform these tasks. In this text, we will focus on biota with sizes smaller than a few millimeters in diameter, which are the most important community affecting the distribution of chemical compounds in the sea. The distinction between dissolved material that is carried with water flow and particulate material that can sink is operationally defined as 0.45 μm, a common pore size for filters. Particulate material actually contains only a small fraction of the total carbon and other chemical elements found in seawater. Even for carbon, which is a very abundant element in particulate material, the >0.45 μm fraction contains ≤1 μmol kg^{-1} C below the euphotic zone relative to the 40–80 μmol kg^{-1} C found as dissolved organic carbon (DOC) and ~2000 μmol kg^{-1} C found as dissolved inorganic carbon (DIC). Suspended particles with a size <20 μm sink very slowly (<1 m d^{-1}) while particles as large a 100 μm in diameter, usually fecal pellets, can sink as rapidly as 100 m d^{-1}.

Plankton are categorized by their size, metabolic function, and internal structure. The size classification consists of *picoplankton* (<2 μm), *nanoplankton* (2–20 μm), *microplankton* (20–200 μm), and *mesoplankton* (>200 μm). The classification by metabolic function distinguishes those species that create organic matter from dissolved inorganic matter, the *autotrophs* (phytoplankton and some bacteria), and those that gain energy and nutrition by consuming previously existing organic matter, the *heterotrophs* (some bacteria and zooplankton). There are a few species that can exist by both autotrophy and heterotrophy; these are called *mixotrophs*. The final distinction is based on whether single celled algae contain a nucleus and internal organelles; those that have none are *prokaryotes,* and those that have them are *eukaryotes*.

1.4.1 Types of Plankton

Bacteria. Bacteria are prokaryotes in the range of 1 μm in size and consist of species that are both autotrophic and heterotrophic. Bacterial concentrations range from 10^5–10^7 cells cm^{-3}

of seawater, with higher concentrations in the surface ocean. The cyanobacteria *Synechococcus* and *Prochlorococcus* are believed to dominate the picophytoplankton in most oceanic regimes. These species are believed to contain most of the green, light-absorbing pigment, chlorophyll, in these regions. Their importance was discovered in the 1970s and 1980s with the development of epifluorescence microscopy as a routine method for studying microscopic living organisms. Species of autotrophic bacteria that generate energy from chemical reactions rather than from sunlight are called "chemoautotrophic" and play an important role as primary producers of organic matter at oxic–anoxic interfaces found in anoxic basins like the Black Sea and at hydrothermal vents. Organic matter production by this energy pathway may have been important for the origin of life, but in today's ocean it represents only a tiny fraction of that produced by photoautotrophy.

Larger prokaryotic bacteria such as *Trichodesmium* form colonies in surface waters and are known to fix N_2 gas into NH_3 in regions of the ocean where other nitrogen compounds are in low concentration. This opportunistic behavior gives the nitrogen fixers an advantage when the inorganic fixed nitrogen compounds (NH_4^+ and NO_3^-) are unavailable (see Section 4.1.3).

Heterotrophic bacteria exist throughout the ocean both as individual entities in seawater ($\approx$80 percent) and attached to particle surfaces and sediments. These bacteria consume dissolved organic matter, because they ingest nutrients solely by dissolved transport across their membranes. Some may secrete enzymes, which break down large molecules so they can be transported across the cell walls. In the euphotic zone of the ocean, heterotrophic bacteria play an important role in organic matter recycling. Below the euphotic zone and in sediments, heterotrophic bacteria are responsible for the vast majority of organic matter respiration.

Phytoplankton. Phytoplankton are unicellular autotrophs that contain chlorophyll. The prokaryotic, autotrophic bacteria discussed above are included in this classification, but it is not limited to them. There are many phytoplankton that are eukaryotes. In addition to the autotrophic bacteria, there are three other principal types of phytoplankton: diatoms, coccolithophorids, and photoautotrophic flagellates. Diatoms are nano to micro in size, but often form colonies that can be greater than 1 mm long. Their primary distinction is that they produce silica frustules. Because of the opaline frustules, diatoms control the dissolved silica cycle in the sea, which is characterized by a biologically active concentration profile with low concentrations in surface waters that increase with depth. These phytoplankton dominate in regions of the ocean where upwelling is important, such as the Southern Ocean and coastal areas. A portion of the silica frustules that leave the surface ocean are preserved in sediments. In areas like the Southern Ocean around Antarctica, the sediments consist predominately of diatom frustules because of their dominance in the surface waters. Observations of diatom frustules in sediments and "diatomaceous earth" deposits for the last 100 million years indicate that diatoms have been important phytoplankton since the Cretaceous Period.

Coccolithophorids are nano-sized plankton that produce calcite ($CaCO_3$) plates called coccoliths that armor their protoplasm. Generally, coccolithophorids are less prevalent than diatoms and tend to be most abundant in subtropical gyres and in warm waters. *Emiliania huxleyi* is the most abundant coccolithophorid. Because their calcareous plates reflect light, blooms of coccolithophorids can be seen from space using satellite images (Brown and Yoder, 1994). The calcitic coccoliths make up the bulk of the $CaCO_3$-rich sediments currently preserved in the ocean, and nearly pure coccolith deposits, such as in the White Cliffs of Dover in England, provide geologic evidence that this group has existed for the last 170 million years.

Photoautotrophic flagellates are pico-nano size plankton usually with two flagella, whip-like appendages that beat within grooves in the cell wall. They are found in most regions of the ocean, but they contain no mineral "shell" and are thus poorly preserved in marine sediments, though the cysts of species that produce them can be preserved. Dinoflagellates are a kind of photoautotrophic flagellate with red to green pigments that can form large blooms in coastal waters that are sometimes called "red tides."

Zooplankton. Zooplankton obtain their energy heterotrophically by grazing both smaller phytoplankton and other zooplankton. They are eukaryotic and exist in sizes from nanometers to centimeters. Food webs that contain pico- and nano-size phytoplankton are generally dominated by nano-size flagellate zooplankton with very fast growth rates. They feed by direct interception, sieving, and filtering. Ciliate zooplankton are a little larger than flagellates and graze on flagellates and small diatoms. Heterotrophic dinoflagellates are larger, nano-micro size heterotrophs, and graze on nanophytoplankton and nanozooplankton.

The only zooplankton of any abundance that form mineral shells (or tests) are the microzooplankton foraminifera, pteropods, and radiolarians, which form tests of $CaCO_3$ (usually calcite for foraminifera and always aragonite for pteropods) and SiO_2 (opal) for radiolarians. Larger zooplankton (copepods, euphausiids, and crustaceans) play an important role in the flux of chemical species to the ocean interior because they create large ($\approx$100 μm) fecal pellets that sink very rapidly ($\approx$100 m d^{-1}) into the deep sea. Sediment trap experiments have shown that fecal pellets often dominate the particle flux from the surface ocean. This can have important consequences for the distribution of metabolic products because their rate of release to the water in dissolved form depends on both the degradation and dissolution rates as well as the sinking velocity. The sinking velocity greatly depends on the density difference between the particles and seawater, so fecal pellets that contain the minerals $CaCO_3$ and SiO_2 sink much faster than those that only contain organic matter.

1.4.2 Marine Metabolism: Estimates of Fluxes

The presence and distribution of phytoplankton in the sea is determined primarily by the abundance of chlorophyll. While this is not a direct measure of the number of cells, because they contain different amounts of chlorophyll under different conditions, it is the most rapid and widely used method of identifying photosynthetically active regions. Since the color of the ocean can be measured using satellites, it is possible to determine the global content of chlorophyll in the surface waters of the sea. The global chlorophyll pattern seen in Fig. 1.17 is strongly related to the wind-driven circulation discussed in Section 1.3.2. Low chlorophyll concentrations are found in the center of the subtropical gyres, where

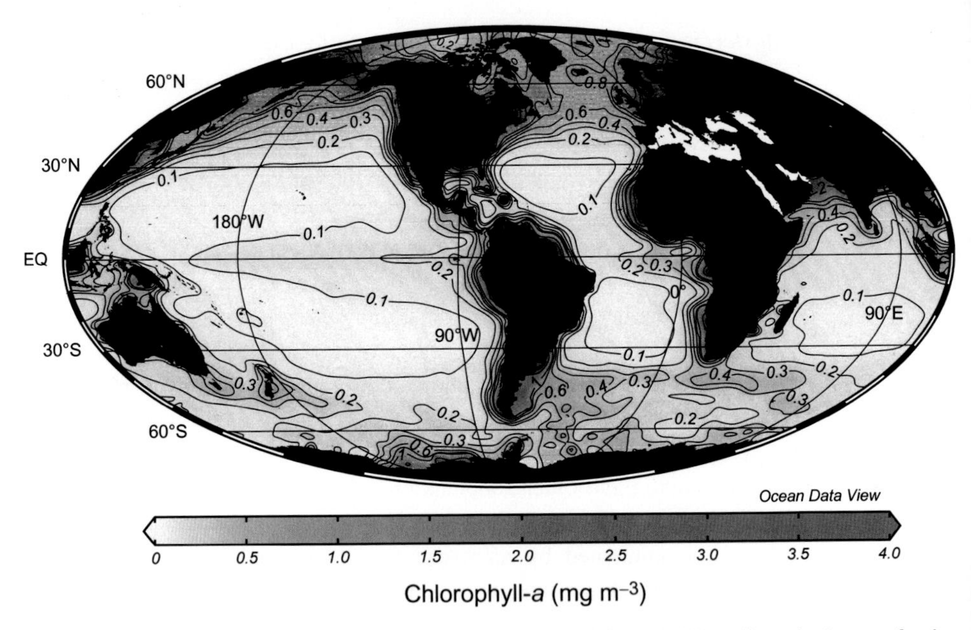

Figure 1.17 Global map of mean annual surface ocean chlorophyll-*a* concentration from MODIS satellite color imagery for the period 2009–2013. (https://oceancolor.gsfc.nasa.gov/data/aqua/). A black and white version of this figure will appear in some formats. For the color version, refer to the plate section.

Ekman-driven downwelling pushes high-nutrient water away from the sunlit surface. High chlorophyll concentrations are found along many continental margins and in the subarctic gyres where Ekman-driven upwelling brings nutrients to the surface.

While one does not expect a direct relationship between the standing stock of chlorophyll and the rate of photosynthesis, there are regional correlations. These relationships have been used with satellite color measurements to estimate global marine primary production. Caveats to keep in mind are that satellites measure color over only one optical depth (about 30 m in the open ocean) and the level to which light penetrates in the open ocean is more like 100 meters (the approximate 1 percent light level), and that chlorophyll and photosynthesis are not uniformly correlated. While biological oceanographers attempt to compensate for this and other factors, there is currently no accurate way to relate surface chlorophyll concentrations to euphotic zone photosynthetic productivity in vast regions of the ocean. Satellite-determined global maps of marine photosynthesis generally underestimate productivity in the vast subtropical gyres compared to direct, shipboard measurements (see Chapter 3).

The biological process that most influences the chemistry of the atmosphere and the sea is the flux of organic matter from the surface to the deep ocean, that which escapes the upper ocean before it is respired by zooplankton and bacteria. This carbon flux is sometimes called the ocean's *biological carbon pump*. The relationship between biological productivity and the carbon pump requires careful definition of frequently used terms that describe biological rates. The transformation of *dissolved inorganic carbon* (DIC) to organic carbon by autotrophs is called *gross primary production* (GPP). Since autotrophs also respire organic carbon back to DIC, *net primary production* (NPP) is the term used to describe the difference between GPP and autotrophic respiration. NPP is the flux that biological oceanographers believe to be most closely related to rates determined by ^{14}C-DIC incubation experiments. Finally, *net community production* (NCP) is the difference between NPP and the respiration by heterotrophs (bacteria and animals). This is the amount of organic carbon produced by photosynthesis that is available to be lost from the surface ocean. On an annual basis, the organic carbon flux out of the upper ocean, the biological pump, should equal the *annual net community production* (ANCP) of organic carbon.

Discussion Box 1.3

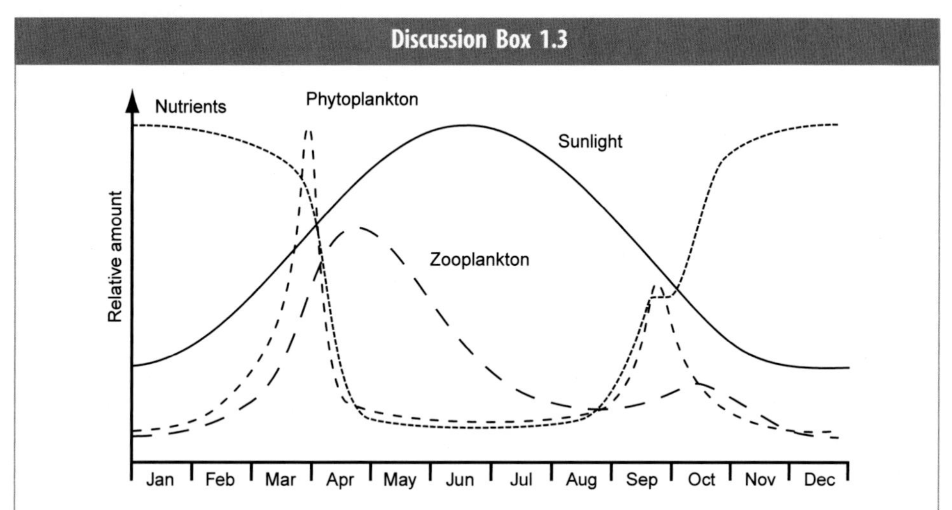

In this section, we have focused on summarizing the types of plankton and their metabolic rates, but we can also think about how these interact with the seasonality of the mixed layer described in the previous section. This figure shows idealized seasonal changes in sunlight, mixed layer nutrient concentrations (nitrate or phosphate for example), phytoplankton biomass, and zooplankton biomass in a region like the North Atlantic that produces large blooms of phytoplankton.

- What controls the variation in sunlight over time?
- Why do nutrients decrease in the spring? Why do they increase in the fall?
- What likely limits the abundance of phytoplankton in the winter? What factors likely cause phytoplankton abundance to decrease after the spring bloom? What causes the secondary increase in phytoplankton abundance in the fall?
- How might the rate at which zooplankton can grow and multiply affect the phytoplankton annual cycle? For example, what if zooplankton multiplied faster? Conversely, what if zooplankton were controlled by a strict annual life cycle that only produced new zooplankton at a specific time of year?

References

Abraham, J. P., M. Baringer, N. L. Bindoff, et al. (2013) A review of global ocean temperature observations: implications for ocean heat content estimates and climate change, *Reviews of Geophysics*, **51**, 450–483, doi:10.1002/rog.20022.

Broecker, W. S., and T.-H. Peng (1982) *Tracers in the Sea*, 690 pp. Palisades: Eldigio Press.

Brown, C. W., and J. A. Yoder (1994) Coccolithophorid blooms in the global ocean, *Journal of Geophysical Research,* **99**, 7467–7482.

Bruland, K. W., R. M. Middag, and M. C. Lohan (2014) Controls of trace metals in seawater, in *Treatise on Geochemistry* (eds. H. D. Holland and K. K. Turekian), pp. 19–51. Oxford: Elsevier.

Emerson, S. R., and J. I. Hedges (2008) *Chemical Oceanography and the Marine Carbon Cycle*, 453 pp. Cambridge: Cambridge University Press.

Hatje, V., C. H. Lamborg, and E. A. Boyle (2018) Trace-metal contaminants: human footprint on the ocean, *Elements*, **14**, 403–408.

IPCC (2013) Climate Change 2013: The physical science basis, Chapter 3: Ocean observations. *Contribution of Working Group I to the Fifth Assessment Report of the Intergovernmental Panel on Climate Change*, (eds. T. F. Stocker, D. Qin, G.-K. Plattner, et al.), 1535 pp. Cambridge: Cambridge University Press.

Jackett, D. R., and T. J. McDougall (1997) A neutral density variable for the world's oceans, *Journal of Physical Oceanography*, **27**, 237–263.

Key, R. M., A. Kozar, C. L. Sabine, et al. (2004) A global ocean climatology: results for the global data analysis project (GLODAP), *Global Biogeochemical Cycles*, **18**, GB4031, doi:10.1029/2004GB002247.

Lamborg, C. L., C. R. Hammerschmidt, K. L. Bowman, et al. (2014) A global ocean inventory of anthropogenic mercury based on water column measurements, *Nature*, **512**, 65–67, doi:10.1038/nature13563.

Lewis E. L. (1980) The Practical Salinity Scale 1978 and its antecedents, *Journal of Oceanic Engineering*, **5**, 3–8.

Libes, S. M. (1992) *An Introduction to Marine Biogeochemistry*, 928 pp. New York: John Wiley and Sons.

Millero, F. J., C.-T. Chen, A. Bradshaw, and K. Schleicher (1980) A new high pressure equation of state for seawater, *Deep-Sea Research A*, **27**, 255–264.

McDougall, T. J., et al. (2010) *The International Thermodynamic Equation of Seawater – 2010: Calculation and Use of Thermodynamic Properties*, Intergovernmental Oceanographic Commission, Manuals and Guides No. 56, 196 pp. UNESCO.

Orians, K. J., and K. W. Bruland (1986) The biogeochemistry of aluminum in the Pacific Ocean, *Earth and Planetary Science Letters*, **78**, 397–410.

Pilson, M. E. Q. (2013) *An Introduction to the Chemistry of the Sea*, 533 pp. Cambridge: Cambridge University Press.

Pinet, P. R. (1994) *Invitation to Oceanography*, 508 pp. Boston: Jones and Bartlett.
Schlitzer, R. (2021) Ocean Data View, https://odv.awi.de
Sverdrup, H. U. (1947) Wind-driven currents in a baroclinic ocean; with application to the equatorial current of the eastern Pacific. *Proceedings of the National Academy of Sciences of the U.S.A.*, **33**, 318–326.
Talley, L. D., G. L. Packard, W. J. Emery, and J. H. Swift (2011) *Descriptive Physical Oceanography: An Introduction*, 560 pp. London: Elsevier.
Thurman, H. (1994) *Essentials of Oceanography*, 7th ed., 550 pp. New York: Macmillan.
Tsunogai, S., M. Nishimura, and S. Nakaya (1968) Calcium and magnesium in seawater and the ratio of calcium to chlorinity as a tracer for water masses, *Journal of the Oceanographical Society of Japan*, **24**, 153–159.

Problems for Chapter 1

1.1. Typical summertime mixed layer temperatures in the central Baffin Bay in the Arctic are a balmy 2°C while salinity is near 31. The freezing temperature for water at this salinity is –1.7°C.

a. How much energy would need to be removed from a 1 m^2 column of the mixed layer (30 m deep) to cool the water from its summertime temperature to its freezing point? If heat is lost from the surface at a rate of 60 W m^{-2}, how long will it take for the mixed layer to reach freezing temperature?

b. After the water reaches freezing temperature, how long will it take for 1 cm of ice to form at the surface given the same cooling rate? How long would it take to freeze a 30 m deep column of water at this cooling rate (if such a thing were possible)?

1.2. Write a recipe for making 1 kg of artificial seawater with salinity = 35, including all ions that have concentrations greater than 10 mmol kg^{-1}. This solution will be made in a N_2-filled glove bag with no CO_2, so that there are no carbon-containing species in the water. The following salts, acids, and water are available: NaCl (*s*), $MgSO_4$ (*s*), $CaCl_2$ (*s*), KOH (*s*), $Mg(OH)_2$ (*s*), HCl (concentration = 12 M; density 1180 kg m^{-3}), and H_2O (density = 998.2 kg m^{-3} at 20°C). Assume all salts and acids dissociate completely. H^+ and OH^- combine to make water in the solution, but there will be an excess of OH^- that need not be considered here. How many grams of each solid and how many mL of HCl and H_2O are needed to make the artificial seawater solution?

When we study the carbonate system (Chapter 5), we'll revisit this problem to consider alkalinity changes by removing the solution from the glove bag and allowing it to exchange gases with the atmosphere.

1.3. The figure to the right shows a depth profile of salinity in the South Atlantic Ocean (25°S, 25°W). A few selected points from this profile are given in the table below.

a. Calculate the expected sodium and calcium concentrations at these five depths in mmol kg^{-1}. What processes have been important to setting the observed salinity values at different depths?

b. At the same time this salinity profile is collected, the concentration of calcium at 1.3 m is accurately measured to be 10.63 mmol kg^{-1}. What could account for the difference in the measured value and that calculated in part a?

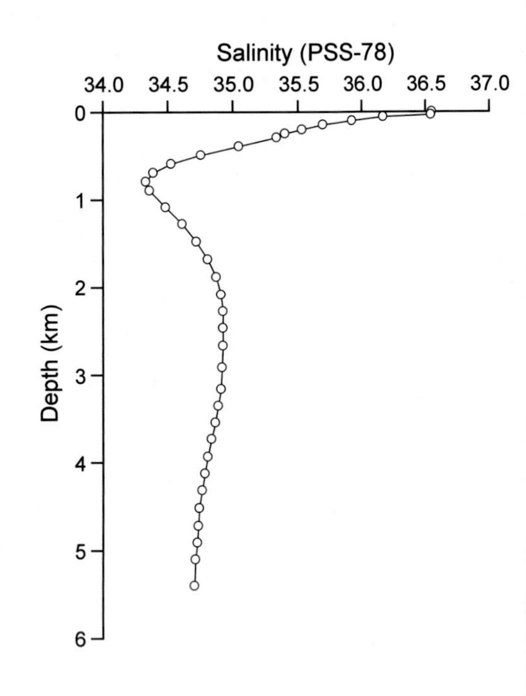

Selected points

Depth (m)	Salinity (PSS-78)
1.3	36.549
391.3	35.051
796	34.331
2 465.5	34.926
5 400.1	34.708

1.4. Determine the fraction (in percent) of N_2 in the total atmosphere/ocean system that is dissolved in the ocean. Do the same calculation for O_2 and for Ar. Assume the whole ocean has a temperature of 20°C and a salinity of 35, and that gases in the ocean are at equilibrium with the atmosphere. (This is not exactly correct, because deep ocean temperatures are much lower than 20°C, and oxygen is depleted in the deep ocean. However, the calculation is still a good first approximation.) Besides the information in this chapter, Appendix A may also be helpful to you.

1.5. The ocean takes up and gives off a large amount of heat every year.

a. Based on the information in Figs. 1.2 and 1.11, estimate the change in heat content (in J m^{-2}) between March and August in the subarctic Pacific Ocean.

b. What average heating rate (in W m^{-2}) is required between March and August to achieve this change?

1.6. The ocean's thermohaline circulation involves the mixing of several major water masses together to form the water found in the deep Pacific. Determine the contribution of each of the three major deep-water masses to water in the deep Pacific based only on the potential temperature and salinity of each. Use your best estimates of temperature and salinity (to the nearest 0.1°C and 0.1 salinity) from Table 1.3 and Fig. 1.7.

2 Geochemical Mass Balance: Chemical Flow across the Ocean's Boundaries

2.1 Major Ion Mass Balance: Weathering and Authigenic Mineral Formation 45
2.1.1 The Source: Weathering and River Fluxes to the Ocean 45
2.1.2 Residence Times of Seawater Constituents 47
Discussion Box 2.1 50
2.1.3 Mackenzie and Garrels Mass Balance 50
2.1.4 The Sinks: Hydrothermal Circulation 56
Discussion Box 2.2 65
2.1.5 The Sinks: Reverse Weathering 65
2.2 Gases in the Ocean–Atmosphere System 68
2.2.1 Air–Sea Chemical Equilibrium: Henry's Law Solubility 69
Discussion Box 2.3 73
2.2.2 Gas Sources and Sinks in the Ocean–Atmosphere System 74
Appendix 2A.1 A Brief Review of Rocks and Minerals 76
Appendix 2A.2 The Meaning of Residence Time 78
Appendix 2A.3 The Kinetics of Air–Sea Gas Exchange 80
2A.3.1 The Gas Exchange Flux Equations 80
2A.3.2 Measurements of Gas Exchange Rates in the Ocean 84
2A.3.3 Gas Transfer Due to Bubbles 87
References 90
Problems for Chapter 2 94

Concentrations both of the elements that make up most of the salinity of the oceans and of the major gases provide clues to the mechanisms that control their sources and sinks. The chemical perspective of oceanography revealed by major element concentrations is about processes that occur across ocean boundaries – weathering reactions on land, authigenic mineral formation in marine sediments, reactions with the crust at hydrothermal areas, and air–sea interaction. The amount of time some of the dissolved constituents remain in solution before they are removed chemically is very long, suggesting the possibility for chemical equilibrium between seawater and the minerals in ocean basins as a controlling factor. Conversely, energy from both the Sun and the Earth's interior is constantly driving seawater constituents away from chemical equilibrium. This chapter is divided into two parts: ocean mass balances controlling the major ions in seawater and the processes controlling the distribution of gases between the atmosphere and ocean. The first section is based on a classic geochemical mass balance of major seawater dissolved ions between river inflow and the concentrations observed in the sea, and how these observations are used to infer mechanisms. The second section introduces how thermodynamic equilibrium and kinetic principles determine the distribution of the major atmospheric gases between the atmosphere and ocean.

2.1 Major Ion Mass Balance: Weathering and Authigenic Mineral Formation

The main mechanisms for delivery of dissolved ions to the ocean are river inflow and atmospheric input. Formation of authigenic minerals (those minerals that form in situ) is the ultimate sink (Fig. 2.1) for these constituents. Authigenesis primarily involves precipitation of plant and animal shells, chemical reactions in sediments, and high-temperature reactions at hydrothermal regions. We begin with a brief review of the chemical reactions influencing the dissolved ion concentrations of rivers and end with an attempt to balance the river sources with plausible sinks for the major seawater ions.

2.1.1 The Source: Weathering and River Fluxes to the Ocean

Based on chemical measurements of river water and of atmospheric particles, river inflow is by far the most important mechanism for the delivery of dissolved major ions and

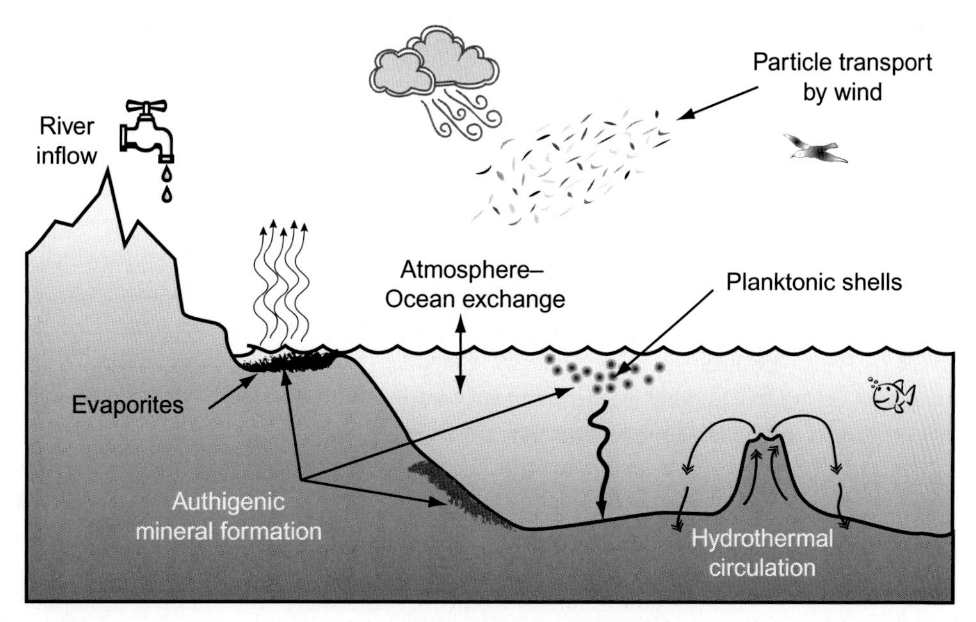

Figure 2.1 A schematic representation of the sources and sinks of the dissolved chemical constituents of seawater. Sources for ions are from rivers and wind, and sinks are authigenic mineral formation. The atmosphere is both a source and sink for gases. Redrafted from Emerson and Hedges (2008).

Table 2.1 The sources of major ions to rivers. The average concentration of dissolved species in rivers is naturally somewhat uncertain. These values and their sources are from Holland (1978). Some of the total river water concentrations differ from those in Table 2.3 where the numbers are from Li (2000).

	Anions			Cations				Neutral
	HCO_3^-	SO_4^{2-}	Cl^-	Ca^{2+}	Mg^{2+}	Na^+	K^+	H_4SiO_4
Source				meq kg^{-1}				mmol kg^{-1}
Atmosphere	0.58	0.09	0.06	0.01	≤0.01	0.05	≤0.01	≤0.01
Rock Types								
Silicates				0.14	0.20	0.10	0.05	0.21
Carbonates	0.31			0.50	0.13			
Sulfates		0.07		0.07				
Sulfides		0.07						
Chlorides			0.16	0.03	≤0.01	0.11	0.01	
Organic C	0.07							
SUM	0.96	0.23	0.22	0.75	0.35	0.26	0.07	0.22

silicates to the ocean. This is not the case for all elements; some trace metals like iron and lead have important sources from atmospheric dust, but our discussion will focus on the flux of major elements to the ocean. The concentration and origin of the major ions in river water is presented in Table 2.1. Weathering of rocks on land is the origin of the cations, Na^+, Mg^{2+}, Ca^{2+}, and K^+, while the source of the anions, Cl^-, SO_4^{2-}, and HCO_3^- is partially from rock weathering and partially from the gases CO_2, SO_2, and HCl that are delivered to the atmosphere via volcanic emissions over geologic time.

Weathering involves the breakdown of rocks on land to create dissolved constituents in solution and altered mineralogy on land, not only creating smaller rock particles but also chemically altering the minerals themselves. To aid those unfamiliar with geology, a very brief review of important rocks and minerals involved in the weathering process is presented in Appendix 2A.1. During weathering, carbon dioxide, which ultimately comes from the atmosphere but is also generated in soils by respiration, reacts with aluminosilicate and carbonate minerals. Some of these reactions are illustrated in Table 2.2 using formulae for pure minerals. Carbonate minerals are represented by calcium carbonate, $CaCO_3$, which is the dominant constituent of limestone rocks on land and carbonate shells produced in the ocean today. Silicates are represented by the igneous minerals potassium feldspar and phlogopite mica, and by the clay mineral kaolinite. Clay minerals are the reaction products of igneous rock weathering. They consist of the fine-grained particles that make up most muds. There are many different igneous and clay minerals and lots of mixed versions between the mineralogically pure endmembers.

The common feature of weathering reactions (Table 2.2) is consumption of CO_2 and the release to solution of HCO_3^- along with the cations of the weathered mineral. During limestone weathering, Ca^{2+} and HCO_3^- enter solution. In the case of silicate weathering involving potassium feldspar and phlogopite mica (Table 2.2), H_4SiO_4, HCO_3^-, Mg^{2+}, and

Table 2.2 Examples of reactions that occur between rocks and in CO_2-rich soil water to form the dissolved composition of river water. (a) Reaction with carbonate rocks. (b and c) Reactions with two examples of silicate rocks.

(a) CO_2 in soils reacts with water to form H^+ that dissolves $CaCO_3$ according to the net reaction (iii):	
i	$CO_2 + H_2O \rightleftarrows HCO_3^- + H^+$
ii	$CaCO_3(s) + H^+ \rightleftarrows HCO_3^- + Ca^{2+}$
iii	$CaCO_3(s) + CO_2 + H_2O \rightleftarrows 2HCO_3^- + Ca^{2+}$
	carbonate rocks *dissolved ions*
(b) CO_2 in soils reacts with water to form H^+ that reacts with potassium feldspar to form the clay mineral (kaolinite) according to the net reaction (vi)	
iv	$2CO_2 + 2H_2O \rightleftarrows 2HCO_3^- + 2H^+$
v	$2KAlSi_3O_8(s) + 2H^+ + 9H_2O \rightarrow Al_2Si_2O_5(OH)_4(s) + 2K^+ + 4H_4SiO_4$
vi	$2KAlSi_3O_8(s) + 2CO_2 + 11H_2O \rightarrow Al_2Si_2O_5(OH)_4(s) + 2K^+ + 2HCO_3^- + 4H_4SiO_4$
	K-feldspar *kaolinite* *dissolved ions*
(c) CO_2 in soils reacts with water to form H^+ that reacts with phlogopite mica to form the clay mineral (kaolinite) according to the net reaction (ix)	
vii	$14CO_2 + 14H_2O \rightleftarrows 14HCO_3^- + 14H^+$
viii	$2KMg_3AlSi_3O_{10}(OH)_2(s) + 14H^+ + H_2O \rightarrow Al_2Si_2O_5(OH)_4(s) + 2K^+ + 6Mg^{2+} + 4H_4SiO_4$
ix	$2KMg_3AlSi_3O_{10}(OH)_2(s) + 14CO_2 + 15H_2O \rightarrow Al_2Si_2O_5(OH)_4(s) + 2K^+ + 6Mg^{2+} + 14HCO_3^- + 4H_4SiO_4$
	phlogopite *kaolinite* *dissolved ions*

K^+ enter solution. Bicarbonate is by far the most abundant anion in most rivers, because it is formed during all weathering reactions. According to Table 2.1, about two-thirds of the HCO_3^- in rivers has an atmospheric origin and the other third comes from carbonate rocks. In the simple $CaCO_3$ dissolution reaction in Table 2.2a, half of the carbon source for HCO_3^- comes from CO_2 and half from $CaCO_3$. In the aluminosilicate reactions (Table 2.2b&c), all the HCO_3^- formed has an atmospheric origin. Combination of these mass balance concepts requires that about half of the HCO_3^- in river water that originally came from the atmosphere did so through aluminosilicate weathering (Fig. 2.2).

2.1.2 Residence Times of Seawater Constituents

The simplest conceptual model of seawater composition would be to assume that seawater contains ions in the same ratios as delivered by rivers, but at higher concentrations due to evaporation. Comparison of the relative ion concentrations in rivers and the ocean (Fig. 2.3) indicates, however, that this is clearly not the case. River water contains mostly dissolved Ca^{2+}, Na^+, and HCO_3^-, while the ocean contains mainly Na^+ and Cl^-. The reasons for the differences are the varying reactivities of the elements in the ocean. Since the predominant removal mechanism of dissolved ions from the sea is in the form of solid authigenic minerals, those elements that readily combine to form an insoluble mineral, or rapidly adsorb to the surface of solids, are lower in concentration than those that persist in solution until high concentrations are reached. Observations of saline lakes and evaporite basins like the Dead Sea and the Great Salt Lake demonstrate that NaCl (salt) does not

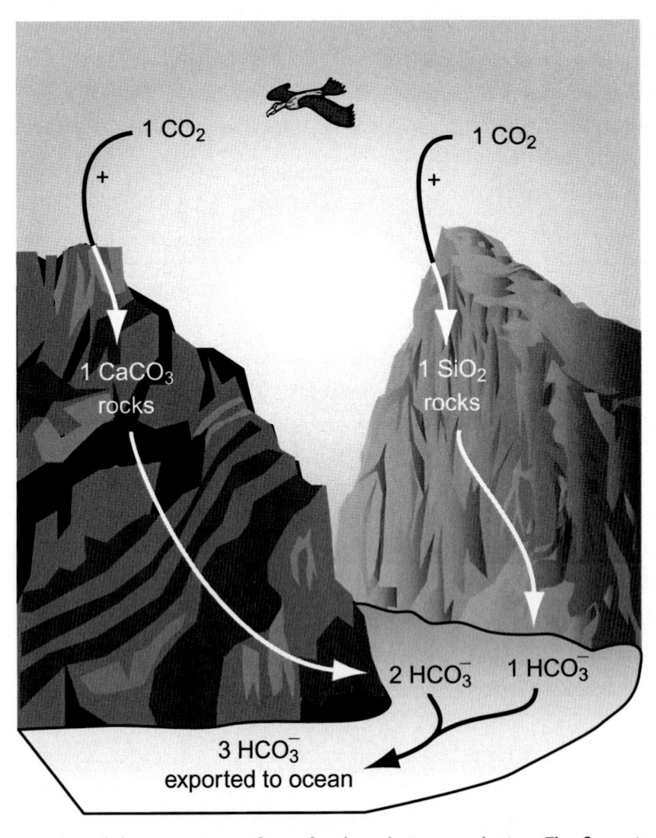

Figure 2.2 A schematic representation of the approximate flow of carbon during weathering. The figure is a representation of the fluxes in Table 2.1, which suggest that about two-thirds of the HCO_3^- produced by weathering via the reactions in Table 2.2 comes from the atmosphere and 1/3 comes from dissolution of $CaCO_3$ rocks. Redrafted from Emerson and Hedges (2008).

precipitate from water except in regions where there are very high evaporation rates and ions become highly concentrated in a brine. Thus, Na^+ and Cl^- are relatively unreactive. On the other hand, the sea floor in some regions of the ocean is almost purely $CaCO_3$, indicating that solid calcite and aragonite readily form at present day seawater concentrations of Ca^{2+} and CO_3^{2-}, making these major ions relatively reactive.

The ratio of the inventory (total amount) of an element in the sea to its inflow (or outflow) rate at steady state (unchanging total mass) is a measure of its reactivity. These two terms combine to define the *residence time* of a dissolved constituent with respect to river inflow, τ_R:

$$\tau_{R_i} = \frac{\text{ocean inventory (mol)}}{\text{river inflow flux (mol yr}^{-1})} = \frac{[C_i]\, V_O}{F_{i,river}}. \tag{2.1}$$

As usual, concentration is indicated by brackets [ ]. The notation V_O is the ocean volume and $F_{i,river}$ is the river inflow flux of the ith constituent. Residence time is the average amount of

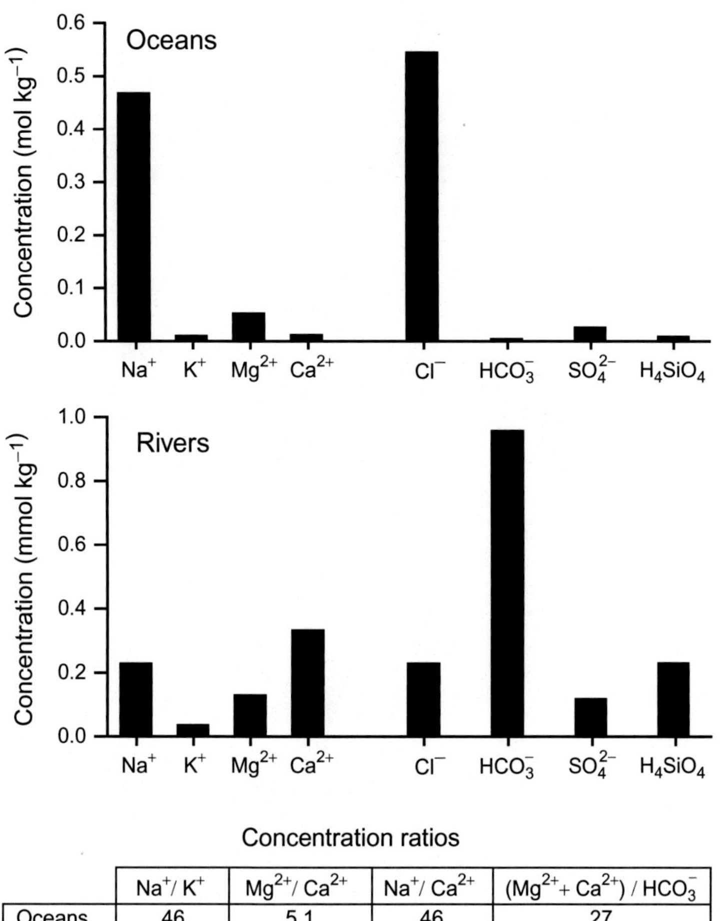

Concentration ratios

	Na^+/K^+	Mg^{2+}/Ca^{2+}	Na^+/Ca^{2+}	$(Mg^{2+} + Ca^{2+})/HCO_3^-$
Oceans	46	5.1	46	27
Rivers	6.0	0.39	0.70	0.48

Figure 2.3 Bar graphs showing the relative concentrations of the major ions in seawater (upper panel) and river water (lower panel). The Table compares concentration ratios in seawater and river water. Redrafted from Emerson and Hedges (2008).

time an ion remains in the ocean, and it is also a measure of the time required to remove a dissolved species from the ocean via chemical reactions. The meaning in mathematical terms is described in more detail in Appendix 2A.2. Residence times of the major constituents in the sea, calculated from their seawater concentrations and global average concentrations in rivers (Table 2.3), vary from ~10^8 y for Cl^- to ~10^5 y for dissolved inorganic carbon $\left(DIC = \left[HCO_3^-\right] + \left[CO_3^{2-}\right] + [CO_2]\right)$. The renewal rate of water via the river inflow and evaporation cycle is on the order of 40 000 y, while the circulation time of the ocean is on the order of 1000 y. Thus, the most reactive of the major ions, HCO_3^- and CO_3^{2-}, reside in the sea only about three times as long as the water, but long enough to circulate through the ocean roughly 100 times before being removed. The least reactive of the major ions, Cl^-, resides in the sea about 2500 times as long as the water. This is why the seawater concentration of DIC is only about three times that of rivers, but the Cl^- concentration is about 2500 times larger.

Discussion Box 2.1

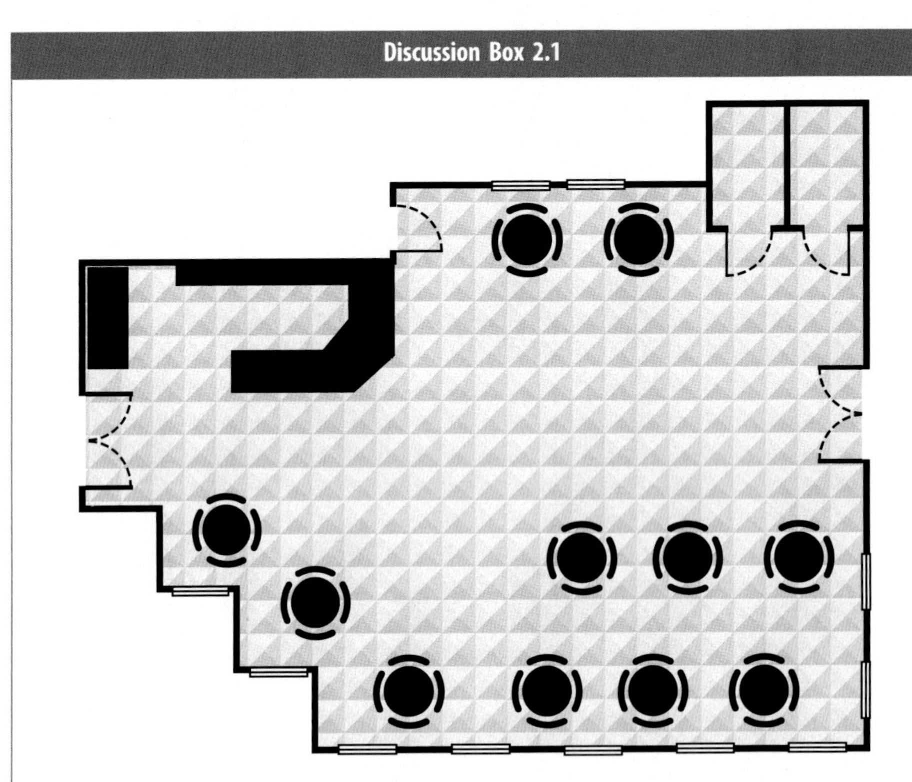

Suppose that your class decides to determine the residence time of people inside the campus coffee shop, with the floor plan shown in the figure. You may assume that you have some modest resources at your disposal for any required equipment.

- What observations do you need to make to determine the residence time of people in the coffee shop? Be specific; what do you observe and record?
- How would you make the calculation? What assumptions do you need to make?
- There are different ways to estimate residence time in a situation like this. What alternate ways can you use to determine the residence time? What observations and assumptions are needed for this alternate method?
- Compare your technique to that discussed in the text for major ions in seawater.

2.1.3 Mackenzie and Garrels Mass Balance

A classic approach to showing the importance of authigenic mineral formation to the removal of major elements from seawater was the chemical mass balance of Mackenzie and Garrels (1966). Their goal was to calculate a rough mass balance of fluxes between

Table 2.3 Residence times of some of the major dissolved elements in the ocean. Global average concentrations in rivers are from Li (2000).

Constituent	Seawater concentration (mmol kg^{-1})	Inventory[a] (10^{18} mol)	River water concentration (μmol kg^{-1})	River inflow[b] (10^{12} mol y^{-1})	τ (10^6 y)
H_2O					0.04
Na^+	469.1	647	231	8.6	75
Mg^{2+}	52.8	72.9	128	4.8	15
Ca^{2+}	10.3	14.2	332	12.4	1.1
K^+	10.2	14.1	38.4	1.4	10
Cl^-	545.9	753	220	8.2	92
SO_4^{2-}	28.2	38.9	115	4.3	9.0
DIC[c]	2.3	3.2	958	35.7	0.1
H_4SiO_4	0–0.2	0.1[d]	158[d]	5.9	0.01

[a] Ocean Mass = 1.38×10^{21} kg
[b] River flow rate = 3.73×10^{16} kg y^{-1} (Dai and Trenberth 2002)
[c] DIC = dissolved inorganic carbon = $HCO_3^- + CO_3^{2-} + CO_2$
[d] H_4SiO_4 values from Tréguer and De La Rocha (2013)

dissolved river inflow and authigenic mineral formation for each major element. They assumed that the composition of rivers had not changed over the longest residence time of the major elements (i.e., that of Cl^-, about 10^8 y) and that the river–ocean system is at steady state over this period of time. This is only a rough calculation, because the time interval is long enough to include very different climates and configurations of the continents and thus probably different weathering and river compositions. Also, many known authigenic minerals are formed intermittently in very restricted locations, like the salt deposits of closed basins and marginal seas. Nevertheless, residence times for all the major ions except Ca^{2+} and DIC are greater than 10^6 y (Table 2.3), so time rates of change will be strongly damped in response to intermittent removal reactions. Mackenzie and Garrels proceeded with the expectation of deriving first-order information about the long-term, marine chemical mass balance.

The first step was to calculate the total mass of each of the major ions discharged by rivers over the last 10^8 y and then remove them from the ocean in a normative way by combining cations and anions to form solids with the stoichiometry of known sedimentary rocks (Table 2.4). Salts found in sedimentary rocks that are made of the major dissolved constituents of river water and seawater are halite (NaCl), calcite ($CaCO_3$), anhydrite ($CaSO_4$), pyrite (FeS_2), and opal (SiO_2) (Table 2.4).

While $CaSO_4$ and FeS_2 are known to be the major sinks of sulfur in sedimentary rocks, their concentrations and accumulation rates are not known well enough to determine burial fluxes. Thus, it was simply assumed that at steady state half of the inventory of SO_4^{2-} is removed to the mineral FeS_2 and half to $CaSO_4$. Although iron is only a trace element in seawater and not listed in Table 2.4, it is the third most abundant element in marine sediments

Table 2.4 (a) A summary of the inventory of major ions in seawater after the normative removal from the ocean by formation of the main sedimentary minerals using the reactions in (b). The table in (a) follows the procedure of Mackenzie and Garrels (1966) but uses river flow rates and concentrations in Table 2.3 and the silica mass balance of Trégeur and De La Rocha (2013).

(a) Inventory

	Major ion →	SO_4^{2-}	Ca^{2+}	Cl^-	Na^+	Mg^{2+}	K^+	H_4SiO_4	HCO_3^-
Mass removed in $10^8 y$ (10^{18}*mol*) →		429	1 238	821	861	477	143	589	3 573
Mineral formed	Moles Removed			*Amount of ion remaining after reaction*					
Pyrite, FeS_2	215[a]	214	1 238	821	861	477	143	589	3 573
Anhydrite, $CaSO_4$	214[a]	0	1 024	821	861	477	143	589	3 573
Calcium carbonate, $CaCO_3$	1 024		0	821	861	477	143	589	1 525
Sodium chloride, NaCl	821			0	40	477	143	589	1 525
Opal, SiO_2	630[b]				40	477	143	0	1 525

[a] Assume half of the SO_4 is removed by pyrite formation and half by $CaSO_4$ formation.
[b] The biogenic opal (SiO_2) burial is taken from Trégeur and De La Rocha, 2013.

(b) Formation reactions

Pyrite	$SO_4^{2-} + 2CH_2O(s) \rightleftarrows S^{2-} + 2CO_2 + 2H_2O$ followed by $Fe^{2+} + S^{2-} + S^0 \rightleftarrows FeS_2$
Anhydrite	$Ca^{2+} + SO_4^{2-} \rightleftarrows CaSO_4(s)$
Calcium carbonate	$Ca^{2+} + 2HCO_3^- \rightleftarrows CaCO_3(s) + CO_2 + H_2O$
Sodium chloride	$Na^+ + Cl^- \rightleftarrows NaCl(s)$
Opal	$H_4SiO_4 \rightleftarrows SiO_2(s) + 2H_2O$

(~6 percent by mass), so it can be assumed that plenty is available as labile iron oxides to supply the Fe necessary to form authigenic pyrite. Since it is assumed the concentrations in the ocean approximate a steady state with respect to inflow and outflow, precipitation of these two minerals is the fate of SO_4^{2-} that flows into the sea via rivers. Sedimentation of the entire inventory of SO_4^{2-} also removes about 15 percent of the calcium ion inventory. Next, the remaining Ca^{2+} is removed as $CaCO_3$, which also removes about 60 percent of the HCO_3^-. Then, Na^+ and Cl^- are precipitated as halite, removing all the Cl^- and about 95 percent of the Na^+. Finally, H_4SiO_4 is removed as opal. Since no additional cations or anions are incorporated into opal, the amount of silicic acid removed is based on estimates of opal accumulation rates in marine sediments. In the original mass balance of Mackenzie and Garrels, the burial rate of opal in marine sediments was not known well; here, values estimated recently by Tréguer and De La Rocha (2013) are used, indicating that inflow and removal are in balance to within the error of the burial estimates.

With the exception of some relatively minor clay mineral exchange and adsorption reactions, the calculation is essentially complete, and a summary of the results (Fig. 2.4) indicates some rather embarrassing conclusions. All the river inflow of Mg^{2+} and K^+ and

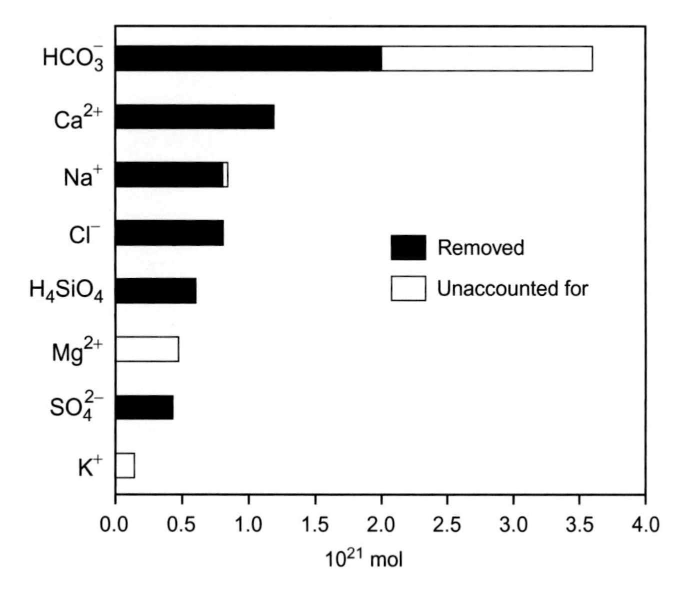

Figure 2.4 A bar graph representing the results of the Mackenzie and Garrels mass balance calculation as redone in Table 2.4. The bar length is the total number of moles of the ion delivered to the ocean in 100 million years. The dark region of the bar is the portion of that flux accounted for by the normative calculation in Table 2.4. Modified from Emerson and Hedges (2008).

about 40 percent of that for HCO_3^- remain unexplained by the formation of the major authigenic minerals!

Even though the HCO_3^- imbalance is not as severe as that for Mg^{2+} and K^+, its excess inflow presents a serious problem in the marine mass balance, because HCO_3^- has the shortest residence time of the major ions, and its concentration is influential in controlling the content of CO_2 in the atmosphere. The global carbon balance between the atmosphere, ocean, and land (Fig. 2.5) is dominated by weathering reactions that consume CO_2 from the atmosphere and by $CaCO_3$ precipitation in the ocean that returns CO_2 to the atmosphere. The first is indicated by the dissolution reactions in Table 2.2a and the latter by $CaCO_3$ precipitation:

$$2\,HCO_3^- + Ca^{2+} \rightleftarrows CaCO_3(s)\downarrow + CO_2(g)\uparrow + H_2O$$

The rate of global production of bicarbonate by weathering can be determined from the flow of HCO_3^- in the world's major rivers. This flux represents a drain of the CO_2 from the atmosphere that must be balanced by resupply to maintain a steady fCO_2 in the atmosphere. (fCO_2 represents the fugacity of CO_2 in the atmosphere, which is similar to the partial pressure but includes the small non-ideality of the CO_2 molecule. See Chapter 5.)

The rate of production of bicarbonate by weathering reactions can be estimated by the flux of HCO_3^- to the ocean via rivers (river flow rate, Q_{river}, times concentration):

$$F_{HCO_3^-,river} = Q_{river}\left[HCO_3^-\right]_{river} = \left(3.7\times10^{16}\ \mathrm{kg\ y^{-1}}\right)\left(9.6\times10^{-4}\ \mathrm{mol\ kg^{-1}}\right)$$
$$= 3.6\times10^{13}\ \mathrm{mol\ y^{-1}}$$

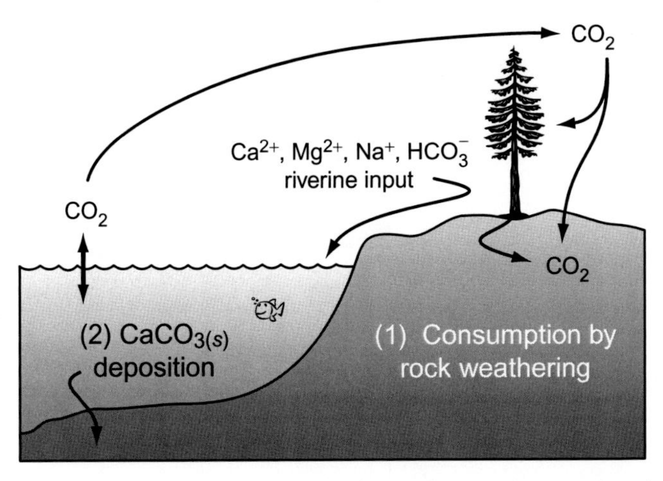

Figure 2.5 A schematic representation of the long-term, global carbon balance between the atmosphere, land, and ocean. Flow (1) represents CO_2 consumption from the atmosphere by weathering and the transport of HCO_3^- to the sea. Flow (2) is the return of CO_2 to the atmosphere by precipitation of $CaCO_3$ in the oceans. Modified from Emerson and Hedges (2008).

And the rate of bicarbonate removal from seawater solution by $CaCO_3$ precipitation (Fig. 2.5) is two times the flux of calcium from rivers. (This is an upper limit for the removal of HCO_3^- by precipitation of $CaCO_3$ because a small amount of the Ca^{2+} from rivers is precipitated as anhydrite in the Mackenzie and Garrels mass balance, Table 2.4.) Using the river flow rate and Ca^{2+} concentration in Table 2.3:

$$F_{\mathrm{CaCO_3,ppt}} = 2\, Q_{\mathrm{river}} \left[\mathrm{Ca}^{2+}\right]_{\mathrm{river}} = 2\left(3.7\times10^{16}\ \mathrm{kg\ y^{-1}}\right)\left(3.3\times10^{-4}\ \mathrm{mol\ kg^{-1}}\right)$$
$$= 2.5\times10^{13}\ \mathrm{mol\ y^{-1}}.$$

There is a bicarbonate imbalance between weathering and marine $CaCO_3$ precipitation of $3.6 - 2.5 \times 10^{13} = 1.1 \times 10^{13}$ mol y^{-1}.

This imbalance stems from the fact that we considered both $CaCO_3$ and silicate rocks during the weathering reactions to create HCO_3^- and remove CO_2 from the atmosphere, but only $CaCO_3$ precipitation to consume HCO_3^- from rivers and return CO_2 to the atmosphere (Figs. 2.2 and 2.5). If there are no other important reactions removing HCO_3^- from seawater, this imbalance would change the ocean's carbonate chemistry and the atmosphere's fCO_2 on a timescale of the residence time of HCO_3^- in seawater (~100 000 years).

The response of atmospheric fCO_2 to the imbalance of bicarbonate sources and sinks to the ocean created by silicate weathering must be considered in the light of the size of the global inorganic carbon reservoirs. There is 50 times more carbon in the dissolved inorganic carbon $\left(\mathrm{DIC} = HCO_3^- + CO_3^{2-} + CO_2\right)$ of seawater than in the atmosphere (see Table 8.1), so the fCO_2 of the atmosphere is in some ways a slave to the concentration of ocean DIC on timescales longer than ocean circulation (~1000 y). Chemical equilibrium between ocean DIC and atm fCO_2 means that processes that alter the DIC of seawater will also change fCO_2^a. The silicate weathering reaction (Eq. vi in Table 2.2) represents a transformation of CO_2 to HCO_3^- without a change in dissolved inorganic carbon in the ocean–atmosphere system – the amount of carbon consumed by CO_2 reaction during silicate weathering is equal to the amount of carbon produced as HCO_3^-. If this reaction is not balanced by a HCO_3^--consuming and CO_2-producing reaction, the fugacity of CO_2 of the ocean and atmosphere will decrease.

The amount of the fCO_2 decrease to be expected based on the inflow–outflow imbalance can be calculated based on marine carbonate chemistry principles. Many readers will not be familiar with details of carbonate equilibrium and some of the terminology of carbonate chemistry, because it is not introduced until Chapter 5. We press on with this discussion anyway to emphasize the magnitude of this geochemical imbalance. One of the end-of-chapter problems in Chapter 5 will return to the question of the geochemical HCO_3^- source and sink imbalance to ensure this discussion is revisited after we study carbonate chemistry.

In carbonate chemistry terms, the exchange of CO_2 for HCO_3^- in the silicate weathering reaction causes an increase in carbonate alkalinity $\left(A_C = HCO_3^- + 2CO_3^{2-}\right)$ but, as stated above, no change in DIC. We show in Chapter 5 (Fig. 5.7) that an increase in alkalinity that is greater than DIC shifts the ocean's carbonate system toward an increase of CO_3^{2-} concentration and a decrease in fCO_2. This change can be explicitly calculated by the carbonate equilibrium reactions, if we know the amount of the alkalinity change. If the imbalance of sources and sinks for HCO_3^- indicated by the Mackenzie and Garrels mass balance persisted for the ocean residence time of HCO_3^-, it would create an increase of alkalinity of $0.8 \times 10^{-3}\,\text{mol kg}^{-1}$ $(1.1 \times 10^{13}\ \text{mol y}^{-1} \times 10^5\ \text{y}/1.38 \times 10^{21}\ \text{kg} = 0.8 \times 10^{-3}\ \text{mol kg}^{-1})$, which is about a 30 percent increase in alkalinity over today's value of $2.42 \times 10^{-3}\ \text{mol kg}^{-1}$ (see Table 5.3). Using the carbonate equilibrium equations described in Chapter 5, one can show that an increase in alkalinity by this amount without a change in DIC (assuming a global mean DIC of $2.09 \times 10^{-3}\ \text{mol kg}^{-1}$) would alter the fCO_2^a in equilibrium with seawater from 290 μatm to 40 μatm. This is a dramatic change, and we know from measurements of fCO_2^a of the ancient atmosphere that nothing of this sort has occurred.

Results of fCO_2^a measurements in atmospheric bubbles of ice cores drilled in the Antarctic ice cap provide a remarkable record of how atmospheric fCO_2^a pressure has changed during the past 800 kiloyears. These data (see Chapter 8, Fig. 8.3) indicate that fCO_2^a has undergone regular ~100 000 y cycles over at least the last 800 ky in which the CO_2 pressure of the atmosphere changed from ~280 μatm during interglacial periods to ~200 μatm during glacials, presumably because of the ocean's response to glacial–interglacial climate changes. However, there is no indication of a longer-term decrease in the mean fCO_2^a over this time period. The relative stability of the glacial–interglacial mean atmospheric fCO_2 over the last hundreds of thousands of years indicates that it is not possible for the ocean-atmospheric carbonate system to be out of balance by the magnitude suggested by the Mackenzie and Garrels mass balance. There must be marine reactions that provide feedbacks to silicate weathering which help stabilize the HCO_3^- concentration in the ocean and the fCO_2^a of the atmosphere on timescales equal to or shorter than the bicarbonate residence time in seawater.

Mackenzie and Garrels (1966) explained the Mg^{2+}, K^+, and HCO_3^- flux imbalance, and the associated carbon cycle problem, by suggesting a major sink for these ions in the form of *reverse weathering* reactions in marine sediments. Reactions of the type in Table 2.2 consume cations, silicate, and bicarbonate if they proceed from right to left, which is in the opposite direction of weathering reactions on land, hence the name "reverse weathering." This solution to the mass balance conundrum has a basis in thermodynamics. Sillen (1967) proposed that reactions driving the major rock-forming minerals and seawater toward thermodynamic equilibrium were the main controls over the composition of seawater, especially seawater pH. The original calculations were handicapped by the lack of necessary thermodynamic data; in particular, the equilibrium constants for important clay mineral reactions were unknown. As the thermodynamic data accumulated, subsequent calculations (e.g., Berner, 1971; Helgeson and Mackenzie, 1970) showed that, among the silicate minerals, Mg- and K-rich clay minerals were thermodynamically stable at the pH of seawater. The implication was that when ion-poor clay minerals like kaolinite are washed into the sea by rivers, they should undergo alteration by consuming cations, particularly K^+ and Mg^{2+}, to form more cation-rich minerals like illite and montmorillonite. Because there was little observational evidence for such reactions at the time, reverse weathering fell out of favor for some time as a mechanism to explain unknown sinks for Mg^{2+}, K^+, and HCO_3^-.

The discovery of hydrothermal vents on the sea floor and the chemical changes brought about by water–rock reactions at high temperature became a more compelling explanation for the shortcomings of the geochemical mass balance. High-temperature reactions were known to be a sink for Mg^{2+} and HCO_3^-, based on laboratory experiments and studies of seawater wells in Iceland, an island born from mid-ocean ridge hydrothermal activity. After measurements were made of the chemistry of high-temperature hydrothermal vents on the sea floor, it was widely assumed that the geochemical mass balance was closed. This conclusion, however, was based on an estimate of the flow rate of seawater through hydrothermal vents, which, it turns out, was not well enough understood to be certain of the flux.

The present status of the geochemical balance between river input to the ocean and removal by heterogeneous chemical reactions is that mechanisms have been discovered that can close the mass balances of Mg^{2+}, K^+, and HCO_3^-, but determining whether the fluxes caused by these processes are great enough to match the inflow from rivers is still uncertain. High-temperature hydrothermal reactions are clearly still important to the geochemical mass balance, but the euphoria that these reactions fully explain imbalances of the required magnitude is gone, because of uncertainties in the rates of flow of seawater through hydrothermal regions. Evidence has been found for low-temperature sedimentary reactions that in some cases resemble "reverse weathering," but determining how widespread these reactions are in ocean sediments is elusive. We present the status of these two most promising solutions to the mass balance problem in the next two sections. Be prepared for a lack of definite answers, as this aspect of the marine geochemistry mass balance is still in flux.

2.1.4 The Sinks: Hydrothermal Circulation

Chemical Changes at High Temperature. Molten basalt nears the sea floor in tectonically active regions, resulting in high-temperature reactions that alter seawater chemistry. This

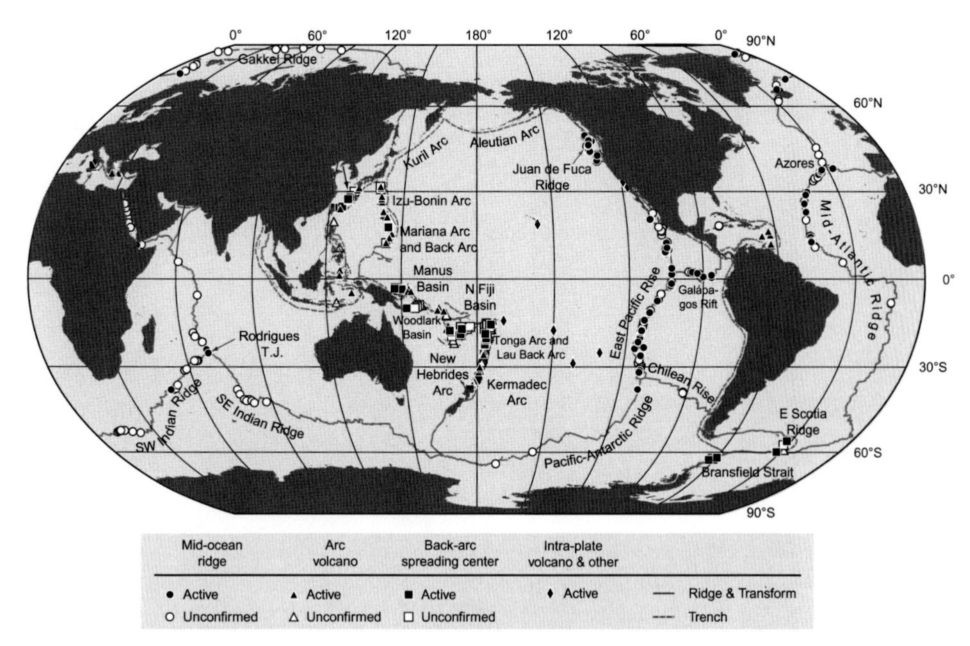

Figure 2.6 Schematic map of global ridge crest and arc systems showing active hydrothermal vents that have been discovered (filled symbols) and those that are known to exist because of characteristic signals in the overlying water column (open symbols). Modified from the International InterRidge program; https://vents-data.interridge.org/maps.

phenomenon is described here by first reviewing the most important chemical changes in the hydrothermal waters circulating in these regions, and then discussing what these changes mean in terms of fluxes to and from the ocean.

Most hydrothermal regions have been discovered at mid-ocean ridge systems where new crust is formed and at island arcs where the crust subducts (Fig. 2.6). In some of these locations, seawater comes in contact with hot, newly formed basalt, causing circulation cells in the vicinity of the heat source in which warm, chemically altered seawater rises to the ocean bottom and fresh, unaltered bottom water flows into the cracked crust to fill the void (Fig. 2.7). Laboratory experiments reveal that when seawater is heated to high temperatures and pressures (e.g., 350°C and 400 atm) in the presence of basalt, Mg^{2+}, SO_4^{2-}, and HCO_3^- are removed, and Ca^{2+} and H_4SiO_4 are released into solution (e.g., Bischoff and Dickson, 1975; Mottl and Holland, 1978). Some of the important reactions are the precipitation of anhydrite ($CaSO_4$), which removes Ca^{2+} from solution (Eq. 2.2), and magnesium-hydroxysulfate-hydrate ($MgSO_4 \cdot (1/4)Mg(OH)_2 \cdot 1/2H_2O$, Eq. 2.3), which removes Mg^{2+} (Li, 2000):

$$Ca^{2+} + SO_4^{2-} \rightleftarrows CaSO_4(s) \tag{2.2}$$

$$\frac{5}{4}Mg^{2+} + SO_4^{2-} + H_2O \rightarrow MgSO_4 \cdot \frac{1}{4}Mg(OH)_2 \cdot \frac{1}{2}H_2O(s) + \frac{1}{2}H^+ \tag{2.3}$$

$$\frac{5}{4}Mg^{2+} + CaSO_4 + H_2O \rightarrow MgSO_4 \cdot \frac{1}{4}Mg(OH)_2 \cdot \frac{1}{2}H_2O(s) + \frac{1}{2}H^+ + Ca^{2+}. \tag{2.4}$$

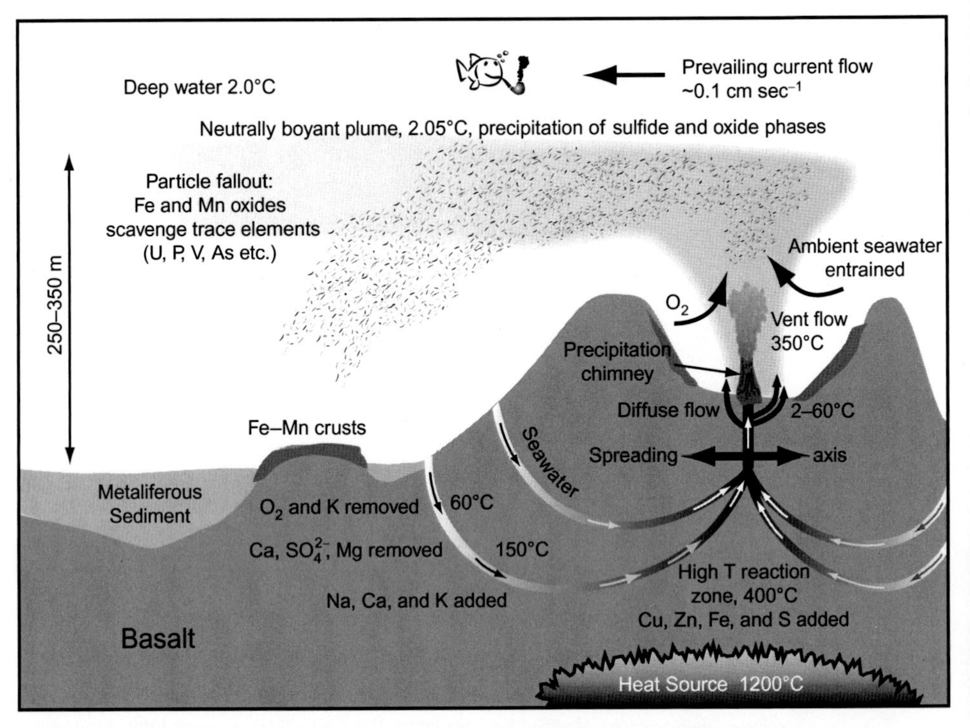

Figure 2.7 Schematic representation of a mid-ocean ridge hydrothermal system and the overlying water. Arrows in the sediments overlying the heat source represent circulation that feeds the high-temperature "black smokers" that outcrop on the sea floor. The shaded region above the sea floor represents the hydrothermal plume that rises and entrains surrounding water until it reaches a density level equal to the surroundings. Dots indicate particles formed from plume metals that originate at the high-temperature source. Modified from German and Seyfried (2014).

Subsequent high-temperature reactions increase the Mg concentration of the authigenic minerals at the expense of Ca, resulting in a release of Ca^{2+} into solution (e.g., Eq. 2.4). The production of hydrogen ions in the above reactions titrates HCO_3^- and CO_3^{2-} to form CO_2 in hydrothermal solutions, resulting in a much lower pH than in normal seawater. Not shown in these equations are the release of H_4SiO_4 to solution by high-temperature silicate reactions and the reduction of SO_4^{2-} to S^{2-} by oxidation of Fe^{2+} to Fe^{3+} within iron-containing silicates.

These general results entered the chemical perspective of oceanography when the first hydrothermal areas on the sea floor were discovered near the Galápagos Islands. John Edmond from MIT did the first detailed sampling of the warm (~20°C) water exiting the hydrothermal vents and found trends that were similar to those found in Icelandic hydrothermal wells and predicted from laboratory experiments (Edmond et al., 1979). Edmond predicted the hydrothermal water endmember concentrations (seawater concentrations in solutions contacting basalt at temperatures exceeding 300°C), and later he and his students measured these concentrations in vents expelling water at these temperatures, sometimes called "black smokers" because of the presence

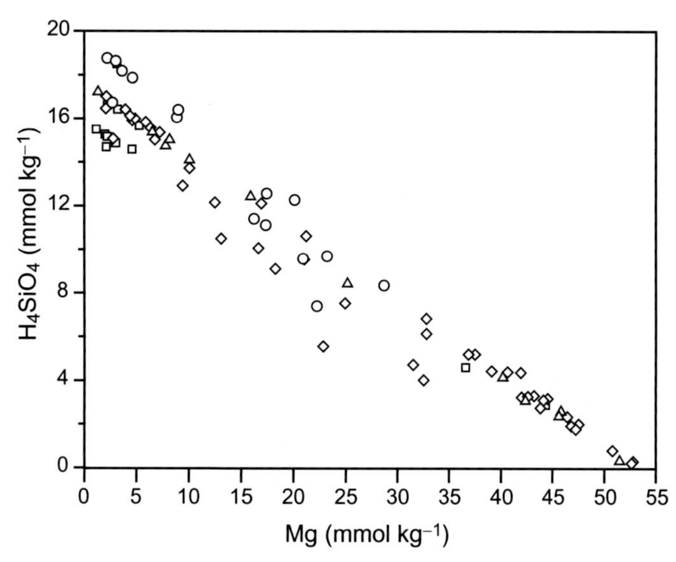

Figure 2.8 H_4SiO_4 and Mg^{2+} concentrations measured in the neck of a "hot smoker" at the 21°N East Pacific Rise hydrothermal area in the North Pacific Ocean. Different symbols indicate data from different vents. The range of concentrations vary from the seawater endmember on the right side with 53 mmol kg^{-1} Mg^{2+} and a few hundred μmol kg^{-1} H_4SiO_4 to the "hydrothermal endmember" with near-zero Mg^{2+} concentration and between 16 and 20 mmol kg^{-1} H_4SiO_4. From Von Damm et al. (1985).

of black iron sulfide particles formed when the waters encounter seawater (e.g., Von Damm et al., 1985).

The high-temperature hydrothermal vent chemical measurements were made with special sampling equipment deployed on submersible vehicles. Titanium syringes were inserted into the hydrothermal water plume, retrieving samples that were a turbulent mixture of the hydrothermal solution with surrounding seawater. An example of the correlation between dissolved Mg^{2+} and H_4SiO_4 in samples from a "black smoker" at 21°N on the mid-Pacific Rise (Fig. 2.8) reveals measurements that reflect a continuous range of mixing between the seawater and hydrothermal endmembers. Since nearly all measurements at 350°C indicate undetectable Mg^{2+} concentrations in the hydrothermal solutions, other constituents are often plotted against the Mg^{2+} concentration so that their hydrothermal endmember concentrations can be identified as the point where $[Mg^{2+}]$ extrapolates to zero. The case in Fig. 2.8 indicates that the hydrothermal endmember has a H_4SiO_4 concentration that is about 100 times that of ocean bottom water in the Pacific. Some of the most notable chemical changes in the hydrothermal endmember solutions from a single vent on the East Pacific Rise are presented in Table 2.5. It should be noted that concentrations of the hydrothermal endmember are variable from vent to vent, partly because of phase separation in those vents that have low enough pressures and high enough temperatures for the solutions to separate into "brine" and "vapor" phases (German and Von Damm, 2003). This complicates interpretation of the data, so they are sometimes

Table 2.5 The chemical composition of hydrothermal endmember water from the Hanging Garden hydrothermal vent at 21°N in the North Pacific Ocean. Hydrothermal endmember/seawater concentration ratios are in the last column. Data are from Von Damm et al. (1985) and are an example of the values determined for the hydrothermal endmember solution.

Chemical species	Concentration (mmol kg^{-1})	Hydrothermal: Seawater
pH (pH units)	3.3	
Alk (meq kg^{-1})	–0.5	
Fe^{2+}	2.4	Very large
Mn^{2+}	0.878	Very large
Si	15.6	98
Li^{+}	1.32	51
Rb^{+}	0.033	25
K^{+}	23.9	2.44
Ca^{2+}	11.7	1.14
Na^{+}	433	0.95
Cl^{-}	496	0.92
Sr^{2+}	0.065	0.75
SO_4^{2-}	0	0
Mg^{2+}	0	0
H_2S	8.7	

presented on a constant Cl^- basis to distinguish phase separation effects from changes due to chemical reactions.

Generally, high-temperature vent fluids are almost totally depleted of Mg^{2+}, SO_4^{2-}, and HCO_3^-. The pH is low enough that the alkalinity is sometimes negative (excess of H^+ over $HCO_3^- + 2CO_3^{2-}$). The vent fluids are enriched in Ca^{2+} and K^+ and extremely enriched in Fe^{2+}, Mn^{2+}, and other trace metals (German and Von Damm, 2003). Thus, hydrothermal solutions are a potential sink for Mg^{2+}, HCO_3^-, and SO_4^{2-}, but they are a source for Ca^{2+} and K^+. Uptake of Mg^{2+} and HCO_3^- at high-temperature vents is in this sense required to provide the missing sink for these elements. Also, the hydrothermal Ca^{2+} source represents an additional sink for HCO_3^- from rivers because hydrothermal Ca^{2+} is introduced to seawater without accompanying HCO_3^-. When the hydrothermal Ca^{2+} combines with HCO_3^- to form $CaCO_3$ via the reverse of reaction (iii) in Table 2.2, riverine-introduced HCO_3^- is removed to the sediments as solid $CaCO_3$ and to the atmosphere as CO_2.

Observations that exacerbate the geochemical mass balance problem, rather than fix it, are the supply of an unnecessary additional sink for SO_4^{2-} ($CaSO_4$ precipitation), and the additional sources for H_4SiO_4 and K^+. An additional sink for SO_4^{2-} does little damage to the marine mass balance in Table 2.4 because it affects the Ca^{2+} mass balance at the level of only about 15 percent even if the hydrothermal sink equals the SO_4^{2-} riverine inflow. The hydrothermal source for silicic acid is impressive at the high-temperature vents

(Fig. 2.8) but is estimated to be only about 10 percent of the riverine inflow rate in recent marine silica mass balances (Tréguer and De La Rocha, 2013). The hydrothermal source for K^+ cannot be rationalized as easily because there is no observed sink in the marine environment for the riverine and hydrothermal sources. Some observations suggest that alkali metals (including K^+) are released from basalts at high temperature and reincorporated into basaltic rock on the sea floor after they have cooled. Thus, K^+ may be, in a sense, "recycled" in the vicinity of hydrothermal vents. The rates of release and incorporation of K^+ by this process are uncertain enough to obscure its importance in the marine K^+ mass balance.

Determining the influence of hydrothermal circulation on the mass balance of chemical constituents in seawater boils down to knowing how much heat and water flows through hydrothermal areas and how important the reactions at high versus low temperature are to the total hydrothermal flux. The rest of this subsection is devoted to explaining how hydrothermal flow rates are determined.

Water Flow through Hydrothermal Systems. A geochemical mass balance approach can be used to determine the global hydrothermal water flow by determining the flow necessary to create a sink equal to the source from rivers. The classic example is the mass balance for magnesium. Since the hydrothermal solution endmember contains essentially undetectable Mg^{2+} (Fig. 2.8, Table 2.5), and, to a first approximation, there is no other significant sink for Mg^{2+} in sedimentary rocks, the marine mass balance for Mg^{2+} is simply that the river inflow equals the hydrothermal removal as seawater circulates through the system:

$$Q_{river}\left[Mg^{2+}\right]_{river} = Q_{hydro}\left(\left[Mg^{2+}\right]_{SW} - \left[Mg^{2+}\right]_{hydro}\right), \tag{2.5}$$

where Q indicates river and hydrothermal flow rates (kg y^{-1}) and brackets as usual indicate concentrations (mol kg^{-1}).

Hydrothermal circulation is viewed as flow of seawater with its ambient concentration into the hydrothermal zone to replace water that is heated and rises to exit on the sea floor (Fig. 2.7). Consequently, the right-hand side of Eq. 2.5 expresses the difference in Mg^{2+} concentration between seawater, $[Mg^{2+}]_{SW}$, and the hydrothermal solution, $[Mg^{2+}]_{hydro}$. In the case of Mg^{2+}, the cool limb takes seawater to the hot zone, and the altered seawater rises with no Mg^{2+} left in solution at all. Since the river inflow of Mg^{2+} is about 4.8×10^{12} mol y^{-1} (Table 2.3), the hydrothermal flow based on this argument is simply:

$$Q_{hydro} = 4.8 \times 10^{12}\ \text{mol y}^{-1}/53\ \text{mol m}^{-3} = 9.1 \times 10^{10}\ \text{m}^3\ \text{y}^{-1} \sim 9.1 \times 10^{13}\ \text{kg y}^{-1} \tag{2.6}$$

For context, the global river flow rate is ~3730×10^{13} kg y^{-1}.

Similar kinds of clever and elegant calculations have been made for helium, strontium, and lithium isotopes as well as germanium/silica ratios (see Elderfield and Schultz, 1996, for a review). However, present geophysical estimates of the hydrothermal water fluxes at high-temperature circulation regions indicate that all of these values are too large!

The key to determining the flow of water through hydrothermal areas using geophysical methods is the relationship between heat and water flux. The total hydrothermal heat flux from mid-ocean ridges is estimated at about 11 TW (TW = terawatts = 10^{12} watts), and ~3 TW of this occurs at the ridge axes due to cooling of newly injected molten basalt during sea floor spreading (e.g., Mottl and Wheat, 1994). A detailed explanation of the source of these numbers is presented in Emerson and Hedges, 2008, section 2.3.2. The heat capacity (C_p) and temperature (T) of the hydrothermal solutions governs the relationship between heat flow and water flow:

$$Q_{\mathrm{H_2O},conv} = \frac{\Sigma q_{conv}}{\Delta T\ C_p} = \frac{\mathrm{J\,s^{-1}}}{\mathrm{K \times J\,g^{-1}\,K^{-1}}} = \mathrm{g\,s^{-1}}, \qquad (2.7)$$

where Σq_{conv} is the convective heat flow at the ridge axis. The heat capacity of seawater is 4.0 J g^{-1} K^{-1} at 5.0°C and 350 bar and 5.8 J g^{-1} K^{-1} at 350°C and 350 bar. If all the axial heat flow of 3 TW were released from black smokers, which are the vents debauching ~350°C water, then the water flow at the axis would be

$$Q_{\mathrm{H_2O},conv} = \frac{\Sigma q_{conv}}{\Delta T\ C_p} = \frac{3 \times 10^{12}\ \mathrm{J\,s^{-1}}}{350\ \mathrm{K} \times 5.8\ \mathrm{J\,g^{-1}\,K^{-1}}} = 1.5 \times 10^{9}\ \mathrm{g\,s^{-1}} = 4.8 \times 10^{13}\ \mathrm{kg\,y^{-1}}. \qquad (2.8)$$

This is only about half the flow calculated from the Mg mass balance in Eq. 2.6.

A more serious problem with the geochemical estimates of the hydrothermal heat flux is that much of the flux occurs at temperatures much lower than that of black smokers. For example, it has been suggested that only about 20 percent of the axial heat flow occurs at high temperatures with the rest occurring at much lower temperatures (German and Seyfried, 2014; Nielsen et al., 2006). The same heat flux at lower temperatures implies, via Eq. 2.8, a much higher water flow and that most of the hydrothermal heat flow occurs at much lower temperatures than exist at ridge crests (Table 2.6).

Table 2.6 Heat and water fluxes associated with sea floor hydrothermal circulation. TW is terawatts (10^{12} watts). For comparison, the convective hydrothermal heat flow determined by heat flow measurements is 11 TW and the water flow from rivers is 3730 × 10^{13} kg y^{-1} (modified from German and Seyfried, 2014).

Location	Water temp. (°C)	Heat flux (TW)	Water flow (10^{13} kg y^{-1})
Axial flow (0–1 Ma*)	All flow at 350°C	2.8	5.6
	20% at 350°C, 80% at 5°C	2.8	375
Off-Axis (1–65 Ma)		7 ± 2	2 000–10 000
Hydrothermal plumes (20% of axial at 350°C diluted by 10^4)		0.56	~11 000

* This value is the age of the crust in millions of years. A sea floor spreading rate of 1 cm y^{-1} would place the 1 Ma age at 10 km on either side of the ridge crest.

The geochemical mass balances were flawed because they assumed that the entire hydrothermal sink or source had the chemistry of the 350°C endmember. The reason for this assumption was simply that almost all the early knowledge of hydrothermal chemistry changes was based on the concentrations of the 350°C endmember. The simple expression in Eq. 2.5 suffers from that assumption. In reality, most of the hydrothermal flux is at low temperatures, and the Mg^{2+} concentration in these waters is not well known.

The most extreme example of low-temperature, off-axis chemical changes comes from hydrothermal plumes that rise above the regions of hydrothermal activity until mixing with surrounding waters cools the plume water to ambient temperatures but leaves the water chemistry still anomalous (Fig. 2.7). As pointed out by German and Seyfried (2014), the dilution of hydrothermal water to plume water can be as large as 10^4 times based on ^3He tracer measurements. Assuming that 20 percent of the axial heat flux is from high-temperature vents that ultimately become plumes and are diluted with surrounding waters by a factor of 10^4 suggests that the flow of deep ocean water through hydrothermal plumes is equivalent to the inflow of water from rivers (Table 2.6), indicating the potential importance of hydrothermal plumes in affecting ocean chemistry.

A very dramatic example of the chemical effects of hydrothermal plumes on ocean chemistry is in relatively recent measurements of Fe, Mn, Al, and ^3He along an E–W section at 10–15°S in the Pacific Ocean (Fig. 2.9). The isotope ^3He is a known indicator of hydrothermal circulation, because this rare isotope of helium is in extremely low concentrations in the atmosphere, and it seeps to the Earth's surface from the interior via volcanoes both on land and in the ocean. Because it is chemically inert, only dilution can change the ^3He concentration as water moves away from hydrothermal areas. The big surprise in these data (Fig. 2.9) is that the ratios of $Fe/^3He$ and $Mn/^3He$ are nearly constant for distances of several thousand kilometers away from the ridge crest, indicating that these chemicals are transported great distances before being chemically removed from solution. Resing et al. (2015) suggest that the refractory nature of iron in seawater is caused by complexation with organic ligands. They argue that, if the data in Fig. 2.9 are typical of hydrothermal vents, up to 20 percent of the iron that supports productivity in Southern Ocean surface waters may have come from hydrothermal regions. The fascinating implication is that biological processes important to the ocean's carbon cycle may be directly influenced by tectonic processes!

The discovery of chemical changes that occur in hydrothermal areas was originally believed to be a very large, previously unrecognized chemical sink that would account for missing sinks in the ocean's chemical mass balance for Mg^{2+}, K^+, and HCO_3^-. As our understanding of the processes at hydrothermal areas deepened, the importance of the most visually dramatic and easily sampled hydrothermal locations (hot smokers) waned. There is little doubt, currently, that hydrothermal processes play an important role in the geochemical mass balance for some of the major ions, and even more for trace metals. However, it has proven difficult to determine the hydrothermal fluxes quantitatively, because lower-temperature circulations probably dominate, and they are too widely dispersed on the ocean floor for their chemistry to be well studied.

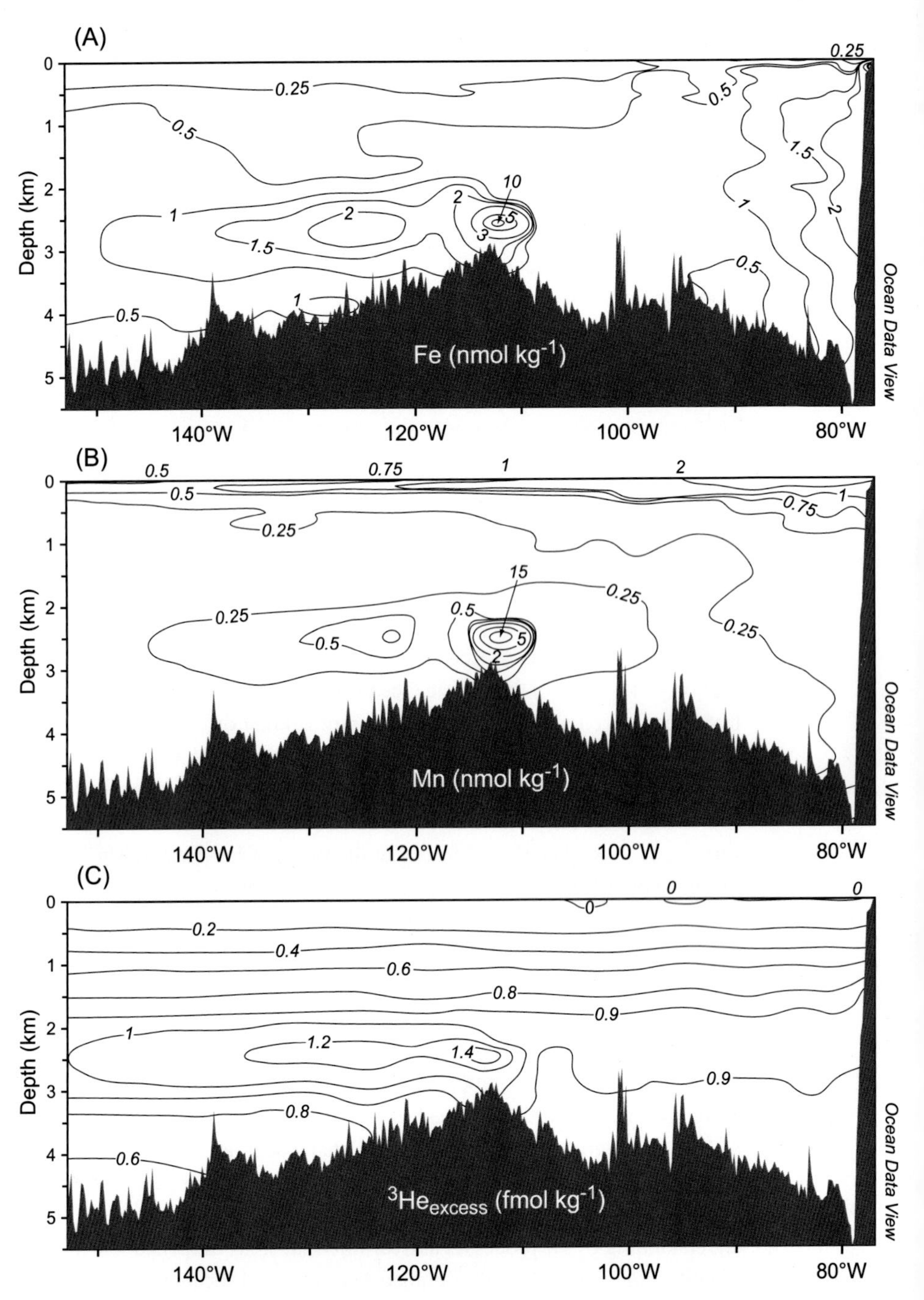

Figure 2.9 Sections of dissolved Fe (A), Mn (B), and excess ^{3}He (C) along GEOTRACES transect GP16 that crossed the East Pacific Rise at 10–15°S. Adapted from Resing et al. (2015).

Discussion Box 2.2

For this discussion, we will consider a simple version of the ocean where the only source of Mg^{2+} is river inflow and the only sink is high-temperature hydrothermal processes that remove 100 percent of the Mg^{2+} from seawater as it flows through the axial hydrothermal system. To start, the system is at steady state.

- Draw a diagram of the important fluxes in the system.
- Now, suppose that the concentration of Mg^{2+} in rivers suddenly doubles and stays at that higher value forever. The rate of water flow out of rivers or through hydrothermal vents is unchanged. What would initially happen (qualitatively) to the average concentration of Mg^{2+} in the ocean?
- What will the average concentration of Mg^{2+} in the ocean be when it reaches steady state again (compared to the concentration before the change in river Mg^{2+})?
- How can you express this system mathematically (i.e., write an equation that would allow you to solve for the average ocean Mg^{2+} concentration)?
- Draw a rough figure showing the evolution of the average ocean Mg^{2+} concentration over time during this scenario.

2.1.5 The Sinks: Reverse Weathering

Recent studies of early diagenesis in marine sediments (chemical and physical changes after deposition) reveal reactions in siliceous deposits that remove Al and cations from pore water to authigenic solid phases. In continental margin sediments, these observations are similar to what might be called reverse weathering, while in deep-sea sediments, observations indicate a more general enrichment of Al in newly formed opaline sediments. Both cases indicate the presence of low-temperature reactions in newly formed marine sediments as suggested by Mackenzie and Garrels (1966).

Detailed studies in deltaic nearshore sediments have demonstrated the existence of rapid diagenesis in which cation-rich aluminosilicates form in rapidly accumulating sediments of the Amazon, Mississippi, and Congo deltas (see Aller, 2014). Pore water observations in these environments (e.g., Fig. 2.10) indicate the removal of potassium from solution during diagenesis. The Cl^- profiles from the same sediment cores demonstrate that potassium is removed by reaction rather than varying merely due to a change in salinity. A generalized equation for these "reverse weathering" reactions was suggested by Aller (2014):

$$\begin{aligned} &\textit{Reactive}\ SiO_2 + Al(OH)_4^- + \left(Fe^{2+,3+}, Mg^{2+}, K^+, Li^+\right) + F^- + HCO_3^- \\ &\quad \rightarrow \textit{aluminosilicates} + CO_2 + H_2O. \end{aligned} \tag{2.10}$$

Michalopoulos and Aller (1995) devised a direct test of these types of reactions by placing seed materials (glass beads, quartz grains, and quartz grains coated with iron oxide) into anoxic Amazon delta sediments. After 12 to 36 months, they observed the formation of K–Fe–Mg-rich clay minerals on the seed materials and suggested that formation of these

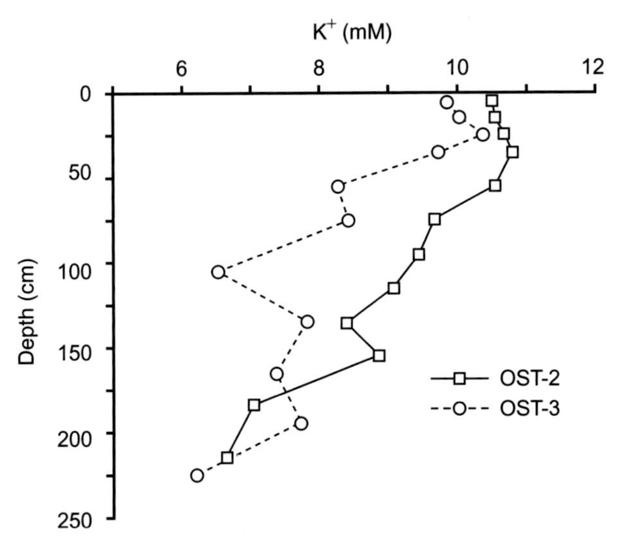

Figure 2.10 Pore water concentrations of potassium in rapidly accumulating deltaic sediments in the Amazon delta. OST-2 and OST-3 indicate two different sites. Redrawn from Aller (2014).

materials in Amazon sediments alone could account for the removal of 10 percent of the global riverine input of K^+. Since environments like the Amazon delta account for ~60 percent of the flux of detrital material to the oceans, the importance of these reactions globally might be much greater than in this single delta alone.

A separate mechanism for reverse weathering-like reactions in marine sediments has been observed in silicate-rich sediments of the deep sea. Only a small fraction of the particulate flux of silica from ocean surface waters is preserved in sediments, but it is still one of the most important components of marine sediments, particularly in the Southern Ocean and equatorial Pacific (Fig. 2.11).

The ocean water column is everywhere undersaturated with respect to opal, meaning that diatom shells reaching the sediments should undergo dissolution until the H_4SiO_4 concentration of the pore waters surrounding them increases to saturation. One of the mysteries of silica diagenesis has been that the saturation H_4SiO_4 concentration obtained by incubating fresh diatom frustules in the laboratory is greater than the asymptotic value observed in sediment pore waters. Experiments using diatoms from Southern Ocean surface waters in stirred flow-through reactors indicate H_4SiO_4 concentrations in chemical equilibrium with opal range between 1000 and 1600 μmol l^{-1} H_4SiO_4 (Van Cappellen and Qiu, 1997a). These values are much greater than those determined in the pore water profiles from the same location at depths where dissolution has ceased (Fig. 2.12C), suggesting that values in sediment pore waters are not at equilibrium with the pure opal produced in the surface waters. A striking clue to the reason for observations like those in Fig. 2.12 is that the asymptotic H_4SiO_4 pore water values are strongly dependent on the amount of detrital (clay) material present in the sediments, i.e., the more detrital aluminosilicates in the sediment, the lower the asymptotic H_4SiO_4 concentration.

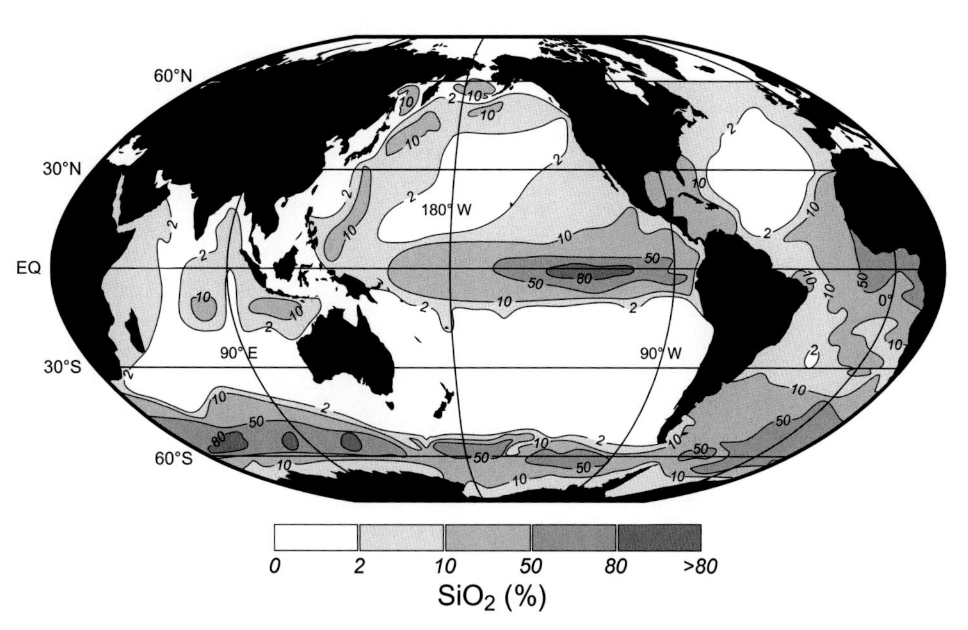

Figure 2.11 The global distribution of biogenic SiO_2 (in wt%) in marine sediments. Redrafted from Broecker and Peng (1982). A black and white version of this figure will appear in some formats. For the color version, refer to the plate section.

Field observations implicating the importance of aluminosilicates to opal diagenesis were followed by laboratory experiments to determine the effect of Al derived from detrital aluminum silicates on the solubility and dissolution kinetics of opal. Dixit et al. (2001) mixed opal-rich (~90 percent SiO_2) sediments from the Southern Ocean with different amounts of either kaolinite or ground basalt in long-term (21-month) batch experiments. The observed concentration of H_4SiO_4 at the end of these experiments was strongly influenced by the presence of the aluminosilicate phase. Values ranged from ~1000 μmol kg^{-1} H_4SiO_4 for nearly pure opal to ~400 μmol kg^{-1} for a 1:4 aluminosilicate:opal mixture – similar to the lowest observations in Fig. 2.12C. An authigenic phase forms in the presence of dissolved Si and Al that is less soluble than pure opal. Detailed observations indicate an Al atom substitutes for 1 in 70 of the Si atoms in opal, decreasing its solubility by about 25 percent.

The flux of dissolved silica from rivers and hydrothermal processes is ultimately balanced by the burial of biogenic opaline silica, bSi, in the form of diatom tests created in the surface ocean. Although the most obvious siliceous sediments are those found in the deep sea, about half of the opal burial takes place in near-shore sediments and deltas because of their rapid sedimentation rates (DeMaster, 2002). The open-ocean opal/aluminum diagenesis studies discussed here reveal the existence of widespread diagenetic alteration of opaline sediments, and the near-shore pore water studies indicate major ion removal from solution as they are incorporated into authigenic solid phases. Both of these studies indicate that low-temperature diagenesis in siliceous sediments is a widespread

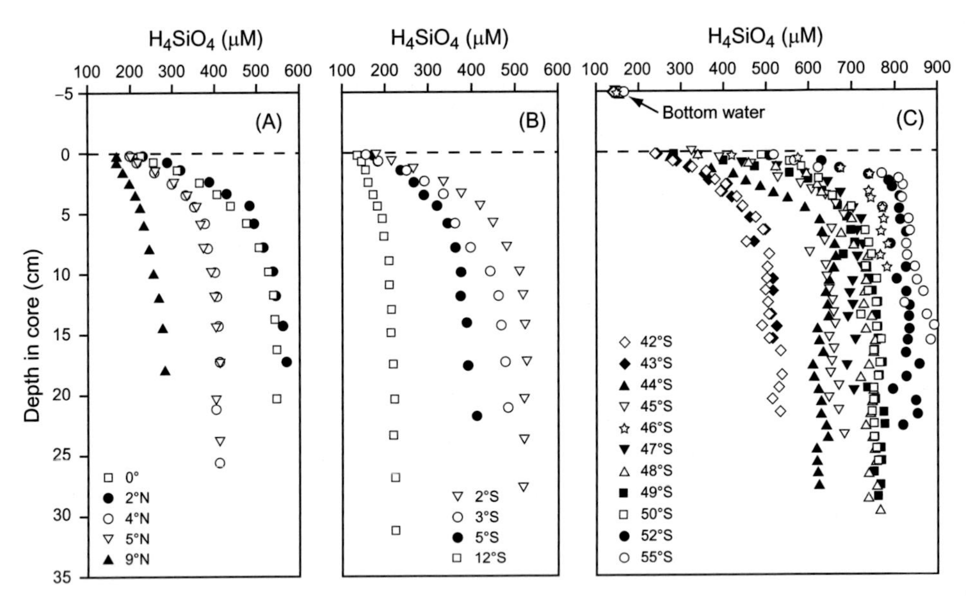

Figure 2.12 Sediment pore water H_4SiO_4 concentrations as a function of depth in sediments from: (A) & (B) a N–S transect along 140°W in the equatorial Pacific (McManus et al., 1995) and (C) a N–S transect through the Indian Ocean sector of the polar front in the Southern Ocean (Rabouille et al., 1997).

process in marine sediments and that "reverse weathering"–like reactions are ubiquitous. Still, it has so far been impossible to be quantitative about the magnitude of the fluxes caused by these reactions.

2.2 Gases in the Ocean–Atmosphere System

The marine mass balance of major ions involves the flow of elements that originate in rivers and exit the ocean as solids in sediments. For most gases, ocean concentrations are dominated by the source–sink exchange between the ocean and atmosphere. The most important control on the concentration of any of the major atmospheric gases in the ocean is thermodynamic (or solubility) equilibrium between the gas in the atmosphere and that in the dissolved phase. Kinetic air–sea gas exchange processes (Appendix 2A.3) determine how rapidly gas concentrations return to equilibrium when perturbed by physical and biological processes.

We begin this section with a brief discussion of gas-water chemical equilibrium, because chemical thermodynamics plays an important role in the air–sea distribution of gases. The fundamentals of thermodynamic chemical equilibrium are described in more detail in the online Chapter 3 of Emerson and Hedges (2008) (www.cambridge.org/emerson-hamme). The kinetics of air–water gas exchange (presented in Appendix 2A.3) is not as formally developed as chemical equilibrium, but is essential for interpreting fluxes of climatically and biologically important gases like CO_2 and O_2 between the atmosphere and ocean.

2.2.1 Air–Sea Chemical Equilibrium: Henry's Law Solubility

The amount of a gas in the atmosphere, or any vapor phase, is most often expressed as a pressure (in units of atmospheres, bars, or pascals). One atmosphere (atm) is equal to 1013.25 millibars (mbar) pressure or 101.325 kilopascals (kPa). The total pressure of gases in the atmosphere, P_{atm}, is equal to the sum of the partial pressures of all the individual gases:

$$P_{atm} = p^a_{N_2} + p^a_{O_2} + p^a_{H_2O} + p^a_{Ar} + p^a_{CO_2} + \ldots, \tag{2.11}$$

where the superscript a indicates atmosphere.

The amount of a gas in the atmosphere can also be quantified by its *mole fraction*, X, which is defined as the number of moles of that gas ($_g$) per total moles of atmospheric gases ($_{atm}$) ($mol_g\ mol_{atm}^{-1}$) in the absence of water vapor. The mole fraction is independent of humidity, which varies geographically and with height in the atmosphere. For an ideal gas, the mole fraction and volume fraction are identical, and the units frequently used for the latter are $cm_g^3\ cm_a^{-3}$ or parts per million by volume (ppmv). The partial pressure of a particular gas, C, and its mole fraction are related to each other via the total atmospheric pressure and the partial pressure of water vapor, $p^a_{H_2O}$:

$$p^a_C = X_C(P_{atm} - p^a_{H_2O}). \tag{2.12}$$

In a dry atmosphere, the partial pressure and mole fraction have the same numerical value when $P_{atm} = 1$. Mole fractions for the major atmospheric gases are presented in Table 2.7. The $p^a_{H_2O}$ at equilibrium with water in the atmosphere varies with temperature and can be the third highest gas partial pressure in the atmosphere! (Saturation equilibrium varies from 0.7 percent of the total atmospheric pressure at 2°C to 4 percent at 30°C; see Appendix C.)

The amount of gas dissolved in water at thermodynamic equilibrium is described by *Henry's law* in which the concentration in the water, $[C]$ (mol kg^{-1} atm^{-1}) and *fugacity*, f, in the gas phase are related via the *Henry's law coefficient*, K_H (mol kg^{-1} atm^{-1}).

$$[C] = K_{H,C} fC. \tag{2.13}$$

The fugacity of a gas is related to partial pressure in the same way as activities are related to concentrations of ions in solution (see online Emerson and Hedges (2008) Chapter 3, www.cambridge.org/emerson-hamme). Interaction of molecules with each other in a real gas diminishes the reactivity of an individual gas slightly, creating an effective partial pressure called the fugacity. As the total gas pressure approaches zero, the pressure and fugacity become equal. In practice, the difference between fugacity and partial pressure for the major gases in the atmosphere is very slight – unlike the differences between activity and concentration for an ion in seawater. Except for CO_2, the fugacity and partial pressure of major atmospheric gases differ by less than 0.1 percent. Klots and Benson (1963) suggest a value for $(p\text{-}f)/p$ for N_2 of 0.014–0.040 percent between 2 and 27°C. Weiss (1974) calculates this value for CO_2 to be much larger but still less than 1 percent (0.30 to 0.44 percent between 0 and 30°C). For this reason, fugacity and partial pressure are frequently used interchangeably.

Gas fugacity is defined for both air and water, such that the concentration of a dissolved gas is at solubility equilibrium (or atmospheric saturation, $[C_{eq}]$) when the atmospheric and water fugacities are equal:

Table 2.7 Solubilities of the major atmospheric gases in seawater ($S_P = 35$) at one atmosphere pressure and 20°C. X_C is the mole fraction in the atmosphere. (Greenhouse gases, CO_2, CH_4, N_2O, and SF_6, are from NOAA's 2019 global means; DMS is from Woodhouse et al., 2010; and the others are from Glueckauf, 1951). $K_{H,C}$ is the Henry's law coefficient. $[C_{eq}]$ is the concentration expected in seawater at saturation equilibrium with the atmosphere. Saturation concentrations and Henry's law coefficients are calculated using equations in Appendix D. β and α are the Bunsen and Ostwald solubility coefficients, respectively (see the footnotes).

Gas	X_C ($mol_g\ mol_{atm}^{-1}$)	$K_{H,C}$ ($mol\ kg^{-1} atm^{-1}$)	$[C_{eq}]$ [a] ($\mu mol\ kg^{-1}$)	β [b] ($cm_g^3\ cm_{sw}^{-3}$)	α [c] ($cm_g^3\ cm_{sw}^{-3}$)
N_2	7.8084×10^{-1}	5.50×10^{-4}	4.198×10^{2}	1.26×10^{-2}	1.36×10^{-2}
O_2	2.0946×10^{-1}	1.10×10^{-3}	2.255×10^{2}	2.53×10^{-2}	2.72×10^{-2}
H_2O	up to 4×10^{-2}				
Ar	9.34×10^{-3}	1.21×10^{-3}	1.108×10^{1}	2.79×10^{-2}	2.99×10^{-2}
CO_2	4.10×10^{-4}	3.24×10^{-2}	1.28×10^{1}	7.44×10^{-1}	7.99×10^{-1}
Ne	1.818×10^{-5}	3.84×10^{-4}	6.83×10^{-3}	8.83×10^{-3}	9.47×10^{-3}
He	5.24×10^{-6}	3.31×10^{-4}	1.70×10^{-3}	7.61×10^{-3}	8.17×10^{-3}
CH_4	1.87×10^{-6}	1.21×10^{-3}	2.22×10^{-3}	2.79×10^{-2}	2.99×10^{-2}
Kr	1.14×10^{-6}	2.19×10^{-3}	2.44×10^{-3}	5.03×10^{-2}	5.40×10^{-2}
N_2O	3.32×10^{-7}	2.33×10^{-2}	7.56×10^{-3}	5.35×10^{-1}	5.74×10^{-1}
Xe	8.7×10^{-8}	3.95×10^{-3}	3.36×10^{-4}	9.08×10^{-2}	9.74×10^{-2}
SF_6	9.96×10^{-12}	1.91×10^{-4}	1.86×10^{-9}	4.39×10^{-3}	4.71×10^{-3}
DMS	up to 3×10^{-10}	5.68×10^{-1}	up to 5×10^{-4}	1.31×10^{1}	1.40×10^{1}

[a] $[C_{eq}] = K_{H,C} fC$; the fugacity is assumed equal to the partial pressure, p, except for CO_2.
[b] $\beta = K_H(RT_{STP})\rho$; where $R = 0.082057$ L atm K^{-1} mol^{-1}; $T_{STP} = 273.15$ K; ρ is the density of seawater (at $T_{Celsius} = 20$°C and $S_P = 35$, $\rho = 1.0248$ kg L^{-1}).
[c] The Ostwald solubility coefficient, $\alpha = K_H RT\rho = \beta(T/T_{STP}) = \beta T/273.15$.

$$fC^a = fC^w, \tag{2.14}$$

where the superscript a indicates air and w indicates the water phase. It can seem a little strange to think about a gas dissolved in water as having what is essentially a partial pressure, so imagine that f^w signifies the fugacity that a gas would have if it were at equilibrium with the water. The Henry's law relationship in Eq. 2.13 makes it possible to determine the fugacity of a gas in water if the concentration is known:

$$[C] = K_{H,C} fC^w. \tag{2.15}$$

A special case of this equation is the relationship between the solubility equilibrium concentration in seawater and the fugacity of a gas in air:

$$[C_{eq}] = K_{H,C} fC^a. \tag{2.16}$$

Chemical oceanographers also refer to the departure of the concentration of a gas C from solubility equilibrium, ΔC, as its "saturation anomaly" or "supersaturation," which is frequently presented as a percentage because the numbers are small. Positive values are supersaturated with respect to atmospheric equilibrium and negative values are undersaturated.

$$\Delta C(\%) = \left(\frac{[C]}{[C_{eq}]} - 1 \right) \times 100 \tag{2.17}$$

Using the Henry's law coefficient can sometimes be tricky, because different units are used in the literature. In this text, we present the Henry's law coefficient in units of moles per kilogram atmosphere (mol kg^{-1} atm^{-1}). Other units that are often used are molar (mol l^{-1} atm^{-1}) and volume fraction (ml_{gas} l_{soln}^{-1} atm^{-1}). The ml_{gas} in the latter is defined at standard temperature and pressure. (STP is 0°C and 1 atm.) Since the volume of a mole of ideal gas at STP is 22.414 liters, there is a uniform relationship between moles and mL (STP). Another potential confusion is that the Henry's law relationship is sometimes referred to as the reciprocal of the value given here, e.g., $1/K_H$. The bulk of the marine literature follows the definition used in Eq. 2.13.

Other solubility coefficients frequently encountered are the *Bunsen coefficient*, β, and the *Ostwald solubility coefficient*, α. These values are directly related to the Henry's law coefficient (see the footnotes in Table 2.7), but they present solubility in different units that have some conceptual advantages. The Bunsen coefficient is defined as the volume of gas at STP dissolved in a unit volume of solvent at some temperature, T, when the partial pressure of the gas is 1 atm. The Ostwald solubility coefficient is the same as the Bunsen coefficient except that the gas volume is at the same temperature as the solvent water rather than STP. These two quantities represent equilibrium between the pure gas and water, where the solvent and solute are presented in the same units. One can envision these constants as air–water partition coefficients for a pure gas at equilibrium with water, when the gas phase is 1 atmosphere and the gas and water reservoirs have equivalent volumes. A tricky thing about Bunsen and Ostwald solubility coefficients is that they represent a volume of gas per volume of solvent (not solution). Since gases increase the volume of the solution when they dissolve into it, a correction on the order of 0.14 percent should be made for this difference (Weiss, 1970). The values presented in Table 2.7 have not been corrected for this effect, and since this is a potential point of confusion, we use the Henry's law coefficient throughout this book.

Henry's law coefficients of the major atmospheric gases and their temperature dependencies are quite variable (Fig. 2.13). Equations used to calculate values for K_H in seawater (Table 2.7) are presented in Appendix D, and MATLAB and Python routines for determining the saturation concentration as a function of temperature and salinity are available at www.cambridge.org/emerson-hamme. Helium is the least soluble noble gas, and its solubility also depends very little on temperature. On the other hand, krypton is almost 10 times as soluble as helium with a much greater temperature dependence. (Xenon is even more soluble and temperature-dependent but not shown in Fig. 2.13 to avoid a y-axis scale that is too large.) One of the most striking observations demonstrated by Fig. 2.13 is that gases are more soluble in cold water than in warm water, exactly opposite to most solids and the solubility of water vapor in air (the humidity). Most of us are familiar with the concept of heating a pot of water to get more salt or sugar to dissolve in it. In that case, the energy

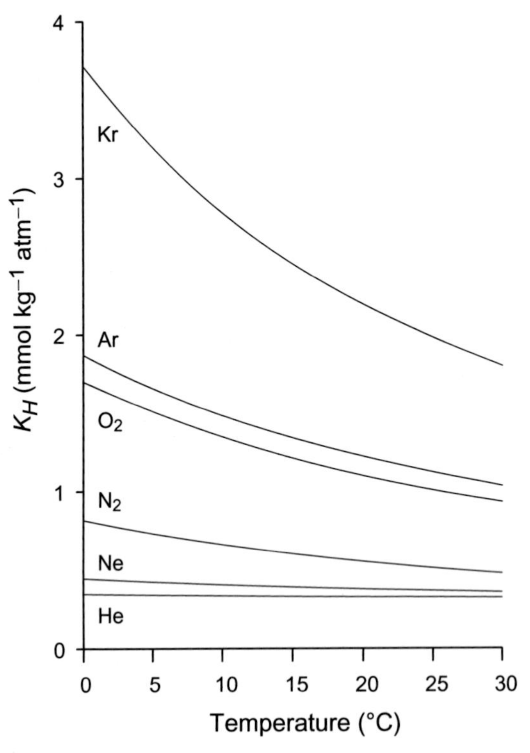

Figure 2.13 The Henry's law coefficients, K_H (mmol kg^{-1} atm^{-1}) of noble gases, O_2, and N_2 in surface seawater with $S_P = 35$ as a function of temperature. The curves are drawn using the relationships in Appendix D. Modified from Emerson and Hedges (2008).

added by heating breaks ionic bonds in the solid, promoting the dissolved phase. For gases, the dissolved phase contains small van der Waals interactions between the gas and surrounding water molecules that stabilize the presence of the dissolved gas. Adding heat acts to break those interactions, promoting the gaseous (bond-free) state.

Another notable aspect of the data in Fig. 2.13 is that gas solubility–temperature dependence is not linear. Thus, mixing between parcels of water of different temperatures at saturation equilibrium with the atmosphere results in a mixture that is supersaturated. This effect has been observed for noble gases in the ocean and has been used to evaluate mixing rates between waters of different densities (e.g., Emerson et al., 2012).

A little-discussed aspect of thermodynamic solubility is that gas solubility decreases with increasing hydrostatic pressure. The effect is fairly large, about 14 percent per 1000 m of water depth (Enns et al., 1965; Taylor, 1978). This pressure effect on solubility has few implications for most oceanographic circumstances, because dissolved gases normally interact with the gas phase only at the water surface where the hydrostatic pressure is zero and the atmospheric pressure is one atmosphere. Once the concentration of gases is set at the air–water interface, it is not changed as the water circulates to depth except by internal ocean processes that produce or consume the gas, like biological respiration of oxygen. Situations where the pressure effect on solubility is relevant include some instruments that are used to determine in situ measurements of gases. A gas tension device (GTD) is an instrument used to determine the total gas pressure in the water using a pressure sensor mounted in an air-filled chamber separated from the water by a gas-permeable membrane. In this case, a gas phase is created artificially at depth, and the pressure effect must be considered when interpreting the results.

Discussion Box 2.3

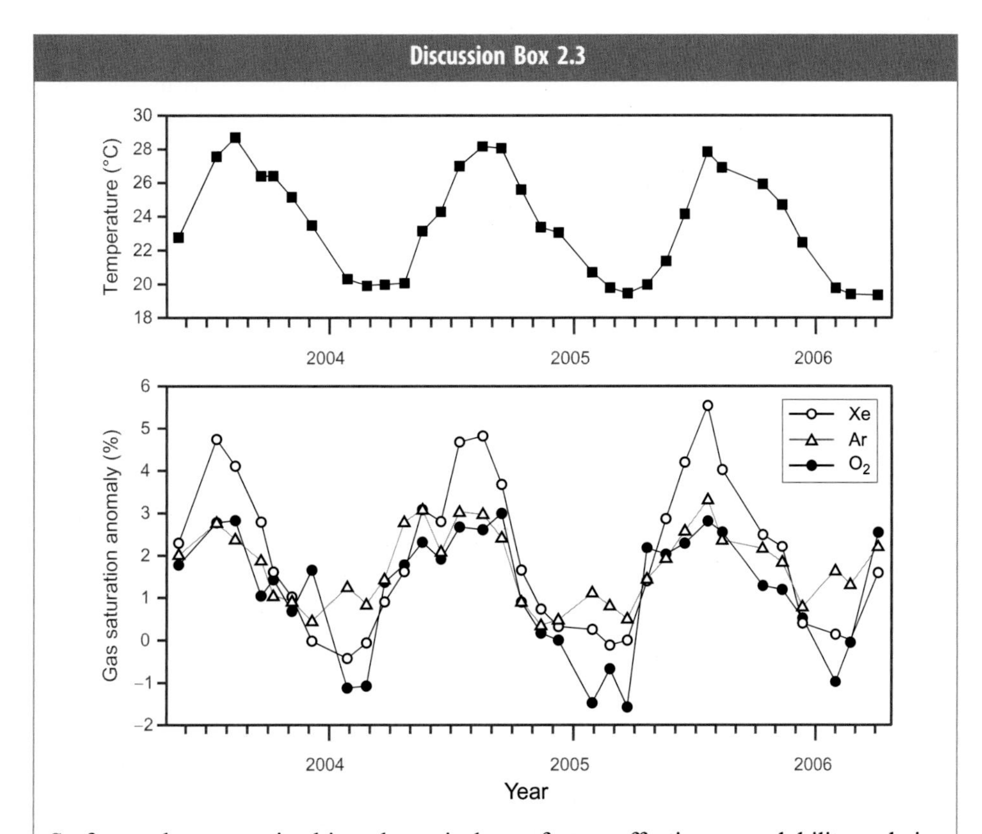

So far, we have examined in a theoretical way factors affecting gas solubility and air–sea gas exchange fluxes (Appendix 2A.3). Now, let's apply this theory to understanding some real observations. This figure shows a three-year time series of surface temperature as well as gas saturation anomalies in the mixed layer for oxygen, argon, and xenon at a station near Bermuda at approximately 32°N, 64°W (data from Stanley et al. 2009).

- First consider the inert gases Ar and Xe. What could explain the periods of high saturation anomaly for these gases? What about the periods of low saturation anomaly?
- What could cause Ar and Xe to behave differently from each other?
- O_2 and Ar have nearly the same physical properties (solubility, diffusion coefficient, etc.). How do these gases compare to each other in summer? In winter? Speculate about what explains differences between them.

2.2.2 Gas Sources and Sinks in the Ocean–Atmosphere System

The residence time for gases in the combined ocean-atmosphere system with respect to processes that produce or consume them (Table 2.8) yields information about the reactivity of the gas, just like the residence times for ions in solution (Table 2.3). The residence times are evaluated by determining the inventory of the gas in both the ocean and atmosphere and dividing by the production or consumption rate of the gas. Atmospheric pressures in Table 2.8 are measured values, and concentrations in the ocean are calculated assuming solubility equilibrium with a standard atmosphere at 2°C. Concentrations of N_2 and Ar in the ocean are observed to be near equilibrium, but O_2 is on average about 50 percent depleted by respiration in the deep ocean. The CO_2 inventory is a special case, because its fugacity is determined by chemical equilibrium with other components of the dissolved inorganic system $\left(DIC = CO_2(aq) + HCO_3^- + CO_3^{2-}\right)$. The total inventory for CO_2 in the ocean is thus the measured concentration of DIC (see Chapter 5).

Gases N_2, O_2, and Ar are very insoluble – the total amount in the atmosphere greatly exceeds that in the ocean, i.e., >99 percent of the gas is in the atmosphere (Table 2.8). The gas concentrations in the ocean for these insoluble gases are controlled by solubility equilibrium with the atmospheric fugacity. The case for CO_2 is the opposite, because CO_2 is in chemical equilibrium with HCO_3^- and CO_3^{2-}, both of which have much higher dissolved concentrations. The amount of carbon in the DIC in seawater is 45 times that of the CO_2 in the atmosphere. On average, the fugacity of CO_2 in the atmosphere is within 10 percent of that in the surface ocean, making fCO_2^a primarily controlled by the processes that control the fCO_2 of ocean surface water.

Residence times for N_2 and O_2 are determined with respect to the most important long-term flux from/to the ocean–atmosphere system – the burial rate of organic matter in

Table 2.8 Residence times of atmospheric gases in the combined ocean/atmosphere system. Mole fractions in the atmosphere are denoted as X_C. Ocean concentrations were calculated based on equilibrium with the atmosphere at 2°C via Henry's law constants except for CO_2. The fCO_2 of the atmosphere and the dissolved inorganic carbon (DIC) content of the ocean are measured values. Fluxes for N_2 and O_2 are calculated from organic carbon (OC) burial rates and N/OC or O_2/OC ratios. The range of OC burial estimates is $0.04–0.13 \times 10^{14}$ mol C y^{-1} (from Berner, 2006, and Hedges and Keil, 1995, respectively). The N/OC ratio is assumed to be 0.03 (Berner, 2006) and the O_2:OC ratio during respiration is assumed to be −1.0.

Gas	X_C mol_g/mol_{atm}	Inventory Atm (10^{18} mol)	Inventory Oce. (10^{18} mol)	Inventory Σ (10^{18} mol)	Oce./Σ	Flux (10^{14} mol y^{-1})	τ (10^6 y)
N_2	0.7808	138	0.8	139	0.006	OC bur. × 0.03 = .0005–.002 (N_2)	700–2 300
O_2	0.2095	37.1	0.4	37.5	0.01	OC bur. × 1.0 = 0.04–0.13	3–9
Ar	0.0093	1.65	0.02	1.67	0.01	^{40}K decay	1800
CO_2	0.0004	0.07	3.20 (DIC)	3.27	0.98	Weathering = 0.36	0.09

marine sediments. The nitrogen sink consists of reduced nitrogen primarily in amino acids of organic matter and thus is determined by the organic C burial rate times the N/C ratio of organic matter in shales and coal deposits ($\Delta N:\Delta C = 0.03$, Berner, 2003). Since roughly one mole of O_2 is produced per mole of organic carbon buried ($\Delta O_2:\Delta C = -1.0$, Hedges and Keil, 1995), this organic matter burial represents the most important long-term flux of oxygen to the atmosphere. There are a range of estimates for the long-term organic carbon burial rate. Using the values suggested by Berner (2003) and Hedges and Keil (1995) as endmembers results in a range of about a factor of 3 for the residence times for oxygen and nitrogen gas in the atmosphere. By these calculations, residence times of N_2 and O_2 are very long – more than a billion years for N_2 and 3–9 million years for O_2.

Since the major fluxes of oxygen to and from the atmosphere are photosynthesis and respiration ($\sim 127 \times 10^{14}$ mol y^{-1} – more than two orders of magnitude greater than organic carbon burial), one could calculate a much shorter oxygen residence time with respect to these fluxes. Photosynthesis and respiration, however, are gross fluxes that are nearly equal, leaving little net production of organic carbon or molecular oxygen. Gross fluxes exchange the oxygen atoms, but do not change the concentration. As an example, the residence time for the isotope ratio of oxygen in the atmosphere is only about 3000 years and is determined by the atmospheric inventory of oxygen divided by the global respiration rate. Because oxygen is fractionated by respiration but not by photosynthesis – the gross flux matters for determining the residence time of the isotope ratio. The residence time for the oxygen inventory is more than 1000 times longer, because it can change only if the rates of photosynthesis and respiration are unequal resulting in net burial of organic C and production of O_2.

Similarly, fluxes of nitrogen fixation and denitrification in the ocean are much greater than the burial rate of organic nitrogen in sediments, but again these are gross fluxes with respect to the N_2 reservoir. The nitrogen fixation/denitrification exchange is between the pools of fixed nitrogen (mostly NO_3^-) and N_2. The inventory of N_2 is four orders of magnitude greater than that for NO_3^- so changes in the nitrogen fixation and denitrification rates do not greatly influence the atmospheric N_2 concentration on these timescales.

The residence time for CO_2 in the atmosphere and ocean system (Table 2.8) is the same as that determined earlier in this chapter for DIC (Table 2.3). The result determined using the organic carbon burial flux, as done for O_2 and N_2 gives a value that is about 10 times longer demonstrating that weathering of inorganic carbonates and silicates creates the dominant flux of carbon to the ocean. Argon gas is inert except for its production by the radioactive decay of potassium–40. An approximate global residence time of ^{40}Ar is the mean life of ^{40}K, which is its radioactive half life (1.2×10^9 y) divided by 0.693. (See Chapter 7.)

Appendix 2A.1 A Brief Review of Rocks and Minerals

To help understand the discussion of weathering, a very brief review of the names and chemical formulae for some of the important rock-forming minerals are presented here. The silicate discussion is mainly from Drever (1982).

Igneous Rocks

The ultimate source of most cations, as well as the silicic acid dissolved in rivers and the ocean, is igneous rock. Granites are light-colored acidic rocks, while basalts are dark colored with high concentrations of metal ions. These rocks originate from deep within the Earth, where at one time they were in a molten state. They are made of minerals like *feldspar*, *mica*, and *quartz*. *Feldspars* are the pink, green, and white minerals visible in granite:

$$\textit{Potassium Feldspar} = KAlSi_3O_8\,\text{(potassium aluminosilicate)}$$
$$\textit{Albite Feldspar} = NaAlSi_3O_8\,\text{(sodium aluminosilicate)}.$$

Micas are the shiny flakes in igneous rocks that catch your eye by reflecting sunlight while you pick your way along a path on a mountain hike:

$$\textit{Phlogopite Mica} = KMg_3AlSi_3O_{10}(OH)_2.$$

Quartz is the glassy-looking parts of granitic rocks:

$$\textit{Quartz} = SiO_2.$$

Clay Minerals

Clay minerals are formed when igneous rocks weather. These minerals are the main constituent of fine grained (<63 μm) particles in mud. In general, these minerals are less cation rich than their igneous precursors. *Kaolinite* has the simplest clay mineral formula because it is pure aluminosilicate. It is the mineral that held the secret to making porcelain, which was greatly valued by the emperors of China before AD 1000 after they discovered how hard and clear kaolinite becomes when heated to 1300–1400°C. Other, more complicated clay minerals, e.g., *illite* and *montmorillonite*, have various amounts of cations added to their structures.

Kaolinite = $Al_2Si_2O_5(OH)_4$(aluminosilicate)
Illite = Similar to phlogopite mica, but less cation rich.
Montmorillonite = $(Na, Ca)(Al, Mg)_6(Si_4O_{10})_3(OH)_6 - nH_2O$

Authigenic Minerals

Minerals that precipitate from solution at Earth's surface temperatures and pressures are termed *authigenic* minerals. These minerals are: the main constituents of shells of plants and animals that live primarily in the ocean surface waters; evaporite minerals that precipitate in places like the Dead Sea and the Great Salt Lake; and clay minerals that form in marine sediments. The main authigenic minerals formed from seawater are

Calcite and Aragonite = $CaCO_3$ (shells of coccolithophores, foraminifera, corals),
Opal = $SiO_2(H_2O)_n$ (shells of diatoms and radiolarians),
Pyrite = FeS_2 (forms in SO_4^{2-}-reducing sediments),
Halite = NaCl (salt, formed in evaporite basins),
Gypsum and Anhydrite = $CaSO_4(H_2O)_2, CaSO_4$ (formed in evaporite basins).

2A.2 Appendix 2A.2 **The Meaning of Residence Time**

Two analogies are presented to explain the meaning of residence time. The simplest is that of filling a reservoir. Consider a child's swimming pool with a volume, V_{pool} (m^3), being filled with a water hose of inflow rate, Q_{in} ($m^3\ hr^{-1}$). How long does it take to fill the pool? From time $t = 0$ to $t = \tau$ the rate of change in volume is

$$dV = Q_{in} dt. \tag{2A.2.1}$$

We define the residence time, τ_{fill}, as the time when the pool starts overflowing, as illustrated in Fig. 2A.2.1A, by

$$\tau_{fill} = \frac{V_{pool}}{Q_{in}}. \tag{2A.2.2}$$

In this analogy, the residence time is the time at which the volume of the system reaches steady state (i.e., the time at which the pool begins to overflow).

Now let's make the analogy a little more realistic. The pool is full, but the water has gotten dirty. It has *E. coli* bacteria of concentration, C (mol m^{-3}), and you want to renew the water in the pool with clean water that has an *E. coli* concentration of C_{in}. At $t = 0$, you begin the inflow again. Your dog is swimming in the pool during renewal to keep the water well mixed. How long will it take to renew the old water with freshwater?

Since the concentration of *E. coli* in the pool asymptotically approaches that of the inflowing water, the pool water will theoretically never be *totally* renewed, so one defines a time, τ_{renew}, when a certain fraction of the old water has been replaced. What is the time τ_{renew}, and how much water has been replaced at this time?

The concentration change from the beginning is now

$$V_{pool}\frac{dC}{dt} = C_{in}Q_{in} - CQ_{in}, \text{ with initial conditions at } t = 0 \text{ of } C = C_0. \tag{2A.2.3}$$

These equations have the solution below that is illustrated in Fig. 2A.2.1B:

$$C - C_{in} = (C_0 - C_{in})e^{-\left(\frac{Q_{in}}{V_{pool}} \cdot t\right)} \tag{2A.2.4}$$

at $t = \tau_{renew}$

$$\frac{C - C_{in}}{C_0 - C_{in}} = e^{-\left(\frac{Q_{in}}{V_{pool}} \cdot \tau_{renew}\right)}. \tag{2A.2.5}$$

Now, if we assume that the renewal residence time is equal to the final, steady-state inventory in the pool divided by the rate of inflow, as in the definition of the residence time in Eq. 2A.2.1, then τ_{renew} is equal to τ_{fill}:

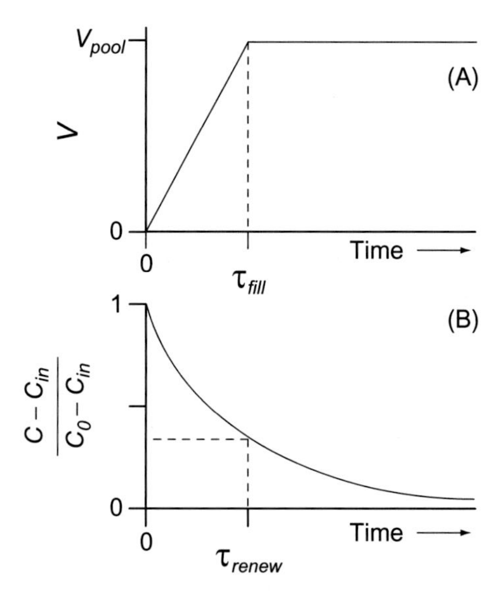

Figure 2A.2.1 (A) The volume of water, V, in the swimming pool as a function of time as defined by the solution to Eq. 2A.2.1. V_{pool} is the final value when the pool is full. (B) The relative concentration of *E. coli* in the pool as a function of time as defined by the solution to Eq. 2A.2.4. C_0 is the concentration of *E. coli* in the pool at time $t = 0$, while C_{in} is the concentration in the water flowing into the pool, such that the relative concentration is one ($C = C_0$) at $t = 0$ and approaches 0 over time ($C = C_{in}$). Residence times, τ, are indicated for both models. Modified from Emerson and Hedges (2008).

$$\frac{V_{pool}C_{in}}{Q_{in}C_{in}} = \frac{V_{pool}}{Q_{in}} = \tau_{renew}. \tag{2A.2.6}$$

Substituting Eq. 2A.2.6 into Eq. 2A.2.5 gives the concentration at the residence time, τ_{renew}, as

$$\frac{C - C_{in}}{C_0 - C_{in}} = e^{-\left(\frac{Q_{in}}{V_{pool}} \times \frac{V_{pool}}{Q_{in}}\right)} = e^{-1} = \frac{1}{e} = \frac{1}{2.7} = 0.37 \tag{2A.2.7}$$

or

$$C = 0.37(C_0 - C_{in}) + C_{in}. \tag{2A.2.8}$$

The residence time (or renewal time) is defined by the e-folding time, $1/e$. In this example, it is the time required for the concentration in the pool to evolve from its initial value (C_0) to a concentration 37 percent from its final value (C_{in}). The process of renewing the water in the pool is 63 percent (1 – 0.37) of the way to completion after one residence time has elapsed.

Appendix 2A.3 The Kinetics of Air–Sea Gas Exchange

The rate at which gases move between the atmosphere and ocean is controlled by environmental conditions, the gas's solubility in water (Table 2.7), and the concentration difference between the atmosphere and ocean. Exchange rates are faster when the air–sea interface is turbulent than when the surface is calm, so exchange rates are usually presented as proportional to wind speed, since wind is the main source of turbulence. Our ability to predict air–sea gas exchange rates is imperfect, and relies on a combination of empirical observations of gas tracers under different air–sea states and on simple models of gas exchange.

2A.3.1 The Gas Exchange Flux Equations

A general formula for air–water gas transfer, $F_{A\text{-}W}$, in nature includes two separate fluxes: one controlled by molecular diffusion at the air–water interface, F_S, and one by bubbles from breaking waves that occur at higher wind speeds and transport gases below the surface, F_B.

$$F_{A-W} = F_S + F_B. \tag{2A.3.1}$$

We present a description of each of these fluxes in this Appendix.

To understand the diffusive flux of a gas between the atmosphere and the ocean one must think about what is happening at very small scales right at the air–water interface. The ocean has a well-mixed layer near the surface, where concentrations, $[C]$, are essentially uniform but not at saturation equilibrium with the well mixed atmosphere above. For all but extremely soluble gases, it is processes on the liquid side of the interface that control the rate of gas transfer. Assuming that the concentration of gas exactly *at* the air–sea interface is in solubility equilibrium with the atmosphere, $[C_{eq}]$, the chemical driving force for the air–sea flux of gas C, $F_{S,C}$, is the difference in concentration between the solubility equilibrium value and the concentration in the bulk ocean surface waters. Factors that affect the rate of gas transfer like individual gas molecular diffusion coefficients and environmental physical conditions like wind speed and air–sea roughness are parameterized by a gas exchange mass transfer coefficient, k_S, which has units of (distance/time). The flux of gases across the interface (mol m^{-2} d^{-1}) is the product of the chemical driving force measured by the air–sea difference in concentration or fugacity, f, and the mass transfer coefficient.

$$\begin{aligned} F_{S,C} &= -k_{S,C}\{[C] - [C_{eq}]\} \\ F_{S,C} &= -k_{S,C}\{[C] - K_{H,C} f C^a\} \\ F_{S,C} &= -k_{S,C} K_{H,C}\{f C^w - f C^a\} \end{aligned} \quad (2A.3.2)$$

The sign of the equations in this text is such that the flux is positive when gas moves from the atmosphere to the ocean (ingassing) and negative when it moves from the ocean to the atmosphere (outgassing).

Values for the gas exchange mass transfer coefficient have been determined by a combination of mechanistic models of processes that occur near the air–water interface, experiments in laboratory wind tunnels, and field observations. Even for the same physical conditions, the mass transfer coefficient varies for different gases because molecular diffusion coefficients are gas dependent. However, the primary physical process that determines the value of the mass transfer coefficient is turbulence near the air–water interface.

Turbulence in the fluid on either side of the air–sea interface is determined by eddies of many sizes (Fig. 2A.3.1). Approaching the interface from below, smaller and smaller eddies with progressively less velocity control water movement. The *diffusive boundary layer* is the layer within which molecular diffusion processes become a more important transport mechanism than the physical water movement by eddies. The transition from

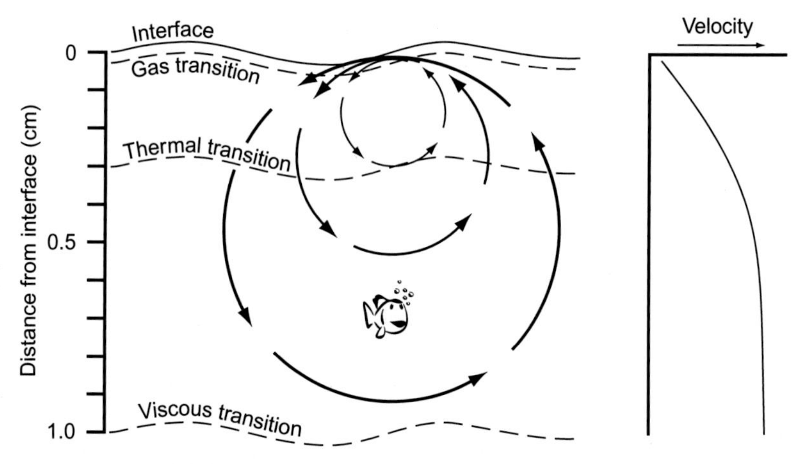

Figure 2A.3.1 A schematic representation of the boundary layers for momentum, heat, and mass near the air–water interface. The velocity of the water and the size of eddies in the water decrease as the air–water interface is approached. The larger eddies have greater velocity, which is indicated here by the length of the arrow in the eddy. Because random molecular motions of momentum, heat, and mass are characterized by molecular diffusion coefficients of different magnitude (0.01 $cm^2\ s^{-1}$ for momentum, 0.001 $cm^2\ s^{-1}$ for heat and 10^{-5} $cm^2\ s^{-1}$ for mass), there are three different distances from the air–sea interface where molecular motions become as important as eddy motions for transport. The scales are called the viscous (momentum), thermal (heat), and diffusive (molecular) boundary layers near the interface. Modified from Emerson and Hedges (2008).

eddy to molecular diffusion for momentum, heat, and gases occurs at different depths because the molecular diffusion of momentum (the kinematic viscosity, $\upsilon \sim 10^{-2}\ \mathrm{cm^2\,s^{-1}}$) is about 10 times greater than that of heat $(D_{\mathrm{heat}} \sim 10^{-3}\ \mathrm{cm^2\,s^{-1}})$ and about 1000 times greater than molecular diffusion of gases $(D_{\mathrm{gas}} \sim 10^{-5}\ \mathrm{cm^2\,s^{-1}})$. The relationships among the distances, X, for the transition from eddy to molecular diffusion for momentum, X_m, heat, X_h, and gas, X_g, scale with the molecular diffusion coefficient via Einstein's diffusion equation: $X = (2Dt)^{1/2}$, where t is the diffusion time. For the same times, t, $X_m/X_h \sim 3.2$ and $X_m/X_g \sim 32$. Assuming the transition from eddy to molecular diffusion for momentum occurs at ~1 cm, the transition for heat occurs at ~3.1 mm and for gases at about 0.3 mm (Fig. 2A.3.1).

Within the viscous boundary layer, momentum flux is controlled by the kinematic viscosity, υ, which is a measure of the resistance of flow to the influence of gravity. "Stickier" or "thicker" substances have higher kinematic viscosity. In this region, eddy mixing of heat and gases decreases to zero as the interface is approached. The fluxes of heat and gases are proportional to their molecular diffusion coefficients because diffusion dominates transport near the interface; however, the flux is inversely proportional to the kinematic viscosity, v, because it is more difficult to diffuse through "stickier" substances. Warmer temperatures act to increase the flux by both increasing diffusion coefficients, as the energy input to random molecular motions increases, and by decreasing kinematic viscosity, as warmer fluids flow more easily. Mass transfer coefficients of gases A and B (for example) follow this proportionality to the power, n, which varies in different environments between 0.5 and 1.0:

$$k_{S,A} \propto \left\{{D_A}/{v}\right\}^n;\ k_{S,B} \propto \left\{{D_B}/{v}\right\}^n. \tag{2A.3.3}$$

The ratio of the kinematic viscosity to diffusion coefficient has been given the name Schmidt number, which has the symbol Sc (see Appendix E):

$$Sc = {v}/{D}. \tag{2A.3.4}$$

Thus, the mass transfer coefficient, k_S, is inversely proportional to the Schmidt number to the power n. For gases A and B:

$${k_{S,B}}/{k_{S,A}} = \left\{{Sc_A}/{Sc_B}\right\}^n. \tag{2A.3.5}$$

Comparison of transfer coefficients of the same gas at different temperatures and/or in different liquids, requires knowledge of the dependence of the gas exchange rate on the kinematic viscosity. However, when comparing the gas transfer coefficients of different gases in the same solution at the same temperature, the dependence on kinematic viscosity plays no role because it is a property of the fluid and cancels. Later we will see that chemical oceanographers have chosen to normalize measured mass transfer coefficients to a Schmidt number for CO_2 at 20°C and $S = 35$ ($Sc_{CO_2(20°C,\, S=35)} = 660$; see Appendix E).

Different conceptual models of gas exchange suggest values for n in Eqs. 2A.3.3 and 2A.3.5 ranging from 0.5 to 1. This n value has been determined experimentally in wind tunnels by measuring the air–water flux of gases with widely different molecular diffusion coefficients at a variety of wind speeds (Jähne et al., 1987). At wind speeds typical of those over the ocean, the experimental results align most closely with $n = 0.5$.

The square root dependence of the mass transfer coefficient on the Schmidt number matches the prediction of a gas exchange model called the Surface Renewal Model, which is presented schematically in Fig. 2A.3.2. This model envisions small parcels of water (order 100 μm diameter) moving up to the air–sea interface and replacing parcels that have

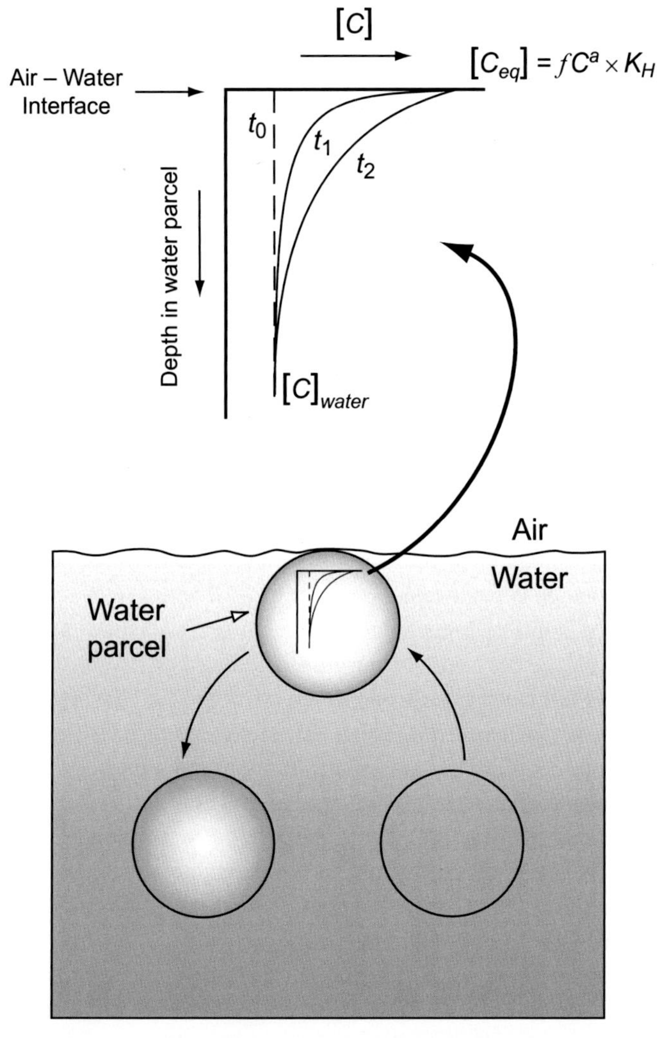

Figure 2A.3.2 A schematic representation of air–sea gas exchange for the Surface Renewal Model in a situation where the gases are undersaturated in the surface mixed layer (lower concentration than expected at equilibrium). Water parcels arrive at the surface with the bulk concentration of the mixed layer, then gases diffuse through the air–sea interface and into the water parcel with the concentration profile at different times shown by t_0, t_1, and t_2. Eventually, the water parcel is replaced by a new parcel from below. The scale of these water parcels is on the order of 100 μm. Modified from Emerson and Hedges (2008).

been directly at the interface for a short time (typically less than a second). While each new parcel remains in contact with the interface, gases diffuse between the water and the air, bringing the concentration in the water parcel toward equilibrium. Figure 2A.3.2 depicts a situation with a mixed layer concentration below saturation equilibrium. During the period of time that the parcel remains at the interface, the concentration in the parcel increases toward equilibrium as gases enter the water by molecular diffusion. The derivation of the model relationship to the square root of the gas molecular diffusion coefficient is presented in Chapter 10.1.3 of Emerson and Hedges (2008).

2A.3.2 Measurements of Gas Exchange Rates in the Ocean

Many different tracers have been used to determine thc gas transfer coefficients for air–sea gas exchange. Natural tracers of gases that exchange between the ocean and atmosphere like ^{14}C of CO_2 and ^{222}Rn were used to derive the first estimates of gas exchange rates (specifically the mass transfer coefficient, k_S.) These radioactive tracers and their applications to air–sea exchange are described in Chapter 7 about radioisotopes. More recently, atmospheric eddy correlation methods have been used, in which rapid measurements of gas concentration in the atmosphere are correlated to the turbulent velocity of air to determine the gas flux at the ocean surface. Dimethyl sulfide (DMS) is ideal for this method because it is a very soluble gas (Appendix D) that is produced in the surface ocean by biological processes but has very low concentrations in the atmosphere. Information from these natural tracers has some drawbacks in that these tracers average over different characteristic times and are difficult to use at high wind speeds where gas exchange is most rapid.

To improve knowledge of the air–sea exchange rate, the oceanography community developed methods for purposely adding gas tracers to the ocean and then determining their fate over time. In these experiments, two tracers, at least one of which must be a gas, are released into the ocean surface and measured as a function of time (Watson et al., 1991). The gas concentration decreases both by dilution due to mixing with tracer-free surrounding waters and by gas flux to the atmosphere. Since there are two tracers and two unknowns, the gas transfer velocity can be determined. Following a tracer patch and measuring the concentration decrease with time has an advantage over the "natural" tracer experiments in that the gas exchange experiment follows the same water parcel rather than measuring concentrations at a stationary location that is likely to incorporate water masses with different gas exchange histories. The gases ^{3}He (a rare isotope of helium) and sulfur hexafluoride (SF_6) are most often used as the tracers in these experiments because they meet the criteria of being detectable at very low levels, of having virtually no natural background, and of being inert in the aquatic environment.

An important caveat of dual tracer experiments is that the exact relationship between the gas exchange rates of the two tracers must be known in order to couple the equations describing dilution and gas exchange. In particular, the exact value of the exponent, n, in the Schmidt number dependency is uncertain because of the effect of bubbles on the two tracers (see later). Laboratory determinations of the effect of bubbles on the Schmidt-number dependence indicate a value of n in the range of 0.44 to 0.50 (Asher and Wanninkhof, 1998), and a field experiment, in which a nonvolatile tracer (bacterial spores) was released in addition to the two gases suggested an optimal value for n of 0.51 (Nightingale et al., 2000). Both of these results indicate that the departure from square root dependency causes less error than uncertainties associated with keeping track of the tracer patch.

Mass transfer coefficients from all dual tracer experiments to date are plotted versus wind speed in Fig. 2A.3.3. The y-axis value, k_{660}, is the value of the mass transfer coefficient determined from the gases SF_6 and ^3He normalized to the Schmidt number for CO_2 in seawater at 20°C ($Sc = 660$), by the relationship in Eq. 2A.3.5. There is a considerable amount of scatter in the wind speed versus k_S relationships, which is probably because wind speed does not adequately represent the turbulent conditions of the sea state. The best fit line of Ho et al. (2011, H11) is presented in the figure along with two other curves that describe values of k_{660} from experiments that had no contribution from bubbles. The line labeled LM86 (Liss and Merlivat, 1986) is from a tracer release experiment in a small sheltered lake by Wanninkhof et al. (1985). The line labeled G-M16 (Goddijn-Murphy et al., 2016) is from eddy correlation measurements of DMS over the ocean.

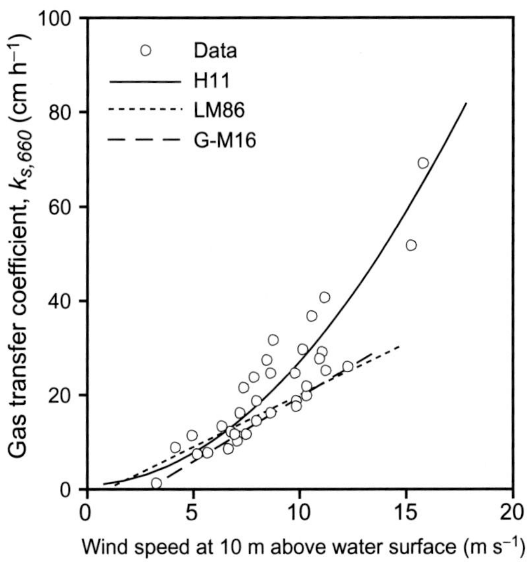

Figure 2A.3.3 Gas transfer coefficients for air–sea gas exchange normalized to a gas of Schmidt number 660, k_{660} (cm hr^{-1}), as a function of wind speed at 10 meters above the water surface. The circles represent a compilation of data from tracer release experiments (from David Ho, personal communication, and in Ho et al., 2011). Lines indicate different predictions: H11 (Ho et al., 2011) is the line drawn through the tracer release data; LM86 is the prediction by Liss and Merlivat (1983) for the intermediate range of wind speeds, which originates from a purposeful gas release experiment in a lake by Wanninkhof et al. (1985); G-M16 (Goddijn-Murphy et al., 2016) is from the results of eddy correlation measurements of DMS in the atmosphere.

Bubble processes have diminishing importance for highly soluble gases because most of the gas is in solution already, and dimethyl sulfide is so soluble in water (Appendix D) that the contribution to DMS exchange from bubbles even at high winds is minor. Equations used by these authors to describe the gas transfer coefficient – wind speed relationships are

Table 2A.3.1 Gas exchange relationships as a function of wind speed that are used in Figs. 2A.3.3 and 2A.3.5. The air–sea flux is conceptually modeled as the sum of diffusive transfer at the air–sea interface, F_s, and the transfer due to bubbles, F_B (see text). Mass transfer coefficients, k_S, are presented in units of cm hr^{-1}. Both k_S and k_P are normalized to a Schmidt number, Sc, of 660 (CO_2 in seawater at 20°C), where $k_{S,C} = k_{S,660}\,(Sc_C/660)^{-0.5}$. Schmidt numbers are presented in Appendix E. Wind speeds^{+} (U_{10}) are values at 10 m height above the air–sea surface, while u_*^a and u_*^w are the friction velocities on the air and water side of the interface, respectively.

(a) Bubble free exchange, F_s (all in units of cm h^{-1})

Relationship	Source
1. $k_{s,660} = 2.72\,U_{10} - 9.20\ (3.5 < U_{10} < 13\ \mathrm{m\,s^{-1}})$	(small lake measurements, Liss and Merlivat, 1986)
2. $k_{s,660} = 2.2\,U_{10} - 3.4$	(DMS atm. eddy correlation, Goddijn-Murphy et al., 2016)
3. $k_{s,660} = 48.7\,u_*^a$	(NOAA COARE, Liang et al., 2013)

(b) Results of purposeful tracer release experiments (all in units of cm h^{-1})

Relationship	Source
1. $k_{s,660} = 0.26\,U_{10}^2$	(Ho et al., 2011)
2. $k_{s,660} = 0.21\,U_{10}^2 + 0.32\,U_{10}$	(Nightingale et al., 2000)

(c) Exchange due to bubbles, $F_B = F_{inj} + F_p$ (Using the model of Liang et al., 2013)

1. Small bubbles that totally collapse:

   $F_{inj} = \beta_{inj} k_{inj} X_C$;

   X_C is the atmospheric mole fraction of gas C

   $k_{inj} = 5.56\left(u_*^w\right)^{3.86}$ (units of mol m^{-2} s^{-1})

   β is a value between 0 & 1 used to calibrate the bubble model

2. Larger bubbles:

   $F_p = \beta_p k_p\left((1+\Delta P)\left[C_{eq}\right] - [C]\right)$

   ΔP = increased hydrostatic pressure; $[C_{eq}]$ = atmospheric saturation concentration

   $k_p = 5.5\left(u_*^w\right)^{2.76}(Sc_C/660)^{-0.67}$ (units of m s^{-1})

   $\Delta P = 1.52\left(u_*^w\right)^{1.06}$

3. Calibration of the L13 bubble model using surface ocean N_2 gas measurements as a function of wind speed (Emerson et al., 2019), Fig 2A.3.5

   $\beta_{inj} = \beta_p = 0.37 \pm 0.14$

$^{+}U_{10}$ and u_* relationships (all in units of m s^{-1}):

$u_*^a = (C_d)^{0.5}\,U_{10}$

$u_*^w = (\rho_a/\rho_w)^{0.5}\,u_*^a = 0.034\,u_*^a$

where $C_d = 0.0012$ (for $U_{10} < 11\ \mathrm{m\,s^{-1}}$);

$C_d = (0.49 + 0.065\,U_{10}) \times 10^{-3}$ (for U_{10} = 11-20 m s^{-1})

ρ_a = density of air and ρ_w = density of seawater

presented in Table 2A.3.1. The curves used by Liang et al. (2013) for the bubble-free k_{660} values are very similar to those presented in the figure, and the earlier gas transfer coefficient – wind speed relationships of Wanninkhof (1992) and Nightingale et al. (2000) are only slightly different from that of Ho et al. (2011). Note that the "no-bubble" lines in Fig. 2A.3.3 (LM86 and G-M16) and the ocean tracer release line (H11) are the same to within the scatter of the data until a wind speed of 8–10 m s^{-1} when the tracer release results begin to become greater as wind speed increases.

The residence time of gases in the surface ocean mixed layer represents the time it would require gas exchange to return the concentration 63 percent of the way to equilibrium after being perturbed by some process. In this case, residence time is the inventory in the mixed layer divided by the air–sea flux:

$$\tau_{gas,ML} = \frac{h[C]_{ML}}{k_S[C]_{ML}}, \tag{2A.3.6}$$

where $[C]_{ML}$ is the mixed-layer concentration (mol m^{-3}) and h is the mixed layer depth (m). The denominator, $k_S[C]_{ML,}$ represents the one-way flux out of the mixed layer. The equation simplifies to h/k_S. (You would get the same result by using the difference between $[C]_{ML}$ and $[C_{eq}]$ in the numerator and the net flux in the denominator.) With a global average k_S for most gases of around 17 cm hr^{-1} or 4 m d^{-1} (Wanninkhof, 2014), gases in a 50 m mixed layer have a residence time of around 12 days (50/4). If a bloom of phytoplankton were to increase ΔO_2 to $+10$ percent (supersaturated) and then suddenly stop producing oxygen, it would take about 2 weeks for gas exchange to draw the ΔO_2 63 percent of the way back toward equilibrium (see Appendix 2A.2).

2A.3.3 Gas Transfer Due to Bubbles

Readers who have seen the surface of the ocean on a windy day will have noticed that breaking waves introduce lots of small air bubbles into the water. The hydrostatic pressure only 1 meter below the ocean surface is about 10 percent of the entire atmospheric pressure, so the pressure inside a bubble at a depth of 1 meter is ~110 percent of that at the ocean surface. Thus, there is a strong tendency for bubbles entrained in the downwelling limb of a breaking wave to lose some or all of their gas to the surrounding fluid by diffusion across the bubble surface, as the higher pressure inside the bubble forces gas to diffuse into the surrounding water. The flux created by bubbles has been described in many ways (e.g., Fuchs et al., 1987; Keeling, 1993; Liang et al., 2013; Wolf, 1997). We adopt the model of Liang et al. (2013) who used the formalism for gas transfer by bubbles of Fuchs et al. (1987) in a high-resolution three-dimensional model of the upper ocean which simulates the creation of bubbles by wave processes.

The bubble model of Fuchs et al. (1987) assumes that the full spectrum of bubbles can be described by a combination of two bubble transfer mechanisms (Fig. 2A.3.4).

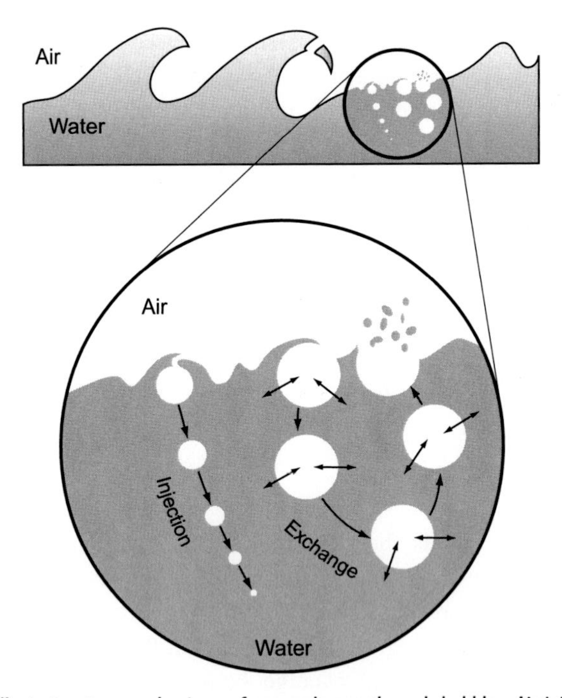

Figure 2A.3.4 A schematic diagram illustrating two mechanisms of gas exchange through bubbles. Air injection (total trapping), described with the empirical constant k_{inj}, indicates smaller bubbles that totally dissolve as they are subducted into the water by a wave. Exchange (partial trapping), described by the empirical constant k_P, describes larger bubbles that are submerged by waves and exchange their contents partially with the surrounding water before they resurface at the interface.

$$F_B = F_{inj} + F_P. \tag{2A.3.7}$$

The first mechanism, F_{inj}, applies to small bubbles, generally <50 μm in diameter, that completely dissolve, injecting their entire contents into the water. This mechanism has been called "air injection" or "total trapping." (The subscript $_{inj}$ refers to injection.) In this case, the flux of gas from the bubble depends only on the total volume of air transferred by these bubbles, which is described by the product of an empirical transfer velocity, k_{inj} (mol m^{-2} s^{-1}) and the mole fraction of the gas in the atmosphere, X:

$$F_{inj} = k_{inj} X_C. \tag{2A.3.8}$$

This flux is a one-way transfer to the ocean and is thus always positive.

The second mechanism, F_P, applies to bubble transfer caused by larger bubbles, generally >50 μm in diameter, that do not fully dissolve but instead exchange some gases across the bubble–water interface and then rejoin the atmosphere. This mechanism is usually called "exchange" or "partial trapping." (The subscript $_P$ refers to partial trapping.)

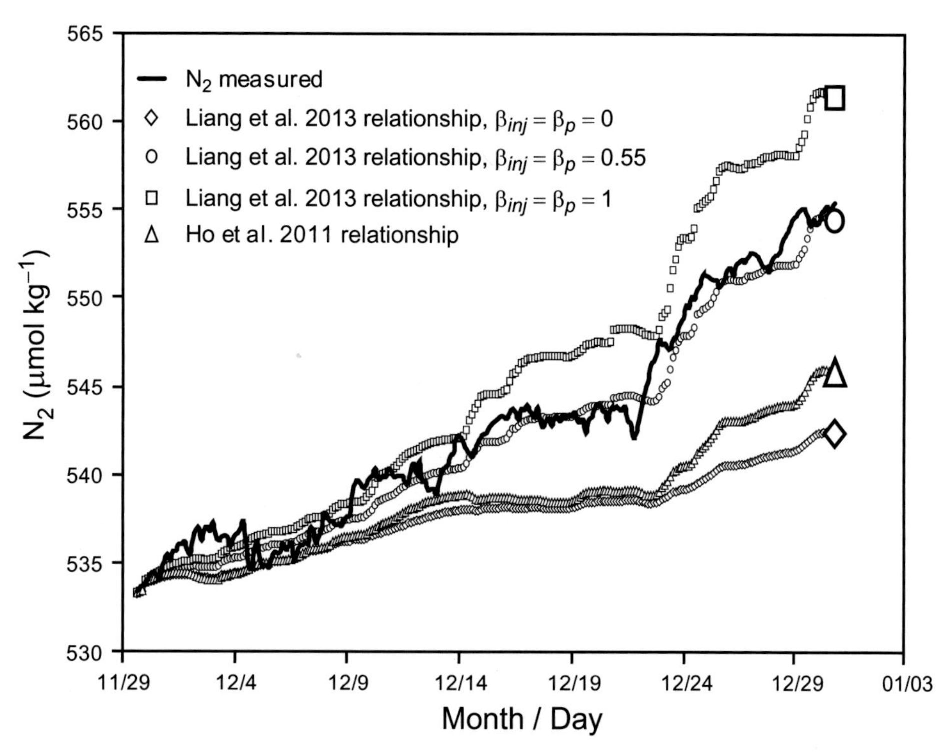

Figure 2A.3.5 The concentration of nitrogen gas in the surface waters as a function of time during one month in the winter of 2007 when T, S, and wind speed were also measured. The black line is the data. The line of triangles is a prediction based on the k_s value of Ho et al. (2011) with no explicit bubble model. The line of diamonds is a prediction using the gas exchange with the bubble model of Liang et al. (2013), but with the bubble flux set to zero. The line of squares is a prediction using the bubble model of Liang et al. (2013) as in their paper. The line of circles is a prediction using the bubble model of Liang et al. (2013) but with bubble strengths (beta values, Table 2A.3.1) that best fit the N_2 data. Redrawn from Emerson et al. (2019).

Conceptually, this process depends on two unknowns: the partial trapping mass transfer coefficient, k_P (m d^{-1}) and the degree of over pressure of the gas in the bubble, ΔP (a fraction like 0.1), caused by hydrostatic pressure and surface tension.

$$F_P = -k_{P,C}\{[C] - (1 + \Delta P)\,[C_{eq}]\}. \tag{2A.3.9}$$

Theories of the process of gas transfer across the bubble interface (e.g., Levich, 1962; Merlivat and Memory, 1983) indicate that the exchange coefficient k_P has the same Schmidt-number dependence as the flux across the air–water interface, except that the exponent, n, is 0.67 instead of 0.5. Bubbles accumulate surfactants on their surface, which creates a smooth laminar layer in which the gas exchange rate is dependent on the molecular diffusion coefficient raised to the two-thirds power rather than one-half (Jähne et al., 1987).

The air–sea gas exchange model for gas C is now

$$F_C = -k_{S,C}([C] - [C_{eq}]) + k_{inj}X_C - k_{P,C}\{[C] - (1 + \Delta P)[C_{eq}]\}, \quad (2A.3.10)$$

where each of the unknowns, k_S, k_{inj}, k_P, and ΔP, have their own individual dependence on wind speed determined by comparison to data (Table 2A.3.1).

The values of the bubble coefficients k_{inj} and k_P in the model of Liang et al. (2013) have been verified using a 10-year-long data set of in situ N_2 measurements in the surface waters of the subarctic Pacific Ocean (Emerson et al., 2019). A plot of the change in N_2 concentration during one of the winter months is presented in Fig. 2A.3.5. During this period, the mixed layer depth is stable at 110 m, N_2 concentration was measured every three hours, and there are high winds. The only process that could change the N_2 concentration was air–sea interaction. Predictions using Eq. 2A.3.10 and the wind speed dependencies in Table 2A.3.1 are presented along with the data in the figure. Comparison of the data with the model indicates that the air–water diffusive exchange (k_s) alone cannot account for the observed N_2 concentration change – an explicit bubble term is required to reproduce the data. However, the Liang et al. (2013) bubble relationship creates a bubble flux that is too large. The bubble exchange coefficients k_{inj} and k_P must be smaller – about one-third of the value suggested by Liang et al. (2013). (The value indicated as β in Fig. 2A.3.5 is 0.55, but this is just one of the 10 month-long periods used to derive the mean and standard deviation for β in Table 2A.3.1c.)

Gas exchange models are evolving at the time of writing this book. For now, it is recommended that a model with explicit bubble processes be used to do gas exchange calculations at high winds (like the model of Liang et al., 2013), but that values of gas transfer coefficients for bubble processes k_{inj} and k_P be reduced to 0.37 of the values suggested in the work of Liang et al. (2013).

References

Aller, R. C. (2014) 8.11 – Sedimentary diagenesis, depositional environments, and benthic fluxes, in *Treatise on Geochemistry*, 2nd ed. (eds. H. D. Holland and K. K. Turekian), pp. 293–334. Oxford: Elsevier.

Asher, W. E., and R. Wanninkhof (1998) The effect of bubble-mediated gas transfer on purposeful dual-gaseous tracer experiments, *Journal of Geophyical Research Oceans*, **103**, 10555–10560.

Berner, R. (1971) *Principles of Chemical Sedimentology*, 240 pp. New York: McGraw-Hill.

Berner, R. A. (2003) The long-term carbon cycle, fossil fuels, and atmospheric composition, *Nature*, **426**, 323–326.

Berner, R. A. (2006) Geological nitrogen cycle and atmospheric N_2 over Phanerozoic time, *Geology*, **34**, 413–415.

Bischoff, J. L., and F. W. Dickson (1975) Seawater-basalt interaction at 200°C and 500 bars: Implications for the origins of sea-floor heavy metal deposits and regulation of seawater chemistry, *Earth and Planetary Science Letters*, **25**, 385–397.

Broecker, W. S., and T.-H. Peng (1982) *Tracers in the Sea*, 690 pp. Palisades: Eldigio Press.

Dai, A., and K. E. Trenberth (2002) Estimates of freshwater discharge from continents: latitudinal and seasonal variations, *Journal of Hydrometeorology*, **3**, 660–687.

DeMaster, D. J. (2002) The accumulation and cycling of biogenic silica in the Southern Ocean: revisiting the marine silica budget, *Deep-Sea Research II*, **49**, 3155–3167.

Dixit, S., P. Van Cappellen, and A. J. van Bennekom (2001) Processes controlling solubility of biogenic silica and pore water build-up of silicic acid in marine sediments, *Marine Chemistry*, **73**, 333–352.

Drever, J. I. (1982) *The Geochemistry of Natural Waters*, 388 pp. Englewood Cliffs: Prentice-Hall.

Edmond, J. M., C. Measures, R. E. McDuff, et al. (1979) Ridge crest hydrothermal activity and the balances of the major and minor elements in the ocean: the Galapagos data, *Earth and Planetary Science Letters*, **46**, 1–18.

Elderfield, H., and A. Schultz (1996) Mid-ocean ridge hydrothermal fluxes and the chemical composition of the ocean, *Annual Review of Earth and Planetary Science*, **24**, 191–224.

Emerson, S. R., and J. I. Hedges (2008) *Chemical Oceanography and the Marine Carbon Cycle*, 453 pp. Cambridge: Cambridge University Press.

Emerson, S., T. Ito, and R. C. Hamme (2012) Argon supersaturation indicates low decadal-scale vertical mixing in the ocean thermocline, *Geophysical Research Letters*, **39**, L18610, doi:10.1029/2012GL053054.

Emerson, S., B. Yang, M. White, and M. Cronin (2019) Air-sea gas transfer: Determining bubble fluxes with in situ N_2 observations, *Journal of Geophysical Research Oceans*, **124**, 2716–2727.

Enns, T., P. F. Scholander, and E. D. Bradstreet (1965) Effect of hydrostatic pressure on gases dissolved in water, *Journal of Physical Chemistry*, **69**, 389–391.

Fuchs, G., W. Roether, and P. Schlosser (1987) Excess ^{3}He in the ocean surface layer, *Journal of Geophysical Research Oceans*, **92**, 6559–6568.

German, C. R., and W. E. Seyfried Jr (2014) 8.7 – Hydrothermal processes, in *Treatise on Geochemistry*, 2nd ed. (eds. H. D. Holland and K. K. Turekian), pp. 191–233. Oxford: Elsevier.

German, C., and K. Von Damm (2003) Hydrothermal processes, in *The Treatise on Geochemistry* (eds. H. D. Holland and K. K. Turekian), pp. 181–222. Amsterdam: Elsevier.

Glueckauf, E. (1951) The composition of atmospheric air, in *Compendium of Meteorology*, pp. 3–10. Boston: American Meteorological Society.

Goddijn-Murphy, L., D. K. Woolf, A. H. Callaghan, P. D. Nightingale, and J. D. Shutler (2016) A reconciliation of empirical and mechanistic models of the air-sea gas transfer velocity, *Journal of Geophysical Research Oceans*, **121**, 818–835, doi:10.1002/2015JC011096.

Hedges, J. I., and R. G. Keil (1995) Sedimentary organic matter preservation: an assessment and speculative synthesis, *Marine Chemistry*, **49**, 81–115.

Helgeson, H., and F. Mackenzie (1970) Silicate-seawater equilibria in the ocean system, *Deep-Sea Research*, **17**, 877–892.

Ho, D. T., R. Wanninkhof, P. Schlosser, et al. (2011) Toward a universal relationship between wind speed and gas exchange: gas transfer velocities measured with $^3He/SF_6$ during the Southern Ocean Gas Exchange Experiment, *Journal of Geophysical Research Oceans*, **116**, C00F04, doi:10.1029/2010JC006854.

Holland, H. D. (1978) *The Chemistry of the Atmosphere and Oceans*, 351 pp. New York: John Wiley.

Jähne, B., K. O. Münnich, R. Bösinger, et al. (1987) On the parameters influencing air-water gas exchange, *Journal of Geophysical Research Oceans*, **92**, 1937–1949.

Keeling, R. F. (1993) On the role of large bubbles in air-sea gas exchange and supersaturation in the ocean, *Journal of Marine Research*, **51**, 237–271.

Klots, C. E., and B. B. Benson (1963) Solubilities of nitrogen, oxygen, and argon in distilled water, *Journal of Marine Research*, **21**, 48–57.

Levich, V. G. (1962) *Physicochemical Hydrodynamics*, 700 pp. Englewood Cliffs: Prentice-Hall.

Li, Y.-H. (2000) *A Compendium of Geochemistry*, 472 pp. Princeton: Princeton University Press.

Liang, J.-H., C. Deutsch, J. C. McWilliams, et al. (2013) Parameterizing bubble mediated air-sea gas exchange and its effect on ocean ventilation, *Global Biogeochemical Cycles*, **27**, doi:10.1002/gbc.20080.

Liss, P. S., and L. Merlivat (1986) Air-sea gas exchange rates: introduction and synthesis, in *The Role of Air-Sea Exchange in Geochemical Cycling*, pp. 113–127. Dordrecht: Springer.

Mackenzie, F. T., and R. M. Garrels (1966) Chemical mass balance between rivers and the ocean, *American Journal of Science*, **264**, 507–525.

McManus, J., D. E. Hammond, W. M. Berelson, et al. (1995) Early diagenesis of biogenic opal: dissolution rates, kinetics, and paleoceanographic implications, *Deep-Sea Research II*, **42**, 871–902.

Merlivat, L., and L. Memory (1983) Gas exchange across an air-water inter-face: experimental results and modeling of bubble contribution transfer, *Journal of Geophysical Research Oceans*, **88**, 707–724.

Michalopoulos, P., and R. Aller (1995) Rapid clay mineral formation in the Amazon delta sediments: reverse weathering and ocean elemental cycles, *Science*, **270**, 614–617.

Mottl, M. J., and H. D. Holland (1978) Chemical exchange during hydrothermal alteration of basalt by seawater, I. Experimental results for major and minor components of seawater, *Geochimica et Cosmochimica Acta*, **42**, 1103–1115.

Mottl, M., and J. Wheat (1994) Hydrothermal circulation through mid-ocean ridge flanks: fluxes of heat and magnesium, *Geochimica et Cosmochimica Acta*, **58**, 2225–2237.

Nielsen, S. G., M. Rehkämper, D. A. Teagle, et al. (2006) Hydrothermal fluid fluxes calculated from the isotopic mass balance of thallium in the ocean crust, *Earth and Planetary Science Letters*, **251**, 120–133.

Nightingale, P. D., G. Malin, C. S. Law, et al. (2000) In situ evaluation of air-sea gas exchange parameterizations using novel conservative and volatile tracers, *Global Biogeochemical Cycles*, **14**, 373–387.

Rabouille, C., J. F. Gaillard, P. Tréguer, and M. A. Vincendeau (1997) Biogenic silica recycling in surficial sediments across the Polar Front of the Southern Ocean (Indian Sector), *Deep-Sea Research II*, **44**, 1151–1176.

Resing, J. A., P. N. Sedwick, C. R. German, et al. (2015) Basin-scale transport of hydrothermal dissolved metals across the South Pacific Ocean, *Nature*, **523**, 200–203.

Sillen, L.G. (1967) Gibbs phase rule in marine sediments, in *Equilibrium Concepts in Natural Water Systems* (ed. R. F. Gould), pp. 57–69. Washington, DC: American Chemistry Society.

Stanley, R. H. R., W. J. Jenkins, D. E. Lott III, and S. C. Doney (2009) Noble gas constraints on air-sea gas exchange and bubble fluxes, *Journal of Geophysical Research Oceans*, **114**, C11020, doi:10.1029/2009JC005396.

Taylor, C. D. (1978) The effect of pressure upon the solubility of oxygen in water, *Archives of Biochemistry and Biophysics*, **191**, 375–384.

Tréguer, P. J., and C. L. De La Rocha (2013) The world ocean silica cycle, *Annual Review of Marine Science*, **5**, 477–501.

Van Cappellen, P., and L. Qiu (1997) Biogenic silica dissolution in sediments of the Southern Ocean. I. Solubility, *Deep-Sea Research II*, **44**, 1109–1128.

Von Damm, K. L., J. M. Edmond, B. Grant, and C. I. Measures (1985) Chemistry of submarine hydrothermal solutions at 21 °N, East Pacific Rise, *Geochimica et Cosmochimica Acta* **49**, 2197–2220.

Wanninkhof, R. (1992) Relationship between wind speed and gas exchange over the ocean, *Journal of Geophysical Research Oceans*, **97**, 7373–7382.

Wanninkhof, R. (2014) Relationship between wind speed and gas exchange over the ocean revisited, *Limnology and Oceanography: Methods*, **12**, 351–362.

Wanninkhof, R., J. R. Ledwell, and W. S. Broecker (1985) Gas exchange-wind speed relation measured with sulfur hexafluoride on a lake, *Science*, **227**, 1224–1226.

Watson, A. J., R. C. Upstill-Goddard, and P. S. Liss (1991) Air–sea gas exchange in rough and stormy seas measured by a dual-tracer technique, *Nature*, **349**, 145–147.

Weiss, R. F. (1970) The solubility of nitrogen, oxygen and argon in water and seawater, *Deep-Sea Research*, **17**, 721–735.

Weiss, R. F. (1974) Carbon dioxide in water and seawater: the solubility of a non-ideal gas, *Marine Chemistry*, **2**, 203–215.

Wolf, D. K. (1997) Bubbles and their role in gas exchange, in *The Sea Surface and Global Change* (eds. P. S. Liss and R. A. Duce), pp. 173–206. Cambridge: Cambridge University Press.

Woodhouse, M. T., K. S. Carslaw, G. W. Mann, et al. (2010) Low sensitivity of cloud condensation nuclei to changes in the sea-air flux of dimethyl-sulphide, *Atmospheric Chemistry and Physics*, **10**, 7545–7559, doi:10.5194/acp-10-7545-2010.

Problems for Chapter 2

2.1. Mercuric chloride is accidentally spilled into a stream that drains into an urban lake. The contaminated stream is the only waterway draining into the lake and it has a volume flow rate of 50 m^3 s^{-1}. A stream drains out of the lake at the same rate. The concentration of mercury ions in the inflowing stream during the spill is 10 μmol L^{-1} and the spill duration is 5 days, after which the mercury ion concentration in the inflowing stream abruptly falls to 1 μmol L^{-1} and remains constant. The initial concentration of mercury ions in the lake is negligible. The lake has a surface area of 1 km^2 and an average depth of 21.5 m.

a. Find the residence time of water in the lake with respect to stream inflow.
b. What is the concentration of mercury ion in the lake 5 days after the start of the spill?
c. What is the concentration of mercury ion in the lake 10 days after the start of the spill?

2.2. An unusual lake in the Sierra Nevada mountains has chemistry that is not changing over time. Its dissolved chemical composition is determined by four processes: river input, $CaCO_3$ precipitation, hydrothermal processes, and gas exchange. The lake has no outlet. Assume that $CaCO_3$ precipitation is the only sink in this lake for bicarbonate ions, and that it precipitates according to the following reaction: $Ca^{2+} + 2HCO_3^- \rightarrow CaCO_3 + CO_2 + H_2O$.

Lake:
$$[Mg^{2+}] = 46 \text{ mmol kg}^{-1}$$
$$[Ca^{2+}] \text{ not measured}$$
$$[HCO_3^-] \text{ not measured}$$
$$\text{Volume} = 10^{12} \text{ kg}$$

River:
$$[Mg^{2+}] = 0.46 \text{ mmol kg}^{-1}$$
$$[Ca^{2+}] = 1.9 \text{ mmol kg}^{-1}$$
$$[HCO_3^-] = 4.2 \text{ mmol kg}^{-1}$$
$$\text{Flow rate} = 10^{10} \text{ kg yr}^{-1}$$

a. If lake water flowing through the hydrothermal vent contains no dissolved Mg^{2+} when it exits the vent, what is the flow rate of water through the hydrothermal vent?
b. Given the amount of bicarbonate entering the lake via the river, how much calcium must be removed from the lake each year?
c. Determine the increase in calcium concentration of the lake water as it passes through the hydrothermal vent. This is the difference in the calcium concentration of the water that exits the vent from the lake water that enters the vent.

2.3. At an open ocean location in winter, the phosphate concentration in the mixed layer is 1.7 μmol kg^{-1}. The mixed layer is 65 m deep. Combining the gradient in phosphate at the base of the mixed layer with an estimate of the eddy diffusivity yields a turbulent mixing flux bringing additional phosphate up into the mixed layer at the rate of 900 μmol m^{-2} d^{-1}. If the system is temporarily in steady state, what is the residence time of phosphate in the mixed layer?

2.4. Weekly measurements of oxygen are being made in a semi-enclosed bay to determine the balance of gas exchange and biological production in the mixed layer. On the first cruise, the average mixed layer depth is 30 m and surface temperature, salinity, and oxygen concentration are 20°C, 35, and 248 μmol kg^{-1}, respectively. Average winds yield a gas transfer coefficient for oxygen in this bay of 4 m d^{-1}. Assume that a strong thermocline at the base of the mixed layer inhibits vertical mixing with waters below the surface.

a. Calculate the surface oxygen concentration in each of the next 7 days if the only process affecting the evolution of oxygen concentrations in the mixed layer is air–sea gas exchange by diffusive processes.
b. Repeat the same calculation but include net biological oxygen production at a rate of 70 mmol m^{-2} d^{-1}.
c. Make a figure showing the evolution in the surface oxygen saturation anomaly, ΔO_2, over time for your two predictions.

2.5. As described in Appendix 2A.3, the rate at which small bubbles add gases to the ocean mixed layer is independent of the gas saturation state. At steady state, the dissolution of small atmospheric bubbles causes the surface ocean to be supersaturated. The flux into the ocean by small dissolving bubbles equals the diffusive flux out of the ocean across the air–sea interface.

a. Calculate the steady-state saturation anomaly induced by small bubbles for the gases: O_2, Ne, and Xe at wind speeds (U_{10}) of 3, 6, 9, 12, and 15 m s^{-1}. Assume a mixed layer depth of 30 m, a temperature of 20°C, and a salinity of 35. Use the gas exchange Eq. 2A.3.10 in Appendix 2A.3 considering only the mass transfer coefficients for small bubbles, k_{inj}, and air–sea exchange, k_S (i.e., $k_P = 0$). Calculate values for k_S and k_{inj} from equations in Table 2A.3.1, (a) Eq. 3 and (c) Eq. 1, respectively, and the values in Appendices D and E.
b. Make a plot of gas saturation anomaly versus wind speed.
c. Explain what creates the pattern that you observe; i.e., what controls the saturation anomaly differences among the gases?

3 Life in the Surface Ocean: Biological Production and Export

3.1 The Chemistry of Life 97
3.1.1 Redox Processes 97
3.1.2 The Main Elements of Organic Matter: C, H, O, N, P 99
Discussion Box 3.1 106
3.1.3 Trace Elements in Organic Matter: Fe, Zn, Mn, Ni, Cu, Co, Cd 107
Discussion Box 3.2 112
3.2 The Flux of Biologically Produced Elements from the Surface Ocean: The Ocean's Biological Pump 113
3.2.1 A Simplified Whole-Ocean Model of the Biological Pump 114
Discussion Box 3.3 118
3.2.2 Particle Fluxes and Thorium Isotope Tracers 119
3.2.3 Upper Ocean Metabolite Mass Balance: O_2, NO_3^-, DIC, DIC+δ^{13}C-DIC 121
3.2.4 O_2/Ar and O_2/N_2 Tracers 126
3.2.5 Comparing Different Methods for Determining ANCP 128
3.3 Global Distributions of Organic Carbon Export 129
3.3.1 Comparing Measured ANCP with Model Predictions 129
3.3.2 The Anthropogenic Influence: Evidence for Changes in Biological Fluxes 133
Appendix 3A.1 Measurement of Net and Gross Biological Production 134
3A.1.1 Net Primary Production Rates 134
3A.1.2 Gross Primary Production Rates 135
References 136
Problems for Chapter 3 141

Patterns of chemical distributions within the ocean are primarily controlled by biological processes and ocean circulation. Sunlight penetrates only the top 100 meters or so of the ocean (the euphotic zone), just 4 percent of the average ocean depth, but supplies the energy for photosynthesis that is the basis for most marine chemical transformations. Much of the organic matter produced by photosynthesis is consumed by bacteria and animals in the euphotic zone, but some also escapes the sunlit upper waters into the dark sphere. Density stratification between the warmer surface mixed layer and colder deep waters helps to keep the metabolic products of photosynthesis and respiration separate.

The dynamics of this physical and biogeochemical mosaic include the transport of nutrients dissolved in deeper waters to the ocean surface where they are incorporated into particulate and dissolved organic matter during photosynthesis. During this process, dissolved oxygen that is produced during photosynthesis is transported by gas exchange from the euphotic zone of the ocean to the atmosphere. The organic matter produced is transported by gravity and mixing into the ocean interior. In deeper waters, net respiration produces high concentrations of dissolved nutrients and inorganic carbon while it simultaneously depletes oxygen concentrations. Interactions of these physical and biological processes in the upper ocean occur on timescales of weeks, to tens, to hundreds of years and are revealed by the vertical concentration profiles of the biologically active chemical compounds. The chemical perspective of oceanography involves using observations of the metabolic products of photosynthesis and respiration to estimate the rates and derive information about the fluxes and mechanisms of ocean processes in this largely unobserved sphere.

The effects of life processes are felt throughout this book. We begin this chapter by introducing the chemist's view of photosynthesis and respiration with the concepts of reduction-oxidation reactions and the chemical compounds that make up organic matter. From there, we introduce a simple model that considers the interaction of biological and physical processes that control organic matter fluxes to the deep ocean. Finally, we present mass balance methods and tracer distributions used to quantify the rate of the export of metabolic products from surface waters.

3.1 The Chemistry of Life

3.1.1 Redox Processes

This chapter is primarily about organic matter reactions, which are almost exclusively oxidation-reduction reactions (*redox*) in which an element changes the deficiency or excess of electrons it contains relative to the number of protons in its nucleus. Redox reactions are of great interest in the study of environmental chemistry because photosynthesis and respiration fuel most life processes. These reactions are categorized energetically using the same chemical equilibrium thermodynamic formalism of free energies of formation and reaction as other chemical reactions. However, in practice redox reactions are rarely at thermodynamic equilibrium, because the Sun's energy is being used to produce well-organized organic compounds that are far from their minimum free energy state. Thermodynamic discussions of redox reactions can be found in Chapter 3 of Emerson and Hedges (2008; www.cambridge.org/emerson-hamme) and many environmental chemistry texts (e.g., Morel and Hering, 1993; Stumm and Morgan, 1995). Here we present an introduction to redox reactions using the examples of photosynthesis and respiration.

Redox reactions are characterized by a simultaneous *reduction* (electron gain) of one element combined with the *oxidation* (electron loss) of another. One example is the abbreviated form for photosynthesis and respiration in which organic matter (OM) is represented in its simplest possible form as the carbohydrate formaldehyde, $CH_2O(OM)$:

$$CH_2O(OM) + O_2 \rightleftarrows CO_2 + H_2O. \quad (3.1)$$

Carbon and oxygen in this reaction exist in two different oxidation states, which we will indicate with Roman numerals. Carbon has an oxidation state of zero (0) in CH_2O and plus four (IV) in CO_2; oxygen has an oxidation state of (0) in O_2 and (–II) in H_2O. The oxidation state of most elements found in nature can be determined by assuming that oxygen has an oxidation state of (–II) in every molecule except O_2 where it is zero (0). Similarly, hydrogen has an oxidation state of (+I) in every molecule except H_2 where it is (0). If reaction 3.1 proceeds from right to left, it represents net photosynthesis in which carbon is reduced and oxygen is oxidized; when it goes from left to right the net effect is that carbon is oxidized (respired) to CO_2 and oxygen is reduced to H_2O. This reaction is traditionally written the other way around (in which photosynthesis proceeds to the right), and we will return to this tradition later in the chapter. Presenting the reaction as in Eq. 3.1, for the time being, will help clarify the concepts of electron donors and acceptors in the following discussion.

Redox reactions can be thought of as consisting of two simultaneous half-reactions in which electrons are gained and lost. In the case of net respiration, i.e., when the reaction in Eq. 3.1 goes from left to right, the carbon-containing compounds lose four electrons:

$$C(0)H_2O + H_2O \rightarrow C(IV)O_2 + 4H^+ + 4e^-. \quad (3.2)$$

It is frequently stated that C donates four electrons and is by definition oxidized from organic matter, CH_2O, to CO_2. In the second half-reaction, oxygen-containing compounds gain 4 electrons:

$$4e^- + 4H^+ + O(0)_2 \rightarrow 2H_2O(-II). \quad (3.3)$$

Each oxygen atom accepts two electrons so O_2 accepts a total of four electrons and is by definition reduced from an oxidation state of (0), O_2, to an oxidation state of (–II), H_2O. The separation of full redox reactions into half-reactions is conceptual, because electrons do not appear in solution, and thus an oxidation cannot occur without a simultaneous reduction. Also, note that the term "oxidation" only indicates an electron transfer reaction in which an element loses electrons. There are many oxidation reactions that have nothing to do with the element oxygen. For example, organic matter oxidation occurs in nature by a variety of electron acceptor reactions that include not only oxic reactions (involving O_2 as in Eq. 3.1), but also a cascade of other anoxic (the absence of O_2) reactions involving NO_3^-, SO_4^{2-}, Fe^{3+}, and Mn^{4+}, to name a few. The relative energetics of these reactions is discussed in Section 4.3 when we describe reactions that occur in anoxic basins and in sediments.

There is a procedure for balancing redox reactions that can be followed as long as you know the molecular formula of the molecules that are being oxidized and reduced. The full redox reaction can be derived from the sum of two half-reactions, as in the example above. First, write the half-reaction with the two different oxidation states of the element that is oxidized on either side of the reaction arrow, then balance the number of oxygen atoms by adding the required number of H_2O molecules to one side. Now balance the hydrogen atoms by adding H^+ and finally add electrons, e^-, to the half-reaction so that the charge is the same on each side of the equation. Follow this procedure for the element that is reduced. Finally, multiply the half-reactions by the stoichiometric coefficient necessary to make the number of electrons in each half-reaction equal so that they cancel when you add the equations – there can be no net change in electrons during the reaction.

3.1.2 The Main Elements of Organic Matter: C, H, O, N, P

Our short-hand formula for organic matter in reactions 3.1 and 3.2 is CH_2O (OM). In reality, organic matter consists of a large variety of compounds with more complicated formulae that exist in both particulate and dissolved form: particulate organic matter (POM) and dissolved organic matter (DOM). The boundary between POM that sinks and DOM that does not has been set at a size of ~0.5 μm, the pore size of the most common filters used in oceanography. As Fig. 3.1 shows, small particles and colloids are therefore classified as DOM. These particles are small enough and their density difference from water low enough that they are transported by the flow of water along with truly dissolved organic matter. Given this operational definition, even some organisms (viruses and some bacteria) are considered DOM! The major constituent of DOM, dissolved organic carbon (DOC), is the largest reduced carbon reservoir in the ocean with concentrations in surface waters that are about 10 times larger than particulate organic carbon (POC). The average marine concentrations of DOC and POC are ~80 and <10 μmol-C kg^{-1}.

Overall, most marine plants and animals are small, on the order of microns, and are either passively drifting (planktonic) or weakly swimming. The concentration of living organisms in the surface ocean is only a fraction of the total organic carbon concentration in DOC and POC. This relatively small standing crop of living particulate carbon is a great contrast to the situation on land where trees and grasses are a much larger fraction of the total carbon. Since photosynthetic rates of marine and land biomass are about the same (net primary production of about 50 Pg-C y^{-1} each; see Chapter 8 about the marine carbon cycle), organic matter residence times (lifetimes) in the surface ocean are much shorter than on land. The average lifetimes of the bacteria and phytoplankton are on the order of hours to days. Although zooplankton have a much broader size range and life span, they still have lifetimes of only days to months. The combined result of this limited mobility and brief life span is that most marine organisms are captive to, and characteristic of, the seawater in which they occur. The exceptions to this generalization are larger animals that can swim appreciable distances moving between water types and sinking particles that can fall through the thermocline into the interior ocean. Because planktonic ocean life is so dynamic, diverse, and endemic, its net effect on the ocean is sometimes best understood by the chemical patterns it creates and responds to, rather than by direct observation. A brief review of the main algal types found in the ocean was presented in Section 1.4, which is a good background for the rest of this chapter.

One of the classic observations oceanographers have made in support of a chemical perspective on ocean processes is that the C:N:P ratios of mixed marine plankton collected

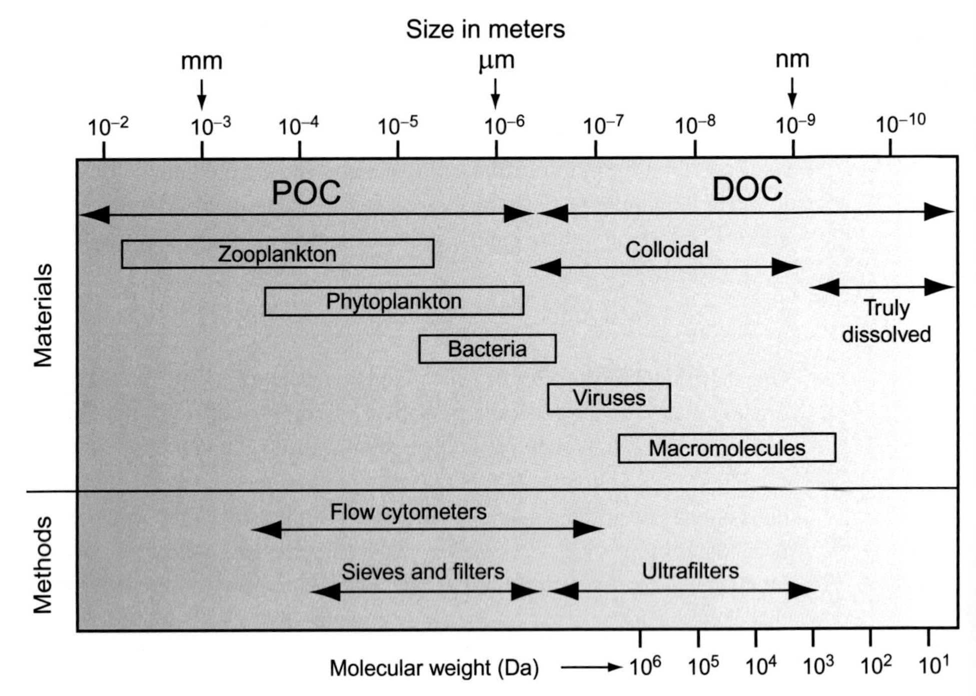

Figure 3.1 Schematic illustration of the sizes of organic matter in seawater using scales of diameter (meters) and molecular mass (one dalton, Da, equals the mass of a proton or neutron). Size ranges are presented for different organic matter types and methods of separation. POC = particulate organic carbon and DOC = dissolved organic carbon. Redrawn from Emerson and Hedges (2008).

by towing nets (~64 µm mesh) through the surface ocean occur at relatively constant values near 106:16:1 on a molar basis. This observation was first published by Alfred Redfield (Redfield, 1958). The N:P ratios used by marine plants and animals are also revealed in the ratios of dissolved inorganic nitrogen (DIN) and dissolved inorganic phosphorus (DIP) in waters below the euphotic zone throughout the world's oceans with a universal slope near 16 (Fig. 3.2).

Following the C:N:P ratios determined for marine organic matter, Redfield, Ketchum, and Richards (1963, hereafter RKR) "fleshed out" the stoichiometry of photosynthesis and respiration as

$$106CO_2 + 16HNO_3 + H_3PO_4 + 122H_2O \rightleftarrows (CH_2O)_{106}(NH_3)_{16}H_3PO_4 + 138O_2. \tag{3.4}$$

The left-hand side of this RKR equation gives the moles of various inorganic nutrients (in their uncharged forms) that are converted to organic matter and molecular oxygen during photosynthesis. This reaction is an elaboration of the simple photosynthesis-respiration reaction presented in Eq. 3.1 only considering the five most abundant elements in organic matter and written so that photosynthesis proceeds from left to right. It is important to recognize that $(CH_2O)_{106}(NH_3)_{16}H_3PO_4$ does not represent a specific

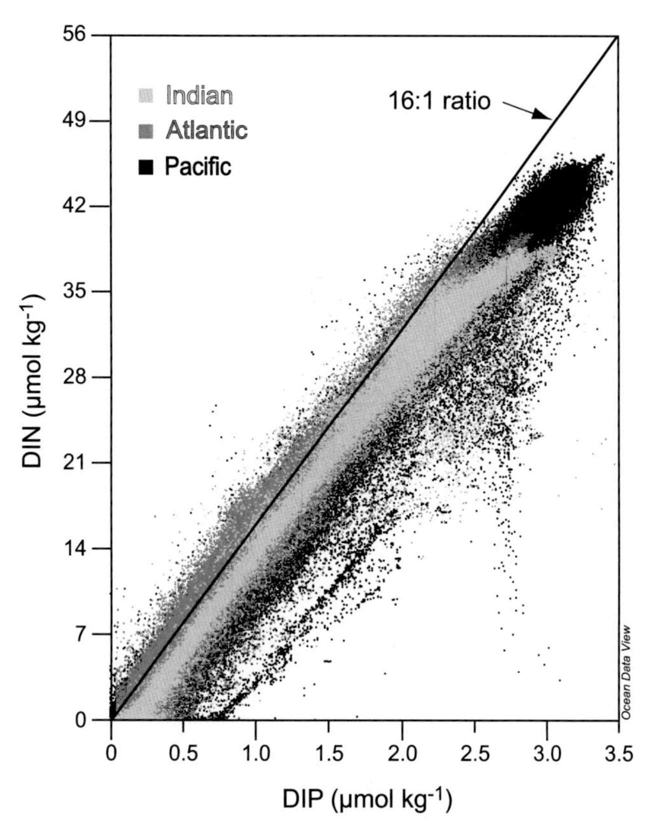

Figure 3.2 The concentration of dissolved inorganic nitrogen and phosphorus (DIN vs. DIP) in the world's oceans. Data are from the CLIVAR and Carbon Hydrographic Data Office (CCHDO, https://cchdo.ucsd.edu). The 16:1 ratio is presented for reference. Note that ratios are slightly higher in the Atlantic than in the Pacific and Indian Oceans because of denitrification reactions in the low-oxygen-containing waters and sediments of the Pacific and Indian Oceans (see Chapter 4). A black and white version of this figure will appear in some formats. For the color version, refer to the plate section.

molecule of organic matter, and we do not think of having a mole of organic matter with this stoichiometry in the same way as we define a mole of the chemical compounds or elements that make up organic matter. For example, if 106 moles of CO_2 underwent photosynthesis according to this reaction, we would say that 106 moles of organic carbon were produced and that 16 moles of organic nitrogen were produced but we would *not* say that a mole of organic matter had been produced.

The number of moles of oxygen produced and water consumed in Eq. 3.4 was estimated entirely theoretically from an assumption that one mole of O_2 is released for every atom of carbon converted into biomass, and two moles for every atom of nitrogen. The reduction half-reactions for Eq. 3.4 are

$$CO_2 + 4H^+ + 4e^- \rightarrow CH_2O + H_2O \tag{3.5}$$

$$NO_3^- + 9H^+ + 8e^- \rightarrow NH_3 + 3H_2O \tag{3.6}$$

and the oxidation half-reaction is

$$2H_2O \rightarrow O_2 + 4H^+ + 4e^-. \tag{3.7}$$

Equations (3.5 and 3.6) were combined in a ratio of 106/16 and the result is added to Eq. 3.7 by adjusting the stoichiometry so that the electrons cancel. This accounts for all the hydrogens that show up in $(CH_2O)_{106}(NH_3)_{16}H_3PO_4$ and the production of 138 moles of O_2. Whereas most of the reactants and products of photosynthesis must travel with the ambient seawater, O_2 can escape to the atmosphere and particulate organic matter can sink. The differential transport of these two chemically extreme products of photosynthesis largely controls the distribution of redox sensitive compounds in the atmosphere and oceans.

The concept of a fixed stoichiometry among the elements C, N, P, and O_2 remains a guide for understanding the chemistry of biological processes in the ocean; however, the values for the stoichiometric coefficients have evolved over time and their geographic constancy in the ocean is currently being challenged. For example, an organic chemist would recognize that the stoichiometric formula given for biomass is impossibly hydrogen rich and suspiciously over packed with oxygen as well. These excesses result in part because the formula $(CH_2O)_{106}(NH_3)_{16}H_3PO_4$ assumes that organic carbon in plankton occurs exclusively in the form of carbohydrate $(CH_2O)_n$, which is the most hydrogen and oxygen rich of all biochemical compounds. In addition, NH_3 carries roughly three times the amount of hydrogen that occurs in protein, where most nitrogen actually occurs in living organisms.

The stoichiometry of real organic matter and the oxygen requirement for its oxidation are demonstrated by plots of the H/C ratio versus O/C ratio of the main components of living organisms compared to the average phytoplankton composition from five ocean samples. As can be seen in Fig. 3.3A, the RKR formula $(C_{106}H_{263}O_{110}N_{16}P)$ has both a higher hydrogen and a higher oxygen content compared to carbon (H/C = 2.45 and O/C = 1) than any of the proteins, lipids, or carbohydrates that make up phytoplankton, and thus is impossible. The proportions of major elements (C, H, O, N, and sometimes S) in a variety of organic molecules and materials that make up living organisms in the marine and terrestrial environment and the oxygen requirement for oxidizing each material are presented in Table 3.1. Note that the RKR H/C ratio is much higher than any of these molecules. A biochemical analysis of marine organic matter types (Table 3.2) demonstrates that the carbon in average marine plankton is roughly 65 percent protein, 19 percent lipid, and 16 percent carbohydrate. The nominal formula $(C_{106}H_{177}O_{37}N_{17})$ corresponds to atomic H/C and O/C ratios of 1.67 and 0.35, respectively. RKR plankton contain too much oxygen because carbohydrates are more O-rich than protein and lipid. The N:C ratio versus the respiration quotient (Fig. 3.3B) $(RQ = \Delta O_2/\Delta C)$ for these plankton samples illustrates that more oxygen is required to oxidize a mole of carbon in true plankton (RQ = 1.44) than in RKR plankton (RQ = 1.3).

So far, we have considered only the elements carbon, nitrogen, and oxygen, which dominate the redox chemistry during photosynthesis and respiration. The ratio of these components to phosphorus, which resides primarily in adenosine triphosphate (ATP), nucleic acids, and lipids, is difficult to determine accurately by measuring the P content of organic components because it is in much lower concentration and measurement error

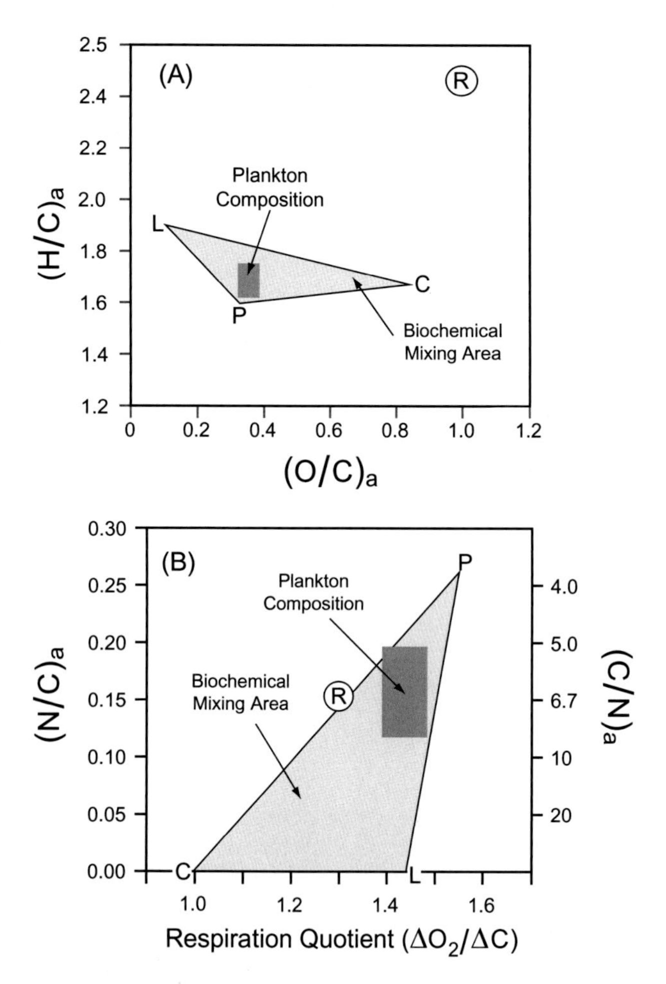

Figure 3.3 H, C, N, and O ratios in organic matter. The triangle in each plot indicates values for proteins, P, carbohydrates, C, and lipids, L, at the apices. The dark rectangle represents and average of values from five plankton samples. R indicates the RKR formula. (A) A van Krevelan plot of the H/C vs. O/C ratios in organic matter on an atomic basis (subscript "a"). (B) The nitrogen/carbon ratio in organic matter, N/C on the left-hand axis and C/N on the right-hand axis, versus the molecular oxygen utilization to organic carbon degradation rate, the respiration quotient ($\Delta O_2/\Delta C$). Redrawn from Hedges et al. (2002).

becomes important. The ratios of C, N, and O to P have been investigated by measuring changes of DIC, DIN, and DIP on constant density surfaces in the aphotic zone of the ocean. These results conclude that a N:P ratio of 16 best fits the data in areas where denitrification is unimportant (Anderson and Sarmiento, 1994). The nitrogen concentration in the analysis by Hedges et al. (2002) was somewhat higher than traditional values (C:N = 106:17 rather than 106:16), probably because several of the samples were from the Southern Ocean where nitrogen fixation may make these data more N-rich than the global average. We assume that the long-standing Redfield stoichiometry of N:P = 16 is appropriate for organic matter respiration in the absence of denitrification.

Table 3.1 C, H, N, and O elemental compositions of common biopolymers and plant materials, and the calculated respiration quotient ($\Delta O_2/\Delta C$ ratio during oxic respiration).

Material	Formula	Example	H/C	O/C	N/C	$\Delta O_2/\Delta C$[a]
Carbohydrate[b]	$C_{100}H_{167}O_{83}$	Cellulose	1.67	0.83	0	1.00
Protein[b]	$C_{100}H_{158}O_{32}N_{26}S_{0.9}$	Net plankton	1.58	0.32	0.26	1.57
Lipid[b]	$C_{100}H_{189}O_{11}$	Oleic acid	1.89	0.11	0	1.42
Chlorophyll[b]	$C_{100}H_{140}O_{9.5}N_{7.5}$	Chlorophyll-a	1.40	0.10	0.08	1.40
Lignin[b]	$C_{100}H_{108}O_{38}$	Gymnosperm	1.08	0.38	0	1.08
Tannin[b]	$C_{100}H_{66}O_{42}$	Angiosperm	0.66	0.42	0	0.96
Marine plankton	$C_{100}H_{167}O_{35}N_{16}S_{0.4}$	Net plankton	1.67	0.35	0.16	1.45
Bacteria	$C_{100}H_{167}O_{35}N_{16}S_{0.4}$	Gram-negative	1.67	0.35	0.16	1.45
Wood	$C_{100}H_{100}O_{50}$	Gymnosperm	1.00	0.50	0	1.00
Tree leaf	$C_{100}H_{100}O_{50}N_{10}$	Angiosperm	1.00	0.50	0.10	1.13

[a] O_2 requirement for total respiration to CO_2, H_2O, and HNO_3 was calculated using the equation:

$$C_\alpha H_\beta O_\gamma N_\delta S_\sigma + \omega O_2 = \alpha CO_2 + \beta H_2O + \delta HNO_3 + \sigma SO_3,$$
$$\text{where } \omega = 1.00\alpha + 0.25\beta + 1.25\delta + 1.5\sigma - 0.5\gamma$$

(Hedges et al., 2002)

[b] Calculated from representative structure (vs. directly measured).

Table 3.2 The four major biochemical compositions (by % carbon) for various common organisms. See Table 3.1 for representative elemental compositions for these biochemical types.

Material	Protein	Carb.[a]	Lipid	Pigment
Bacteria	55–70	3–10	5–20	2–5
Phytoplankton	25–50	5–50	5–20	3–20
Zooplankton	45–70	3–5	5–20	1–5
Vascular plant	2–5	37–55	<3	5–20
Wood	<1	40–80	<3	0

[a] Carb. = carbohydrates, mainly polysaccharides including neutral, basic, and acidic sugars.

The C:N (or C:P) ratio of marine plankton is more difficult to determine by analysis of DIC:DIN (or DIC:DIP) changes along constant density surfaces because: (a) relative changes in DIC are small with respect to the high background value (~2000 μmol kg^{-1}), (b) shallow subsurface regions, where the largest respiration rates are found, are contaminated by fossil fuel CO_2, and (c) in order to determine changes due to respiration one must know the concentrations of DIP, DIN, and DIC in the water when it left the surface because starting values (*preformed values*; see Chapter 4) have to be subtracted from those measured in the interior to determine the changes due to respiration.

Attempts to determine the C:P ratio during respiration by removing contamination by anthropogenic CO_2 using methods described in Chapter 8 indicate a C:P ratio that is in the vicinity of 120 (Anderson and Sarmiento, 1994). The data in these studies, however, come from ocean depths greater than 400 meters, and it has been shown (Körtzinger et al., 2001; Shaffer et al., 1999) that the C:P ratio in degrading organic matter increases from values of about 100 at 100 meters to values between 120 and 130 at depth. In the face of these ambiguities, we adopt the long-standing C:P value of 106:1 in this book.

An improved RKR stoichiometry can be calculated from the formula of Anderson (1995):

$$C_\alpha H_\delta O_\chi N_\beta P_\varphi + \gamma O_2 \rightarrow \alpha CO_2 + \beta HNO_3 + \varphi H_3PO_4 + 0.5 \cdot (\delta - \beta - 3\varphi) H_2O \quad (3.8)$$

and

$$\gamma = \alpha + 0.25\delta - 0.5\chi + 1.25\beta + 1.25\varphi. \quad (3.9)$$

If we use the stoichiometry of the phytoplankton analyzed by Hedges et al. (2002), but corrected for N by increasing the C:N ratio from 106:17 to 106:16, and stick with the traditional C:P ratio of 106:1, the coefficients for Eqs. 3.8 and 3.9 are α, δ, χ, β, and $\varphi =$ 106, 179, 38, 16, and 1, respectively. Using these values in the above equation results in an oxygen:phosphate stoichiometry, $\Delta O_2{:}\Delta P$, of –153 and thus a molar ratio of oxygen change to carbon change ($\Delta O_2{:}\Delta OC$, respiration quotient) of 1.44. The improved RKR equation corresponding to this analysis becomes

$$106CO_2 + 16HNO_3 + H_3PO_4 + 80H_2O \rightarrow C_{106}H_{179}O_{38}N_{16}P\,(OM) + 153O_2, \quad (3.10)$$

where again, $C_{106}H_{179}O_{38}N_{16}P$ represents ratios in bulk organic matter, not a specific molecule. Based on the plankton study of Hedges et al. (2002) and the analysis of Anderson (1995), formulas in the range $C_{106-120}H_{170-180}O_{35-45}N_{14-18}P$, which require an RQ of 1.42–1.46 appear to be realistic for the mean stoichiometry of marine organic matter. We use an RQ value of 1.44 in the calculations of this book.

Adopting a single mean stoichiometry for organic matter in the ocean is tempting, and true to a first order, but geographic variability in these ratios has recently been observed. Martiny et al. (2013) found a systematic latitudinal variation in the C:N ratio of suspended particulate organic matter with C:N ratios higher than 106:16 in subtropical locations and lower than 106:16 in high-latitude regions. This result is consistent with variations in the surface ocean dissolved organic matter C:N:P ratios (Abell et al., 2000; Letscher and Moore, 2015).

DeVries and Deutsch (2014) took a closer look at changes in oxygen and DIP along isopycnal surfaces in the ocean thermocline using a data-constrained model of ocean circulation, which identifies the surface origin of ocean thermocline waters so that changes due to respiration could be isolated from preformed values (concentrations that the water had at the surface before being carried into the interior). They coupled the model with global average measurements (climatological values) of the DIP and O_2 concentration and were able to determine the respiration stoichiometries in shallow regions of the thermocline where most of the respiration takes place (100–300 m). Their results indicate an astounding variation of the ratio of O_2 consumption to the production of DIP. The $\Delta O_2{:}\Delta DIP$ values vary from –80 in the high latitudes to –180 in the subtropics. They attribute this variability to the range of C:P values in organic matter produced in the surface ocean (higher C:P in the organic matter requires more oxygen to respire per P produced). The global mean value for ΔO_2:ΔDIP in their study was –150:1, which is very close to the value of –153:1 determined from the mean organic matter analysis above. The ratios represented in Eq. 3.10 thus seem to be an accurate assessment for global mean values, but the assumption of regional constancy is less viable.

Discussion Box 3.1

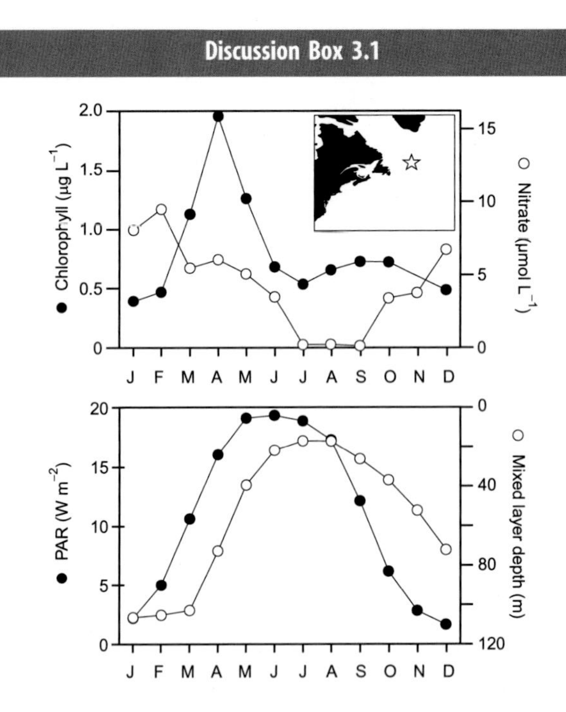

The chapter so far has focused on the reactants necessary to the photosynthetic creation of organic matter. This recipe also suggests what factors can prevent or limit efficient photosynthesis.

This figure shows average seasonal cycles of surface chlorophyll concentration, nitrate concentration, photosynthetically active radiation (PAR), and mixed layer depth at 50°N, 46°W in the North Atlantic. Chlorophyll data from the MODIS Aqua satellite retrieved from NASA's OceanColor website, nitrate data from the World Ocean Atlas 2013, PAR data from Bishop et al. (1997), and mixed layer depth data from the Argo program via Holte et al. (2017).

- What do these data suggest about what limits productivity in the winter at this location? In the summer?
- What initiates the spring increase in phytoplankton?
- Are high surface nutrient concentrations an indicator of ideal conditions for phytoplankton growth? Why or why not?
- Will the addition of anthropogenic carbon to the surface ocean stimulate ocean productivity at this location?

3.1.3 Trace Elements in Organic Matter: Fe, Zn, Mn, Ni, Cu, Co, Cd

The requirement for metals in enzymes that control many different types of metabolic processes has been known for a long time (Harvey, 1945). However, the concentrations of these trace metals in seawater are very low, and many are complexed by dissolved organic compounds, which can inhibit their reactivity. Thus, the role of metals in photosynthesis and respiration is only now being quantified. The first-row transition metals (manganese, Mn; iron, Fe; cobalt, Co; nickel, Ni; copper, Cu; and zinc, Zn) and the second-row element cadmium, Cd, are essential for the metabolism, growth, and reproduction of all marine phytoplankton. The required concentrations or "quotas" of metals in phytoplankton are driven largely by the biochemical demands of the photosynthetic machinery and/or the makeup of the many different proteins required for specific metabolic purposes (Bruland et al., 2014; Morel et al., 2014; Twining and Baines, 2013). Iron (Fe) is the only trace metal that has been shown unequivocally to limit the rate of photosynthesis in the sea, and by some estimates (Moore et al., 2013) does so in as much as 30–40 percent of the surface ocean. Other metals have been shown to stimulate growth in bottle experiments (Mn, Zn, Co, and Cu) but are probably more important in regulating species composition of phytoplankton communities, because of the differences in cellular quotas required for growth among species (Sunda, 2012).

In phytoplankton, iron is required for carbon fixation, because of its essential role in electron transport in photosystems I and II, for the reduction of nitrate and nitrite to make them bioavailable, and for chlorophyll synthesis. Photosynthesis is typically iron limited in so-called HNLC (high-nutrient, low-chlorophyll) regions of the surface ocean that have high concentrations of the macronutrients NO_3^- and DIP year-round (the Equator, the Antarctic, and North Pacific; roughly 30 percent of the ocean's surface). High macronutrient concentrations in surface waters normally do not indicate an ideal situation for phytoplankton growth but rather that something else is limiting productivity, preventing the drawdown of the nutrients. Iron limitation has been shown experimentally in these HNLC areas by seeding the surface ocean with Fe and monitoring the biological response (e.g., de Baar et al., 2005). A consistent observation in these experiments is that larger diatoms grow more quickly than the other soup of phytoplankton after Fe addition. The reason is believed to stem from the physics of nutrient supply, diatoms' ability to store Fe, and the role of grazers. Because large diatoms have a relatively low surface area to volume ratio compared to picoplankton and nutrient uptake is limited by diffusion across the cell wall, they require higher Fe concentrations in the water to maintain the supply necessary for growth. Many species of diatoms also have the ability to produce ferritin and other compounds to store Fe within their cells, which allows them to continue to grow after added Fe has dropped back to ambient concentrations. Finally, the growth of picoplankton is often kept in check by small grazers that can rapidly reproduce, while larger diatoms can, at least initially, escape grazing pressure because zooplankton capable of eating them need more time to grow and reproduce.

Perhaps the most extreme example of a process requiring Fe is nitrogen fixation occurring in specialized plankton able to catalyze this reaction (the transformation of

N_2 gas to organic N). Enzymes responsible for N_2 fixation are very iron-rich, and the use of N_2 as a nitrogen source can require up to 10 times more Fe than growth on NH_4^+. In regions of the ocean where the source of iron is limited (areas far from continents and the flux of iron-rich dust) and the availability of fixed nitrogen (NO_3^- and NH_4^+) in surface waters is low, the rate of nitrogen fixation can be a limiting factor in phytoplankton growth. In this sense, photosynthesis rates in these regions can be viewed as being co-limited by iron and nitrogen (Sunda, 2012).

Manganese (Mn) by many assessments has the second highest quota in marine phytoplankton. Manganese occurs in the water-oxidizing complex of photosystem II (Eq. 3.7) and is thus essential for photosynthesis. Manganese also occurs in the enzyme superoxide dismutase, an antioxidant that removes toxic superoxide radicals $\left(O_2^-\right)$ produced by photosynthesis. The concentration of manganese is rarely limiting to photosynthesis, because dissolved forms available to phytoplankton exist in surface waters. Very little total dissolved Mn is complexed by organic ligands (scc later) and the oxidized inorganic form of manganese (Mn(IV)) is photoreduced in surface waters to the highly soluble and biologically available Mn(II) species.

Zinc (Zn) is the predominant metal in the enzyme carbonic anhydrase, which catalyzes the transformation of HCO_3^- to CO_2. This mechanism is important in the sea, because the pool of HCO_3^- contains 100 times more carbon than the pool of CO_2, which is usually the form of inorganic carbon that is reduced to organic carbon during photosynthesis. Diatoms and some cyanobacteria also use carbonic anhydrase to concentrate CO_2, and it has been observed that in some cases both Co and Cd can substitute for Zn in the carbonic anhydrase enzyme. Zn is also an important component of the enzyme alkaline phosphatase, which is necessary for phytoplankton to be able to use dissolved organic phosphorous (DOP) during growth.

The known roles of other metals in cell biochemistry are not certain but suspected to be quite specific. Copper (Cu) is used in both photosynthesis and respiratory electron transfers and is present in the enzyme superoxide dismutase. It is necessary for the growth of cyanobacteria at concentrations available in seawater but can be toxic at higher concentrations. Nickel (Ni) is believed to be associated primarily with the enzyme urease, which is required for phytoplankton to utilize urea as a nitrogen source. Cobolt (Co) plays an important role in the growth of cyanobacteria; both *Prochlorococcus* and *Synechococcus* have Co requirements that cannot be replaced by Zn. Co is also incorporated into vitamin B12, but this is unlikely to represent a large portion of cellular Co.

Trace metal requirements for some marine eukaryotes have been presented with a Redfield-like stoichiometry (Morel et al., 2014, from the data of Ho et al., 2003):

$$(C_{106}N_{16}P)_{1000}Fe_8Mn_4Zn_{0.8}Cu_{0.4}Co_{0.2}Cd_{0.2}. \tag{3.11}$$

While this stoichiometry provides an order of magnitude image of metal requirements, one can challenge it on a number of grounds. For example, organisms in laboratory cultures may take up metal concentrations based on their availability in the media, and

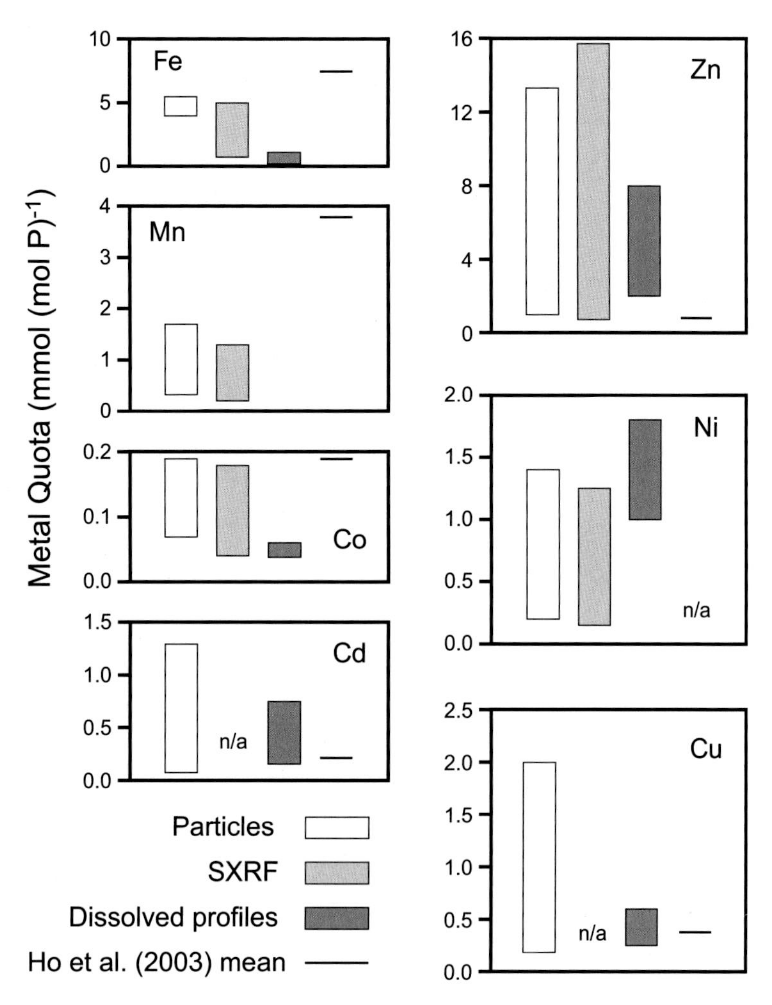

Figure 3.4 Observed trace metal : DIP concentration ratios (mmol metal : mol P) in marine particulate matter and dissolved in seawater (quotas). Note that the scales of the axes vary. "Particles" indicates concentrations measured by traditional methods on particles filtered from seawater samples; "SXRF" indicates concentrations determined by Synchrotron radiation x-ray fluorescence on individual particles; "Dissolved profiles" indicates dissolved metal/DIP ratios along isopycnal surfaces in the ocean; and "Ho et al. (2003) mean" indicates values from the laboratory cultured plankton used to derive the stoichiometry in Eq. 3.11. Modified from Twining and Baines (2013).

recent ocean observations indicate a great deal of stoichiometric plasticity in observed metal requirements. In Fig. 3.4, measurements derived from laboratory cultures (Eq. 3.11) are compared with concentrations determined in particles sampled from the ocean and dissolved metal : DIP ratios in the upper water column of the ocean (<800 meters). Observations indicate that the quotas determined in these different ways vary by more than a factor of 5 for Fe, Mn, Zn, Ni, Co, and Cd and by a factor of

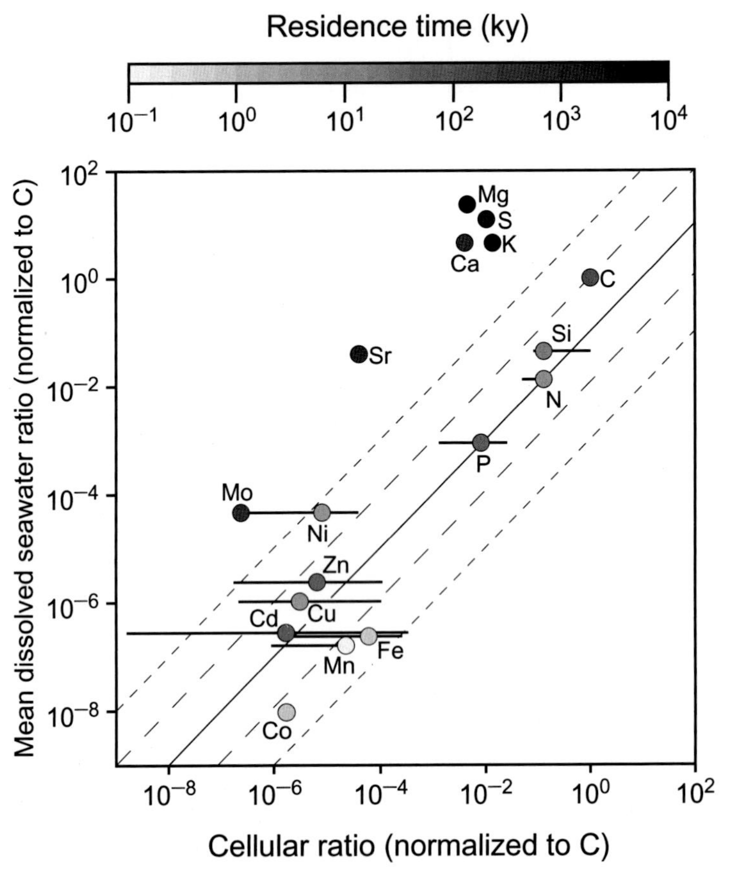

Figure 3.5 A log–log plot of metal:carbon ratios of elements dissolved in seawater versus the quotas determined in plankton samples. The solid line runs through values for nitrogen and phosphorus. Dashed lines indicate one and two orders of magnitude departure from the solid line. Shading distinguishes element residence times in kyr. Modified from Moore et al. (2013).

10 for Zn. Perhaps the greatest inconsistency with the stoichiometry in Eq. 3.11 is that Zn quotas, while quite variable, are of the same magnitude as those for Fe. Twining and Baines (2013) suggest a hierarchy of trace mental quotas determined from their measurements in marine plankton of

$$Fe \sim Zn > Mn \sim Ni \sim Cu >> Co \sim Cd.$$

A very crude picture of the role of biological processes in influencing the concentration of elements in the sea comes from comparing the measured intracellular concentration with that found in seawater. This analysis has only recently become possible because of the growing number of contamination-free trace metal measurements. A log–log plot comparing the concentration of elements relative to carbon (element:C ratio) in ocean phytoplankton and dissolved in seawater (Fig. 3.5 from Moore et al., 2013) illustrates in a very general sense which elements are in excess or depleted in seawater relative to cellular quotas. Carbon is used for normalization because it is by far the most abundant element in organic matter. Horizontal bars indicate the range of element ratios found in ocean plankton and hence the plasticity of the quotas discussed above. Dissolved seawater concentrations are average values, which for nutrient-like elements are close to the deep-water value not the value found in surface water. Use of the average concentrations rather than surface water values focuses the interpretation of the figure on timescales that are equal to or longer than the circulation time of the ocean. The solid line in Fig. 3.5 is drawn through the concentrations of N and P, which are the macronutrients that limit plankton growth in most of the ocean. Dashed lines on either side of the N and P line indicate one and two orders of magnitude departure from a perfect correlation with N and P. Elements with seawater concentrations above the line, like the major ions Mg, Ca, K, and S, have concentrations in seawater that are in great excess relative to values in plankton cells. The concentration of these elements is not very dependent on biological processes – geological factors discussed in Chapter 2 are more likely to influence their abundance in seawater. Elements that are near the line are influenced by biological processes. Elements that are below the dashed line, like Fe, Mn, and Co in the lower left corner, are removed from seawater by processes with timescales that are shorter than biological processes, like adsorption to particles (a process called "scavenging," see later). Residence times, indicated by shading of the points, range from less than 1000 years for elements that are strongly active biologically or readily adsorbed to greater than one million years for the major elements. The residence time (see Chapter 2) is a complex function of the availability of the element in rocks, its solubility in water, and biological processes.

One of the most important observations about the speciation of metals in seawater has been the role of chelation or complexation (we use these terms interchangeably) of metals by organic ligands (organic molecules that bond strongly with metals in solution). Studies of metals in seawater using electrochemical techniques were developed to an art by Ken Bruland and students (see Bruland and Rue, 2001, for a description of one of the methods) and have demonstrated that most metals, particularly bioactive ones, are chelated with strong metal-binding organic ligands.

The metal concentrations presented in the periodic table of Chapter 1 and the metal:C ratios in Fig. 3.5 refer to the "total" metal concentration dissolved in seawater – that which is measured in an unfiltered, acidified water sample. (Measurements on filtered samples indicate that the "total" values are predominantly dissolved.) The dissolved concentration is often categorized into two general pools: (a) the portion chelated with organic compounds and (b) the portion present as the free metal, Me^{n+}, plus concentrations complexed with inorganic anions like Cl^-, CO_3^{2-}, and/or OH^-. Inorganic complexes are categorized with the free metal because they are difficult to distinguish experimentally and are less strongly bound to the metals than the organic complexes.

Analysis of seawater samples by electrochemical methods (summarized by Morel et al., 2014) suggest that greater than 99 percent of Fe^{3+} in surface waters is complexed with strong Fe^{3+}-binding ligands which exist at subnanomolar to nanomolar concentrations in excess of the dissolved Fe; greater than 99 percent of dissolved Cu exists as organic complexes; approximately 98 percent of Zn in surface seawaters is complexed

Discussion Box 3.2

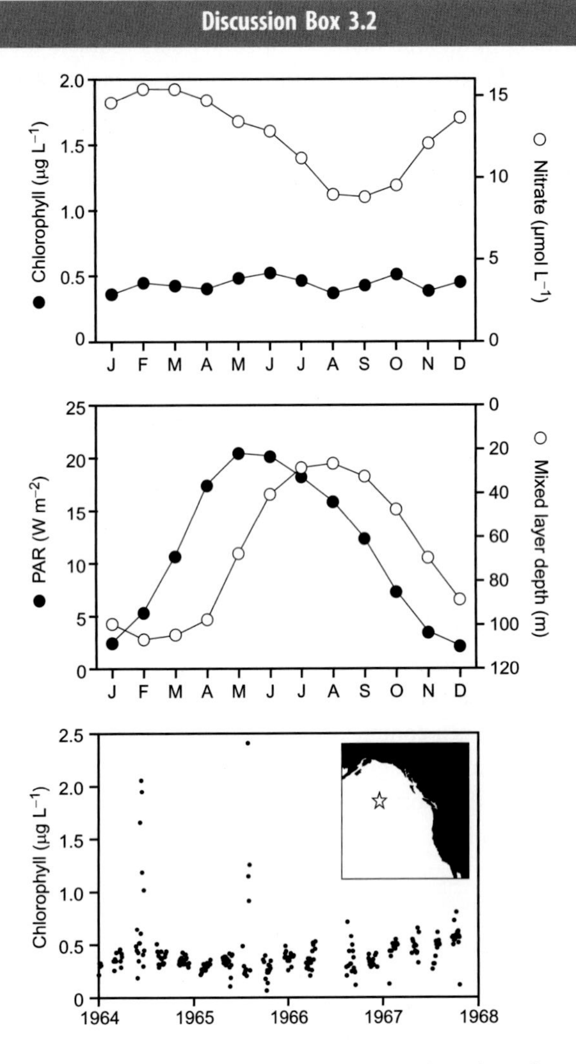

The top two panels in this figure show average seasonal cycles of surface chlorophyll concentration, nitrate concentration, photosynthetically active radiation (PAR), and mixed layer depth at Ocean Station Papa (50°N, 145°W) in the North Pacific from the same sources as Discussion Box 3.1. The lower panel shows a time series of surface chlorophyll measurements in the 1960s when the station was frequently occupied by the Canadian Weathership program (Fisheries and Oceans Canada).

- What do these data suggest about what limits productivity at this location?
- Compare these seasonal cycles in the North Pacific to those in the North Atlantic (Discussion Box 3.1). What might cause the similarities and differences you observe?
- How do the high frequency chlorophyll measurements in the 1960s compare to the average annual cycle? What could explain the deviations?

with organic ligands; greater than 90 percent of cobalt exists complexed to strong Co-binding organic ligands; and about 80 percent of Cd in surface waters is complexed with organic ligands. Among the micro-nutrient trace metals, Ni and Mn are the exceptions with regard to complexing by ligands: only about 30 percent of the total dissolved Ni and very little of the dissolved Mn is complexed. Identification of the chemical compounds that make up the metal-binding ligands is currently an area of active research. While metal–ligand interactions are not yet very quantitatively understood at the writing of this book, two general characteristics are that measurable concentrations of the ligands exist throughout the water column and that, so far at least, they seem to be very metal specific.

Trace metal nutrients are taken up into phytoplankton cells by specialized transport proteins on the cell membrane. Binding of dissolved metals with these proteins is thought to be limited to the concentration of "free" metal ions in solution. Thus, chelation by organic complexes is generally believed to decrease metal bioavailability making chemical speciation extremely important to photosynthesis (Sunda, 2012). However, ligand-bound trace metals are available to some plankton. For example, organic compounds called siderophores are believed to be produced by bacteria specifically to enable the organism's acquisition of iron. Electrochemical analytical methods have revealed that siderophores produced by bacteria for Fe acquisition have similar stability constants to those of Fe-binding ligands observed in surface seawater (Barbeau et al., 2001, 2003) suggesting that at least some metal-binding ligands were once membrane-bound metal acquisition proteins. A further twist is that many free metal ions are highly insoluble and easily adsorbed to particles sinking through the water column. Metal-binding ligands may help to maintain higher total metal concentrations while lowering free metal concentrations. Thus, the role of organic metal complexation in the uptake of trace metals by phytoplankton is complicated – some organic compounds render metals unavailable whereas others enhance metal uptake and solubility.

3.2 The Flux of Biologically Produced Elements from the Surface Ocean: The Ocean's Biological Pump

Determining fluxes, the transfer of chemical compounds or fluid between different parts of the environment, represents a challenge that is different from measuring concentrations because there are no certified standards for a quantity like a flux. To quantify rates, the element of time must be introduced either by incubations in the laboratory or at sea or by a model of concentration distributions that includes information about rates of transport. Because there are no standards, the only method for assessing accuracy of fluxes is to compare different flux methods with each other until a consensus is obtained (see Section 3.2.5). We present and compare several of the most important methods here.

Clear definitions of terms used frequently to describe marine biological productivity and the flux of organic matter from the surface ocean are required to understand the interdisciplinary nature of studies of the marine biological pump. The transformation of dissolved inorganic carbon (DIC) to organic carbon by autotrophs (mainly organisms performing photosynthesis) is called gross primary production (GPP). Since autotrophs also respire organic carbon back to DIC, net primary production (NPP) is the term used to describe the difference between GPP and autotrophic respiration. NPP is not directly measurable by current methods but is believed to be most closely represented by rates determined by ^{14}C-DIC incubations. Some of the details of how GPP and NPP are determined in the ocean are presented in Appendix 3A.1. The rarest flux, the one that escapes all respiration in the upper ocean, is net community production (NCP), the difference between NPP and heterotrophic respiration. This is the amount of organic carbon produced by photosynthesis that is available to escape the surface ocean.

On an annual basis, the organic carbon flux out of the upper ocean, the biological pump, should equal the annual net community production (ANCP) of organic carbon. These terms are frequently used interchangeably. One has to be careful in the definition of the "upper ocean," though, because light levels and mixed layer depths vary seasonally in most regions. Since the biological pump is a measure of the organic carbon that escapes the upper ocean on an annual basis, the "upper ocean" is defined here as the depth of the winter mixed layer. This definition avoids including in the biological pump (or ANCP) organic matter that escapes the surface ocean in summer, when the mixed layer depth is shallow, but then degrades and is returned to dissolved form in the mixed layer when it deepens in winter.

We focus our attention on ANCP in the rest of this chapter because this is the flux that controls the observed large-scale chemical distribution deeper in the ocean, and it helps regulate the partial pressure of CO_2, the most critical greenhouse gas, in the atmosphere (see the example in Chapter 8). This biologically driven marine flux is an important thread connecting the physical, chemical, and biological processes in the ocean to the global carbon cycle and thus climate.

This section begins with a simple two-layer model of the ocean-wide mean value of the biological pump to emphasize the relationship between physical mixing and biological processes. We then move on to upper ocean observations used to determine the global distribution of the biological pump and compare these values with estimates from remote sensing (by satellites) and global circulation models.

3.2.1 A Simplified Whole-Ocean Model of the Biological Pump

The distribution of chemical constituents in the ocean provides insight into biological processes. So far, we have considered the biological uptake of many elements by plankton in the surface ocean and their release back to dissolved form during respiration. We can now combine those reactions with the physical sinking of particles and circulation of ocean water between the surface and deep in a quantitative way. A construct historically used by chemical oceanographers to model large-scale circulation and biological fluxes involves dividing the ocean into well mixed reservoirs (boxes). The simplest of these is the two-layer ocean that was employed effectively by W. S. Broecker and others primarily in the 1970s and 1980s (e.g., Broecker, 1971; Broecker and Peng, 1982). This model, depicted in Fig. 3.6, divides the ocean into an upper layer where both photosynthesis and respiration

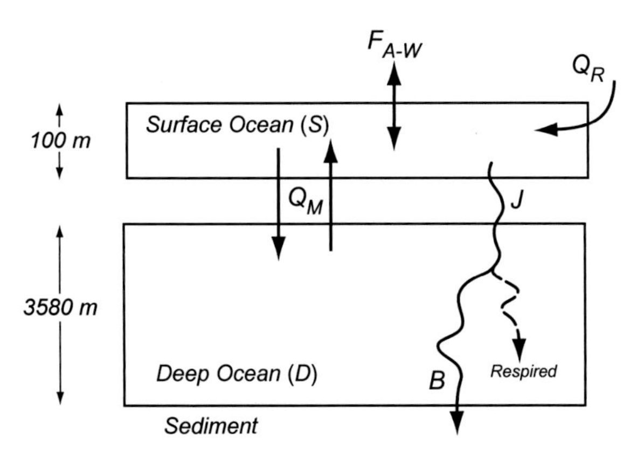

Figure 3.6 A schematic representation of the two-layer ocean including the terms used in the mass balances for the surface (*S*) and deep (*D*) reservoirs. Q_R and *B* are the river water inflow rate and sediment burial rate. $F_{A\text{-}W}$ is the air–water gas exchange rate, which affects only gases. Exchange between the surface and deep reservoirs is via sinking particles, *J*, and water transport, Q_M.

occur and a deep layer where only respiration occurs. We set the dividing line between the layers at ~100 m, the approximate depth of the 1 percent light level above which most photosynthesis occurs and also the mean depth of the winter mixed layer in most of the ocean. The rest of the ocean, including the thermocline, is in the deep layer, which is on average 3580 m thick.

We introduce the two-layer model, because it can provide fabulous first-order insight into nutrient distributions and biological export from the euphotic zone. However, it is important to keep in mind that the model assumes that the surface and deep ocean are each well mixed and homogeneous, so it can give misleading results about processes that occur on timescales that are shorter than whole-ocean mixing or space scales smaller than indicated by the boxes. For this reason, more complex models are now used for research purposes. In Chapter 8, we will introduce a three-box ocean and atmosphere model that separates the surface ocean into high and low-latitude boxes.

In this version of the two-layer model, we consider river inflow to be the only way most elements enter the ocean and burial of particles in sediments, B (mol y^{-1}), to be the only way elements are removed. Water flows between the two layers (upward/downward advection or mixing), and particles sink from the surface layer into the deep, J (mol y^{-1}). We neglect hydrothermal fluxes and atmospheric deposition, which have less importance for the major nutrients. For now, we will not consider gas exchange between the ocean and atmosphere; however, we show this flux in the model diagram ($F_{A\text{-}W}$) because we will add it in later versions (Chapter 8). Water flowing into the ocean via rivers with a flow rate of $Q_R(\mathrm{m^3\,y^{-1}})$ carries element concentrations for component C of $[C]_R(\mathrm{mol\,m^{-3}})$, while upward and downward advection, both Q_M ($\mathrm{m^3\,y^{-1}}$), carry the element concentrations of the deep and surface reservoirs respectively, $[C]_D(\mathrm{mol\,m^{-3}})$ and $[C]_S(\mathrm{mol\,m^{-3}})$. The sinking particle flux (J) is not carried by water movement, and so cannot be split into a water transport and concentration term. The changes in surface and deep concentrations with respect to time are:

$$V_S\left(\frac{d[C]_S}{dt}\right) = Q_R[C]_R + Q_M([C]_D - [C]_S) - J \qquad (\text{mol y}^{-1}), \tag{3.12}$$

$$V_D\left(\frac{d[C]_D}{dt}\right) = Q_M([C]_S - [C]_D) + J - B \qquad (\text{mol y}^{-1}), \tag{3.13}$$

where V_S and V_D are the volumes of the surface and deep reservoirs (m^3), respectively. At steady state, the changes with respect to time are zero, so inputs equal outputs for both boxes and for the ocean as a whole. In this case, the flow of species C from rivers is equal to its burial rate in sediments:

$$\frac{d[C]}{dt} = 0 \text{ and } Q_R[C]_R = B. \tag{3.14}$$

Since there are only two reservoirs, Eqs. 3.12 and 3.13 have interrelated terms and are not independent, meaning that in this model we can solve for only one unknown.

In order to gain quantitative insight about fluxes between the surface and deep reservoirs, the mixing rate, Q_M, must be estimated. This has been done using measurements of natural carbon-14 in the dissolved inorganic carbon (DIC) of surface and deep waters. ^{14}C is produced in the atmosphere and it decays in the deep ocean with a half-life of 5700 years. We demonstrate its utility for determining ocean circulation later in Chapter 7 on radioactive tracers; for now, we will adopt a mixing rate, Q_M, that is consistent with a 1000-year residence time with respect to circulation of water in the deep ocean box. The water residence time in the deep ocean (τ_M) is equal to the ocean volume divided by the mixing rate:

$$Q_M = {}^{V_D}\!/_{\tau_M} = \frac{1.31 \times 10^{18} (\text{m}^3)}{1000 \text{ y}} = 1.31 \times 10^{15} \text{ m}^3 \text{ y}^{-1}. \tag{3.15}$$

Before using the model to calculate the sinking particle flux (the biological pump), we use it to make a few general statements about biogeochemical dynamics in the ocean. First, the mixing rate, Q_M, is about 35 times greater than the inflow rate from rivers (3.73×10^{13} m^3 y^{-1}, Appendix A):

$$\frac{Q_M}{Q_R} = \frac{1.31 \times 10^{15}}{3.73 \times 10^{13}} = 35. \tag{3.16}$$

Water circulates on average about 35 times between the surface and deep ocean before it evaporates. This is consistent with the mean residence time of water with respect to river inflow of ~40 000 y calculated in Chapter 2 (Table 2.3).

We can also evaluate the importance of the mixing (upwelling) versus river inflow to the delivery of nutrients to the surface ocean and to the fraction of the particle flux that is buried (see Broecker and Peng, 1982 for more details). Phosphorus will be used rather than nitrate as the limiting nutrient in this illustration to avoid the complexities of nitrogen fixation and denitrification; however, phosphate and nitrate are related stoichiometrically in most of the ocean (Fig. 3.2), so nitrate could also be used. The average dissolved inorganic phosphorus concentrations in rivers, $[DIP]_R$ is 1.3 μmol kg^{-1} (Meybeck, 1979). The mean concentrations in the deep sea and surface ocean are $[\text{DIP}]_D = 2.3$ μmol kg^{-1} and $[\text{DIP}]_S =$ 0.0–1.3 μmol kg^{-1}, respectively. (The latter value is the range for average subtropical ocean to high-latitude surface waters, Toggweiler and Sarmiento, 1985.) The model phosphorus fluxes are river input (or sediment burial):

$$\begin{aligned} Q_R[\text{DIP}]_R = B &= \left(3.73 \times 10^{13}\ \text{m}^3\ \text{yr}^{-1}\right)\left(1.3\ \mu\text{mol kg}^{-1}\right)\left(1000\ \text{kg m}^{-3}\right) \\ &= 4.8 \times 10^{10}\ \text{mol yr}^{-1} \end{aligned} \tag{3.17}$$

upwelling to the surface layer:

$$\begin{aligned} Q_M[\text{DIP}]_D &= \left(1.31 \times 10^{15}\ \text{m}^3\ \text{yr}^{-1}\right)\left(2.3\ \mu\text{mol kg}^{-1}\right)\left(1025\ \text{kg m}^{-3}\right) \\ &= 3.1 \times 10^{12}\ \text{mol yr}^{-1} \end{aligned} \tag{3.18}$$

where 1000 kg m^{-3} and 1025 kg m^{-3} are the densities of freshwater and seawater, respectively. The flux of particles sinking into the deep ocean is

$$\begin{aligned} J_P &= Q_M\left([\text{DIP}]_D - [\text{DIP}]_S\right) + Q_R[\text{DIP}]_R \\ &= \left(1.31 \times 10^{15}\ \text{m}^3\ \text{yr}^{-1}\right)\left(2.3\ \mu\text{mol kg}^{-1} - \left(0.0\ \text{to}\ 1.3\ \mu\text{mol kg}^{-1}\right)\right)\left(1025\ \text{kg m}^{-3}\right) \\ &\quad + 4.8 \times 10^{10}\ \text{mol yr}^{-1} \\ &= 1.4 - 3.1 \times 10^{12}\ \text{mol yr}^{-1}. \end{aligned} \tag{3.19}$$

These results imply that phosphorus taken up by biological organisms in the surface ocean is 65 times more likely to come from upwelling than from rivers ($3.1 \times 10^{12} / 4.8 \times 10^{10}$), indicating that ocean circulation is far more important in regulating biological productivity than river inflow. We can also compare the sinking particle flux, J_P, to the total input flux to the surface layer, $Q_M[\text{DIP}]_D + Q_R[\text{DIP}]_R$, to see that between 50 and 100 percent of DIP reaching the surface layer is removed via biological incorporation into sinking particles. Considering the deep sea, only 1 in 30 to 65 (1–3 percent) of P atoms that sink to the deep ocean in biological particles are permanently buried in the sediments, while the rest are respired in the deep and ultimately recycled back to surface waters. This results in a residence time for phosphorus with respect to burial of 30 000–65 000 y – 30 to 65 times the ocean circulation rate. If we were to track a typical P atom over time, we would observe that after flowing into the ocean with a river, this P atom was incorporated into a particle that sank into the deep ocean then was respired back to a dissolved form and mixed back to the surface ocean an average of 30-65 times before escaping respiration to be permanently incorporated into the sediments. Of the time that the P atom does spend in the ocean, 99 percent is spent in the dark deep sea and less than 1 percent is the sunlit surface waters, because of the very different size of these layers.

From the fluid transport rates and inorganic carbon concentrations, we can also derive an estimate of the flux of carbon from the upper ocean at steady state:

$$\begin{aligned} J_C &= Q_M\left([\text{DIC}]_D - [\text{DIC}]_S\right) + Q_R[\text{DIC}]_R \\ &= \left(1.31 \times 10^{15}\right)\left(2269 - 1941\ \mu\text{mol kg}^{-1}\right)\left(1025\ \text{kg m}^{-3}\right) \\ &\quad + \left(3.73 \times 10^{13}\ \text{m}^3\ \text{yr}^{-1}\right)\left(958\ \mu\text{mol kg}^{-1}\right)\left(1000\ \text{kg m}^{-3}\right) \\ &= 4.8 \times 10^{14}\ \text{mol y}^{-1}. \end{aligned} \tag{3.20}$$

The net flux of DIC from the deep ocean to the surface waters via mixing again dominates the calculation and must be balanced by the particulate carbon flux to the deep (the biological pump). The calculated value for J_C is 0.48×10^{15} mol yr^{-1} or 5.7×10^{15} g-C y^{-1} (~5 Pg-C y^{-1}). This is an upper limit for the *organic carbon* flux as calculated from this model, because the surface to deep DIC gradient is actually controlled by both organic C and $CaCO_3$ carbon fluxes, so our calculation of J_C must be a combination of organic and inorganic carbon export. Averaged over the ocean surface, this estimate of the biological pump is 1.3 mol-C m^{-2} y^{-1}, which are the units we will use in the following discussion of observational and model-determined organic carbon export from the upper ocean. Notice that from just the average surface and deep concentrations and the ocean's mixing rate, we have derived an estimate of the global rate of organic carbon export or ANCP! We shall see that the value from the two-layer model underestimates values from upper ocean metabolite mass balance and global circulation models by about a factor of two, yet it demonstrates the power of chemical distributions to yield estimates of important fluxes. The probable reason for the underestimate by the two-layer model is that the observed DIC gradients at the base of the ocean's mixed layer are steeper than a linear extrapolation between the surface ocean and deep ocean averages.

Discussion Box 3.3

The two-layer model is not only useful for determining quantitative fluxes in the ocean, but also serves as a mental model of how the ocean would react to various changes. Any change in concentrations or water flow rates will temporarily throw the ocean system out of balance, but mechanisms exist that will slowly return the ocean back toward steady state.

For example, consider a situation in which global fertilizer use permanently increases the concentrations of nutrients in rivers to double their preindustrial value, an extreme scenario but instructive. In some parts of the ocean, surface nutrients are always drawn down to near zero concentrations, while in other areas surface nutrient concentrations are controlled by the availability of trace metals or sunlight. For this example, let's suppose that these other controls do not change such that the surface nutrient concentrations remain the same after the increased river concentration as they were during preindustrial times (you may use a value of 0.5 μmol DIP kg^{-1} for any calculations). We will also assume that the fraction of the surface-to-deep particle flux that is buried (B/J) does not change.

- What fluxes in the two-layer model will *initially* increase in this scenario? Consider what happens just after the change occurs, when the ocean is not in steady state.
- How does the input rate of nutrients to the whole ocean *initially* compare to the output rate via burial in sediments? What does this imply about the whole ocean concentration of nutrients? In which layer does a change occur?
- Qualitatively, how does this change in nutrient concentrations affect upwelling/downwelling fluxes? How do those changes affect export? How does this lead the ocean to return to steady state?
- Quantitatively, what are the particle and burial fluxes of P (J_P and B_P) at the new steady state?

3.2.2 Particle Fluxes and Thorium Isotope Tracers

The most obvious method for determining the flux of particles in the ocean is to catch them as they sink using sediment traps, which are rain-gauge-like collectors of particles that have been deployed in the ocean hundreds of times. Although these deployments have revealed a wealth of information about the nature of sinking particles, there are good reasons to expect problems in determining the true falling flux of particles of different sizes and densities using a hollow cylinder in a moving fluid. Fluxes measured by shallow sediment traps over an annual cycle are frequently different from fluxes estimated by other methods (see Section 3.2.5).

The most important contribution of sediment trap studies to the chemical perspective of oceanography is the data they provide about the nature of particles that leave the upper ocean and the rate at which they degrade and dissolve as they sink. Correlations between organic and inorganic material in particles that are captured in sediment traps and on large-water-volume filters indicate that the mechanism(s) that control(s) particulate organic and inorganic matter fluxes from the surface ocean are similar (Lam et al., 2011, Sanders et al., 2010.) It has been suggested that the correlation implies that shells of $CaCO_3$ (and opal) act as ballast, by increasing the density of particles so that they sink faster, aiding the transport of POC to depth (Armstrong et al., 2002; Klaus and Archer, 2002) because of the differences in density between inorganic and organic particulate material.

While correlation does not necessarily indicate causation, a simple calculation of the weight percent of the three main constituents of the particulate flux demonstrates the origin of the ballast hypothesis. Based on worldwide gradients of DIC and alkalinity in the top of the permanent thermocline, about 6 percent of the carbon that leaves the upper ocean is $CaCO_3$ (Sarmiento et al., 2002), and the SiO_2 to $CaCO_3$ molar ratio found in sediment traps is about 2 (Broecker and Peng, 1982). Thus, to a first approximation, for each 100 atoms of sinking carbon six of them are carbonate carbons and there are an additional 12 atoms of silicon from opal shells. Given that organic matter contains approximately 55 wt% C and is associated with 12 wt% C in $CaCO_3$ and 45 wt% Si in SiO_2 (assuming 10% water); average marine particles contain roughly 62, 17, and 21 wt% of organic matter, calcium carbonate, and opal, respectively.

Using average densities (g cm^{-3}) of ~2.5 for calcite and opal, and ~1.1 for organic matter, and ~1.0 for seawater, mineral and organic matter will have relative density differences from water of ~1.5 and ~0.1 g cm^{-3}, respectively. According to Stokes' law, the sinking rate of a spherical particle, ω, is directly proportional to its density difference from the ambient fluid and its radius squared: $\omega = 2(\rho_p - \rho_f)\, gr^2/9\mu$, where ρ_p and ρ_f are the densities of the particle and fluid, respectively, g is the gravitational acceleration, r is the particle radius, and μ is the dynamic viscosity of the fluid. Under these conditions, a pure inorganic particle will sink 15 times faster than an organic particle of the same dimensions. If inorganic particles act as ballast for sinking organic matter (Fig. 3.7), the respiration rate profile in the water column is substantially deepened. Without ballast, almost no POM would make it below 200 m based on these calculations (compare the two curves in Fig. 3.7)! While this is a compelling argument for the importance of ballast,

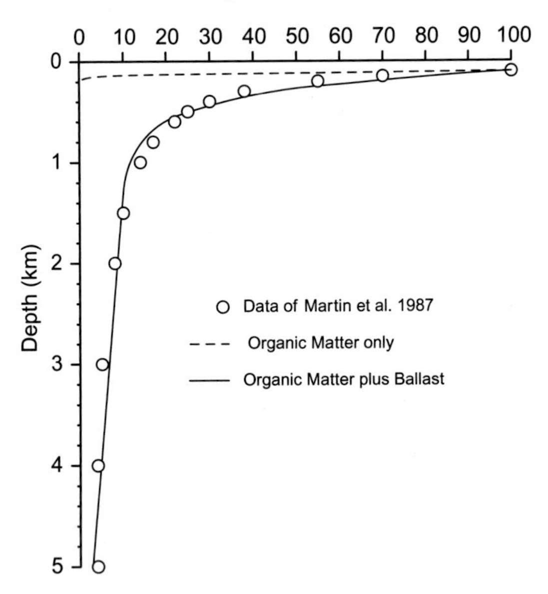

Figure 3.7 The percent particulate organic matter remaining below 100 meters assuming a simple steady-state model where the organic carbon flux out of the surface ocean, $s(d[C]/dz)$, is equal to the first-order organic carbon respiration rate, $k[C]$, at all depths. The sinking velocity, s(m d^{-1}), is determined from Stokes' law for particles that start at 100 m with a diameter of 125 μm (see text). Lines are drawn for the cases in which the particles are pure organic matter (dashed line) and a mixture of 60 percent organic and 40 percent mineral matter by weight (solid line). Circles are the sediment trap measurements of Martin et al. (1987). Lines are from Michael Peterson, unpublished results. Modified from Emerson and Hedges (2008).

observations that inorganic and organic concentrations in particles are correlated may also simply mean that the mechanisms that control the fluxes are similar, particularly in shallow regions of the aphotic zone (Sanders et al., 2010).

The symbols in Fig. 3.7 are data from sediment traps deployed in the mid 1980s by John Martin and colleagues (Martin et al., 1987) and the associated depth curve is often referred to as the "Martin curve." There are many versions of this curve, certainly varying from place to place in the ocean (Buesseler and Boyd, 2009), and the mechanisms controlling the shape are uncertain. This early work, however, established the general concept that the processes of sinking and organic matter respiration conspire to result in an exponential decrease in particulate organic matter flux with an *e*-folding depth of 150–200 meters. By "*e*-folding depth" we mean the depth below 100 m (the winter mixed layer depth) at which the particulate organic matter concentration reaches $1/e$ (0.37) of the value it had at 100 m.

An important chemical tracer for determining the flux of particles from the surface ocean is the difference in activity between the isotope thorium-234, ^{234}Th (half-life = 24.1 days), and its radioactive parent uranium-238, ^{238}U (half-life = 4.5 billion years), in surface waters of the ocean. Details of this method are presented in Section 7.3.3 and Figs. 7.10 and 7.11; we introduce the method here because it has become one of the standard bearers for determining particulate carbon export.

Uranium is a conservative element in the ocean – its ratio to salinity is everywhere constant within the measurement error. This means that the dissolved concentration of ^{238}U is easily predicted. ^{238}U decays to ^{234}Th, which is highly reactive to particles and readily adsorbed, which then removes ^{234}Th from the surface ocean as the particles sink creating a deficiency of ^{234}Th near the surface compared to a situation with no particles. Because the vast majority of particles in the surface ocean are biologically produced and because the ^{234}Th half-life is of the same magnitude as the residence time of these particles in the surface ocean, the deficiency of ^{234}Th activity compared to the activity of its parent, ^{238}U, is proportional to the flux of particulate thorium from the surface ocean. The method is an effective gauge of particle transport dynamics, and there have been hundreds of ^{238}U-^{234}Th activity profiles measured in the upper ocean. However, there are some downsides to the method that limit its application. First, the method requires shipboard sampling, with little prospect of developing an automated in situ method, so an annual time series of measurements is achievable only at ocean sites dedicated to monthly visits by ships. Second, determining particulate organic carbon flux from the activity deficiency requires knowing the organic C/^{234}Th ratio of the particles involved in the flux, which is variable in time and space and must also be measured. Finally, the ^{238}U-^{234}Th activity deficiency is sensitive only to particle flux and thus misses export due to dissolved organic matter.

Correcting this last issue by estimating the flux of dissolved organic carbon (DOC) from the surface ocean requires accurate estimates of the change in DOC concentration with depth below the top ~100 meters of the ocean. This has been accomplished on a global scale by Dennis Hansell and Craig Carlson, with an example presented in Fig. 3.8. The change in concentration between surface waters and values in the top of the thermocline, is about 30 μmol kg^{-1} (from ~75 μmol kg^{-1} in ocean surface waters to about 45 μmol kg^{-1} at ~500 m). The $1/e$ depth of this "semi-refractory" category of DOC is similar to that for organic matter particles – a few hundred meters, and the fraction of organic matter export that is due to DOC has been determined to be between 15 and 25 percent globally (Hansell et al., 2009; Letscher and Moore, 2015). The role of DOC in respiration can be determined by comparing DOC and oxygen utilization in subsurface waters (see Section 4.2.4).

3.2.3 Upper Ocean Metabolite Mass Balance: O_2, NO_3^-, DIC, DIC+δ^{13}C-DIC

Biological processes result in a measurable change in metabolic products in the upper ocean; thus, one can use observed changes in the concentrations of metabolites like O_2, NO_3^-, DIC, and the stable isotopes of DIC to derive rates of carbon export (ANCP) from the upper ocean. An example of simultaneous, in situ measurements of the upper ocean concentrations of O_2, NO_3^-, and DIC (Fig. 3.9) from the subarctic Pacific illustrates the summertime drawdown of DIC and of NO_3^- as well as oxygen supersaturation in the mixed

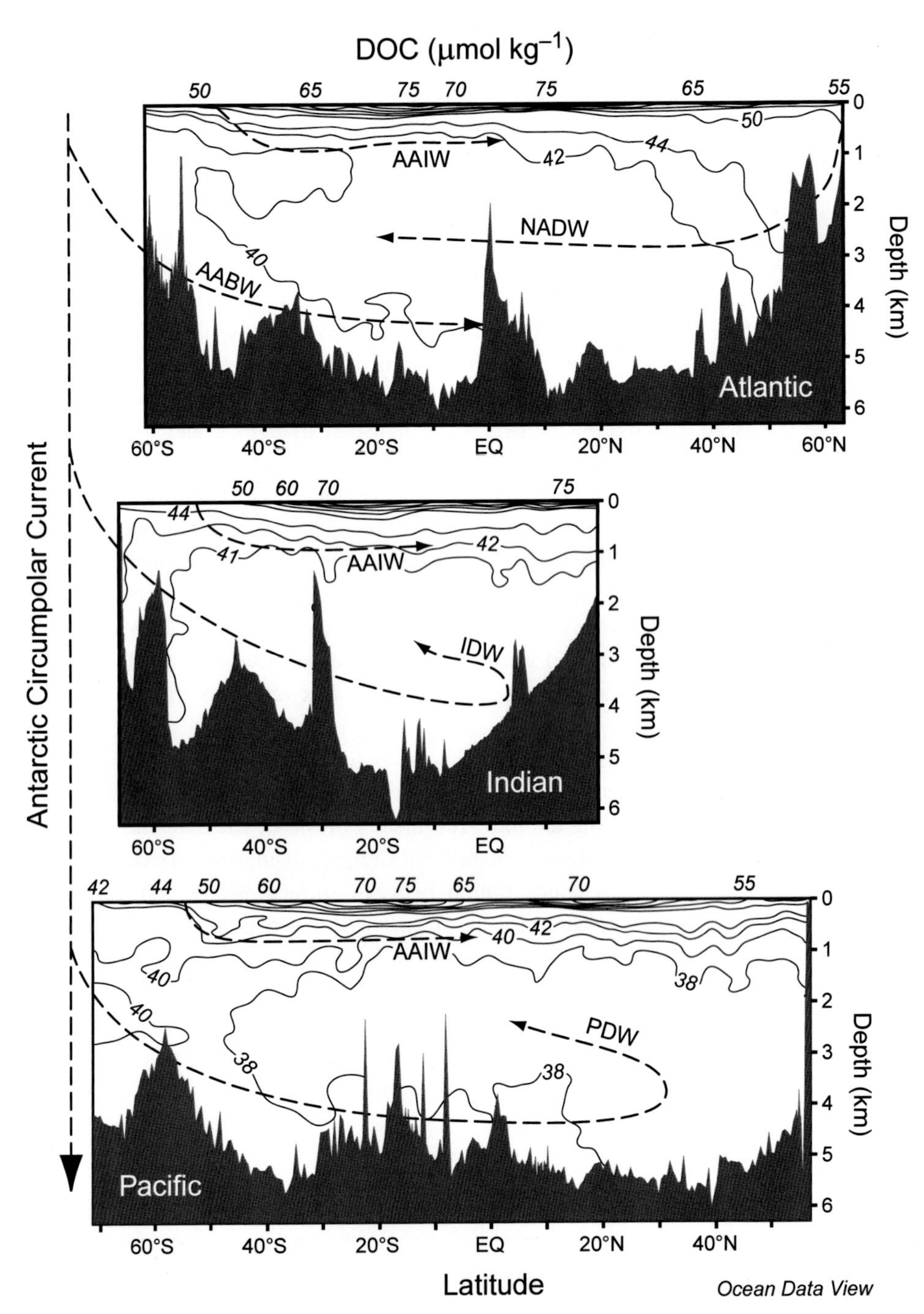

Figure 3.8 Cross sections of dissolved organic carbon (DOC, μmol kg^{-1}) in the Atlantic, Indian and Pacific Oceans. Surface values are 65–75 μmol kg^{-1} which decrease to roughly half this value in the upper pycnocline. Data are from Climate Variability and Predictability (CLIVAR) Repeat sections P16N & P16S for the Pacific, A16N & A16S for the Atlantic, and I09N & I08S for the Indian Ocean. (Data are compiled on the Ocean Carbon Data System, OCADS website: www.ncei.noaa.gov/access/ocean-carbon-data-system/.)

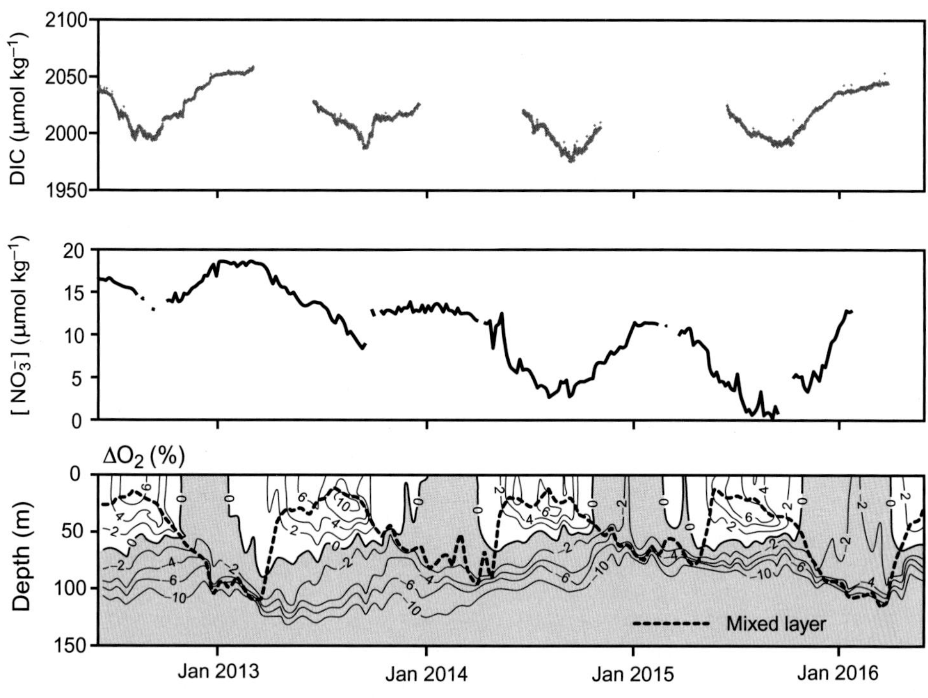

Figure 3.9 A four-year time series of DIC, NO_3^-, and oxygen saturation anomaly (ΔO_2) at Ocean Station Papa (OSP). NO_3^- and DIC data in the top two panels are mixed layer values measured by profiling floats or on a surface mooring, and the oxygen saturation anomaly ΔO_2 (%) versus depth in the third panel is determined from O_2 data measured by profiling floats. The data emphasize the seasonality of these metabolic products at this location. Redrawn from Yang et al. (2018).

layer, all of which demonstrate the influence of biological processes. In the wintertime, surface NO_3^- and DIC increase due to mixing with deeper waters, while the oxygen saturation anomaly is driven close to atmospheric saturation because of low biological production and vigorous gas exchange by high winds. We describe in the following paragraphs how these tracers are used to determine biologically induced fluxes from the upper ocean.

Interpreting time series of measurements in terms of ANCP requires an upper ocean mass balance to evaluate the roles of biological and physical processes like gas exchange, mixing, and advection. The mass balance differs from the two-layer model of Section 3.2.1 in that we consider only the surface layer, and we add terms for horizontal mixing and for air–sea gas exchange. The change in concentration, ($[C]$, mol m^{-3}), integrated to the winter mixed layer depth, h (m), with respect to time, t, is equal to the physical fluxes, F (mol m^{-2} d^{-1}), and the biological production term, J (mol m^{-3} d^{-1}):

$$\frac{d}{dt}\int_{z=0}^{z=h}[C]dz = F_{A-W} + F_z + F_x + \int_{z=0}^{z=h} J\,dz. \tag{3.21}$$

The dominant physical fluxes are: air–sea gas exchange, $F_{A\text{-}W}$, vertical exchange across the base of the winter mixed layer, F_z, and the horizontal advective fluxes, F_x. Integration is done to the depth of the winter mixed layer rather than the seasonally changing mixed layer depth or the euphotic zone depth, because the flux we are after is that which leaves the upper ocean to the interior over an annual period. Each of the terms describing physical processes are expanded to show their dependence on both transport coefficients and concentration gradients in Table 3.3. The different metabolic tracers of ANCP have unique sensitivities to the physical processes, which are indicated qualitatively in the lower half of Table 3.3 and in the following discussion.

Table 3.3 Mass balance equations for determining ANCP from time series measurements of the concentrations of O_2, NO_3^-, or DIC.

(a) Eq. 3.22 states that the temporal change in tracer concentration, $[C]$, (mol m^{-3}) in the mixed layer is equal to processes of gas exchange with the atmosphere, G, transport by vertical and horizontal motions, V and H, respectively, and biological production, J. Equation 3.22 is integrated with respect to depth between $z = 0$ and the depth of the winter mixed layer, $z = h$ (m) to derive Eq. 3.23. Details about the gas exchange formulation are presented in Appendix 2A.3 of Chapter 2, and the definitions of equation terms are in the table's footnote.

(b) Relative sensitivity of the determination of ANCP, $\int_{z=0}^{z=h} J\,dz$, to the physical transport terms on the right side of Eqs. 3.22 and 3.23. The non-steady-state term on the left side of Eqs. 3.22 and 3.23 is not listed in (b) because it is measured and usually near zero for an annual cycle. A dash (–) indicates little or no dependence, a plus (+) indicates moderate dependence (10–30% of the ANCP), and two pluses (++) indicate strong dependence (40–90% of the ANCP). The superscript (*) indicates that although the DIC + δ^{13} C-DIC mass balance depends strongly on advection and mixing, these terms do not need to be determined separately (see text).

(a)

$$\frac{dC}{dt} = G + V + H + J \qquad (\text{mol m}^{-3}\,\text{d}^{-1}) \tag{3.22}$$

$$\int_{z=0}^{z=h} \frac{d[C]}{dt} dz = -k_s([C_{ml}] - [C_{eq}]) + F_B + w([C_h] - [C_{ml}]) + K_z\left(\frac{d[C]}{dz}\right)_h + u\int_{z=0}^{z=h} \frac{d[C]}{dx} dz + \int_{z=0}^{z=h} J dz \qquad (\text{mol m}^{-2}\,\text{d}^{-1}) \tag{3.23}$$

(b) Tracer	$F_{A\text{-}W}$	F_z	F_{xy}	Reference
O_2	++	+	-	Bushinsky and Emerson (2015)
NO_3^-	-	++	++	Archer et al. (1993)
DIC	+	++	++	Fassbender et al. (2016)
DIC + δ^{13}C – DIC	+	++*	++*	Quay and Stutzman (2003)

k_s = gas exchange mass transfer coefficient (m d^{-1}); $[C_{ml}]$ = concentration in the mixed layer (mol m^{-3}); $[C_{eq}]$ = concentration at saturation equilibrium with the atmosphere (mol m^{-3}); F_B = bubble air–sea flux (mol m^{-3} 2d^{-1}); w = vertical velocity (m d^{-1}); $[C_h]$ = concentration at depth h (mol m^{-3}); K_z = eddy diffusion coefficient at depth h (m^2 d^{-1}); u = horizontal advection velocity (m d^{-1}).

The main terms in the upper ocean mass balance for oxygen are production by biological processes and exchange with the atmosphere. ANCP estimates from oxygen are only weakly dependent on horizontal advection, because air–sea exchange has a residence time for oxygen in the mixed layer of only a few weeks, so oxygen is dominated by local processes. (A water parcel with a horizontal advection rate of 5 cm s^{-1} covers only about 60 km or ~½ degree in 2 weeks.) Vertical mixing can be of order 30 percent importance in regions where the oxygen minimum is strong and close to the upper ocean (Bushinsky and Emerson, 2015), but is less important in other locations.

There has been a major effort in chemical oceanography to determine the rate of air–sea gas exchange as a function of wind speed over the ocean (see Appendix 2A.3). From this work, gas exchange mass transfer coefficients can be evaluated from remotely sensed estimates of wind speed. Thus, oxygen gas exchange fluxes can be determined as long as one has accurate measurements of the air–sea pO_2 difference. Recent advances in the accuracy of oxygen sensors that can be deployed in situ on a variety of platforms like profiling floats have made this possible (Bushinsky and Emerson, 2015). A caveat here is that oxygen air–sea exchange is sensitive to bubble processes caused by breaking waves at wind speeds greater than 8–10 m s^{-1}, because it is a very insoluble gas. Bubble processes are less well understood than diffusive exchange at the air–sea interface, and this uncertainty adds error to determining ANCP from oxygen mass balance in subarctic and polar regions where wind speeds are high in the wintertime (see Appendix 2A.3).

The upper ocean mass balance of NO_3^- has been used to determine summertime NCP in high-latitude regions where there is measurable NO_3^- in surface waters year-round (Plant et al., 2016; Wong et al., 2002). Since there is no air–sea exchange of NO_3^-, the mass balance consists of the change with time being equal to NCP and mixing processes. Because it is difficult to determine mixing and advection rates over the period of an entire year, NCP has usually been determined from NO_3^- concentration change during the summer when the thermocline is strong and exchange with the deeper ocean is assumed to be negligible.

The upper ocean mass balance of dissolved inorganic carbon $\left(\mathrm{DIC} = \left[\mathrm{HCO_3^-}\right] + \left[\mathrm{CO_3^{2-}}\right] + [\mathrm{CO_2}]\right)$ is dependent on the physical processes of both gas exchange and water transport. Because only a small fraction of DIC in seawater is the gas CO_2 ($DIC/CO_2 \sim 2000/10 = 200$) and because the different species are in chemical equilibrium, the gas exchange residence time for CO_2 is roughly 10 times longer than for oxygen (see Chapter 8) – more like a year instead of a few weeks. This makes the DIC mass balance sensitive to all terms in Equation 3.21: horizontal and vertical fluxes as well as air–sea exchange, and they must be evaluated to determine ANCP (Fassbender et al., 2016).

Summertime NCP values determined by mass balance of NO_3^-, DIC, and O_2 from data in Fig. 3.9 are 8, 12, and 11 mmol C m^{-2} d^{-1}, respectively. Horizontal processes are ignored in these calculations. Biological fluxes determined from NO_3^- and O_2 are converted to carbon using Redfield ratios for the fluxes of nitrate, $J_C = J_N(\Delta C : \Delta N)$ where $\Delta C : \Delta N = 106/16$, and oxygen, $J_C = J_{O_2}(\Delta C : \Delta O_2)$ where $\Delta C : \Delta O_2 = 106/153$. The agreement in ANCP calculated from the O_2 and DIC mass balances reinforces confidence in the accuracy of the calculation of biological processes by these tracers, because different terms in their mass balances dominate. The lower result determined from the NO_3^- data may mean horizontal processes are an important flux for this mass balance even in summer at this location.

A promising twist for improving the utility of DIC mass balance as an ANCP tracer is to measure both the bulk carbon and its stable isotope, DIC and the δ^{13}C-DIC. δ notation for isotopes will be introduced in Chapter 6; suffice it to say, for now, that δ^{13}C is a measure of the ratio $^{13}C/^{12}C$ relative to a standard. This technique has been demonstrated at the ocean time series locations (Brix et al., 2004; Gruber et al., 1998; Quay and Stutzman, 2003). The residence time with respect to gas exchange of δ^{13}C-DIC is even longer than DIC (multiple years) because it involves exchanging the ^{13}C isotope through the entire DIC pool rather than a chemical equilibrium between CO_2, HCO_3^-, and CO_3^{2-}. Thus, the temporal and areal footprint of the δ^{13}C-DIC tracer approaches basin scales making vertical and horizontal transport the main terms in the upper ocean mass balance with gas exchange also playing a significant role (Table 3.3). Seasonal changes in the upper ocean DIC + δ^{13}C-DIC are illustrated in Chapter 6 (Fig. 6.10), where details of the application of stable isotopes to determining ANCP are discussed.

The consistency of the O_2 versus the DIC + δ^{13}C-DIC tracers has been recently tested by comparing ANCP determined from both tracers using data from subtropical regions of the Atlantic, Pacific, and Indian Oceans (Yang et al., 2019). ANCP determined at the same locations by both tracers confirmed that they result in the same ANCP in the Northern Hemisphere subtropical gyres $\left(\sim 2 \text{ mol C m}^{-2}\text{yr}^{-1}\right)$ even though they integrate over very different timescales. Both tracers resulted in lower ANCP in the Southern Hemisphere subtropical regions, and the results there were not as consistent between the two tracers as in the north.

3.2.4 O_2/Ar and O_2/N_2 Tracers

One way of getting around some of the problems associated with determining the impact of physical processes on the oxygen mass balance is to measure the concentration of an inert gas like argon or nitrogen along with the concentration of oxygen. The biologically inert gas can be used to empirically correct for some of the physical processes that affect oxygen. The main physical properties of a gas that influence transfer between the air and water are its solubility in water and its molecular diffusion coefficient (see Appendix 2A.3). These properties along with the kinematic viscosity of water depend on temperature (and salinity to a lesser extent).

If the two gases measured are oxygen and argon, which have nearly the same solubility and molecular diffusion coefficients (see Appendices D and E), two of the physical processes that cause the oxygen gas concentration to depart from saturation equilibrium with the atmosphere (temperature change and bubble processes) are removed by writing the gas exchange flux in terms of the differences in the gas saturation anomalies, $\Delta C = [C]/\left[C_{eq}\right] - 1$. Effectively, Ar is being used as a biologically inert oxygen, and some of the physical influences on the oxygen mass balance discussed in Section 3.2.3 can be ignored. In cases where vertical and horizontal fluxes are negligible, the biological oxygen flux becomes even simpler:

$$J_{O_2} = k_{s,O_2}(\Delta O_2 - \Delta Ar)[O_2]_{eq}, \tag{3.24}$$

where k_{s,O_2} is the gas exchange mass transfer coefficient for oxygen (see Appendix 2A.3).

The use of the oxygen and argon pair to determine net biological oxygen production was done first by making measurements at ocean time series sites (Hamme and Emerson, 2006; Spitzer and Jenkins, 1989; Stanley, 2007), where a full annual cycle of the saturation anomalies of oxygen and argon were determined. There are now creative methods of expanding the reach of dual gas measurements beyond ship-occupied time-series sites. First, O_2/Ar ratio measurements can be made continuously by portable mass spectrometers after extraction of the gas from seawater (e.g., Cassar et al., 2009). Basin-scale measurements of the O_2/Ar ratio in ocean surface waters have been determined in this way by installing instruments in commercial container ships and tapping into the surface waters used to cool the engines as they ply the ocean for international trade. If enough cruises along similar routes are occupied, one derives a data set that is expansive geographically and covers all seasons as illustrated for the North Pacific in Fig. 3.10. Palevsky et al. (2016) used these data to determine the ANCP in the surface ocean across the North Pacific. The ΔO_2/Ar ratio plotted in this figure is nearly identical to the (ΔO_2 – ΔAr) difference in Eq. 3.24. Results of this study indicate that ANCP in the western region of the North

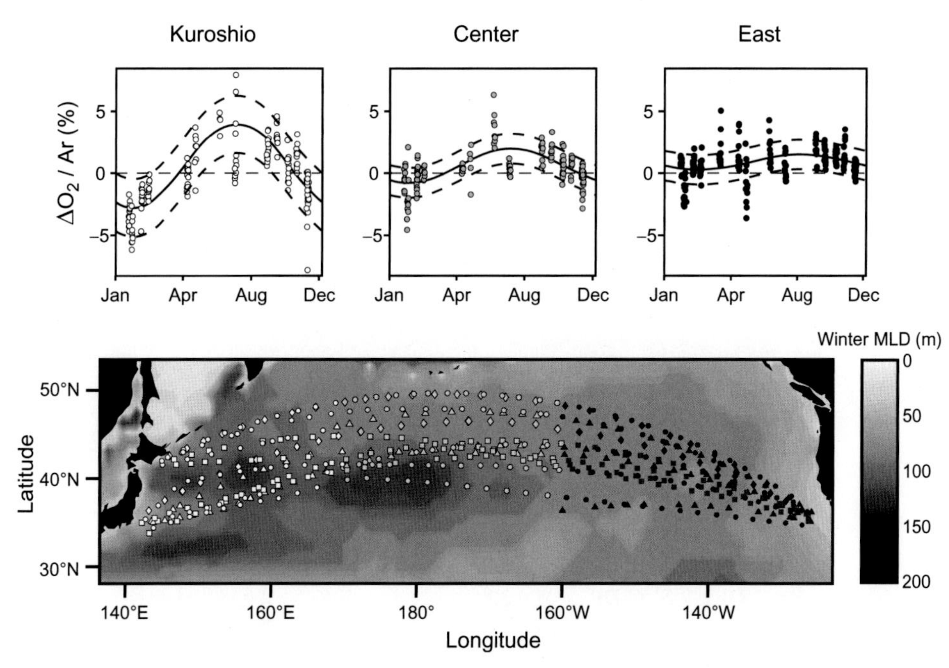

Figure 3.10 The oxygen/argon saturation anomaly in the surface waters of the North Pacific Ocean $\Delta(O_2/Ar)(\%) = \{(O_2/Ar)_{ml}/(O_2/Ar)_{eq}) - 1\} \times 100$ at different times of the year, determined by repeated measurements at the locations indicted on the map. Seawater samples were measured in the engine room of container ships on repeated transects shown by different symbols. Gray scale on the map indicates the depth of the surface winter mixed layer. Redrafted from Palevsky et al. (2016).

Pacific is lower than in the other two regions and that the depth of the winter mixed layer plays a strong role in determining the ANCP value.

A second in situ method for determining ANCP from dual gas measurements at the other extreme in space and time are the measurements of oxygen and nitrogen gas on surface moorings. The concentration of nitrogen gas can be determined by in situ measurements of total gas pressure and oxygen concentration. Gas pressure is determined using a device called a gas tension device (GTD). With simultaneous measurements of total gas pressure and oxygen partial pressure, one derives accurate measurements of both oxygen and nitrogen gas (Emerson et al., 2002). Determining the net biological oxygen production using this gas pair is not as simple as for O_2 and Ar because N_2 is about half as soluble as O_2, which makes N_2 more sensitive to bubble processes. Emerson and Stump (2010) demonstrate the determination of ANCP by this method at Ocean Station Papa using the full simultaneous equations for two gases rather than the simplifications in Eq. 3.24.

3.2.5 Comparing Different Methods for Determining ANCP

At the beginning of our discussion of methods for quantifying ANCP, we stressed that there are no standards for determining fluxes and that accuracy must be assessed by comparing different approaches. An example of this is presented in Table 3.4, where we compile ANCP and particle export rates determined at three prominent ocean time-series sites using the different approaches discussed in this section. ANCP results from the metabolite mass balances of the O_2/Ar ratio, DIC + δ^{13}C-DIC, O_2, and the oxygen utilization rate (OUR). The last of these methods is described Section 4.2.3. The differences among these methods are fairly small, only a $\pm$ 20–25 percent variation, which is no

Table 3.4 ANCP (mol-C m^{-2} y^{-1}) determined by different methods at three time-series locations: the Bermuda Atlantic Time-series Study (BATS), the Hawaii Ocean Time series (HOT), and Ocean Station Papa (OSP). ANCP values for the three locations are for a single year of measurements. These values except those for O_2 are from the compilation by Emerson (2014). O_2 values are from Yang et al. (2018). "Mean" values are the mean and standard deviation of the "Mass Balance" values. Sediment trap results are the mean and standard deviation of annual values between 1988 and 2008 (Church et al., 2013 for BATS and HOT; Timothy, 2013 for OSP). Thorium isotope results are from Benitez-Nelson et al. (2002, HOT) and Charette et al. (1999, OSP).

	BATS	HOT	OSP
Mass Balance			
O_2/Ar	3.9, 5.6	2.7, 1.4, 3.3	1.6, 2.5
DIC + δ^{13}C-DIC	3.8	2.5, 2.3, 3.1	
O_2		2.4	2.2
OUR	4.9		
Mean	$(4.5 \pm 0.9, n = 4)$	$(2.5 \pm 0.6, n = 7)$	$(2.1 \pm 0.5, n = 3)$
Particle Export			
Sediment Traps	0.8 $\pm$ 0.3	0.9 $\pm$ 0.5	0.5 $\pm$ 0.4
^{234}Th-^{238}U		1.5	1.8

greater than most assessments of the error of the individual methods. Mean values at HOT and OSP are the same within the measurement scatter, but values at BATS are nearly twice those of the other two time-series sites. Higher values at BATS are to be expected because this location is in a region of very deep mixed layers in winter, which resupply the upper ocean with limiting nutrients (Jenkins and Doney, 2003), while the other two sites are located within the subtropical and subarctic gyres where there is a pycnocline at about 110 meters that limits the depth of the winter mixed layer.

Sediment trap results at all three of these time-series locations have been determined at 150 to 200 meters nearly monthly for over a 10-year period. The particulate organic matter fluxes are distinctly lower than values determined from the metabolite mass balances. These values are also lower than particle flux estimates from the ^{234}Th method at HOT and OSP, indicating that the sediment traps are probably under-collecting the particle flux. Even the thorium-derived particle flux at HOT is significantly lower than the ANCP determined by the mass balance methods. The reason for this difference at HOT is consistent with the large contribution of DOC to the export of carbon in this region of the subtropical gyre (at least equal to the particulate flux, Emerson 2014; see Section 4.2.4).

3.3 Global Distributions of Organic Carbon Export

Global estimates of the ANCP or net biological carbon export can at present be determined using two types of models: satellite estimates of net primary production (NPP) multiplied by separate estimates of the NCP/NPP ratio to determine a global distribution of ANCP, and global circulation models with embedded ecosystem formulations that provide net organic carbon export. These methods have not been thoroughly verified by observations so far because observational estimates in many regions have only recently become available. Here, we briefly describe the two global methods and make a first attempt at comparing these results with observations that exist to date.

3.3.1 Comparing Measured ANCP with Model Predictions

The ocean's color and optical backscatter can be determined by satellites, from which it is possible to estimate the chlorophyll content and particle concentration in the sea over one optical depth, about the first 30 m of surface waters. These two parameters are measures of the presence and distribution of phytoplankton in the sea. While they are not a direct measure of the number of phytoplankton cells or of phytoplankton carbon (because cells contain different amounts of chlorophyll under different conditions), they are the most rapid and widely used method of identifying the presence of photosynthetic organisms. Satellite observations of chlorophyll and optical backscatter have been used to determine the rate of ocean primary production by comparing chlorophyll concentration and particulate carbon content with in situ determinations of primary production from ^{14}C incubations (e.g., Behrenfeld et al., 2006; Westberry and Behrenfeld, 2014). These relationships have been used with satellite color measurements to estimate global marine primary production. One caveat to keep in mind is that a variable fraction of the total primary production, particularly in the subtropical gyres, occurs deeper than satellites can detect. While biological oceanographers attempt to compensate for this and other factors, the accuracy of converting surface chlorophyll and particle concentrations to photosynthetic productivity in vastly different regions of the ocean is very tricky. Satellite-based NPP estimates are multiplied by NCP/NPP ratios determined either from ecosystem models or empirically from ocean flux measurements to yield global maps of NCP. An example of a map of these estimates is presented in Fig. 3.11A.

The other global estimate of ANCP derives from ocean global circulation models with an embedded ecosystem formulation. A summary of biological carbon export from four of these types of models was presented by Laufkötter et al. (2016) and an example of one is presented here in Fig. 3.11B. Notice that both global maps in Fig. 3.11 suggest the lowest values are in the subtropical regions ($<1.0 - 1.5$ mol C $m^{-2}y^{-1}$), and the highest values are in the equatorial regions (>4 mol C m^{-2} y^{-1} in the Eastern Pacific) and high northern latitudes (particularly in the GCM, >3 mol C m^{-2} y^{-1}; Fig. 3.11B). To a first approximation the highest values are found in the regions of high nutrient concentration (see Fig. 4.2).

A compilation of values of ANCP determined by different mass balance based techniques from observations (O_2, O_2/Ar, DIC+δ^{13}C-DIC, and OUR), is presented in Fig. 3.12 (Quay et al., 2020). Because both observation and model results have stronger north-south than east–west gradients, the observational results have been averaged zonally (along latitude bands). There are distinct differences between the global models and the observations. First, observations are much more homogeneous than the models suggest – mean values of the zonally averaged values in the North Pacific and North Atlantic are 2.6 ± 0.6 and 2.6 ± 0.4, respectively. Second, the Northern Hemisphere trends in the models that suggest higher values in the high-latitude regions are opposite the observations in both the Pacific and Atlantic Oceans, which indicate distinctly lower ANCP in the high latitudes. Finally, one of the most obvious trends in the observations is a much lower ANCP in the Southern Hemisphere subtropical Pacific than in the Northern Hemisphere subtropical Pacific (<1 and 2.5 mol-C m^{-2} y^{-1}, respectively). Model values in the subtropics are uniformly low (~1.0 mol-C m^{-2} y^{-1}).

The observations suggest geographical variations of ANCP do not follow surface nutrient distributions. Possible reasons for the differences between the model results and observations are: (a) the plasticity of the C:nutrient ratios in phytoplankton (higher C:P or C:N ratios in the subtropics and lower at high latitudes), which would weaken the dependence of carbon export on the local availability of limiting nutrients (DeVries and Deutsch, 2014; Teng et al., 2014); (b) horizontal transport of nutrients from the Equator and subtropical–subarctic boundary to the subtropical oceans (e.g., Letscher et al., 2016; Williams and Follows, 1998) could result in lower ANCP on a per m^2 basis in subtropical regions of greater size (like the S. Pacific); (c) finally, the observed ANCP is lowest in the Southern Hemisphere subtropical regions that are bereft of atmospheric iron input (Jickells and Moore, 2015), indicating that iron may be an important factor limiting or co-limiting ANCP. Clearly, we have just scratched the surface of verifying the global

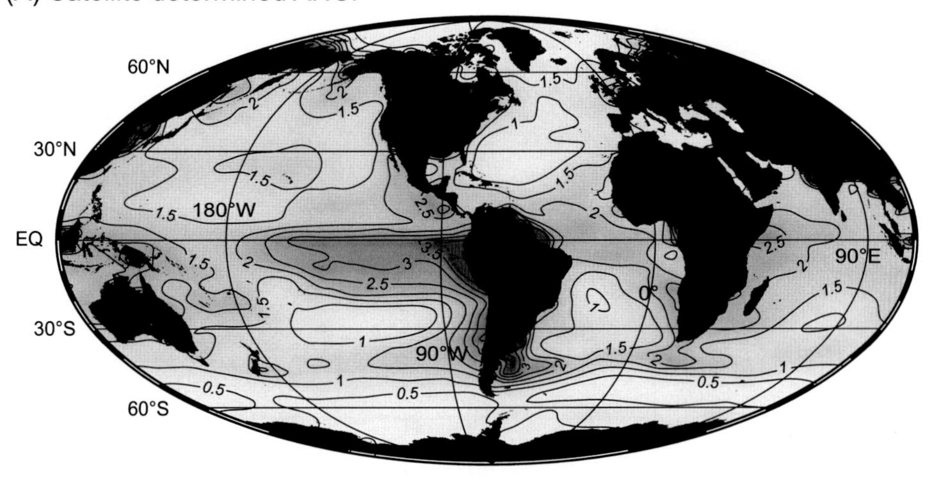

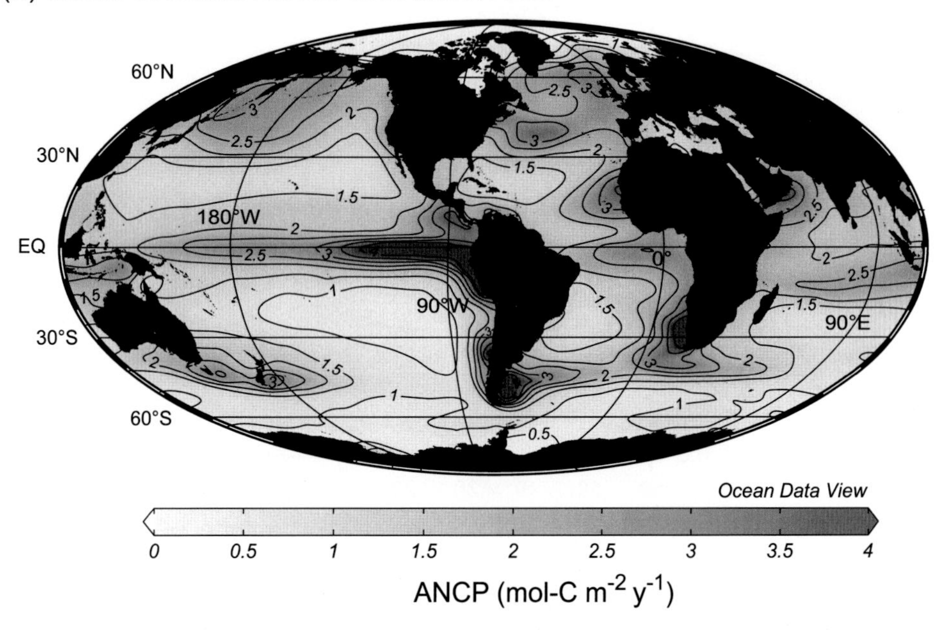

Figure 3.11 Global maps of the biological pump or particulate organic carbon flux from the upper ocean from separate sources. (A) Annual carbon export determined from the relationship C export = NPP (C export/NPP). NPP was determined from satellite-measured ocean color and optical back scatter from 2015 and the CbPM algorithm (Westberry et al., 2008). The export/NPP ratio was determined from the ecosystem model of Laws et al (2011). (B) Carbon export determined from an ocean GCM with the BEC ecosystem model (from Laufkötter et al., 2016). A black and white version of this figure will appear in some formats. For the color version, refer to the plate section.

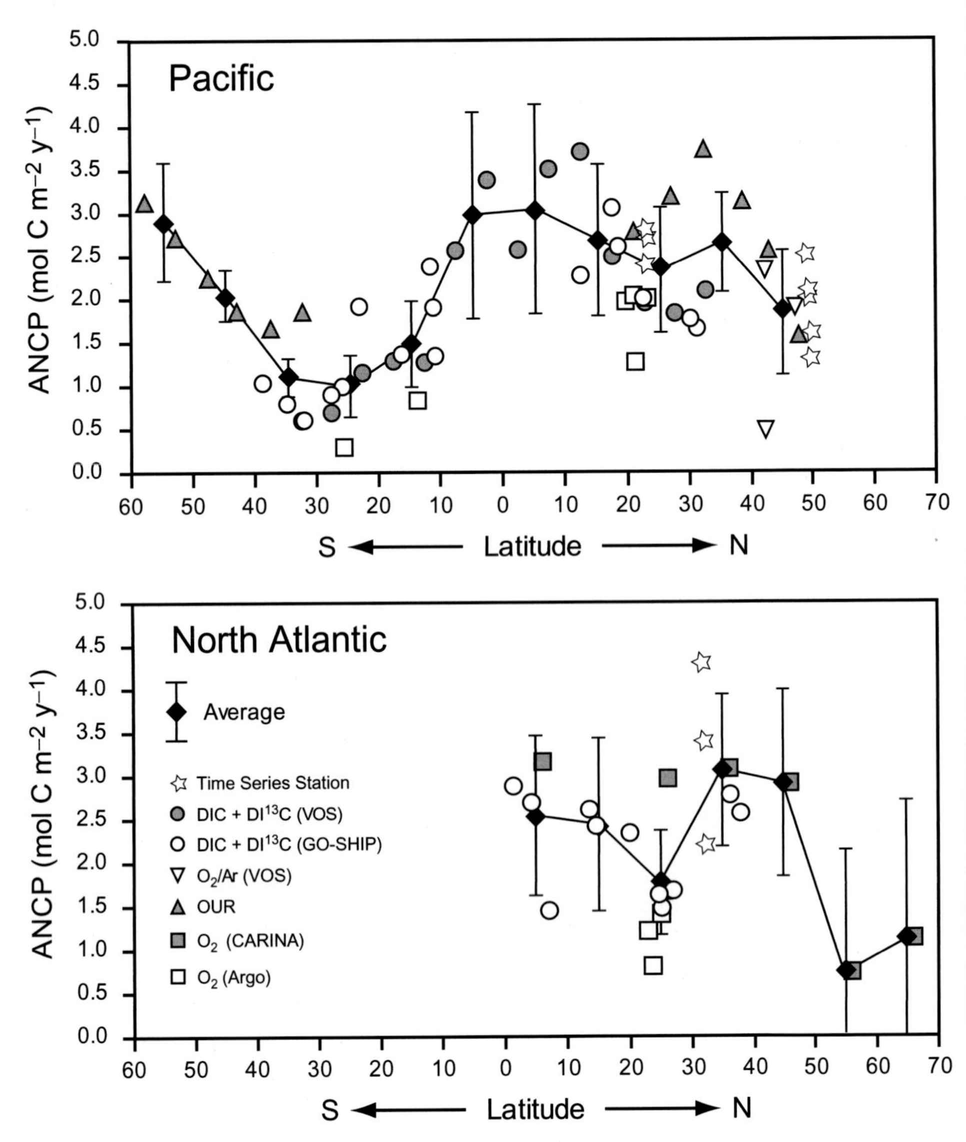

Figure 3.12 Zonally averaged ANCP observations as a function of latitude in the Pacific and N. Atlantic Ocean. Filled diamonds connected with a solid line are the average of separate measurements indicated by the symbols for 10-degree latitude bands. Redrawn from Quay et al. (2020).

distribution of ANCP determined from satellite and global models. More complete coverage of observations will help develop new, improved theories about the geographic variability of ANCP. A deeper mechanistic understanding should enhance the accuracy of global circulation models, and aid the prediction of how climate change will impact ANCP.

3.3.2 The Anthropogenic Influence: Evidence for Changes in Biological Fluxes

It is expected that there will be a feedback between climate change and ocean productivity and organic carbon export. Models suggest that a warmer, more stratified ocean should have a reduced flux of nutrients to the euphotic zone, decreasing primary production in nutrient-limited subtropical regions (e.g., Sarmiento et al., 2004). In higher latitudes, however, an earlier onset of spring stratification could lengthen the productive season, creating an increase in production. The only observations capable of determining global, long-term (decadal or greater) trends in productivity are satellite-based ocean color measurements. The most widely studied to date is the Sea-viewing Wide Field-of-view Sensor (SeaWiFS), which operated continuously from 1997 to 2007. Observations during this time indicate that there is a correlation between increased sea surface temperature and lower primary production determined from ocean color in the subtropical oceans (Behrenfeld et al., 2006), and that the area of the oligotrophic gyres may have increased during this time (Polovina et al., 2008).

There is no doubt that the ocean is warming in response to global climate change due to fossil fuel burning (e.g., Abraham et al., 2013). However, the problem of determining whether global warming is accompanied by a change in ocean chlorophyll concentration or biological productivity is that these changes could be small relative to the mean, and it is anticipated that there may be productivity variations in response to natural global temperature oscillations like ENSO and the Pacific Decadal Oscillation (PDO). One approach to determining how well oceanographers can separate the ecosystem response to climate change from natural variations is by determining changes in ocean productivity and chlorophyll concentration in biogeochemical global models in response to long-term trends in greenhouse gas emissions. By including some assumptions about future atmospheric CO_2 increases due to fossil fuel burning, the duration of the global warming trend and the potential ecological feedback can be projected into the future. Henson et al. (2010) have demonstrated by this approach that detection of climate-change-driven trends in these models requires a time series of about 40 years in length compared to currently available satellite observations of half this. According to this analysis, satellite data of sufficient length to determine ocean ecosystem feedbacks will be available only by the mid-twenty-first century if global satellite color measurements continue. Thus, while there are observed correlations between sea surface temperature and ocean color trends, most of which suggest a decrease in ocean productivity in response to global warming (at least in the subtropical oceans), it is not yet possible to determine whether these trends are part of natural variations or anthropogenic influences.

3A.1 Appendix 3A.1 Measurements of Net and Gross Biological Production

There is a rich history of rate measurements that have been developed to determine the relationships among gross primary production (GPP), net primary production (NPP), and net community production (NCP). Traditionally, the first two of these have relied on incubation experiments using stable or radioisotopes. Here we describe the methods used to determine GPP and NPP to help clarify their relationship to NCP, described in the main text of this chapter. We begin by describing NPP because it has a long history in oceanography.

3A.1.1 Net Primary Production Rates

The rate of primary production in the sea is most widely determined by incubating samples with ^{14}C-enriched DIC (^{13}C can also be used) and then measuring the amount of the isotope that is taken up into the particulate organic carbon. Water from various depths in the euphotic zone is collected into clear plastic bottles, spiked with isotopes, and attached at different depths to a drifting mooring over a period of 12–24 h or kept onboard ship under controlled temperature and light conditions. As production proceeds, the phytoplankton in the bottle take up the enriched inorganic carbon, incorporating it into their cells, where it can be measured at the end of the experiment after separating the cells by filtration. It is assumed that ^{14}C uptake measures only phytoplankton metabolism because most marine zooplankton are not abundant enough or rapid enough grazers to greatly affect the results over a one-day period. A complication arises in defining whether GPP or NPP or something in between is measured in the incubation experiments because photosynthetic organisms are both autotrophic and heterotrophic. While the net result is autotrophic, they both create organic matter (during the day) and respire organic matter (during both day and night). In spite of these caveats, the ^{14}C incubation approach is the most used direct estimate of the flux of carbon through marine autotrophic organisms. Recent syntheses of modern ^{14}C productivity measurements suggest that they measure a flux closer to net than gross primary production (Marra and Barber, 2004).

Methods have also been devised, using similar incubation techniques with ^{15}N-enriched nitrogen isotopes, to determine the uptake of nitrogen compounds (NO_3^-, NH_4^+, and urea) by plankton during production. The N cycle is discussed in Chapters 4 and 6. Briefly, phytoplankton are able to take up NO_3^-, NH_4^+, or urea to fuel primary production, while respiration releases either NH_4^+ or urea. During photosynthesis NH_4^+ and urea are preferred by phytoplankton, because they do not need to be reduced before incorporation into organic compounds. Because of this, they are taken up immediately by autotrophs when released during respiration in the euphotic zone, and they do not accumulate above very low concentrations in the water. In the dark ocean, NH_4^+ is rapidly oxidized to NO_3^-, but this reaction is inhibited by light and thought to occur at only slow rates in the euphotic zone. Because of this, any NO_3^- taken up by plankton in the surface waters is assumed to have been transported from the deeper layers, while any NH_4^+ or urea taken up is assumed to have come from respiration in the surface waters. Dugdale and Goering (1967) called rate measurements from $^{15}NO_3^-$ uptake *new production* to distinguish it from recycled production, quantified by $^{15}NH_4^+$ (and ^{15}N-urea) uptake.

At steady state, over an annual cycle, the rate of NO_3^- uptake by phytoplankton should equal the flux of NO_3^- into the euphotic zone, which must also equal the particulate and dissolved organic N flux out of the euphotic zone. The fraction of NO_3^- mixed into the euphotic zone of the ocean to the total dissolved nitrogen taken up there, termed the *f-ratio*, varies from about 0.1, in regions where there is a well-developed food web to recycle phytoplankton production, to near 1.0 in bloom regions where grazers have not become well established (Eppley and Peterson, 1979). In a steady-state ocean, new production is related via the C:N ratio in organic matter to the flux of organic carbon from the upper ocean and the annual net community production (ANCP), discussed earlier in the chapter.

3A.1.2 Gross Primary Production Rates

The same type of incubation experiment described using ^{14}C has been developed using ^{18}O as a tracer. In this case, seawater enriched in $H_2{}^{18}O$ (^{18}O is a rare isotope of oxygen, see Chapter 6) is incubated in a gas-tight, quartz or glass incubation flask (Bender et al., 1987). Photosynthesis transforms oxygen in H_2O to molecular O_2 (Eq. 3.3). After a period of 12 or 24 hours, the $^{18}O/^{16}O$ ratio of O_2 dissolved in the water is measured. Because O_2 and H_2O do not exchange isotopes inorganically, the only way for the ^{18}O spike to be transferred to O_2 is via photosynthesis. The extent of enrichment of the isotope ratio in O_2 ($^{18}O^{16}O/^{16}O_2$) is a measure of gross oxygen production (GOP) or in carbon equivalent terms gross primary production (GPP). Unlike incubations with ^{14}C labeled DIC, there is little ambiguity about whether the result is closer to GPP or NPP. Both autotrophic and heterotrophic respiration during the incubation period have little influence on the $^{18}O/^{16}O$ ratio in the O_2 produced by gross photosynthesis.

The $H_2{}^{18}O$ incubation method for determining GOP was first used extensively in the Joint Global Ocean Flux Study (JGOFS) program in the 1980s, and has had an important influence on the interpretation of the more traditional ^{14}C incubation measurements. Marra (2002) summarized years of open-ocean simultaneous measurements of ^{14}C-PP (mol-C kg^{-1} d^{-1}) and ^{18}O-GOP (mol-O_2 kg^{-1} d^{-1}) during the JGOFS cruises and demonstrated that they have a ratio of about 2.7 $\left(\text{mol-O}_2\ \text{kg}^{-1}\text{d}^{-1}/\text{mol-C kg}^{-1}\text{d}^{-1}\right)$. More recent, less extensive measurements in different environments have sometimes observed values that are higher than this (see Juranek and Quay, 2013, for a summary).

It is possible to directly compare GOP rates determined by $H_2{}^{18}O$ incubations with measurements of oxygen production using in situ distributions of the three natural oxygen isotopes of the O_2 in ocean surface waters (Luz and Barkin, 2000). The "triple oxygen isotope method" is described in Chapter 6 about stable isotopes, and the reader is referred there for details of the method. Briefly, comparison of gross oxygen production

(GOP) using both incubations and the triple oxygen isotope method indicated that the in situ method gives results that range from equal to twice as high as the in vitro methods (see Juranek and Quay, 2013).

References

Abell, J., S. Emerson, and P. Renaud (2000) Distributions of TOP, TON and TOC in the North Pacific subtropical gyre: implications for nutrient supply in the surface ocean and remineralization in the upper thermocline, *Journal of Marine Research*, **58**, 203–222.

Abraham, J. P., M. Baringer, N. L. Bindoff, et al. (2013) A review of global ocean temperature observations: implications for ocean heat content estimates and climate change, *Reviews of Geophysics*, **51**, 450–483, doi:10.1002/rog-20022.

Anderson, L. A. (1995) On the hydrogen and oxygen content of marine phytoplankton, *Deep-Sea Research I*, **42**, 1675–1680.

Anderson, L. A., and J. L. Sarmiento (1994) Redfield ratios of remineralization determined by nutrient data analysis, *Global Biogoechemical Cycles*, **8**, 65–80.

Archer, D., S. Emerson, J. Powell and C. S. Wong (1993) Numerical hindcasting of sea surface pCO_2 at Weathership Station Papa, *Progress in Oceanography*, **32**, 319–351, doi:10.1016/0079-6611(93)90019-A.

Armstrong, R. A., C. Lee, J. I. Hedges, S. Honjo, and S. G. Wakeham (2002) A new mechanistic model for organic carbon fluxes in the ocean based on the quantitative association of POC with ballast minerals, *Deep-Sea Research II*, **49**, 219–236.

Barbeau, K., E. L. Rue, K. W. Bruland, and A. Butler (2001) Photochemical cycling of iron in the surface ocean mediated by microbial iron (III)-binding ligands, *Nature*, **413**, 409–413.

Barbeau, K., E. L. Rue, C. G. Trick, K. W. Bruland, and A. Butler (2003) Photochemical reactivity of siderophores produced by marine heterotrophic bacteria and cyanobacteria based on characteristic Fe (III) binding groups, *Limnology and Oceanography*, **48**, 1069–1078.

Behrenfeld, M. J., R. T. O'Malley, D. A. Siegel, et al. (2006) Climate-driven trends in contemporary ocean productivity, *Nature*, **444**, 752–755, doi:10.1038/nature05317.

Bender, M. L., K. Grande, K. Johnson, et al. (1987) A comparison of four methods for the determination of planktonic community metabolism, *Limnology and Oceanography*, **32**, 1085–1098.

Benitez-Nelson, C., K. O. Buesseler, D. M. Karl, and J. Andrews (2002) A time-series study of particulate matter export in the North Pacific Subtropical Gyre based on ^{234}Th:^{238}U disequilibrium, *Deep-Sea Research Part I*, **48**, 2595–2611.

Bishop, J. K. B., W. B. Rossow, and E. G. Dutton (1997) Surface solar irradiance from the International Satellite Cloud Climatology Project 1983–1991, *Journal of Geophysical Research – Atmospheres*, **102**, 6883–691.

Brix, H., N. Gruber, and C. D. Keeling (2004) Interannual variability of the upper ocean carbon cycle at Station ALOHA near Hawaii, *Global Biogeochemical Cycles*, **18**, GB4019, doi:10.1029/2004GB002245.

Broecker, W. S. (1971) A kinetic model for the chemical composition of seawater, *Quaternary Research*, **1**, 188–207.

Broecker, W. S., and T. H. Peng (1982) *Tracers in the Sea*, 690 pp. Palisades: Eldigio Press.

Bruland K. W., and Rue E. L. (2001) Analytical methods for determination of concentrations and speciation of iron, in *The Biogeochemistry of Iron in Seawater* (eds. D. R. Turner and K. A. Hunter), pp. 255–289. Chichester: Wiley.

Bruland, K. W., R. M. Middag, and M. C. Lohan (2014) *Controls of trace metals in seawater*, in *Treatise on Geochemistry* (eds. H. D. Holland and K. K. Turekian), pp. 19–51. Oxford: Elsevier.

Buesseler K. O., and P. W. Boyd (2009) Shedding light on processes that control particle export and flux attenuation in the twilight zone of the open ocean, *Limnology and Oceanography*, **54**, doi:10.4319/lo.2009.54.4.1210.

Bushinsky, S. M., and S. R. Emerson (2015) Marine biological production from in situ oxygen measurements on a profiling float in the Subarctic Pacific Ocean, *Global Biogeochemical Cycles*, **29**, doi:10.1002/2015GB005251.

Cassar, N., B. A. Barnett, M. L. Bender, et al. (2009) Continuous high-frequency dissolved O_2/Ar measurements by equilibrator inlet mass spectrometry, *Analytical Chemistry*, **81**, 1855–1864.

Charette, M. A., S. B. Moran, and J. K. Bishop (1999) ^{234}Th as a tracer of particulate organic carbon export in the subarctic northeast Pacific Ocean, *Deep-Sea Research II*, **46**, 2833–2861.

Church, M. J., M. W. Lomas, and F. Muller-Karger (2013) Sea change: charting the course for biogeochemical ocean time-series research in a new millennium, *Deep-Sea Research II*, **93**, 2–15, doi:10.1016,j.dsr2.2013.01.035.

de Baar, H. J. W., P. W. Boyd, K. H. Coale, et al. (2005) Synthesis of iron fertilization experiments: from the Iron Age in the Age of Enlightenment, *Journal of Geophysical Research Oceans*, **110**, C09S16, doi:10.1029/2004JC002601.

DeVries, T., and C. Deutsch (2014) Large-scale variations in the stoichiometry of marine organic matter respiration, *Nature Geoscience*, **7**, 890–894.

Dugdale, R. C., and J. J. Goering (1967) Uptake of new and regenerated forms of nitrogen in primary productivity 1, *Limnology and Oceanography*, **12**, 196–206.

Emerson, S. (2014) Annual net community production and the biological carbon flux in the ocean, *Global Biogeochemical Cycles*, **28**, 14–28, doi:10.1002/2013GB004680.

Emerson, S., and J. Hedges (2008) *Chemical Oceanography and the Marine Carbon Cycle*, 453pp. Cambridge: Cambridge University Press.

Emerson, S., and C. Stump (2010) Net biological oxygen production in the ocean – II: Remote in situ measurements of O_2 and N_2 in subarctic Pacific surface waters, *Deep-Sea Research I*, **57**, 1255–1265, doi:10.1016/j.dsr.2010.06.001.

Emerson, S., C. Stump, B. Johnson, and D. Karl (2002) Autonomous determination of oxygen and nitrogen concentrations in the Ocean, *Deep-Sea Research I*, **49**, 941–952.

Eppley, R. W., and B. J. Peterson (1979) Particulate organic flux and planktonic new production in the deep ocean, *Nature*, **282**, 677–680.

Fassbender, A. J., C. L. Sabine, and M. F. Cronin (2016), Net community production and calcification from 7 years of NOAA Station Papa Mooring measurements, *Global Biogeochemical Cycles*, **30**, 250–267, doi:10.1002/2015GB005205.

Gruber, N., C. D. Keeling, and T. F. Stocker (1998), Carbon-13 constraints on the seasonal inorganic carbon budget at the BATS site in the northwestern Sargasso Sea, *Deep-Sea Research I*, **45**, 673–717, doi:10.1016/S0967-0637(97)00098-8.

Hamme, R. C., and S. R. Emerson (2006) Constraining bubble dynamics and mixing with dissolved gases: Implications for productivity measurements by oxygen mass balance, *Journal of Marine Research*, **64**, 73–95.

Hansell, D. A., C. A. Carlson, D. J. Repeta, and R. Schlitzer (2009) Dissolved organic matter in the ocean: a controversy stimulates new insights, *Oceanography*, **22**, 202–211.

Harvey, H. W. (1945) *Recent Advances in the Chemistry and Biology of Sea Water*, 164 pp. Cambridge: Cambridge University Press.

Hedges, J. I., J. A. Baldock, Y. Gelinas, et al. (2002) The biochemical and elemental compositions of marine plankton: a NMR perspective, *Marine Chemistry*, **78**, 47–63.

Henson, S. A., J. L. Sarmiento, J. P. Dunne, et al. (2010) Detection of anthropogenic climate change in satellite records of ocean chlorophyll and productivity, *Biogeosciences*, **7**, 621–640.

Ho, T., A. Quigg, Z. V. Finkel, A. J. et al. (2003) The elemental composition of some marine phytoplankton, *Journal of Phycology*, **39**, 1145–1159, doi:10.1111/j.0022-3646.2003.03-090.x.

Holte, J., L. D. Talley, J. Gilson, and D. Roemmich (2017) An Argo mixed layer climatology and database, *Geophysical Research Letters*, **44**, 5618–5626, doi:10.1002/2017GL073426.

Jenkins, W. J., and S. C. Doney (2003) The subtropical nutrient spiral, *Global Biogeochemical Cycles*, **17**, doi:10.029/2003GB002085, 2003.

Jickells, T., and C. M. Moore (2015) The importance of atmospheric deposition for ocean productivity, *Annual Review of Ecology, Evolution, and Systematics*, **46**, 481–501, doi:10.1146/annrev-ecolsys-112414-054118.

Juranek, L. W., and P. D. Quay (2013) Using triple isotopes of dissolved oxygen to evaluate global marine productivity, *Annual Review of Marine Science*, **5**, 503–524.

Klaus, C., and D. E. Archer (2002) Association of sinking organic matter with various types of mineral ballast in the deep sea: implications for the rain ratio, *Global Biogeochemical Cycles*, **16**, 1116, doi:10.1029/2001GB001765.

Körtzinger, A., J. I. Hedges, and P. D. Quay (2001) Redfield ratios revisited: Removing the biasing effect of anthropogenic CO_2, *Limnology and Oceanography*, **46**, 964–970.

Lam, P. J., S. Doney, and J. K. B. Bishop (2011) The dynamic ocean biological pump: insights from a global compilation of particulate organic carbon, $CaCO_3$, and opal concentration profiles from the mesopelagic, *Global Biogeochemical Cycles*, **25**, doi:10.1029/2010GB003868.

Laufkötter, C., M. Vogt, N. Gruber, et al. (2016) Projected decreases in future marine export production: The role of the carbon flux through the upper ocean ecosystem, *Biogeosciences*, **1**, 4023–4047, doi:10.5194/bg-13-4023–2016.

Laws, E. A., E. D'Sa, and P. Naik (2011) Simple equations to estimate ratios of new or export production to total production from satellite-derived estimates of sea surface temperature and primary production, *Limnology and Oceanography Methods*, **9**, 593–601, doi:10.4319/lom.2011.9.593.

Letscher, R. T., and J. K. Moore (2015) Preferential remineralization of dissolved organic phosphorus and non-Redfield DOM dynamics in the global ocean: Impacts on marine productivity, nitrogen fixation, and carbon export, *Global Biogeochemical Cycles*, **29**, 325–340, doi:10.1002/2014GB004904.

Letscher, R. T., F. Primeau, and J. K. Moore (2016) Nutrient budgets in the subtropical ocean gyres dominated by lateral transport, *Nature Geoscience*, **9**, 815–819, doi:10.1038/NGEO2812.

Luz, B., and E. Barkin (2000) Assessment of oceanic productivity with the triple-isotope composition of dissolved oxygen, *Science*, **288**, 2028–2031.

Marra, J. (2002) Approaches to the measurement of plankton production, in *Phytoplankton Productivity: Carbon Assimilation in Marine and Freshwater Ecosystems* (eds. P. J. le B. Williams, D. N. Thomas, and C. S. Reynolds), pp. 78–108. Hoboken: Wiley-Blackwell.

Marra, J., and R. T. Barber (2004) Phytoplankton and heterotrophic respiration in the surface layer of the ocean, *Geophysical Research Letters*, **31**, doi:10.1029/2004GL019664.

Martin, J. H., G. A. Knauer, D. M. Karl, and W. W. Broenkow (1987) VERTEX: carbon cycling in the northeast Pacific, *Deep-Sea Research A*, **34**, 267–285.

Martiny, A. C., C. T. Pham, F. W. Primeau, et al. (2013) Strong latitudinal patterns in the elemental ratios of marine plankton and organic matter, *Nature Geoscience*, **6**, 279–283.

Meybeck, M. (1979) Concentrations des eaux fluviales en éléments majeurs et apports en solution aux océans, *Revue de Géologie Dynamique et de Géographie Physique Paris*, **21**, 215–246.

Moore, C. M., M. M. Mills, K. R. Arrigo, et al. (2013) Processes and patterns of oceanic nutrient limitation, *Nature Geoscience*, **6**, 701–710.

Morel, F. M. M., and J. G. Hering (1993) *Principles and Applications of Aquatic Chemistry*, 608 pp. New York: Wiley.

Morel, F. M. M., A. J. Milligan, and M. A. Saito (2014) Marine bioinorganic chemistry: the role of trace metals in oceanic cycles of major nutrients, Vol. 8, *Treatise on Geochemistry* (eds. H. D. Holland and K. K. Turekian), pp. 123–150, doi:10.1016/B978-0-08-095975-7.00605-7. Oxford: Elsevier-Pergamon.

Palevsky, H. I., P. Quay, D. E. Lockwood, and D. P. Nicholson (2016) The annual cycle of gross primary production, net community production, and export efficiency across the North Pacific Ocean, *Global Biogeochemical Cycles*, **30**, doi:10.1002/2015GB005318.

Plant, J. N., K. S. Johnson, C. M. Sakamoto, et al. (2016) Net community production at Ocean Station Papa observed with nitrate and oxygen sensors on profiling floats, *Global Biogeochemical Cycles*, **16**, doi:10.1002/2015GB005349.

Polovina, J. J., E. A. Howell, and M. Abecassis (2008) Ocean's least productive waters are expanding, *Geophysical Research Letters*, **35**, doi:3610.1029/2007GL031745.

Quay, P., and J. Stutzman (2003) Surface layer carbon budget for the subtropical N. Pacific: $\delta^{13}C$ constraints at station ALOHA, *Deep-Sea Research I*, **50**, 1045–1061.

Quay, P., S. Emerson, and H. Palevsky (2020) Regional pattern of the ocean's biological pump based on geochemical observations, *Geophysical Research Letters*, **47**, doi:10.1029/2020GL088098.

Redfield, A. C. (1958) The biological control of chemical factors in the environment, *American Scientist*, **46**, 205–221.

Redfield, A. C., B. H. Ketchum and F. A. Richards (1963) The influence of organisms on the composition of sea water, in *The Sea*, Vol. 2 (ed. M. N. Hill), pp. 99; 26–77. New York: Interscience.

Sanders, R., P. J. Morris, A. J. Poulton, et al. (2010) Does a ballast effect occur in the surface ocean? *Geophysical Research Letters*, **37**, doi:10.1029/2010GL042574.

Sarmiento, J. L., J. Dunne, A. Gnanadesikan, et al. (2002) A new estimate of the $CaCO_3$ to organic carbon export ratio, *Global Biogeochemical Cycles*, **16**, doi:10.1029/2002GB001010.

Sarmiento, J. L., R. Slater, R. Barber, et al. (2004) Response of ocean ecosystems to climate warming, *Global Biogeochemical Cycles*, **18**, GB3003, doi:3010.1029/3003GB002134.

Shaffer, G., J. Bendtsen, and O. Ulloa (1999) Fractionation during remineralization of organic matter in the ocean, *Deep-Sea Research I*, **46**, 185–204.

Spitzer, W. S., and W. J. Jenkins (1989) Rates of vertical mixing, gas exchange and new production: estimates from seasonal gas cycles in the upper ocean near Bermuda, *Journal of Marine Research*, **47**, 169–196.

Stanley, R. H. R. (2007) A determination of air-sea gas exchange and upper ocean biological production from five noble gases and tritigenic helium-3, PhD thesis, Woods Hole Oceanographic Institution.

Stumm, W., and J. J. Morgan (1995) *Aquatic Chemistry: Chemical Equilibria and Rates in Natural Waters*, 3rd ed., 1040 pp. New York: Wiley.

Sunda, W. G. (2012) Feedback interactions between trace metal nutrients and phytoplankton in the ocean, *Frontiers in Microbiology*, **3**, 204, doi:10.3389/fmicb.2012.00204.

Teng, Y-C., F/ W. Primeau, J. K. Moore, M. W. Lomas, and A. C. Martiny (2014) Global-scale variations of the ratios of carbon to phosphorus in exported marine organic matter, *Nature Geoscience*, **7**, 895–898, doi:10.1038/NGE02303.

Timothy, D. A., C. S. Wong, J. E. Barwell-Clarke, et al. (2013) Climatology of sediment flux and composition in the subarctic Northeast Pacific Ocean, *Progress in Oceanography*, **116**, 95–129, doi:10.1016/j.pocean.2013.06.017.

Toggweiler, R., and J. Sarmiento (1985) Glacial to Interglacial changes in atmospheric carbon dioxide: The critical role of the ocean surface water in high latitudes, in *The Carbon Cycle and Anthropogenic CO_2* (eds. E. Sundquist and W. S. Broecker), pp. 163–184. Washington, DC: American Geophysical Union.

Twining, B. S., and S. B. Baines (2013) The trace metal composition of marine phytoplankton, *Annual Review of Marine Science*, **5**, 191–215.

Westberry, T. K., and M. J. Behrenfeld (2014) Oceanic net primary production, in *Biophysical Applications of Satellite Remote Sensing*, pp. 205–230. Heidelberg: Springer.

Westberry, T., M. J. Behrenfeld, D. A. Siegel, and E. Boss (2008), Carbon-based primary productivity modeling with vertically resolved photoacclimation, *Global Biogeochemical Cycles*, **22**, doi:10.1029/2007GB003078.

Williams, R. G., and M. J. Follows (1998) The Ekman transfer of nutrients and maintenance of new production over the North Atlantic, *Deep-Sea Research I*, **45**, 461–489.

Wong, C. S., N. A. D. Waser, Y. Nojiri, et al. (2002) Seasonal cycles of nutrients and dissolved inorganic carbon at high and mid latitudes in the North Pacific Ocean during the *Skaugran* cruises: determination of new production and nutrient uptake ratios, *Deep-Sea Research II*, **49**, 5317–5338.

Yang, B., S. R. Emerson, and M. A. Peña (2018), The effect of the 2013–2016 high temperature anomaly in the Subarctic Northeast Pacific (the "Blob") on net community production, *Biogeosciences*, **15**, 6747–6759, doi:10.5194/bg-15-6747.

Yang, B., S. R. Emerson, and P. D. Quay (2019) The subtropical oceans' biological carbon pump determined from O_2 and DIC/DI^{13}C tracers, *Geophysical Research Letters*, **46**. doi:10.1029/2018GL081239.

Problems for Chapter 3

3.1. A patch of newly upwelled water is identified off the coast of northwestern Africa and marked with a tracer (SF_6, for example) so that it can be tracked. (For this problem, assume that the water doesn't mix significantly with different water masses beneath or next to it.) The initial concentrations measured in the water are 1.6 μmol kg^{-1} phosphate, 20.2 μmol kg^{-1} nitrate, and 2200 μmol kg^{-1} inorganic carbon (DIC).

a. A bloom begins in the upwelled water and draws down the nutrients. Calculate the final concentrations of phosphate, nitrate, and inorganic carbon in the water after the bloom ceases due to nutrient limitation.

b. At this point, dust from the Sahara falls on the water parcel, stimulating nitrogen fixation, which adds just enough ammonia to the water parcel to support any additional possible productivity. For this new situation, calculate the final concentrations of phosphate, nitrate, and inorganic carbon in the water after the bloom ceases.

c. At this point, the water parcel moves downward into the ocean interior, and respiration becomes the dominant process. The oxygen concentration when the water left the surface was 250 μmol kg^{-1}. When the phosphate concentration again reaches 1.6 μmol kg^{-1}, what will be the concentrations of nitrate, inorganic carbon, and oxygen?

3.2. Imagine a well-mixed, 50 L seawater aquarium in the shape of a perfect cube. New water flows into the aquarium with a nitrate concentration of 20 μmol L^{-1} at a rate of 10 L d^{-1}, while water from the aquarium flows out at the same rate. The system is at steady state, and phytoplankton in the aquarium photosynthesize.
   a. What phosphate concentration should the inflowing water have to allow the phytoplankton to make maximal use of the nitrate?
   b. The nitrate concentration of the outflowing water is 1 μmol L^{-1}. What is the rate of organic carbon and oxygen production in the aquarium?
   c. The inflowing water is at equilibrium with air and has an oxygen concentration of 230 μmol L^{-1}. Gases exchange across the upper surface of the aquarium. If the gas transfer coefficient is 1 m d^{-1}, what is the concentration of O_2 in the aquarium?

3.3. During an open ocean iron fertilization experiment, 2100 kg of $Fe(SO_4)•7H_2O$ is mixed into acidified seawater and released into the surface ocean as the ship moves, to create a patch of iron-fertilized water approximately 75 km^2. The mixed layer is 30 m on average during the experiment. At the end of the 25-day experiment, iron concentrations have dropped to background levels again.
   a. Based only on the amount of iron added, what should have been the concentration of iron in the patch if there were no reactions?
   b. An NCP rate of 30 mmol-C m^{-2} d^{-1} is measured in the patch over the 25-day experiment. Estimate how much iron was consumed by biological productivity during the experiment.
   c. What fraction of the iron added is consumed by biological productivity? What process could account for the rest of the iron loss?

3.4. Use the two-box model depicted in Fig. 3.6 to determine the effect of productivity and mixing on the flux of particulate organic matter from the surface to deep ocean. Assume that the concentrations in the deep ocean are constant and independent of changes in surface ocean concentrations and mixing rates (true on timescales of hundreds to thousands of years).
   a. Using a mean surface phosphate concentration of 0.5 μmol kg^{-1} and other values given in the chapter, calculate the flux of phosphorus from the surface to the deep by sinking particles.
   b. Vary the surface ocean phosphate concentration from zero to the deep ocean value, and calculate how the particle flux would change. Make a plot of particle flux versus surface phosphate concentration.
   c. Keep the surface ocean phosphate concentration at the current value but vary the vertical mixing rate from zero to double its present value. Make a plot of particle flux versus mixing rate.
   d. Comment on your plots. What do they imply about the potential effect of changes in phytoplankton nutrient uptake efficiency or stratification on ocean productivity rates?

3.5. Suppose global warming increases the temperature of the surface ocean, leading to permanently increased stratification, and that this extra stratification causes the vertical mixing rate in the ocean to halve. Using a two-box ocean model consider the very long timescale effects if such a change were to persist for hundreds of thousands of years. In this scenario, assume that the average surface phosphate concentration is unaffected by the decreased mixing rate, and that the fraction of particulate matter entering the deep ocean that is ultimately buried does not change.

a. Use the two-layer model to determine the new steady state that the ocean reaches in many thousands of years. Calculate the burial flux of phosphorus at the sea floor, the particulate flux of phosphorus from the surface to the deep box, and the phosphate concentration of the deep box at the new steady state.

b. Qualitatively describe the evolution of the ocean from the steady state before the change in vertical mixing to the new steady state that you solved for in part a. How does each change affect other concentrations or fluxes in the model, and then how do those changes initiate further changes? What causes the system to come back to steady state?

c. Mathematically express the evolution of the phosphate concentration of the deep box over time from the old to new steady state by writing a differential equation for the change in concentration with time. (All the terms in your equation other than the deep box phosphate concentration should be constant with time.) Solve your equation and plot the phosphate concentration of the deep box versus time.

3.6 An autonomous profiling float in the subarctic North Pacific measures nitrate data over a spring and summer deployment period. At the beginning of the spring, the surface nitrate concentration is 15 μmol kg^{-1} and the mixed layer depth is 100 m. Six months later, the surface nitrate concentration is 8 μmol kg^{-1} and the mixed layer depth is 30 m. The subsurface nitrate concentration at 150 m is 20 μmol kg^{-1} and does not change appreciably over the summer. The gradient in nitrate concentrate from the base of the mixed layer to 150 m is approximately constant in time and linear with depth.

a. Make a drawing of the nitrate concentration profiles at the beginning and end of the measurement period.

b. What is the NCP over the spring and summer based only on the change in the nitrate inventory (to 100 m)? Convert and express your answer in mmol-C m^{-2} d^{-1}.

c. Suppose that the vertical mixing coefficient at the base of the mixed layer is 2×10^{-5} m^2 s^{-1} and is constant over the year. How does including this flux affect the calculation of NCP, quantitatively? Think carefully about the signs of the terms in the mass balance.

d. Average horizontal currents in this region are 4 km d^{-1}. Suppose that the nitrate concentration 100 km upstream of the float location is approximately 0.4 μmol kg^{-1} higher than at the float. How does including this flux affect the calculation of NCP, quantitatively?

4 Life in the Deep Ocean: Biological Respiration

4.1 Respiration below the Euphotic Zone 144
4.1.1 Oxygen Concentrations and Apparent Oxygen Utilization (AOU) 147
4.1.2 Nutrient Concentrations and Preformed Nutrients 149
4.1.3 Nitrogen and Phosphorus Cycles 152
Discussion Box 4.1 155
4.2 Respiration Rates 156
4.2.1 Oxygen Utilization Rates (OUR) 156
4.2.2 Interaction of Respiration Rate and Age 159
4.2.3 Relationship between OUR and the Biological Pump 160
4.2.4 Respiration of Particulate and Dissolved Organic Carbon (POC & DOC) 162
4.2.5 Benthic Respiration 162
Discussion Box 4.2 165
4.3 Respiration in the Absence of Oxygen 166
4.4 Anthropogenic Influences 170
References 172
Problems for Chapter 4 174

While photosynthesis and the production of organic matter dominate the surface ocean, respiration is the process that most influences the chemical perspective of oceanography in the deep ocean. For a chemist, respiration is simply photosynthesis (Eq. 3.10) run in reverse, with O_2 consumed while CO_2, nutrients, and energy are liberated. The flux of organic matter out of the surface ocean (the biological pump) is the driving force for the respiration reactions that occur in the deeper ocean. The delicate balance between the supply of organic matter via the biological pump and the renewal of water by mixing with ocean surface waters determines the distribution of metabolites below the euphotic zone. In this chapter, we explore how these processes interact to produce the complex and fascinating distributions of O_2 and nutrients in the deep sea.

4.1 Respiration below the Euphotic Zone

Since we know the approximate global ratio of the metabolic products P, N, C, and O_2 (improved RKR Eq. 3.10), it is possible to compare the global O_2 demand in the ocean

Table 4.1 Relative availability of the major bioactive elements and O_2 versus their use by average marine organisms. Seawater concentrations are the mean values of dissolved P, N, and C in the deep ocean, and the dissolved O_2 value at equilibrium with the atmosphere ($T = 2°C$, $S_P = 35$). Availability ratio refers to the seawater concentrations of N, C, and O_2 relative to P. Use ratio refers to the relative stoichiometry of P, N, C, and O_2 during respiration.

Dissolved bioactive compound	Seawater concentration ($\mu mol\ kg^{-1}$)	Availability ratio	Use ratio	$\frac{Availability}{Use}$
P	2.3	1	1	1
N	34.5	15	16	0.9
C	2275	989	106	9.3
O_2	330	143	153	0.9

below the euphotic zone with the supply of organic matter and oxygen from above. For this comparison (Table 4.1), imagine bringing a cubic meter of average deep water to the surface ocean and following the fate of the P, N, C, and O_2 through a photosynthesis, export, and respiration cycle. When the water arrives at the surface, P, N, and C are in the relative ratios found in deep-sea water, the "availability ratios" in Table 4.1. During photosynthesis, organic matter is created and exported to the deeper ocean. Surface water downwells carrying an O_2 concentration equal to the atmospheric saturation value at the surface temperature. During respiration, P, N, and C are released back to the water via respiration, and O_2 is simultaneously depleted in the ratios indicated as the "use ratios" in the table. From the ratios on the right of Table 4.1, it is evident that the availability and use of both dissolved N and P are balanced. Either of these macro nutrients has the potential to be limiting for primary production, with bioavailable N (mainly NO_3^-) being slightly more in deficit. There is a comparatively large surplus of inorganic carbon, such that, when all N and P have been stripped from the water to form biomass in the euphotic zone, only about 10 percent of the total DIC is removed.

The somewhat unexpected result from this thought experiment is that cold seawater at saturation does not contain quite enough dissolved O_2 versus the amount required to completely oxidize the biomass made from the dissolved nutrients. This would suggest that the deep ocean should be largely anoxic, because seawater downwelling at cold (~2.0°C) high latitudes carries just enough O_2 to meet the demand from respiring organic matter. There is clearly a mismatch between this calculation and the real ocean in that the average O_2 concentration in the deep sea is ~150 $\mu mol\ kg^{-1}$ instead of near zero (Fig. 4.1). This inconsistency was first noted by Redfield (1958), and the reason for it is that surface ocean nutrient concentrations are not uniformly near the detection limit. In polar waters in the subarctic regions and Southern Ocean, nutrients that upwell are not totally removed by biological processes. A map of surface ocean DIP (Fig. 4.2) illustrates that polar surface waters in the subarctic regions and Southern Ocean have high nutrient concentrations, up to half the values in the deep ocean. Thus, high nutrient concentrations in the deep ocean do not require proportionately high levels of oxygen consumption. In other words, not all of the ~2.5 $\mu mol\ kg^{-1}$ of DIP in the deep Pacific Ocean arrived there via respiration of organic matter – some was mixed or advected into the deep ocean from surface waters in

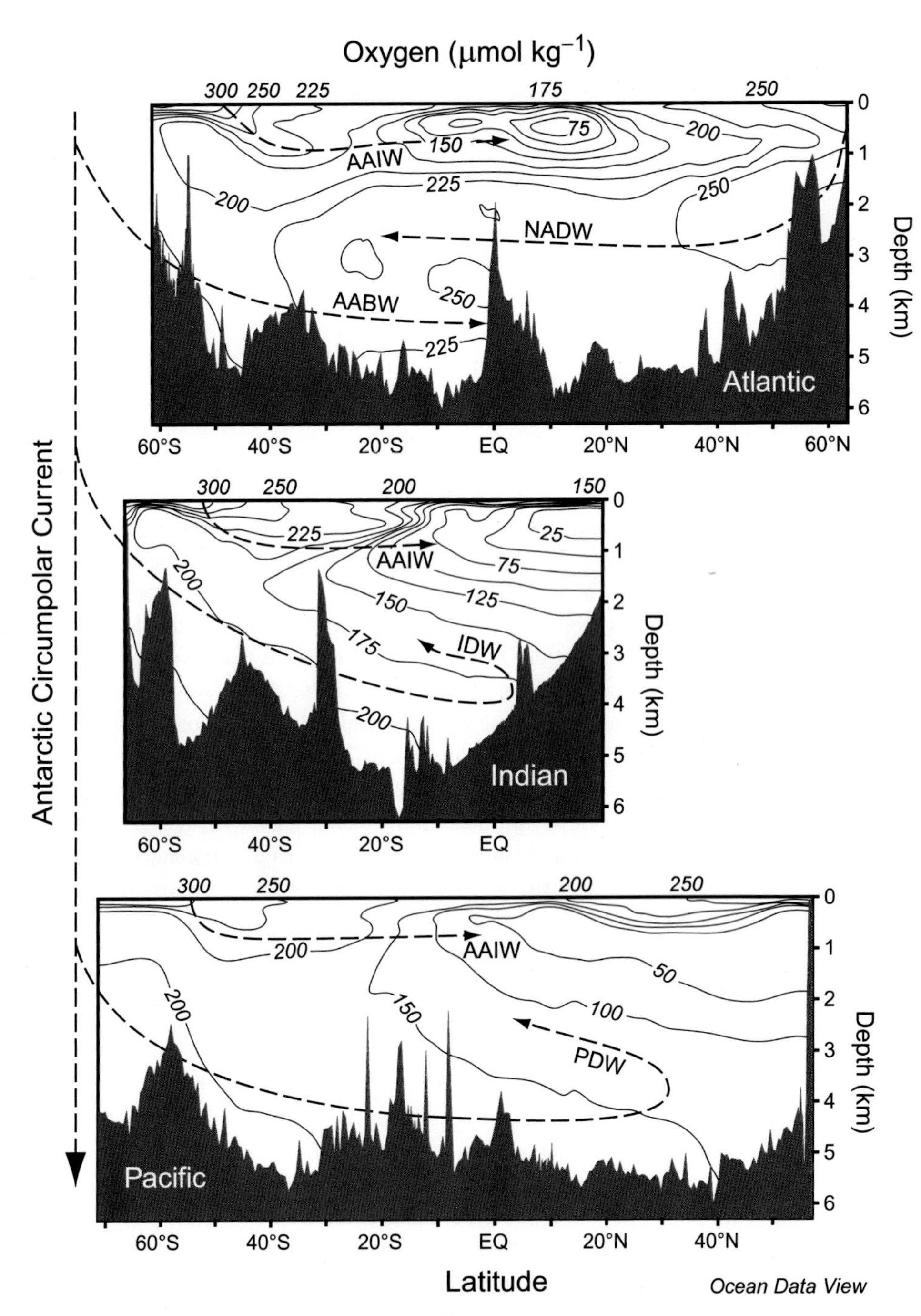

Figure 4.1 Cross sections of oxygen concentration (O_2, μmol kg^{-1}) in the Atlantic, Indian, and Pacific Oceans. Data are from Climate Variability and Predictability (CLIVAR) Repeat sections P16N & P16S for the Pacific, A16N & A16S for the Atlantic, and I09N and I08S for the Indian Ocean. (Data are compiled on the Ocean Carbon Data System, OCADS website: www.ncei.noaa.gov/products/ocean-carbon-data-system/)

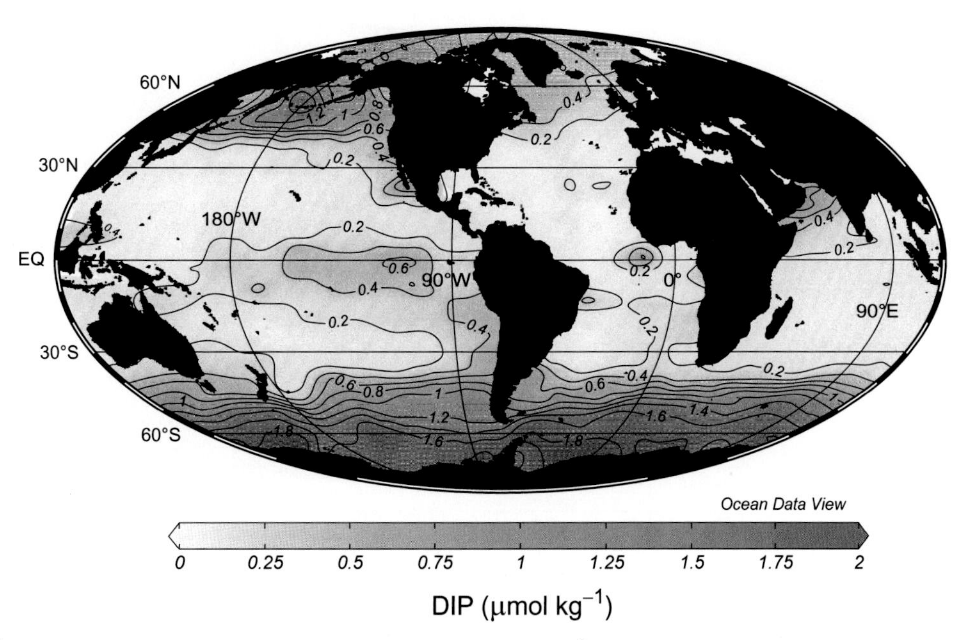

Figure 4.2 The concentration of dissolved inorganic phosphorus (DIP, μmol kg^{-1}) in surface waters of the ocean. Notice elevated concentrations at high latitudes and the Equator. Bottle data are from CCHDO (CLIVAR and Carbon Hydrographic Data Office, (https://cchdo.ucsd.edu). A black and white version of this figure will appear in some formats. For the color version, refer to the plate section.

areas that have high nutrient concentrations. This is also why it is not possible to match both the nutrient and O_2 distributions between the warm surface and deep ocean using a simple two-layer ocean like the one in Fig. 3.6 if one assumes the surface nutrients are near zero. Using the mean ocean phosphate concentration in the two-layer model equations, Redfield ratios for the organic matter flux (J), and *no surface water nutrients* causes the O_2 content of the deep ocean to be completely depleted. (Try it!) The model is too simple to reproduce actual ocean measurements because it does not take into consideration elevated nutrient concentrations in high-latitude waters, and produces too large an organic matter flux. Slightly more complicated models with another surface water reservoir are able to avoid this problem at the expense of creating a more complicated model (see Chapter 8 on the Carbon Cycle).

4.1.1 Oxygen Concentrations and Apparent Oxygen Utilization (AOU)

Oxygen in the ocean does not continually decrease with depth, but instead has a much richer pattern. A typical dissolved O_2 profile exhibits a minimum that is positioned shallower than 1 km (Fig. 4.1). In some regions of the ocean with sluggish circulation and a strong biological pump, O_2 is reduced all the way to zero in these mid-depth minima creating an oxygen deficient zone (ODZ) (Fig. 4.3). The largest of these ODZs are in the North Indian and the Eastern Tropical North and South Pacific. The main processes that contribute to the global large-scale pattern in oxygen concentration are the rapid and efficient respiration of

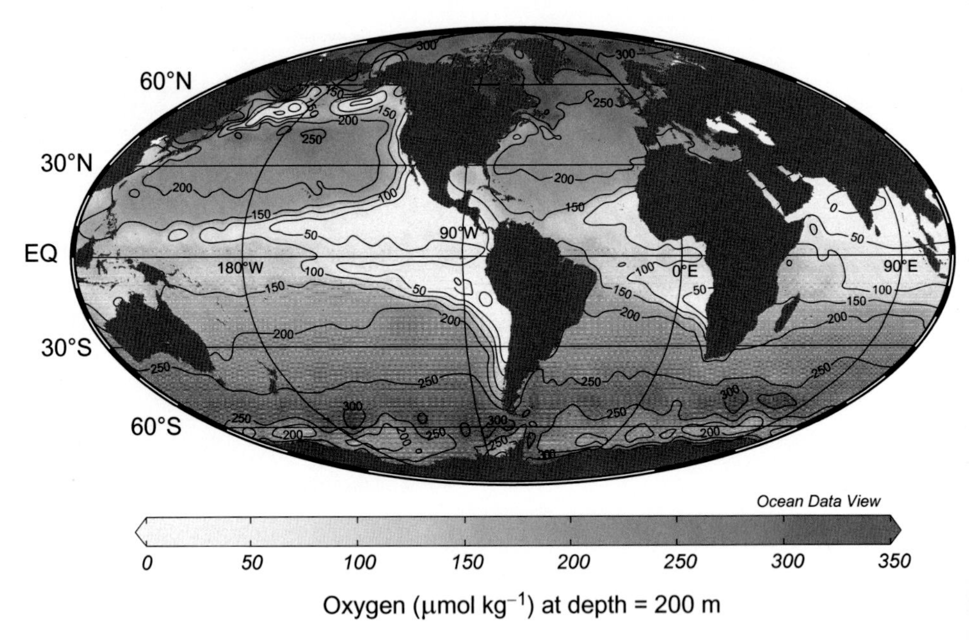

Figure 4.3 The oxygen concentration in the world's ocean at 200 meters depth. Data are from the Global Ocean Data Analysis Project version 2 (GLODAP v2) at the Ocean Data View (ODV) web site (https://odv.awi.de). A black and white version of this figure will appear in some formats. For the color version, refer to the plate section.

organic matter transported from the euphotic zone and the circulation of water from the surface (where it acquires O_2 from the atmosphere) into the deeper ocean. Oxygen concentrations increase again from the mid-depth minima toward a relative high at the bottom. Waters traveling at the ocean's deepest depths have spent most of their history since their last contact with the atmosphere at these deep depths where respiration rates are very slow due to low concentrations of POC and DOC, allowing oxygen concentrations to remain relatively high.

If we assume that the O_2 content of water leaving the surface ocean is in equilibrium with the atmosphere, then the amount of O_2 that has been consumed by organic matter respiration in a water sample can be estimated from its temperature, salinity, and O_2 concentration. The difference between the O_2 the water had at equilibrium with the atmosphere, $[O_2]_{eq}$, and the measured O_2 value is called the apparent oxygen utilization (AOU):

$$\mathrm{AOU} = [\mathrm{O_2}]_{eq} - [\mathrm{O_2}]. \tag{4.1}$$

Sections of AOU in the major oceans resemble those for O_2, but with the opposite sign. However, profiles of AOU and O_2 are not exact mirror images of each other. Surface O_2 concentrations vary by 150 μmol kg^{-1} from the Equator to the polar regions, primarily due to the temperature dependence of the saturation equilibrium value. The advantage of using AOU is that it removes this temperature effect, making it possible to quantify and compare the extent of respiration in waters with very different temperatures and surface origins. In some places, like the Labrador Sea and Southern Ocean south of the polar front, surface oxygen concentrations can be 5 to 10 percent undersaturated in winter (Ito et al., 2004; Wolf et al., 2018), but generally the assumption of saturation equilibrium is not a serious error.

4.1.2 Nutrient Concentrations and Preformed Nutrients

Consumption of oxygen during respiration of organic matter also results in the release of nutrients and inorganic carbon back to the dissolved phase. Because of this, nutrient distributions in the ocean are somewhat similar to the inverse of oxygen distributions (Fig. 4.4), but with important differences. Nutrients are normally low at the ocean surface where oxygen is high. In the deep sea, subsurface nutrient maxima form for the same reason as the oxygen minimum, a combination of respiration rates and the history of the water as it traveled from the surface. However, the subsurface nutrient maximum tends to be broader (extend deeper) than the oxygen minimum. This is because (as demonstrated at the start of this chapter) respiration is not the only source of nutrients to the deep sea.

Using AOU one can determine the fractions of dissolved N and P concentrations that are derived directly from respiration versus those that were present in the water mass when it initially sank from the surface. The latter are termed "preformed nutrients" and are independent of the amount of respiration that has occurred since downwelling. This is the portion of the deep nutrient concentration that was mixed or advected into the deep ocean rather than introduced from respired organic matter. Note that the residence time of DIP in the ocean is on the order of 20 000–100 000 years (Paytan and McLaughlin, 2007), so any individual P atom is likely to have participated in multiple photosynthesis-respiration and ocean circulation cycles. The concept of preformed nutrients applies to its source over the most recent ocean circulation cycle. Preformed and respiration-derived nutrient concentrations are separated by assuming that the ratio of nutrient and oxygen change during respiration, ΔNut.:ΔO_2, is a known constant (the modified RKR ratios discussed in Chapter 3) and that AOU is a measure of the O_2 consumed during respiration. Subtracting the product of the AOU and the respiration ratio from the measured value yields the preformed nutrient concentrations:

$$\mathrm{DIP}_{pre} = \mathrm{DIP} - r_{\mathrm{P:O_2}}\left(\Delta \mathrm{O}_{2,resp}\right) = \mathrm{DIP} - r_{\mathrm{P:O_2}}(\mathrm{AOU}) \quad (4.2)$$

$$\mathrm{DIN}_{pre} = \mathrm{DIN} - r_{\mathrm{N:O_2}}\left(\Delta \mathrm{O}_{2,resp}\right) = \mathrm{DIN} - r_{\mathrm{N:O_2}}(\mathrm{AOU}). \quad (4.3)$$

The concept of nutrient and O_2 change along a surface of constant density in the ocean is illustrated in Fig. 4.5. Recall from Section 1.3.3 that denser waters subduct, flowing from their surface outcrop equatorward under less dense waters. As they leave the surface, these waters carry O_2 concentrations near saturation equilibrium with the atmosphere and may carry non-zero nutrient concentrations (preformed). As a water parcel moves along a constant density surface away from its outcrop and into the upper thermocline, respiration consumes oxygen increasing AOU and dissolved nutrients in the water mass. At every point in the ocean interior within this same water mass, preformed nutrient concentrations should be constant and can be calculated if one knows the temperature and salinity (for determining $[O_2]_{eq}$), nutrient, and O_2 concentrations.

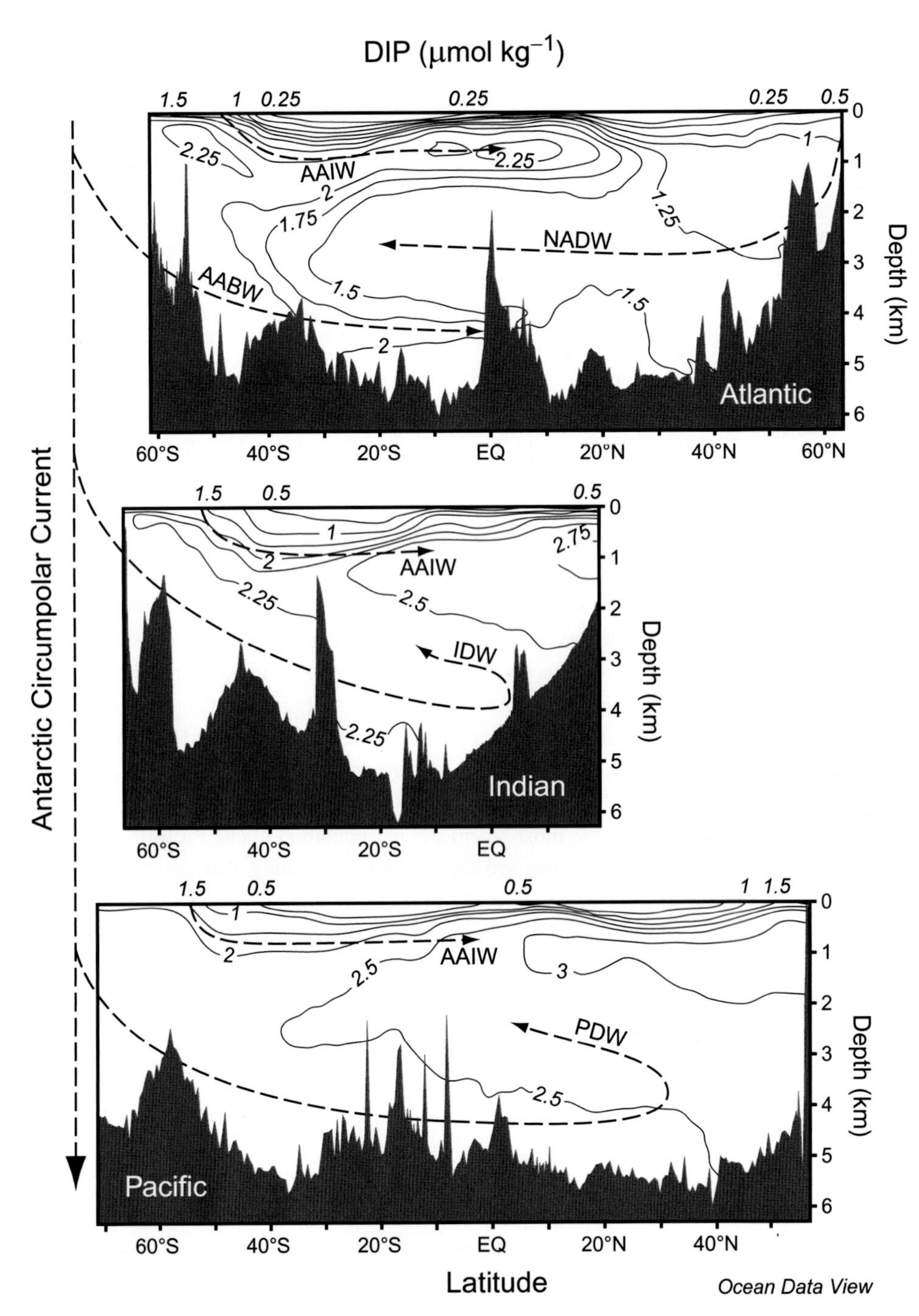

Figure 4.4 Cross sections of dissolved inorganic phosphate concentration (DIP, μmol kg^{-1}) in the Atlantic, Indian, and Pacific Oceans. See the caption to Fig. 4.1 for the data source.

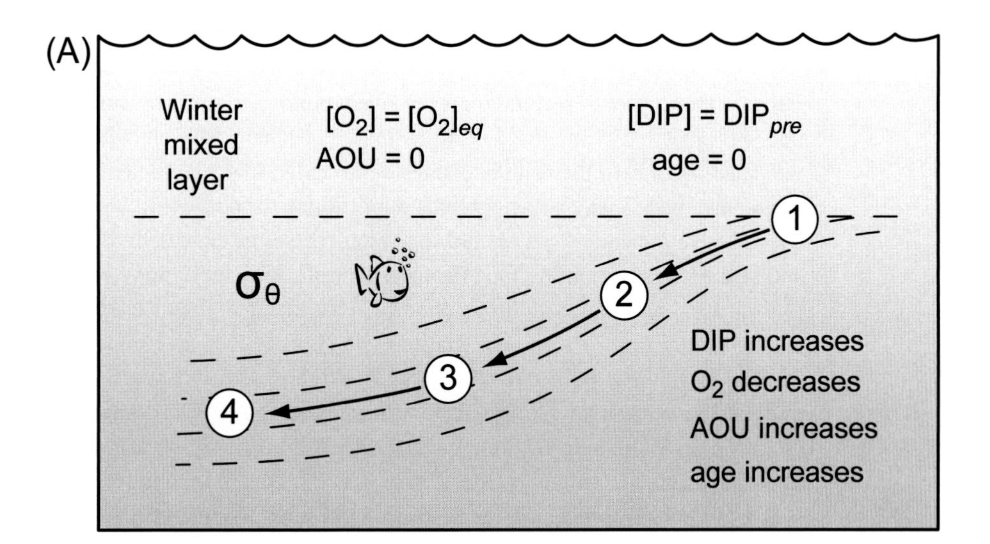

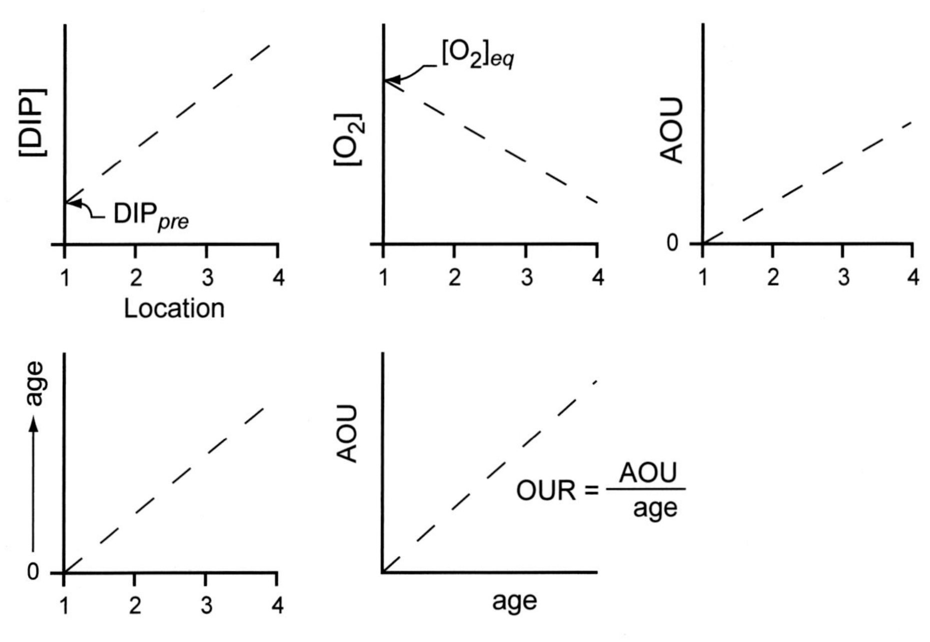

Figure 4.5 (A) Schematic cross section of the upper ocean, illustrating trends on isopycnal surfaces for DIP, O_2, AOU, and age (t = time since the water mass was last at the surface). (B) Schematic plots of DIP, O_2, AOU, and age versus locations (1), (2), (3), and (4) in (A); and a plot of AOU versus age used to derive oxygen utilization rates (OUR). Preformed DIP (constant at all locations) is indicated by the arrow. Modified from Emerson and Hedges (2008).

The preformed nutrient concentrations determined from Eqs. 4.2 and 4.3 are characteristic of specific water masses that formed at the surface and flowed into the interior ocean. Because the impact of respiration is removed, preformed nutrients are close to conservative tracers – just like temperature and salinity – whose value anywhere in the ocean is a weighted average of the original values of the individual water masses that mixed together. There have been many oceanographic applications for nutrient-like conservative and semi-conservative tracers – from identifying water masses as classically done using temperature and salinity, to determining the effects of denitrification on observed nitrate concentrations, to evaluating how much anthropogenic CO_2 has changed the measured DIC concentration. In one of the first of these applications, W. S. Broecker determined the origin of Deep Pacific Ocean water (Broecker et al., 1985). He showed that the main sources of deep water in the Pacific Ocean are North Atlantic Deep Water (NADW), Antarctic Intermediate Water (AAIW), and Antarctic Bottom Water (AABW), all of which are at least partially homogenized in the Circumpolar Deep Water (CDW). Determining how much of each of these source waters contributes to Pacific deep water using endmember mixing of the conservative properties temperature and salinity is not possible because salinities of the endmembers are not sufficiently different. Since concentrations of DIP are well above detection limits in high-latitude surface waters, and very different among the three sources, Broecker et al. (1985) used the preformed nutrient phosphate (which he called PO or PO_4*) along with temperature to determine the importance of the three sources. More recent evaluations using vastly more data and tracers have improved these estimates (Johnson, 2008).

Another application of the preformed nutrient concept is in correcting measured DIC values for anthropogenically introduced CO_2 (Gruber et al., 1996). In this case, the value of preformed DIC is evaluated, and its changes due to organic matter degradation and $CaCO_3$ dissolution are calculated from changes in AOU, NO_3^-, and Alk. These values are then compared with measured DIC values, and the excess is attributed to the concentration of anthropogenic CO_2 that has penetrated the ocean. Details of this procedure are described in the chapter on the carbon cycle (Chapter 8).

4.1.3 Nitrogen and Phosphorus Cycles

Another important application of the concept of combining the concentrations of different metabolic products is that of comparing DIN to DIP using the RKR ratio to determine the sources and sinks of bioavailable nitrogen to the ocean. Ocean cycling of nitrogen and of phosphorus are different in important ways, as illustrated schematically in Fig. 4.6. Both bioavailable N and P are supplied to the ocean by rivers and removed by sediment burial (not shown in Fig. 4.6) and cycled between the euphotic zone and deeper waters by photosynthesis, sinking, respiration, and upwelling. The nitrogen cycle, however, is more complex, because DIN exists in multiple forms and there are additional important sources and sinks.

Phytoplankton are capable of utilizing DIN in many different forms, primarily nitrate (NO_3^-) and ammonium (NH_4^+) but also nitrite (NO_2^-) and urea ($CO(NH_2)_2$). The different bioavailable DIN forms are collectively referred to as fixed nitrogen. Respiration of organic matter produces ammonium, but, in the presence of oxygen, bacteria oxidize ammonium to

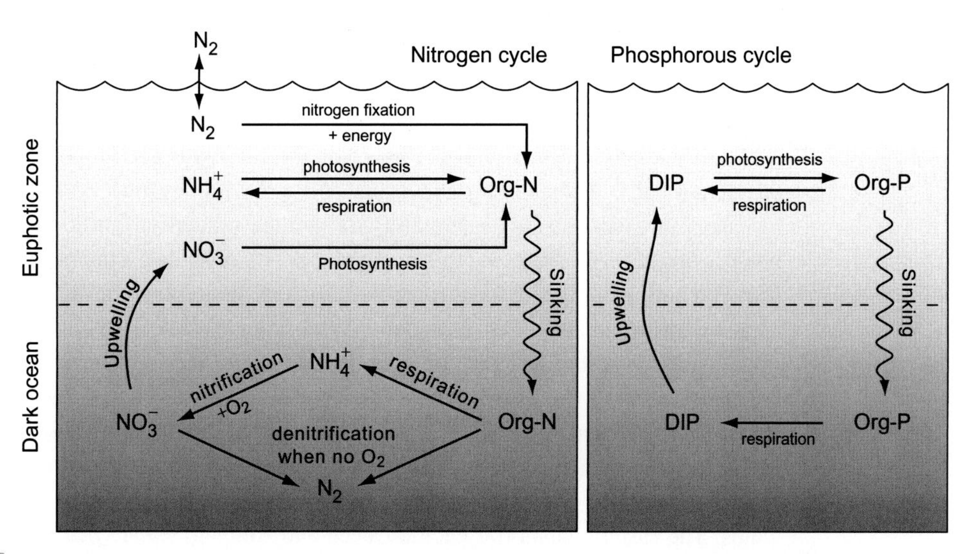

Figure 4.6 Schematic representations of the nitrogen and phosphorus cycles in the ocean indicating the important reactions and transport mechanisms that transform N and P among different compounds, phases, and locations. ("Sinking" refers to both particle settling and mixing of dissolved organic matter.) For simplicity, inflow from rivers and burial in sediments are not shown.

nitrate via a reaction called nitrification. This reaction is inhibited by light, so nitrification primarily occurs below the euphotic zone and, except in locations with negligible oxygen or high ammonium production rates, nitrate is the dominant form of DIN in the deep ocean.

Dissolved N_2 gas exists at high concentrations in the ocean $(400-600\ \mu mol\ kg^{-1})$ but is unavailable to most plankton as a nutrient source. There are, however, some phytoplankton that are able to transform N_2 into bioavailable, i.e., fixed, N through a reaction called nitrogen fixation. This reaction is ecologically favorable in regions of the ocean with low surface nutrients but sufficient iron, which is required by enzymes that facilitate the reaction. The N fixed by these organisms is a significant global source of DIN to the ocean, about double that supplied by rivers (Gruber, 2008). The N cycle also has an additional sink compared with P. In the absence of oxygen, some bacteria can utilize nitrate as an oxidant during organic matter respiration (see Section 4.3), a process called denitrification that produces N_2. Several other processes occurring in anoxic waters also transform fixed N to bio-unavailable N_2, including anammox, a process in which ammonium is oxidized by nitrite. These N_2 producing reactions represent a significant sink for DIN from the ocean that is about 15 times larger than that of the sedimentary burial of N in organic matter (Gruber, 2008). All this makes the whole ocean residence time for fixed N less than 3000 y, far shorter than that for phosphorus. Thus, fixed-N changes are much more likely than P to have been of importance to millennial-scale changes in ocean productivity.

The difference in chemical complexity of these two nutrients has been used to develop the tracer, N*, which is the concentration of fixed N in ocean waters that is either missing because of denitrification or in excess because of nitrogen fixation:

$$N^* = DIN - 16\,DIP, \tag{4.4}$$

where 16 is the whole-ocean Redfield ratio $r_{N:P} = 16$ (see Fig. 3.2). The assumption here is that the N:P ratio during photosynthesis and respiration is constant in the ocean, although we know the ratio must vary locally based on the variations in C:P ratios shown at the end of Section 3.1.2. Values lying above the 16:1 ratio line in Fig. 3.2 have a positive N*, revealing an excess of DIN relative to DIP and the influence of nitrogen fixation. Values lying below the 16:1 ratio line have a negative N*, revealing a deficit of DIN relative to DIP and the influence of denitrification. By this definition, the global average N* is negative, approximately –3.5 μmol kg^{-1} (Deutsch and Weber, 2012).

The global distribution of N* on a density surface of the shallow thermocline (Fig. 4.7) indicates near zero values in the North Atlantic and very negative values in the regions of oxygen deficient zones, namely the eastern equatorial Pacific and Arabian Sea (compare to Fig. 4.3). North Atlantic surface waters are known to be regions of strong nitrogen fixation, supported by high iron fluxes from dust blown westward from the Sahara by the trade winds. The eastern equatorial Pacific Ocean and Arabian Sea are places with low enough oxygen (<10 μmol kg^{-1}) for denitrification to take place in the water column. Note that an N* value different from zero does not indicate that these reactions are actively happening at the location of the measurement. Surpluses and deficits in N can be transported by ocean currents to regions far from their origin.

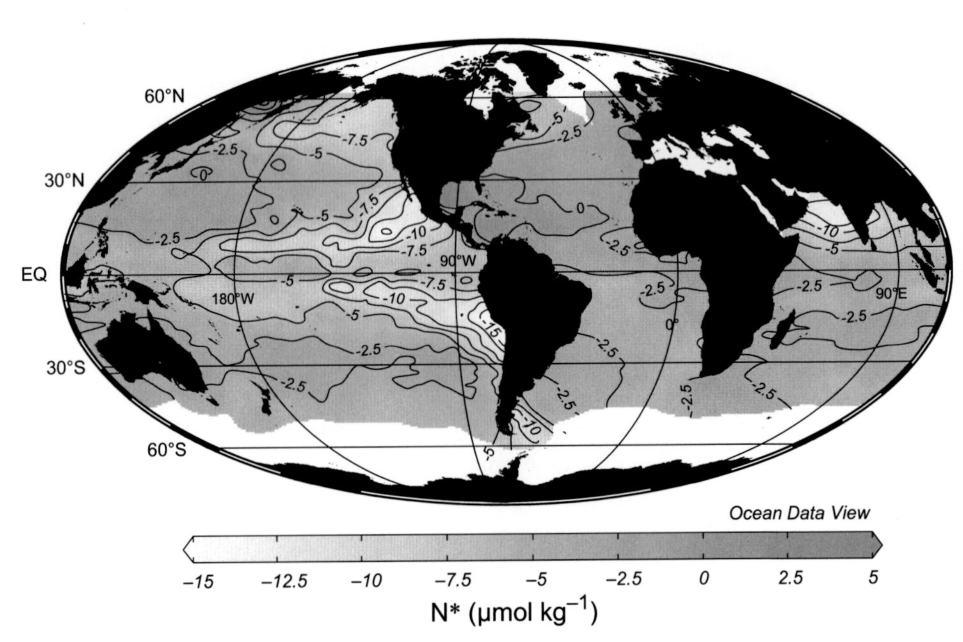

Figure 4.7 N* (DIN – 16*DIP in units of μmol kg^{-1}) on potential density surface $\sigma_\theta = 26.0$ in the world's ocean. White areas at high latitudes indicate regions without water of this density. Calculated from data in the World Ocean Atlas (WOA) 2013. A black and white version of this figure will appear in some formats. For the color version, refer to the plate section.

Discussion Box 4.1

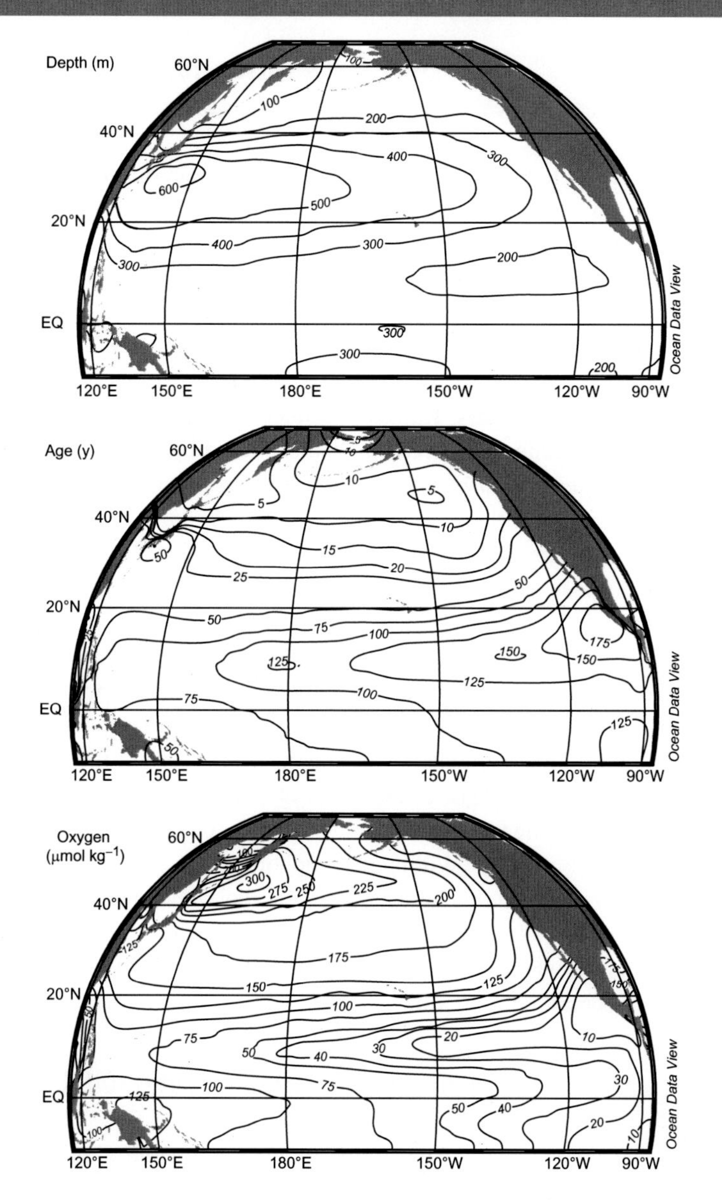

These figures show properties in the North Pacific on the potential density surface $\sigma_\theta = 26.2$. The upper panel shows the mean depth (m) of this isopycnal. The middle panel shows the time (y) since the water was last in contact with the surface (based on CFC-11 measurements, see Section 4.2.1, calculation courtesy of Andrew Shao). The lower panel shows oxygen concentrations (μmol kg^{-1}).

- Where does this isopycnal outcrop? How does the depth of the isopycnal relate to the circulation patterns explored in Chapter 1 (Figs. 1.14 and 1.15)?
- How does the time it takes water to move along this isopycnal relate to the flow path of the water? Which regions have the most sluggish circulation? Why?
- Estimate the equilibrium oxygen concentration for this isopycnal.
- How does oxygen change as the water moves along this isopycnal? What explains the location of the lowest oxygen concentrations?

4.2 Respiration Rates

A rough idea of how respiration decreases as a function of depth in the ocean can be inferred from the flux of particulate organic material caught in sediment traps deployed in the ocean (Section 3.2.2). Figure 3.7 depicts the fraction of organic carbon caught in sediment traps relative to the value at 100 m, the approximate base of the winter mixed layer. The respiration rate of sinking particles is equal to the decrease in flux at each subsequently deeper sediment trap. These data reveal that the vast majority of organic carbon exported from the euphotic zone is respired in the upper few hundred meters. Respiration rates become slower with depth until reaching very low rates in the deep ocean as demonstrated by the much more gradual decrease in organic matter caught by deep sediment traps between 1 and 5 km. While sediment trap studies highlight the general trend of exponentially decreasing respiration rates as a function of depth, other methods of determining the respiration rate are more accurate, because sediment traps suffer from collection biases in trapping particles sinking through a moving ocean.

At the ocean surface where all metabolic rates are high, respiration rates can be measured by incubating seawater samples at in situ temperatures in "dark" bottles, covered in opaque material to block out the light, and measuring oxygen depletion after about one day. Data from such experiments carried out in the open ocean indicate that respiration rates are highest in the euphotic zone, on the order of 1 μmol $kg^{-1}d^{-1}$, and decrease to about one tenth of this value at depths of a few hundred meters (Williams and Purdie, 1991, Fig. 4.8). Within the euphotic zone, these high respiration rates are typically overwhelmed by photosynthesis, so that drawdown of oxygen is only observed in the dark ocean. Note that the incubation approach is only accurate in the upper few hundred meters, where respiration rates are high enough to cause measurable changes in oxygen concentration over a day.

4.2.1 Oxygen Utilization Rates (OUR)

The most accurate means of determining the rate of respiration in the upper ocean – the oxygen utilization rate (OUR) – is by combining AOU with the amount of time that a water mass has been away from the surface ocean mixed layer. This relationship is illustrated in the bottom row of Fig. 4.5 (OUR = AOU/age) and requires a method of determining the amount of time that has passed since the water was last at the surface (the age). There are two tracers of water mass age that work well over the timescale of a few decades that characterizes the age of the upper ocean where most organic matter is degraded: the $^3H - {}^3He$ pair and chlorofluorocarbons (CFCs). Both tracers are gases and stem from anthropogenic activities, either by nuclear weapons testing, ^3H (tritium), or industrial processes, CFCs. They entered the upper ocean via rainfall for tritium or via gas exchange with the atmosphere for CFCs and are at present working their way into the

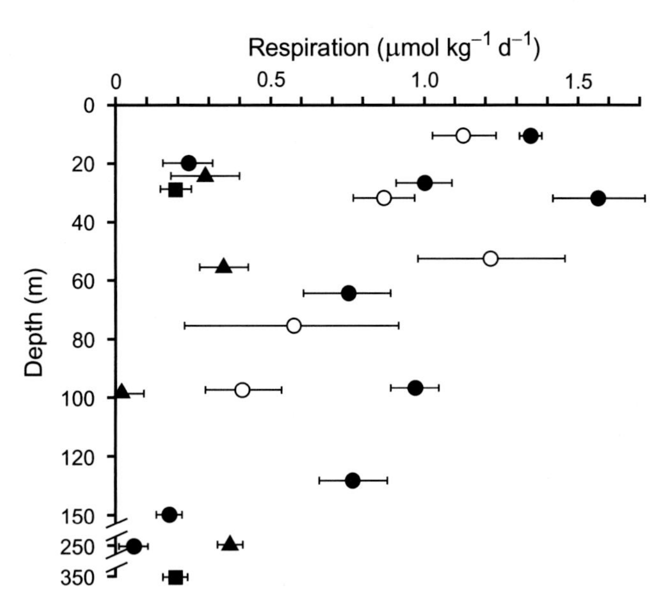

Figure 4.8 Measurements of in vitro changes in oxygen concentration in dark bottles as a function of water depth in the subtropical N. Pacific Ocean. Samples of seawater were incubated either in situ or on the deck of the ship for one day. Different symbols indicate different water casts. Error bars represent the standard error of the mean oxygen change at that depth. Redrawn from Williams and Purdie (1991).

thermocline. The concentrations of these tracers in a transect through the mid-Pacific Ocean (Fig. 4.9) indicate how lines of equal concentration follow density horizons from the more polar regions through the subtropical thermocline and shoal in the region of the Equator.

The systematics of using the ^{3}H–^{3}He pair for determining water mass age are described in Chapter 7. Briefly, ^{3}H injected into the atmosphere from weapons testing in the 1960s and 1970s was incorporated into water vapor and quickly rained out of the atmosphere to the surface ocean. The ^{3}H decays to ^{3}He with a half-life of 12.5 y. Because there is almost no ^{3}He in the atmosphere, that which is produced by ^{3}H decay in the ocean mixed layer escapes to the atmosphere. ^{3}He produced below the ocean mixed layer in the upper ocean thermocline remains dissolved in the water. By measuring both the ^{3}H and ^{3}He content, one can determine from the buildup of ^{3}He the amount of time a water parcel has been out of contact with the atmosphere (the ventilation time).

Man-made chlorofluorocarbon (CFC) concentrations increased in the atmosphere until they were banned in 1987 due to their role in ozone depletion in the stratosphere. Their observed atmospheric history is used to compute the expected surface ocean concentration at equilibrium with the atmosphere at any time in the past. The CFC concentrations measured in the ocean thermocline then identify the time at which this water mass was last at the surface and thus its ventilation age.

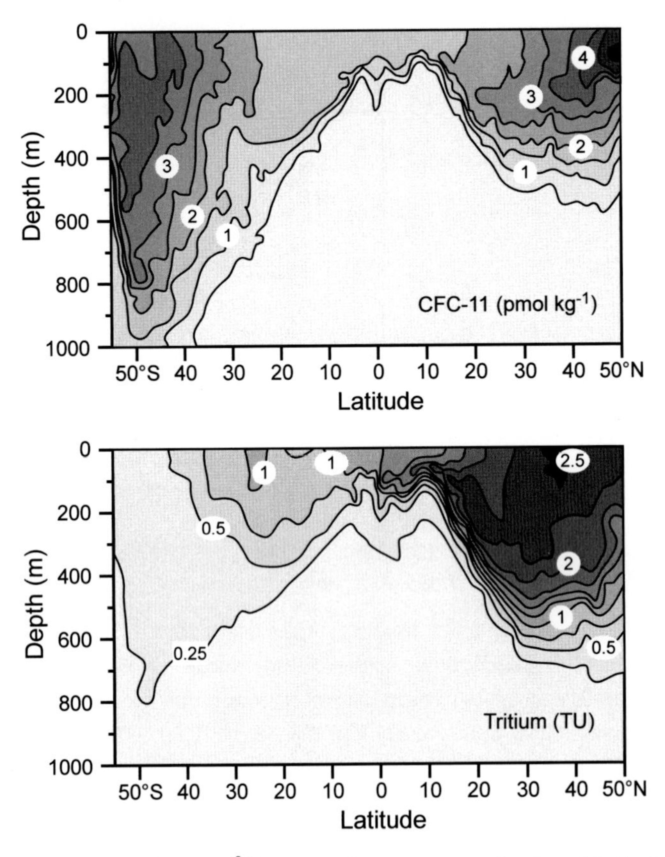

Figure 4.9 Chlorofluorocarbon-11 (CFC-11) and tritium (^{3}H) concentration sections as a function of depth along 135°W in the Pacific Ocean. Both tracers have an anthropogenic origin and have been introduced to the atmosphere in the last ~70 years. Tritium concentrations are in Tritium Units (1 TU = 1 ^{3}H per 10^{18} hydrogen atoms). Numbers indicate the values along individual contours. Redrawn from Jenkins (2002).

Once the ventilation time is determined by either the ^{3}H–^{3}He or CFC method, this time (age) can be plotted against AOU to determine the OUR. In practice, mixing with other waters is important to this calculation, and the three-dimensional distributions of ^{3}H, ^{3}He, CFCs, and O_2 must be considered along with equations that include advection and mixing. William Jenkins (Jenkins, 1998) showed the utility of the ^{3}H–^{3}He pair for determining OUR in the eastern subtropical Atlantic Ocean and later illustrated that the more careful solution considering the three-dimensional nature of ocean transport gives lower values at depth and a more realistic distribution of OUR as a function of depth (Fig. 4.10). This result is probably the best estimate of the depth dependence of respiration in the upper thermocline of the ocean, and its exponential shape has a striking similarity to the sediment trap results (Fig. 3.7). An exponential regression through the data in Fig. 4.10 indicates a scale height of 165 m. This is the depth interval required for the rate at the base of the euphotic zone (~100 m) to decrease to $1/e$ (37 percent) of its

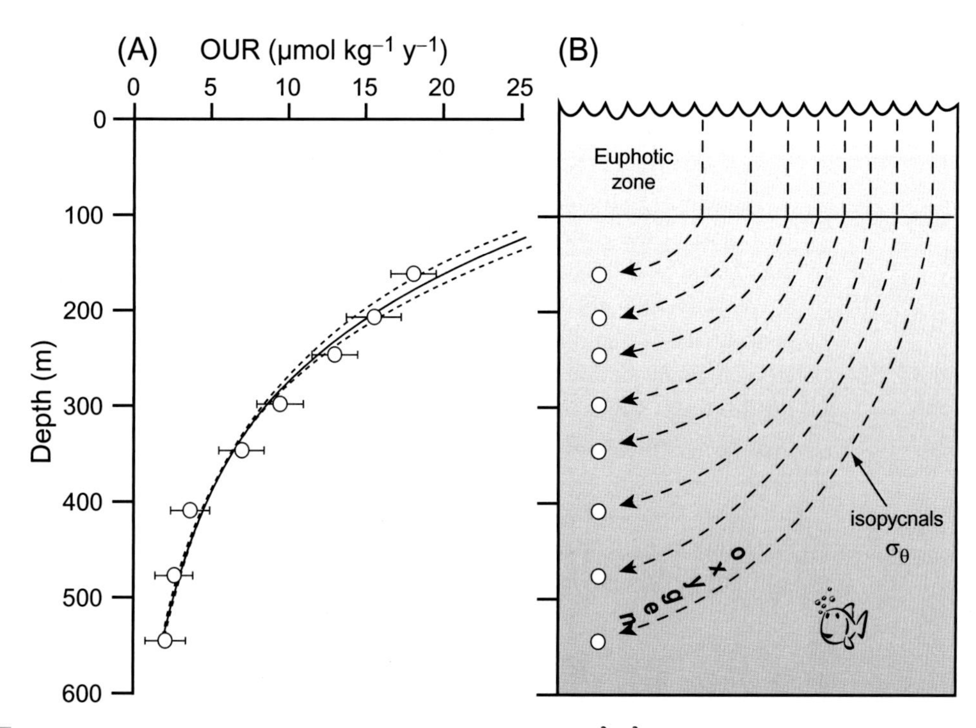

Figure 4.10 Oxygen utilization rate (OUR) versus depth determined by measuring ^{3}H, ^{3}He, and AOU on isopycnal surfaces in the eastern subtropical North Atlantic Ocean. The OUR results in (A) are redrawn from Jenkins (1998). (B) A schematic diagram indicating the pathway of constant potential density surfaces (isopycnals) from the surface ocean into the thermocline. Because more organic matter degrades on surfaces closest to the euphotic zone, the OUR rates are greater on the shallower isopycnals. However, all density surfaces originate at the surface of the ocean so even the deeper isopycnals in (A) have been influenced to some extent by shallow respiration.

value. Most of the organic carbon that exits the euphotic zone is respired in the upper 200 m of the pycnocline.

4.2.2 Interaction of Respiration Rate and Age

Besides respiration rate, the other most important control on oxygen concentrations in the ocean (Fig. 4.1) is the age of the water, i.e., how long the water has been isolated from the atmosphere. We have just shown how shallow water mass ages can be determined from ^{3}H–^{3}He or CFCs. Such measurements demonstrate that the "youngest" waters are near the surface and that age increases as waters move along isopycnals into the interior ocean. Deep-water mass ages are determined from the distribution of natural radiocarbon, ^{14}C, which enters the ocean at the surface through atmospheric exchange and decays in the interior ocean with a half-life of 5700 years (see Section 7.2.1). Those measurements indicate that the youngest deep water is found in the North Atlantic, where North Atlantic

Deep Water (NADW) forms. The thermohaline circulation moves NADW southward in the Atlantic and Antarctic Bottom Water northward in all three ocean basins, filling the deep oceans with mixtures of the major source water masses. The oldest water in the oceans is found at 2000–2500 m depth in the North Pacific, significantly deeper than the oxygen minimum (Fig. 4.1) but not at the very bottom.

Together, respiration rate and water mass age explain AOU distributions and therefore most of the pattern in oxygen concentrations in the deep ocean. Respiration rates are highest near the base of the euphotic zone, but waters there have been at the surface so recently that not enough time has passed for respiration rates to reduce oxygen concentrations significantly. Where waters are very old in the deep ocean, moderate oxygen concentrations may still be found because respiration rates are so low at these depths that oxygen is not reduced to low values despite the long time over which oxygen consumption has occurred. The lowest oxygen concentrations occur at depths where respiration rates and water mass age are both significant. An analogy for the processes that control the oxygen minimum zone is the time required for a journey. Traveling at high speed for a very short time or at a very low speed for a long period will not get you as far as medium speed for a medium amount of time.

4.2.3 Relationship between OUR and the Biological Pump

The biological pump or ANCP measurements discussed in Chapter 3 are linked to the respiration rate discussed here (Fig. 4.10), because the flux of organic matter from the upper ocean must be approximately equal, at steady-state, to the depth-integrated respiration rate divided by the ΔO_2:ΔOC ratio during respiration. In a one-dimensional ocean, the flux of organic carbon from the surface ocean (F_C) is represented by

$$F_C = \int_{z=100m}^{z=\infty} \text{OUR}\, dz \Big/ r_{O_2:C} \quad (\text{mol-C m}^{-2}\,\text{y}^{-1}), \tag{4.5}$$

where $r_{O_2:C}$ is the ΔO_2:ΔOC stoichiometric ratio of 1.45.

Estimates of the depth integrated OUR in the North and South Pacific thermocline at a longitude of about 135°W have been determined using CFC-derived ages and AOU measurements. The results are presented here (Fig. 4.11) in terms of the biological carbon pump or ANCP necessary to create the measured OUR (mol-C m^{-2} y^{-1}, Sonnerup et al., 2013, 2015). The calculated rates of the biological pump from these data fall in the range of 2.5 ± 1.0 mol-C m^{-2} y^{-1}, which is consistent with the biological pump estimates determined from metabolite mass balances in the upper ~100 meters of the surface ocean in Chapter 3 (see Fig. 3.12). Sonnerup's results in Fig. 4.11 represent a richer portrait of meridional variability than the sparse upper ocean data. The general trends in the North Pacific indicate that the highest values occur in the region of the subtropical/subarctic boundary where the flux is about 3.5 mol-C m^{-2} y^{-1}. The biological pump decreases from the gyre boundary to the north and to the south, reaching values about 2 mol-C m^{-2} y^{-1} in the subarctic and subtropical N. Pacific. A decreasing biological pump from the subtropics

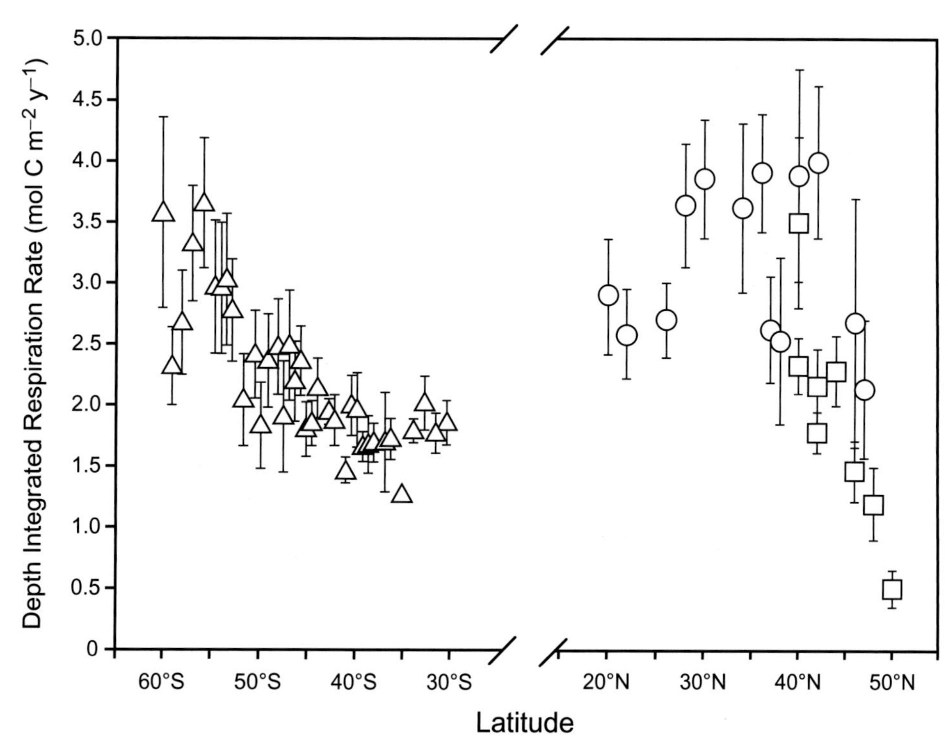

Figure 4.11 Depth integrated respiration rate determined from water mass CFC ages and AOU in the shallow thermocline (~100–500 m) of the Eastern Pacific Ocean. Triangles, circles, and squares are from different cruises. Whiskers represent the error estimated for the CFC ages. Redrawn from Sonnerup et al. (2013 and 2015).

to the subarctic N. Pacific is in sharp contrast to the values suggested by models of the satellite data and by GCMs (Fig. 3.11). There are no flux estimates from the OUR data in the vicinity of the Equator, because it is a region where strong upwelling makes it difficult to determine ventilation age. The biological pump in the South Pacific is highest (~3.0 mol-C m^{-2} y^{-1}) in the region poleward of 50°S and decreases northward to a value of about 1.5 mol-C m^{-2} y^{-1} in the subtropical South Pacific. The trend of decreasing biological pump from high latitudes to the subtropics in the South Pacific follows the trends in Fig. 3.11, but the satellite-derived results suggest much lower values of carbon export in the subtropics than observed (Fig. 4.11).

There are a couple of important simplifications made here. First, OUR should be integrated from 100 meters to the entire ocean depth, but in practice OUR need only be integrated to about the base of the thermocline, because the lion's share of organic matter respiration takes place in the upper several hundred meters of the aphotic zone. The other approximation is that Eq. 4.5 assumes the ocean is one-dimensional. In practice, waters mix along isopycnals (Figs. 4.5 and 4.10), but particles sink vertically. This means that the OUR measured at a given location along an isopycnal's surface is the result of respiration that occurred during the water's transit from the surface to where the oxygen and age tracer are measured. OUR measured at any location is a product of respiration that occurred over a range of locations, not just the waters above the measurement location.

4.2.4 Respiration of Particulate and Dissolved Organic Carbon (POC & DOC)

The flux of organic carbon from the euphotic zone of the ocean to the deep occurs through both particulate and dissolved organic carbon phases (POC and DOC, respectively); POC sinks while DOC is made up of molecules and particles/organisms small enough to remain suspended in the water. While the particulate flux dominates on average (total global rates are estimated to be 70–80 percent particle flux, Doval and Hansell, 2000; Hansell et al., 2009; Letscher and Moore, 2015), its degree of dominance varies dramatically in different regions of the ocean. The DOC:POC ratio of organic carbon export can be determined by comparing AOU to the change in DOC in the upper few hundred meters of the aphotic zone. Assuming that the ΔO_2:ΔDOC ratio during respiration is 1.45, one can calculate what fraction of the AOU change is caused by DOC degradation; the rest of the AOU change is assumed to be caused by particle degradation. The two principal ocean time-series sites (the Bermuda Atlantic Time-series Study, BATS, and the Hawaii Ocean Time-series, HOT) both feature repeated seasonal measurements of DOC. In the 100 – 250 m depth range, about 60 to 70 percent of the AOU at HOT is created by degradation of DOC, whereas this value is only about 15 percent at BATS (Fig. 4.12). This difference indicates that that organic carbon delivery to the ocean interior varies dramatically over different regions. DOC cycling between the surface and upper thermocline is probably more prominent in subtropical regions because small picoplankton, native to these areas, depend on dissolved organic matter cycling. While BATS, on the edge of the North Atlantic subtropical gyre, is not a typical subtropical ocean location because of the influence of the Gulf Stream and the very deep mixed layers in winter (Jenkins and Doney, 2003).

Since non-refractory DOC generally degrades more rapidly than POC, the ratio of DOC versus POC export influences the characteristic organic carbon degradation depth – making it shallower for greater DOC export. Models show that shallower organic matter degradation results in higher oxygen concentrations in thermocline waters, because shallower depths are more readily ventilated with atmospheric O_2 (Kwon et al., 2009). Thus, the types of primary producers present in ocean surface waters may influence the maintenance of the ocean's oxygen minima. By this mechanism, ecosystem changes caused by global warming could also have consequences for oxygen distributions in the deeper ocean.

4.2.5 Benthic Respiration

What little is known about respiration rates in the ocean below 1000 m has been inferred from measuring particle fluxes in deep sediment traps and by the study of O_2

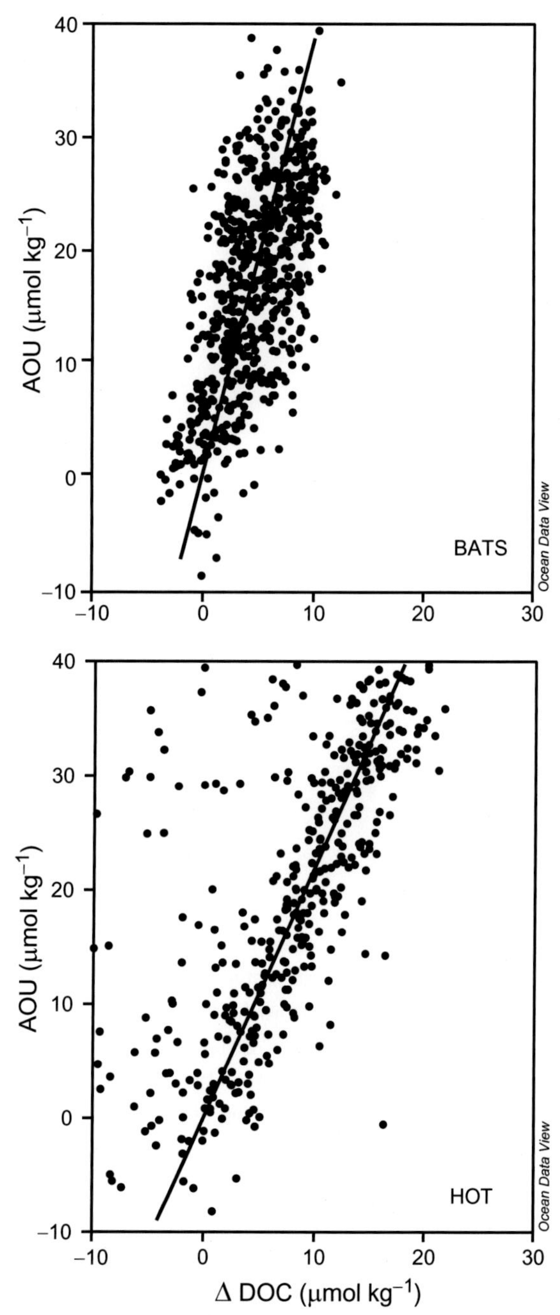

Figure 4.12 AOU versus ΔDOC in water samples in the depth range 100–250 m at two time-series stations: BATS (Bermuda Atlantic Time-series Study) and HOT (Hawaii Ocean Time-series). ΔDOC is the difference between the surface DOC and the measured value. The fraction of the AOU created by degradation of DOC can be determined by the slope of the line assuming the ΔO_2:ΔC ratio during respiration is –1.45. The AOU created by degradation of DOC at BATS is 13 percent of the total respiration, while at HOT it is 66 percent. Redrafted from Emerson (2014).

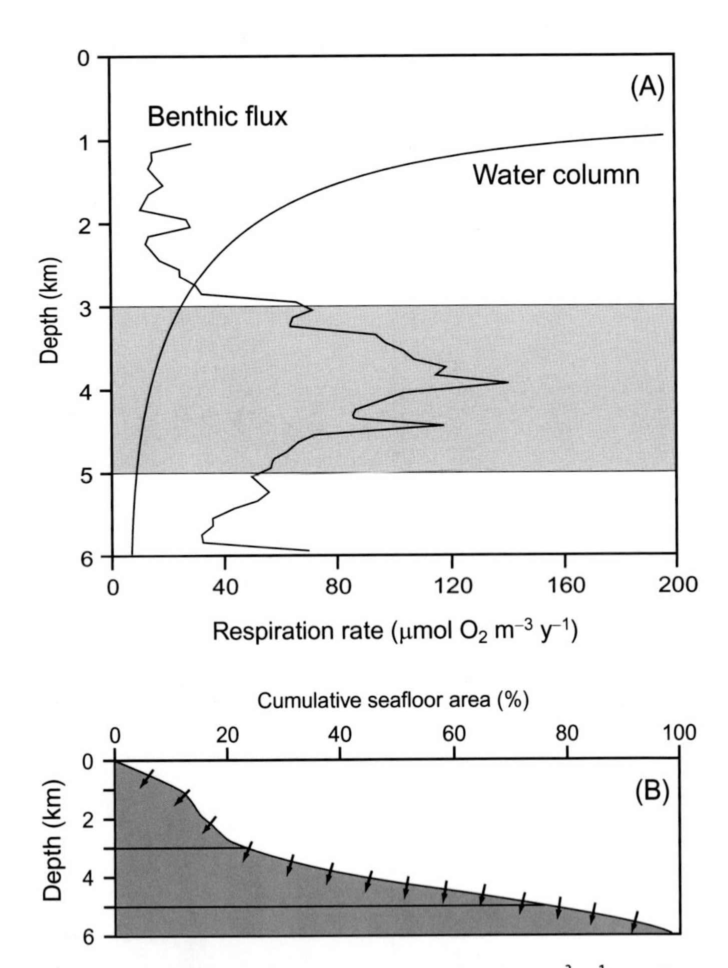

Figure 4.13 (A) Average water column and benthic oxygen consumption rates (μmol O_2 m^{-3} y^{-1}) as a function of depth. The water column curve is from sediment trap and OUR determinations, and the benthic curve is from benthic flux and pore water measurements. The benthic fluxes are normalized to the volume of water exposed per unit of sediment area and indicate that below 3000 m the respiration contribution from sediments is greater than that in the water column. Redrawn from Jahnke and Jackson (1987). (B) The hypsometric curve shows the proportion of sea floor area that exists at different depths and indicates that the region between 3000 and 5000 m depth has the greatest sea floor area to ocean volume ratio, which is indicated by the shaded region in (A). Modified from Emerson and Hedges (2008).

consumption at the sediment–water interface. Most of the particles that escape shallow respiration to reach the deep ocean contain mineral tests or are fecal pellets of zooplankton. These particles transit the water column fairly quickly (weeks), so that some material relatively rich in organic matter reaches the ocean floor. Deep ocean water column respiration rates are estimated from the change in organic matter flux between different sediment trap depths.

Discussion Box 4.2

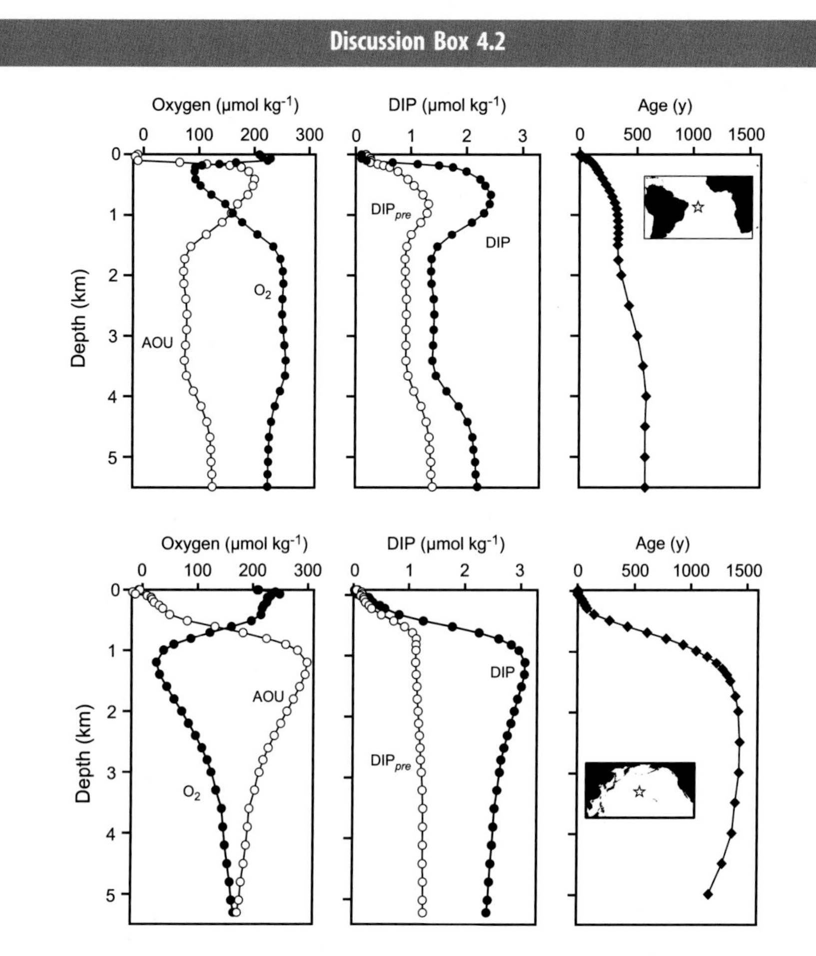

These figures show depth profiles of oxygen and AOU, phosphate and preformed phosphate, and water mass age at 10°S, 25°W in the South Atlantic (upper figures) and at 31°N, 179°E in the North Pacific (lower figures). Data come from the WOCE survey with age estimates from Gebbie and Huybers (2012).

- What controls the surface values of O_2 and AOU? What does a negative AOU indicate? What controls the surface values of DIP and preformed DIP?
- How does the depth of the maximum age compare to the depth of the oxygen minimum? Draw an approximate depth profile of respiration rate. Why is the oxygen minimum at the depth that it is (i.e., not shallower or deeper)?
- How does the shape of the total DIP profile compare to that of AOU for the same locations? How does the difference between preformed DIP and total DIP compare to AOU? What do these comparisons tell you about what controls the shape of the DIP profiles?
- What explains the shape of the preformed DIP profiles?
- AOU and O_2 are not quite mirror images of each other. What causes the subtle differences, particularly in the upper 1 km?

Two methods are used to study benthic respiration in the sediments: placing chambers on the sea floor and measuring the decrease in O_2 in the chambers and measuring O_2 gradients in sediment pore waters to calculate the flux of oxygen into the sediments by molecular diffusion. Benthic respiration rates determined by these methods have been compiled as a function of ocean depth and compared to water column respiration rates, determined from sediment traps, to determine the oxygen consumption that can be attributed to each. Jahnke and Jackson (1987) showed that below about 3 km the amount of O_2 consumed at the sediment–water interface dominates respiration in the deep ocean (Fig. 4.13). The reason for this is that the hypsometry or shape of the deep sea is such that the surface area to volume ratio increases rapidly below about 3 km (e.g., the ocean has a relatively flat bottom).

4.3 Respiration in the Absence of Oxygen

Roughly 90 percent of the organic matter that exits the euphotic zone is respired in the water column. Of the ~10 percent of the organic carbon flux that reaches the sea floor, only about 10 percent of that escapes oxidation and is permanently buried in the sediments. In the sediments of many regions, respiration of organic matter consumes all available oxygen, because the supply of oxygen is limited by slow molecular diffusion from the overlying water through the spaces between sediment grains (pore waters). Oxygen is also fully consumed in the water columns of some coastal fjords and basins where ventilation of deep waters is blocked by a sill or in regions of the open ocean with high respiration rates and sluggish circulation (Fig. 4.3).

As we discussed at the beginning of Chapter 3, oxygen accepts electrons from organic matter during respiration (Eq. 3.3). There are a number of other oxidants in addition to oxygen that are thermodynamically capable of oxidizing organic matter (i.e., the net reaction results in a decrease in free energy). There are many possible oxidizing compounds (see, e.g., Stumm and Morgan, 1995), but the electron acceptors that are most important in the ocean environment, because of their high concentrations, are: $O(0)_2$, $N(V)O_3^-$, Mn(IV), Fe(III), $S(VI)O_4^{2-}$, and organic matter itself during fermentation. (Roman numerals after the above elements indicate oxidation state.) Standard free energy changes for the oxidation half-reaction for organic matter, and the reduction half-reactions for the above electron accepters, are tabulated in Emerson and Hedges (2008, Chapter 3, Table 3.7).

Studies of pore waters in deep-sea sediments (e.g., Froelich et al., 1979) and anoxic basins (e.g., Reeburgh, 1980) have shown that there is an ordered sequence of redox reactions in which the reactions that produce the most energy occur first, and the zones of different reactions are physically separated. From these observations, one can sketch the order and shape of reactant profiles actually observed in sediment pore water chemistry (Fig. 4.14) using the stoichiometry of these different reactions (Table 4.2).

It was a surprise to environmental chemists that the order of this reaction sequence follows the free energy yield of each reaction because there is little relationship between

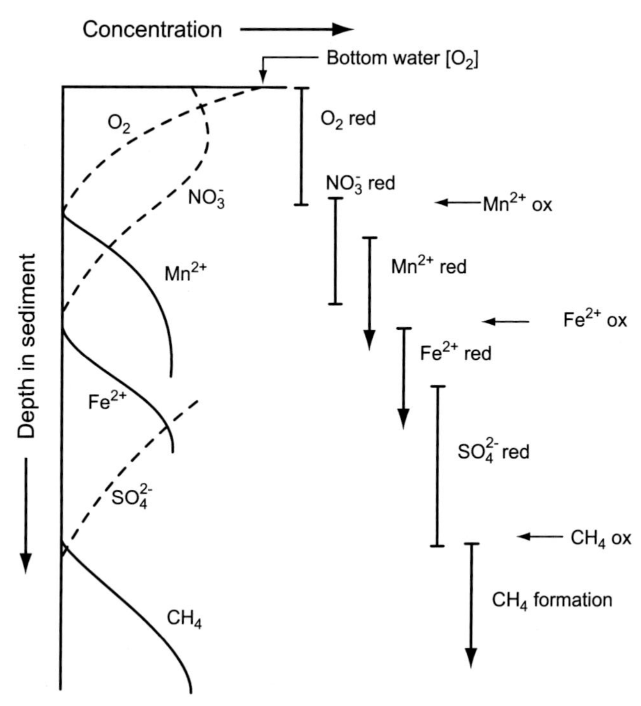

Figure 4.14 A schematic representation of pore water profiles that have been observed in deep-sea and near-shore sediments showing the sequential use of electron acceptors during organic matter respiration. Sediments at the base of this column are the oldest and have already experienced all the conditions above them. Modified from Froelich et al. (1979).

the free energy yield and the rate of a reaction in inorganic chemistry. However, all the organic matter respiration reactions are catalyzed by bacteria that harvest the energy produced for their own cellular processes. The reason that this reaction sequence follows the free energy yield is probably that bacteria are energy opportunists. Specific species of bacteria are able to catalyze only one or at most a very few of these reactions. The bacteria that can catalyze the most energetically favorable reaction, given the mix of available compounds present, outcompete the rest, leading to an ecological community dominated by organisms capable of that reaction. As each oxidant is consumed, the ecological community shifts to a mix of species able to carry out the next most energetically favorable reaction.

The schematic in Fig. 4.14 shows all electron acceptors in a single sequence. This is rarely observed in the environment because regions with abundant bottom water oxygen and moderate organic matter flux to the sediments (i.e., the deep ocean) frequently run out of reactive organic carbon before sulfate reduction becomes important, so pore-water profiles in these regions show no sulfate depletion or methane enrichment. In near-shore environments, where there is sufficient organic matter flux to the sediments to activate sulfate reduction and deplete sulfate in pore waters, zones of oxygen, nitrate, and Mn^{4+}

Table 4.2 Stoichiometry of organic matter (OM) oxidation reactions, listed in the order of most to least energetically favorable in the marine environment. Stoichiometric values are based on the need for 484 e^- to oxidize organic matter containing 106 C with the modified RKR stoichiometry in Chapter 3 and without nitrification.

Redox process	Reaction
Aerobic respiration	$C_{106}H_{179}O_{38}N_{16}P$ (OM) + 153 O_2 → $106CO_2 + 80H_2O + 16NO_3^- + HPO_4^{2-} + 18H^+$
Nitrate reduction (denitrification)	$C_{106}H_{179}O_{38}N_{16}P(OM) + 96.8NO_3^- + 110.8H^+ \rightarrow 48.4N_2 + 106CO_2 + 112.4H_2O + 16NH_4^+ + HPO_4^{2-}$
Manganese reduction	$C_{106}H_{179}O_{38}N_{16}P(OM) + 242\,MnO_2(s) + 498\,H^+ \rightarrow 242Mn^{2+} + 106CO_2 + 306H_2O + 16NH_4^+ + HPO_4^{2-}$
Iron reduction	$C_{106}H_{179}O_{38}N_{16}P(OM) + 242\,Fe_2O_3(s) + 982\,H^+ \rightarrow 484Fe^{2+} + 106CO_2 + 548H_2O + 16NH_4^+ + HPO_4^{2-}$
Sulfate reduction	$C_{106}H_{179}O_{38}N_{16}P(OM) + 60.5\,SO_4^{2-} + 74.5\,H^+ \rightarrow 60.5HS^- + 106CO_2 + 64H_2O + 16NH_4^+ + HPO_4^{2-}$
Methane production (methanogenesis)	$C_{106}H_{179}O_{38}N_{16}P(OM) + 57\,H_2O + 14\,H^+ \rightarrow 45.5CO_2 + 60.5CH_4 + 16NH_4^+ + HPO_4^{2-}$

reduction can be very thin, so that the changes in these species appear to overlap and obscure each other. The sequence of oxidation reactions is sometimes categorized into oxic diagenesis (O_2 reduction), followed by suboxic diagenesis (NO_3^- and Mn^{4+} reduction), and then anoxic diagenesis (Fe^{3+} reduction, SO_4^{2-} reduction, and methane formation). We do not use this terminology here, because the definition of suboxic is ambiguous. We recommend referring to these reactions using the true meaning of the terms – oxic for O_2 reduction and anoxic for the rest (anoxic-NO_3^- reduction, anoxic-Mn^{4+} reduction, and so forth).

The full, layered sequence of reactions is also rarely observed in water columns that are anoxic, like the Black Sea and many coastal fjords. Because mixing processes in the water column are much faster than molecular diffusion, the oxic and anoxic electron acceptors in the waters of these basins can be gathered closely together to form an oxic–anoxic interface where O_2 and H_2S nearly or actually do overlap. An example is the water column of Saanich Inlet, a fjord near Victoria, British Columbia, Canada, where water is trapped behind a sill and oxygen is totally depleted in the deeper waters over much of the year. Depth profiles of seven redox couples measured simultaneously in Saanich Inlet are plotted in Fig. 4.15. In general, the oxic form of the redox couples (O_2, NO_3^-, IO_3^-, Cr^{4+}, As^{5+}) exists above ~130 meters where the waters have measurable O_2 and the reduced form (NH_4^+, I^-, Cr^{3+}, As^{3+} along with H_2S, Mn^{2+}, Fe^{2+}) is most abundant below the depth where oxygen can be detected.

Vertical mixing in the water column or diffusion in pore waters transports reduced species upward into more oxidized regions and transfers oxidized species into the anoxic waters below. Reduced redox species can react with oxidized ones yielding energy that is used by bacteria to produce organic matter by a process called

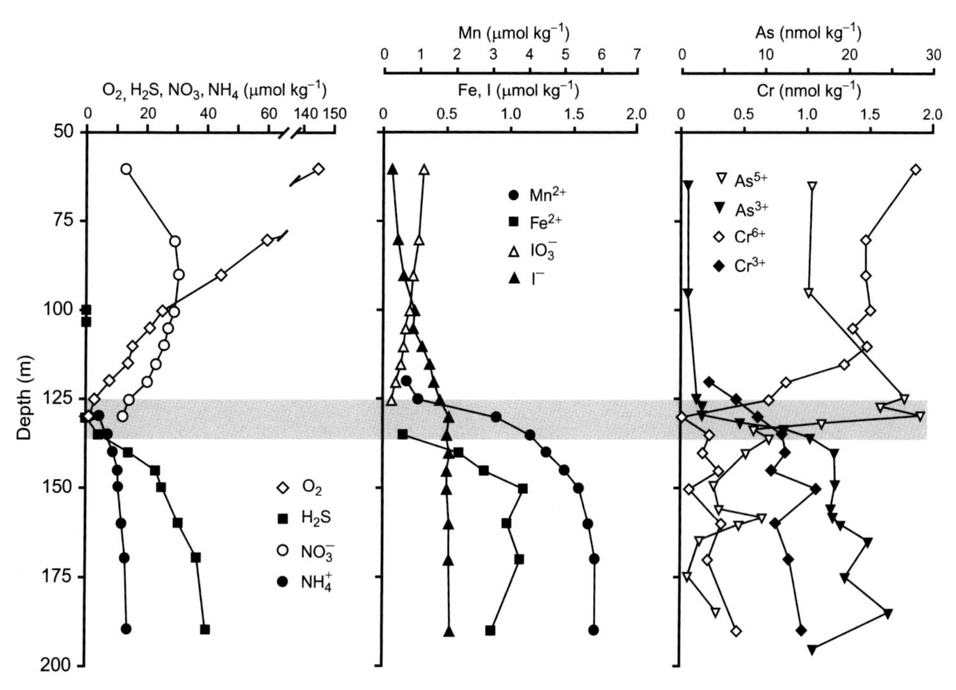

Figure 4.15 Concentrations of redox species O_2, NO_3^-, NH_4^+, H_2S, Fe^{2+}, Mn^{2+}, IO_3^-, I^-, Cr^{3+}, Cr^{6+}, As^{3+}, and As^{5+} with depth in the water column of Saanich Inlet near Victoria, BC. The redox species distribution indicates that oxidized forms are not found in waters containing H_2S and no O_2, but some reduced forms persist in O_2-containing waters. Modified from Emerson et al. (1979) and Peterson and Carpenter (1983).

chemosynthesis. One example of this in Saanich Inlet is the reduced form of Mn, Mn^{2+}, which is transported upward and oxidized to Mn^{3+} or Mn^{4+} by oxygen in the shallower water column. Nitrification, where NH_4^+ is oxidized by O_2 to NO_3^- (see Section 4.1.3), is another example of a redox reaction that fuels organic matter production. Many examples of these reactions are known to occur in the environment where oxidized and reduced species overlap, including the oxidation of Fe^{2+} by Mn^{4+} and the oxidation of methane by sulfate. Microorganisms capable of catalyzing these reactions are currently being discovered and cultured. Note, however, that the kinetics of these reactions vary greatly. Many of the reduced species in Fig. 4.15, have measurable concentrations in the oxic water (above 130 m), because the oxidation rates of these reduced species (Mn^{2+}, Cr^{3+}, and I^-) are slow compared with transport. In contrast, oxidation rates for Fe^{2+} are much faster, and very little overlap occurs between Fe^{2+} and O_2. The persistence of reduced species in oxic water, even though they are thermodynamically unstable, shows that many dissolved chemical species are not at chemical equilibrium in the environment.

4.4 Anthropogenic Influences

One of the main themes of this chapter has been that the delicate balance between organic matter respiration in the ocean thermocline and mixing of oxic waters from above (ventilation) creates a mid-depth minimum in oxygen concentration throughout most of the major ocean basins (Figs. 4.1 and 4.3). Because these processes of respiration, circulation, and ventilation operate over just decades in the upper ocean thermocline (see Discussion Box 4.1), measurements of the concentration of oxygen in the thermocline over the period of modern oceanography have the potential to record changes in these processes. Accurate oxygen measurements have been made in some locations in the Northern Hemisphere since the mid 1950s. In locations with a long-term time series, oxygen concentration measurements show a significant decline in the ocean's thermocline (e.g., Whitney et al., 2007 for the North Pacific; Stendardo and Gruber, 2012 for the North Atlantic; and Stramma et al., 2008 for the ocean's oxygen minimum zones in general). These records indicate different rates of ocean deoxygenation in different regions, but with most areas experiencing a downward trend, at least over the past 30 years. One of the longest and best constrained records is from the subarctic North Pacific, where oxygen in the thermocline is declining at a rate of about 0.5 $\mu mol\ kg^{-1}$ per year (Fig. 4.16). Global models that include warming trends have long predicted that the ocean should become more stratified in response to a warming Earth, which would reduce contact between thermocline waters and the atmosphere (ventilation). However, a recent assessment of the trends in the coastal environment of the eastern North Pacific (Crawford and Peña, 2016) suggests that, for this environment at least, the more recent decreases were preceded by gradually increasing oxygen concentrations. Whether or not the open ocean observations can be attributed to a natural decadal-scale cycle, as appears to be the case in the coastal environment, or to a long-term decrease in O_2 concentration in response to global warming remains to be seen with future measurements.

Observations also show declining oxygen concentrations in continental shelf environments near the mouth of major rivers but for very different reasons. These coastal "dead zones" are an unequivocal indicator of human influence on the ocean. One of the most striking examples is the large anoxic region in the Gulf of Mexico near the mouth of the Mississippi River (Fig. 4.17). In this area and in other coastal regions, fertilizer from croplands has been washed into the river and carried to the coastal environment. The increased nutrient loading of this low-density freshwater spreading out over the seawater causes huge blooms in the surface water, and the greater biological productivity drives an increased export of organic matter out of the near-shore euphotic zone. The increased supply of organic matter to deeper waters in turn fuels higher respiration rates that consume oxygen. The stratification caused by the freshwater reduces vertical mixing that might otherwise supply oxygen to these bottom waters. While these instances of eutrophication are specific to the coastal environment near rivers, they can affect large areas and do threaten the integrity of fish populations and thus the food supply from the ocean.

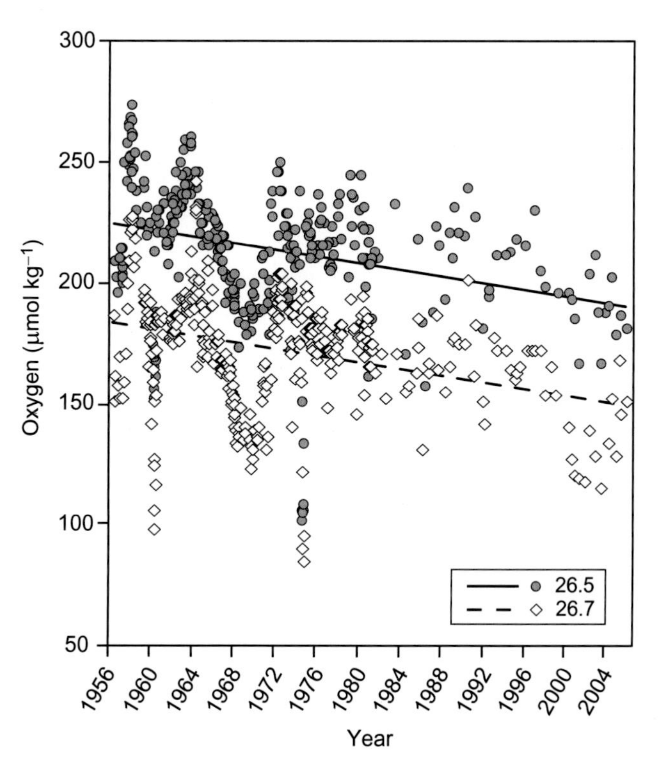

Figure 4.16 A time series of oxygen concentration on potential density surfaces $\sigma_\theta = 26.5$ and 26.7 (average depths of ~140 and ~168 m, respectively) at Ocean Station Papa (OSP, 50°N, 145°W) from 1956 to 2006. Despite the large variability, there is a clear trend of decreasing concentration with time. Data from Whitney et al. (2007).

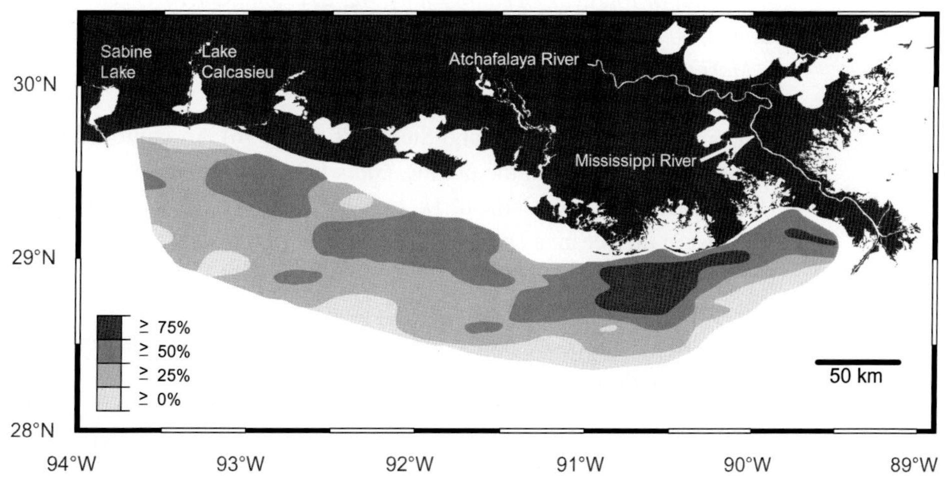

Figure 4.17 Spatial distribution of the percent of time that the coastal area experiences hypoxic conditions (defined as oxygen concentrations below 60 μmol kg^{-1}) in the northern Gulf of Mexico during the time period 1985–2001. The extent and frequency of hypoxic conditions in this area have increased since 1950 as the flux of nutrients carried to the ocean via the Mississippi River has increased threefold. Modified from Rabalais et al. (2002).

References

Broecker, W. S., T. Takahashi, H. J Simpson, and T-H. Peng (1985) Sources and flow patterns of deep ocean waters as deduced from potential temperature, salinity and initial phosphate concentration, *Journal of Geophysical Research Oceans*, **90**, 6925–6939.

Crawford, W. R., and M. A. Peña (2016) Decadal trends in oxygen concentration in subsurface waters of the northeast Pacific Ocean, *Atmosphere-Ocean*, **54**, 171–192, doi:10.1080/07055900.2016.1158145.

Deutsch, C., and T. Weber (2012) Nutrient ratios as a tracer and driver of ocean biogeochemistry, *Annual Review of Marine Science*, **4**, 113–141.

Doval, M. D., and D. A. Hansell (2000) Organic carbon and apparent oxygen utilization in the western South Pacific and the central Indian Oceans, *Marine Chemistry*, **68**, 249–264.

Emerson, S. (2014) Annual net community production and the biological carbon flux in the ocean, *Global Biogeochemical Cycles*, **28**, 14–28, doi:10.1002/2013GB004680.

Emerson, S., and J. Hedges (2008) *Chemical Oceanography and the Marine Carbon Cycle*, 453 pp. Cambridge: Cambridge University Press.

Emerson, S., R. E. Cranston, and P. S. Liss (1979) Redox species in a reducing fjord: equilibrium and kinetic considerations, *Deep-Sea Research A*, **26**, 859–878.

Froelich, P., G. P. Klinkhammer, M. L. Bender, et al. (1979) Early oxidation of organic matter in pelagic sediments of the eastern equatorial Atlantic: Suboxic diagenesis, *Geochimica et Cosmochimica Acta*, **43**, 1075–1090.

Gebbie, G., and P. Huybers (2012) The mean age of ocean waters inferred from radiocarbon observations: sensitivity to surface sources and accounting for mixing histories, *Journal of Physical Oceanography*, **42**, 291–305, doi:10.1175/JPO-D-11-043.

Gruber, N. (2008) The marine nitrogen cycle: overview and challenges, in *Nitrogen in the Marine Environment*, 2nd ed. (eds. D. Capone, D. Bronk, M. Mulholland, and E. Carpenter), pp. 1–50. New York: Elsevier.

Gruber N., Sarmiento J. L., and Stocker T. F. (1996) An improved method for detecting anthropogenic CO_2 in the oceans, *Global Biogeochemical Cycles*, **10**, 809–837.

Hansell, D. A., C. A. Carlson, D. J. Repeta, and R. Schlitzer (2009) Dissolved organic matter in the ocean: a controversy stimulates new insights, *Oceanography*, **22**, 202–211.

Ito, T., M. J. Follows, and E. A. Boyle (2004) Is AOU a good measure of respiration in the oceans? *Geophysical Research Letters*, **31**, doi:10.1029/2994GL020900.

Jahnke, R. J., and G. A. Jackson (1987) Role of sea floor organisms in oxygen consumption in the deep North Pacific Ocean, *Nature*, **329**, 621–623.

Jenkins, W. J. (1998) Studying subtropical thermocline ventilation and circulation using tritium and ^3He, *Journal of Geophysical Research Oceans*, **103**, 5,817–15,831.

Jenkins, W. J. (2002) Tracers of ocean mixing, in *Treatise on Geochemistry*, Vol. 6 (ed. H. Elderfield), pp. 223–246. New York: Elsevier.

Jenkins, W. J., and S. C. Doney (2003) The subtropical nutrient spiral, *Global Biogeochemical Cycles*, **17**, 1110, doi:10.1029/2003GB002085.

Johnson, G. (2008) Quantifying Antarctic Bottom Water and North Atlantic deep water volumes, *Journal of Geophysical Research Oceans*, **113**, C05027, doi:10.1029/2007/JC004477.

Kwon, E. Y., F. Primeau, and J. L. Sarmiento (2009) The impact of remineralization depth on the air-sea carbon balance, *Nature Geoscience*, **2**, 630–635, doi:10.1038/NGE0612.

Letscher, R. T., and J. K. Moore (2015) Preferential remineralization of dissolved organic phosphorus and non-Redfield DOM dynamics in the global ocean: Impacts on marine productivity, nitrogen fixation, and carbon export, *Global Biogeochemical Cycles*, **29**, 325–340.

Paytan, A., and K. McLaughlin (2007) The oceanic phosphorus cycle, *Chemical Reviews*, **107**, 563–576.

Peterson, M. L., and R. Carpenter (1983) Biogeochemical processes affecting total arsenic and arsenic species distributions in an intermittently anoxic fjord, *Marine Chemistry*, **12**, 295–321.

Rabalais, N. N., R. E. Turner, and W. J. Wiseman Jr (2002) Gulf of Mexico hypoxia, aka "The dead zone," *Annual Review of Ecology and Systematics*, **33**, 235–263.

Redfield, A. C. (1958) The biological control of chemical factors in the environment, *American Scientist*, **46**, 205–221.

Reeburgh, W. S. (1980) Anaerobic methane oxidation: rate depth distributions in Skan Bay sediments, *Earth and Planetary Science Letters*, **47**, 345–352.

Sonnerup, R. E., S. Mecking, and J. L. Bullister (2013) Transit time distributions and oxygen utilization rates in the Northeast Pacific Ocean from chlorofluorocarbons and sulfur hexafluoride, *Deep-Sea Research I*, **72**, 61–71, doi:10.1016/j.dsr.2012.10.013.

Sonnerup, R. E., S. Mecking, J. L. Bullister, and M. J. Warner (2015) Transit time distributions and oxygen utilization rates from chlorofluorocarbons and sulfur hexafluoride in the Southeast Pacific Ocean, *Journal of Geophysical Research Oceans*, **120**, 3761–3776, doi:10.1002/2015JC010781.

Stendardo, I., and N. Gruber (2012) Oxygen trends over five decades in the North Atlantic, *Journal of Geophysical Research Oceans*, **117**, 11004, doi:10.1029/2012JC007909.

Stramma, L., G. C. Johnson, J. Sprintall, and V. Mohrholz (2008) Expanding oxygen-minimum zones in the tropical oceans, *Science*, **320**, 655–658, doi:10.1126/science.1153847.

Stumm, W., and J. J. Morgan (1995) *Aquatic Chemistry: Chemical Equilibria and Rates in Natural Waters*, 3rd ed., 1040 pp. New York: Wiley.

Whitney, F. A., H. J. Freeland, and M. Robert (2007) Persistently declining oxygen levels in the interior waters of the eastern subarctic Pacific, *Progress in Oceanography*, **75**, 179–199, doi:10.1016/j.pocean.2007.08.007.

Williams, P. J. leB. and D. A. Purdie (1991) In vitro and in situ derived rates of gross production, net community production and respiration of oxygen in the oligotrophic subtropical gyre of the North Pacific Ocean, *Deep-Sea Research A*, **38**, 891–910.

Wolf, M. K., Hamme, R. C., Gilbert, D., Yashayaev, I., and Thierry, V. (2018) Oxygen saturation surrounding deep water formation events in the Labrador Sea from Argo-O_2 data, *Global Biogeochemical Cycles*, **32**, doi:10.1002/2017GB005829.

Problems for Chapter 4

4.1. The table below shows temperature, salinity, and age data at different depths from a station in the Pacific as well as globally averaged OUR estimates for the same depths.

a. Calculate the expected dissolved oxygen concentrations based on the data in the table. Make plots showing depth profiles of the OUR, water mass age, the calculated oxygen concentration, and any other relevant properties.

b. Explain why the distribution of oxygen varies in the way that it does with depth. Your explanation should include answers to the following questions: What processes are responsible for high or low values at the surface? What processes are responsible for maxima or minima in the profiles? How do those processes interact with each other to create the maxima/minima? Why aren't those maxima/minima at shallower or deeper depths?

Depth (m)	pot T (°C)	Salinity (PSS-78)	OUR (μmol kg^{-1} y^{-1})	Water mass age (y)
144	14.329	34.5192	20.00	2
591	6.193	34.0100	2.60	65
991	3.384	34.3234	0.90	330
1 484	2.517	34.5416	0.20	1 400
2 468	1.510	34.6449	0.12	2 000
5 393	1.048	34.6916	0.10	1 750

4.2. Add oxygen gas (O_2) to the two-box model. Air–sea gas exchange keeps the average oxygen concentration of the surface ocean close to equilibrium, which can be calculated using a temperature of 2°C and a salinity of 35. The mass balance for oxygen gas is different from the previous elements we have looked at, because it is not carried to the deep ocean by particles, but is consumed in the deep ocean by respiration of those particles. Use values from Chapter 3 for necessary constants like the deep ocean DIP concentration and flow rates.

a. Starting with the mass balance for phosphate, determine the average concentration of oxygen in the deep ocean for three different surface water DIP concentrations of 0, 0.5, and 2.3 μmol kg^{-1}.

b. Compare your answer to the deep ocean oxygen concentrations implied in Table 4.1 and observed in Fig. 4.1. What accounts for the differences?

4.3. Below is a plot of oxygen and phosphate versus latitude on an isopycnal with a potential density of 1025.5 kg m^{-3} along a north–south transect at ~150°W in the North Pacific.

a. Calculate the change in oxygen and the change in phosphate as water moves from the outcrop region near 32°N down along the isopycnal to 10°N. Be clear about the sign of the change.

b. This isopycnal has a mean potential temperature of 15.5°C and salinity of 34.7. Calculate the AOU at the outcrop region (32°N) and at 10°N. Compare the AOU change from 32°N to 10°N to the oxygen change calculated in part a.

c. Based on the AOU change determined in part b, calculate the expected contribution of respiration to the increase in phosphate as the water travels from 32°N to 10°N. Compare this expected phosphate change to the actual phosphate change determined in part a. What could account for any differences?

d. The CFC-11 data indicate that it takes water about 25 years to travel from 32°N to 10°N along this isopycnal. Calculate the OUR on this isopycnal.

e. Nitrate along this same isopycnal changes from 2 μmol kg^{-1} at 32°N to 17 μmol kg^{-1} at 10°N. Calculate N* at the 32°N outcrop and at 10°N. Explain what likely causes such values.

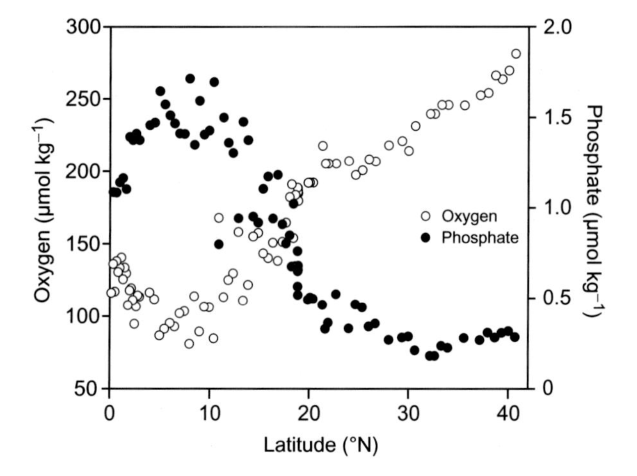

4.4. The following are placed into a sealed reaction chamber: 4.0 mmol of organic carbon in the form of labile organic matter with the usual marine (improved RKR) stoichiometry, 600 μmol of MnO_2, 400 μmol of Fe_2O_3, some bacteria, and 5 kg of seawater with the following concentrations: 260 μmol of kg^{-1} O_2, 30 μmol kg^{-1} NO_3^-, and 28.2 mmol kg^{-1} SO_4^{2-}. The container is sealed so that there is no air inside and placed on a stir plate in a dark room. Quantitatively describe what happens in the container over time to the concentrations of each species named above. Include a calculation of how much organic matter (in terms of total mmol of organic carbon) is still present at each stage of the experiment.

4.5 Apply the rules of balancing redox reactions (end of Section 3.1.1) to other important marine oxidants and reactions presented in this chapter.

a. First, for NO_3^-, for MnO_2, for Fe_2O_3, and for SO_4^{2-}, write half-reactions for the reduction of each oxidant. Then, write balanced whole reactions for the oxidation of CH_2O (simplified organic carbon) by each of these oxidants, showing how to combine half-reactions to achieve the complete whole reaction.

b. Write balanced whole reactions for the oxidation of Mn^{2+} by O_2 and for the oxidation of NH_4^+ to NO_3^- by O_2 (nitrification), showing how to combine half-reactions to achieve the complete whole reaction.

4.6 The decrease in oxygen concentration in waters with a potential density of 1026.7 kg m^{-3} has been measured at Ocean Station P (OSP) at 50°N 145°W in the subarctic NE Pacific to be 0.68 μmol kg^{-1} y^{-1} over the 50 years from the mid-1950s to the mid-2000s (Fig. 4.16). In the 1950s, these waters had an average oxygen concentration of 180 μmol kg^{-1}, temperature of 4.35°C, and salinity of 33.69. The average transport time from the ventilation region to OSP on this isopycnal is ~7 y, while the distance traveled is approximately 4500 km. To evaluate the potential causes of ocean deoxygenation, calculate how large a change would be required in different processes for each to be responsible for the entire observed oxygen decline.

a. First, suppose that increasing water temperature is the only factor driving change (i.e., assume that the water in the ventilation region remains at equilibrium with the atmosphere and that the AOU remains constant). Calculate how much warming would be necessary to cause the entire observed change in oxygen concentration over 50 y. Compare your value to the observed warming rate of 0.012°C y^{-1} in these waters.

b. Second, suppose instead that changing ocean circulation rates are responsible for the change in oxygen concentration. Calculate the velocity of the water as it moves along this isopycnal from the values above. Calculate how much greater the age (travel time from the ventilation region to OSP) of the water would need to be to cause the entire observed change in oxygen concentration over 50 y. What would the change in velocity of the water need to be? Would the circulation rate need to slow down or speed up?

c. Third, suppose instead that changing OUR is responsible for the change in oxygen concentration, i.e., that there are changes in the carbon export rate and the biological pump. Determine the present OUR from the surface outcrop, where it is in equilibrium with the atmosphere, to OSP and calculate how large a change in OUR would be necessary to cause the entire observed change in oxygen concentration over 50 y. Would biological productivity need to increase or decrease to cause such a change?

5 Marine Carbonate Chemistry

5.1 Acids and Bases 178
5.1.1 The Chemical Equilibrium Constant 178
5.1.2 Hydrogen Ion Exchange 180
5.1.3 Acids and Bases in Seawater 182
5.1.4 The Alkalinity of Seawater 187
Discussion Box 5.1 191
5.2 Calculating Carbonate Equilibria and pH 191
5.3 Processes that Control Alkalinity and DIC of Seawater 194
5.3.1 Terrestrial Weathering and River Inflow 194
5.3.2 Alkalinity and DIC Changes within the Ocean 195
Discussion Box 5.2 204
5.4 Mechanisms of Calcium Carbonate Dissolution and Burial 205
5.4.1 Thermodynamic Equilibrium 206
5.4.2 The Kinetics of $CaCO_3$ Dissolution 211
5.5 Anthropogenic Influences 214
Discussion Box 5.3 218
Appendix 5A.1 Carbonate System Equilibrium Equations in Seawater 220
References 221
Problems for Chapter 5 223

Carbonate chemistry is a central component of the chemical perspective of oceanography, primarily because it controls the acidity of seawater and governs the global carbon cycle. Within the mix of acids and bases in the Earth-surface environment, the carbonate system is the primary buffer for the acidity of seawater, which means that the carbonate system prevents seawater acidity from changing rapidly or by large amounts. Ocean acidity in turn determines the reactivity of chemical compounds and solids. The dissolution of the most prevalent authigenic mineral in the environment, $CaCO_3$, is a major source for dissolved carbon to the ocean over long timescales (Chapter 2), and its formation is a major sink. And finally, the carbonate system of the ocean exerts a dominant control over the pressure of carbon dioxide in the atmosphere and determines the rate at which the ocean absorbs anthropogenic CO_2, both of which play a primary role in regulating the temperature of the planet.

The dissolved compounds that make up dissolved inorganic carbon (DIC $= CO_2(aq)$, H_2CO_3, HCO_3^-, and CO_3^{2-}) are in chemical equilibrium on timescales longer than a few minutes. Although the heterogeneous equilibrium between carbonate solids and dissolved carbonate ions is not as rapid, to a first approximation $CaCO_3$ is found in marine sediments that are bathed by waters that are saturated or supersaturated thermodynamically and absent where waters are undersaturated.

Great strides have been made in understanding the thermodynamic equilibrium between carbonate species through the evolution of analytical techniques for measuring carbonate system constituents and thermodynamic equilibrium constants. During the first major global marine chemical expedition, Geochemical Sections (GEOSECS) in the 1970s, marine chemists achieved accuracies for dissolved inorganic carbon, DIC, and alkalinity of ±0.5–1 percent, and for the fugacity of CO_2, fCO_2^{sw}, of ±20 percent. The pH (the negative log of the hydrogen ion concentration) was a qualitative property at this time, because its accuracy was uncertain when measured by glass electrodes, which could not be adequately standardized. By the time of the chemical surveys of the 1980s and 1990s, the World Ocean Circulation Experiment (WOCE) and the Joint Global Ocean Flux Study (JGOFS), the accuracy of carbonate system measurements had improved dramatically to near the standards of today. Part of the improvement was due to new methods such as coulometry for DIC and spectrophotometry for pH. Another important advance was the development of certified, chemically stable DIC and alkalinity standards that resulted from both greater community organization and the where-with-all to make stable standards. These advances made it possible to determine DIC and alkalinity to within several tenths of one percent and fCO_2^{w} to within a couple of microatmospheres. This improvement in techniques has been accompanied by efforts to improve the accuracy of the equilibrium constants governing transformations among species.

Homogeneous chemical equilibria of reactions between carbonate species in water are the primary focus of the first section of this chapter. There are occasions when the kinetics of the reactions between CO_2 and H_2O or OH^- ions are important, and this subject is discussed in Emerson and Hedges (2008). We then move on from carbonate equilibria to discuss processes that control the alkalinity and DIC of seawater and then to the relationship between dissolved carbonate chemistry and the presence of $CaCO_3$ in ocean sediments. Finally, we demonstrate how the anthropogenic absorption of CO_2 into the Earth's carbon cycle has influenced the fCO_2^{sw} and pH of ocean surface waters.

5.1 Acids and Bases

5.1.1 The Chemical Equilibrium Constant

A detailed description of the thermodynamics behind chemical equilibrium constants is presented in online Chapter 3 of Emerson and Hedges (2008; www.cambridge.org/emerson-hamme). We present a short introduction here because equilibrium constants are fundamental to the discussion of carbonate chemistry. The free energy of reaction, ΔG_r, for the generalized reaction in which reactants B and C are altered to products D and E (lowercase letters represent stoichiometric coefficients)

$$bB + cC \rightleftarrows dD + eE \tag{5.1}$$

is

$$\Delta G_r = \Delta G_r^0 + RT\ \ln \frac{(D)^d(E)^e}{(B)^b(C)^c}, \tag{5.2}$$

where R is the universal gas constant = 0.008314 kJ mol^{-1}K^{-1} (see Appendix B) and T is the temperature in kelvin (K = °C + 273.15). The standard free energy of reaction, ΔG_r^0, is determined from the sum of the standard free energies of formation for each of the reactants and products. These values have been determined in the laboratory for every dissolved ion, compound, and solid form of each element and are listed in tables of thermodynamic properties. Parentheses in the last term on the right-hand side of Eq. 5.2 indicate *activities* of the compounds in solution, and the quotient of activities raised to the power of their reaction stoichiometries is referred to as the *ion activity product*, Q. Converting the natural logarithm to base 10 ($\ln(z) = 2.30\ \log_{10}(z)$) gives

$$\Delta G_r = \Delta G_r^0 + 2.3RT\log_{10}Q. \tag{5.3}$$

A negative ΔG_r value indicates that the reaction can occur spontaneously as written, whereas a positive value indicates that the spontaneous reaction of reactants to products is energetically impossible. A ΔG_r value of zero indicates that the chemical system is in a state of *equilibrium*, where no free energy is available and no *net* reaction possible. In this case, the ion activity product is equal to a constant value that is referred to as the thermodynamic equilibrium constant, K. Thus,

$$\Delta G_r^0 = -2.3RT\log_{10}K \text{ at } \Delta G_r = 0. \tag{5.4}$$

This relationship is used to determine equilibrium constants for reactions in nature.

A useful measure of whether a reaction can spontaneously proceed to the right is the ratio Q/K, which we designate with the Greek letter Ω. This value designates whether there is a deficit ($\Omega < 1$) or surplus ($\Omega > 1$) of products versus reactants in the chemical system. The three conditions that can be determined from thermodynamic equilibrium calculations (supersaturation, equilibrium, and undersaturation) and the corresponding relationships of Q, K, and Ω, are summarized in Table 5.1. The main caveat to keep in mind when making predictions for the feasibility of a chemical reaction is that a thermodynamically favored reaction may not in fact occur in nature because the reaction rate is too slow, and reaction rate kinetics are frequently not predictable. Thus, the most definitive application of the thermodynamic equilibrium is to identify chemical reactions that cannot occur spontaneously as written at the given reactant and product activities or concentrations.

Table 5.1 Relationships among ΔG_r, Q, K, and Ω values

State	ΔG_r	Q vs. K	Ω	Prediction
Reactant excess	<0	$Q < K$	<1	Reaction *might* occur as written
At equilibrium	0	$Q \equiv K$	1	No net reaction can occur
Product excess	>0	$Q > K$	>1	Reverse reaction *might* occur as written

The thermodynamic equilibrium constant, K, is defined above as the quotient of the activities at equilibrium of the individual species or solids raised to the power of their reaction stoichiometries. Activity is a measure of the true reactivity of a chemical species, which is different from its actual concentration because of interfering interactions with other ions in the solution. The activity of a chemical species, C, is related to its concentration, [C], via an activity coefficient, γ_C:

$$(\mathrm{C}) = \gamma_{\mathrm{C}}[\mathrm{C}]. \tag{5.5}$$

Thus,

$$K = \frac{(\mathrm{D})^d(\mathrm{E})^e}{(\mathrm{B})^b(\mathrm{C})^c} = \frac{\gamma_{\mathrm{D}}{}^d[\mathrm{D}]^d\gamma_{\mathrm{E}}{}^e[\mathrm{E}]^e}{\gamma_{\mathrm{B}}{}^b[\mathrm{B}]^b\gamma_{\mathrm{C}}{}^c[\mathrm{C}]^c}. \tag{5.6}$$

In concentrated solutions, like seawater, activities are normally less than concentrations (with activity coefficients less than 1), because interactions among the reacting ions and other ions in solution reduce interactions that create reactions. (These concepts are discussed in detail in online Chapter 3 of Emerson and Hedges (2008; www.cambridge.org/emerson-hamme). However, since the relative concentrations of the major ions in seawater are nearly constant (Chapter 1), oceanographers determine equilibrium constants in laboratory seawater solutions with chemistries that represent more than 99 percent of the ocean. The equilibrium constants derived in these experiments are called *apparent* equilibrium constants and are indicated as K'. The apparent equilibrium constant of Eq. 5.1 is related to the concentrations in solution instead of the activities:

$$K' = \frac{[\mathrm{D}]^d[\mathrm{E}]^e}{[\mathrm{B}]^b[\mathrm{C}]^c} = K\frac{\gamma_{\mathrm{B}}{}^b\gamma_{\mathrm{C}}{}^c}{\gamma_{\mathrm{D}}{}^d\gamma_{\mathrm{E}}{}^e}. \tag{5.7}$$

Apparent equilibrium constants are particularly useful because concentrations can be directly measured with greater accuracy than the activity coefficients can be determined. Apparent equilibrium constants for many reactions have been measured as a function of temperature, salinity, and pressure in the seawater medium. With this approach one forgoes attempts to understand the interactions occurring among the ions in solution for a more empirical, but also more accurate, description of chemical equilibria.

5.1.2 Hydrogen Ion Exchange

Whether a specific acid-base pair is important in regulating the pH of seawater depends on the concentrations and equilibrium constants that govern its reactions. Here, we review acid-base equilibria (hydrogen ion exchange) and its relation to pH ($\mathrm{pH} = -\log[\mathrm{H}^+]$) for a generic acid (HBa) and its conjugate anion base (Ba^-). For this system, the hydrogen ion exchange reaction, the apparent equilibrium constant associated with it, and the definition of the total concentration of the protonated and unprotonated forms in solution, $\mathrm{Ba_T}$, are

$$\mathrm{HBa} \rightleftarrows \mathrm{H}^+ + \mathrm{Ba}^-, \tag{5.8}$$

$$K' = \frac{[H^+] \times [Ba^-]}{[HBa]}, \tag{5.9}$$

$$Ba_T = [HBa] + [Ba^-]. \tag{5.10}$$

Combining Eqs. (5.9) and (5.10) results in expressions for the concentration of the acid, HBa, and its conjugate base, Ba^-, as functions of the apparent equilibrium constant, K', and the hydrogen ion concentration, $[H^+]$:

$$[HBa] = \frac{Ba_T \times [H^+]}{K' + [H^+]} \text{ or } \log[HBa] = \log Ba_T + \log[H^+] - \log(K' + [H^+]) \tag{5.11}$$

and

$$[Ba^-] = \frac{Ba_T \times K'}{K' + [H^+]} \text{ or } \log[Ba^-] = \log Ba_T + \log K' - \log(K' + [H^+]). \tag{5.12}$$

Equations are presented in logarithmic form because the relationships are valid for very large concentration ranges and the chemical species central to discussion in nature is the hydrogen ion $[H^+]$, which is presented as a logarithmic value ($pH = -\log[H^+]$). A plot of these logarithmic equations (Fig. 5.1), for an arbitrary choice for Ba_T and K', illustrates that the concentration of the acid, HBa, dominates below $pH = pK'$ (on the acidic side), and that the concentration of the conjugate base, Ba^-, dominates above $pH = pK'$ (the basic side). ($pK' = -\log_{10}(K')$ in the same way that $pH = -\log_{10}[H^+]$.) At a pH equal to pK', $[H^+] = K'$, and the concentrations of the acid and basic forms are equal, $[HBa] = [Ba^-]$.

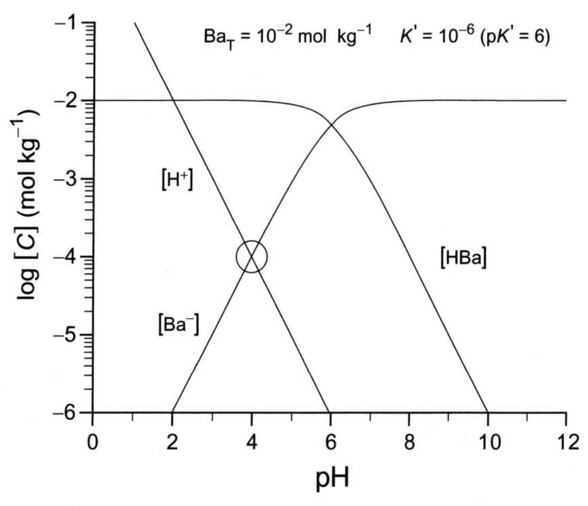

Figure 5.1 Concentrations of the acidic [HBa] and basic $[Ba^-]$ forms of a generic acid, with total concentration $Ba_T = 10^{-2}$ mol kg^{-1} and an equilibrium constant $K' = 10^{-6}$, as a function of pH. The concentrations are equal at the point where $pH = pK'$. When the criteria of charge balance between the ions in solution (H^+ and Ba^-) is included in the equations, the system is defined at a single pH where $[H^+] = [Ba^-]$ indicated by the circle. Both axes are logarithmic scales. Modified from Emerson and Hedges (2008).

Table 5.2 Compounds that exchange protons in the pH range of seawater. Equilibrium constants are for a temperature of 25°C and a salinity of 35 from the equations in Appendix F and from Millero (1995) for nitrogen and sulfur species. An asterisk (*) indicates the concentration is in the μmol kg^{-1} range and variable. ($pK' = -\log_{10}K'$)

		Concentration		
Species	Reaction	(mol kg^{-1})	$-\log_{10}C_T$	pK'
H_2O	$H_2O \rightleftarrows OH^- + H^+$			13.2
DIC	$CO_2 + H_2O \rightleftarrows HCO_3^- + H^+$	$\approx 2.04 \times 10^{-3}$	2.69	5.85
	$HCO_3^- \rightleftarrows CO_3^{2-} + H^+$			8.97
B	$B(OH)_3 + H_2O \rightleftarrows B(OH)_4^- + H^+$	4.16×10^{-4}	3.38	8.60
Si	$H_4SiO_4 \rightleftarrows H_3SiO_4^- + H^+$	*	*	9.38
P	$H_3PO_4 \rightleftarrows H_2PO_4^- + H^+$	*	*	1.61
	$H_2PO_4^- \rightleftarrows HPO_4^{2-} + H^+$	*	*	5.96
	$HPO_4^{2-} \rightleftarrows PO_4^{3-} + H^+$	*	*	8.79
SO_4^{2-}	$HSO_4^- \rightleftarrows SO_4^{2-} + H^+$	2.824×10^{-2}	1.55	1.00
F	$HF \rightleftarrows F^- + H^+$	6.8×10^{-5}	4.17	2.64
Anoxic Water				
N	$NH_4^+ \rightleftarrows NH_3 + H^+$	*	*	9.19
HS^-	$H_2S \rightleftarrows HS^- + H^+$	*	*	6.98

The final constraint in the chemical equilibria equations is that of charge balance, which in this simple solution containing only this one acid-base pair involves the only two ions:

$$0 = [H^+] - [Ba^-]. \tag{5.13}$$

This equation constrains the system to a single location on the plot (where the lines for these two concentrations cross in Fig. 5.1), which uniquely fixes the pH and concentrations of acids and bases in the system. In this simple system, the solution is acidic (pH = 4) because the concentration of the hydrogen ion and anion must be equal. As we will see, the situation for seawater is more complicated.

These simple equations and ideas provide the basis for describing the carbonate system in terms of the fCO_2^{sw}, DIC, pH, and alkalinity of seawater. We will build up a plot similar to that in Fig. 5.1 for the important acids and bases in seawater. These are listed along with their concentrations and apparent equilibrium constants in Table 5.2. We will then demonstrate how the additional constraint of alkalinity (related to the charge balance or buffer capacity) helps determine the pH of seawater.

5.1.3 Acids and Bases in Seawater

Factors that determine the importance of an acid-base pair in controlling the pH of seawater are its total concentration and pK' (Table 5.2). Carbonic and boric acids are the most concentrated hydrogen ion exchangers with pK' values near seawater pH. However, there are other, familiar strong acids in seawater like HCl and H_2SO_4, which play almost no role in regulating the pH because their pK' values are so low. The reason the ocean has high pH (~8.2), near the range of carbonate species pK's, is because of the abundance of carbon in the Earth system and the high reactivity of CO_2 (the residence time of HCO_3^- in the ocean is 0.1 My while that for SO_4^{2-} is 10 My and that for Cl^- is 100 My). Rocks are basic, so reactions with carbonic acid (weathering) produce river water with near neutral pH, far higher than the pK' of strong acids like HCl.

Carbonic Acid. Inorganic carbon exists in four distinct forms in water: the gas in solution or aqueous carbon dioxide, $CO_2(aq)$, and the three products of hydration reactions which are carbonic acid H_2CO_3, bicarbonate HCO_3^-, and carbonate CO_3^{2-}. Concentrations have units of moles kg^{-1}. Chemical equilibria among these species in seawater are described by the apparent constants which have units necessary to make the dimensions of the equilibrium expressions correct:

$$H_2CO_3 \rightleftarrows CO_2(aq) + H_2O \qquad K'_{CO_2(aq)} = \frac{[CO_2(aq)]}{[H_2CO_3]}, \tag{5.14}$$

$$H_2CO_3 \rightleftarrows HCO_3^- + H^+ \qquad K'_{H_2CO_3} = \frac{[HCO_3^-] \times [H^+]}{[H_2CO_3]}, \tag{5.15}$$

$$HCO_3^- \rightleftarrows CO_3^{2-} + H^+ \qquad K'_2 = \frac{[CO_3^{2-}] \times [H^+]}{[HCO_3^-]}, \tag{5.16}$$

where the equilibrium constant, K'_2, indicates the second dissociation constant of carbonic acid (Appendix F). Because only a few tenths of 1 percent of the neutral dissolved carbon dioxide species exists as H_2CO_3 at equilibrium, these neutral species are usually combined and represented with either the symbol $[CO_2]$ or H_2CO_3*. We use the former here:

$$[CO_2] = [CO_2(aq)] + [H_2CO_3] \tag{5.17}$$

Equations (5.14) and (5.15) can be combined to eliminate $[H_2CO_3]$ and give a new composite first dissociation constant of CO_2 in seawater; K'_1:

$$CO_2 + H_2O \rightleftarrows HCO_3^- + H^+ \qquad K'_1 = \frac{[HCO_3^-][H^+]}{[CO_2]}. \tag{5.18}$$

The relationship among the equilibrium constants, K'_1, $K'_{CO_2(aq)}$, and $K'_{H_2CO_3}$ is derived by combining Eqs. (5.14), (5.15), and (5.17):

$$K'_1 = \frac{K'_{H_2CO_3}}{K'_{CO_2(aq)} + 1}. \tag{5.19}$$

K'_1 is the value measured by laboratory experiments (Appendix F), so analytical measurements and theoretical equilibrium descriptions are consistent. Because $K'_{CO_2(aq)} \gg 1$ (the thermodynamic value for $K'_{CO_2(aq)}$ is 350 to 990, Stumm and Morgan, 1996), we can make a further approximation that

$$K'_1 \approx \frac{K'_{H_2CO_3}}{K'_{CO_2(aq)}}. \tag{5.20}$$

At equilibrium, the gaseous CO_2 in the atmosphere, expressed in terms of the fugacity, fCO_2^a (in atmospheres, atm), is related to the aqueous CO_2 in seawater, $[CO_2]$ (mol kg^{-1}), via the Henry's law coefficient, K_H (mol kg^{-1} atm^{-1}; see Chapter 2):

$$K_{H,CO_2} = \frac{[CO_2]}{fCO_2^a}. \tag{5.21}$$

Fugacity is related to partial pressure in a gas in the same way that activity and concentration are related in solutions – interactions among the gases inhibit the full partial pressure from being realized. Weiss (1974) has shown that fCO_2 is only 0.3–0.5 percent different from its partial pressure, pCO_2, for a temperature range of 0–30°C. Because the differences are small, fugacity and partial pressure are sometimes used interchangeably.

With the above equilibria, we are now prepared to define the total concentration of dissolved inorganic carbon and construct a diagram of the variation of the carbonate species concentrations as a function of pH (Fig. 5.2). For simplicity, we begin by assuming that the water is isolated from the atmosphere, so Eq. 5.21 is not necessary to describe the chemical equilibria in this example. The total concentration, C_T, of inorganic carbon in seawater is the dissolved inorganic carbon (DIC) (sometimes called the total CO_2 or ΣCO_2). DIC is the sum of the concentrations of the dissolved inorganic carbon species:

$$\mathrm{DIC} = \left[HCO_3^-\right] + \left[CO_3^{2-}\right] + [CO_2]. \tag{5.22}$$

Since DIC is a total quantity, it has the advantage that it is independent of temperature and pressure unlike the concentrations of its constituent species. For example, as a water sample is brought from the deep ocean to the surface, the concentrations of individual carbonate species will change because temperature and pressure changes affect the equilibrium constants, but its DIC will remain constant. Experimentally, DIC is determined by acidifying the sample, so that all the HCO_3^- and CO_3^{2-} react with H^+ to become CO_2 and H_2O, and then measuring the amount of evolved CO_2 gas. To create a plot of the concentrations of the three dissolved carbonate species as a function of pH, we assign the DIC a value near its average in seawater $\left(\mathrm{DIC} = 2000\ \mu\mathrm{mol\ kg}^{-1}\right)$. Combining Eq. 5.22 with Eqs. 5.16 and 5.18 yields separate equations for the carbonate species as a function of the equilibrium constants, DIC, and pH:

$$[CO_2] = \frac{\mathrm{DIC}}{1 + \frac{K_1'}{[H^+]} + \frac{K_1'K_2'}{[H^+]^2}}, \tag{5.23}$$

$$\left[HCO_3^-\right] = \frac{\mathrm{DIC}}{\frac{[H^+]}{K_1'} + 1 + \frac{K_2'}{[H^+]}}, \tag{5.24}$$

$$\left[CO_3^{2-}\right] = \frac{\mathrm{DIC}}{1 + \frac{[H^+]^2}{K_1'K_2'} + \frac{[H^+]}{K_2'}}. \tag{5.25}$$

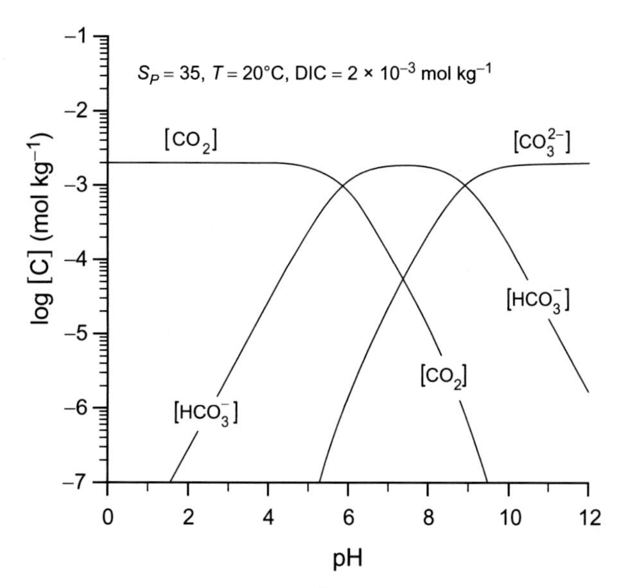

Figure 5.2 Concentrations of the carbonate species CO_2, HCO_3^-, and CO_3^{2-} as a function of pH at a DIC of 2.0×10^{-3} mol kg^{-1}, a temperature of 20°C, and a salinity of 35. (pK' values determined using the equations in Appendix F.)

The plot in Fig. 5.2 demonstrates the relative importance of the three species of DIC in seawater as a function of pH. At $\mathrm{pH} = \mathrm{p}K_1' = 5.85$ the concentrations of CO_2 and HCO_3^- are equal and at $\mathrm{pH} = \mathrm{p}K_2' = 8.97$ the concentrations of HCO_3^- and CO_3^{2-} are equal. At the pH of surface ocean waters of about 8.2, this diagram shows that the dominant carbonate species is HCO_3^-. We now know the ratios of carbonate species that make up DIC as a function of pH (Fig. 5.2), but we have not yet solved for the pH of seawater. Where the solution lies on this plot requires the consideration of charge balance, or the alkalinity, just as in the simple acid-base example. We will define alkalinity soon, but first let's consider the only other acid-base pair of significance in seawater.

Boric Acid. The carbonate system and boric acid are by far the most important contributors to the acid-base chemistry of seawater (Table 5.2). Borate, however, is far less reactive in seawater than carbon. Inorganic carbon is involved in all metabolic processes and varies in concentration geographically, while borate is conservative and maintains a constant ratio to salinity. This means that boric acid can be included in our description of seawater acid-base chemistry without additional measurements other than salinity. The equilibrium reaction and total boron, B_T, equations are:

$$\mathrm{B(OH)_3 + H_2O \rightleftarrows B(OH)_4^- + H^+} \qquad K_B' = \frac{[\mathrm{B(OH)_4^-}]\,[\mathrm{H^+}]}{[\mathrm{B(OH)_3}]}, \tag{5.26}$$

$$\mathrm{B_T} = [\mathrm{B(OH)_3}] + [\mathrm{B(OH)_4^-}]. \tag{5.27}$$

The equations for the boron species as a function of pH and K'_B (Appendix F) are thus:

$$[B(OH)_3] = \frac{B_T \times [H^+]}{[H^+] + K'_B}, \tag{5.28}$$

$$[B(OH)_4^-] = \frac{B_T \times K'_B}{[H^+] + K'_B}. \tag{5.29}$$

We can now construct a figure of both carbonate and borate species as a function of pH (Fig. 5.3), which reveals why boric acid plays a role as a pH buffer in seawater. The two species that exchange hydrogen ions (boric acid and the borate ion) have nearly equal concentrations at a pH between 8 and 9. One does not need a graph to determine this, since the two species that exchange hydrogen ions have equal concentration when the pK' = pH, which in this case is pH = 8.6 (Table 5.2). The species most involved in the buffer system of seawater are those that have curved lines in the pH range of 8 to 9 (Fig. 5.3). These are the points where seawater is most buffered, i.e., the regions where adding (or subtracting) H^+ ion is least likely to change the seawater pH because much of the added H^+ would combine with the basic species rather than increase the $[H^+]$ of the solution. The rest of the acids in seawater with pK' values in the vicinity of 8 to 9, silicic acid and phosphoric acid (Table 5.2), are less important because they have low and variable concentrations (1–200 μmol kg^{-1}), but they must be considered to have a complete representation of the acid-base components of seawater.

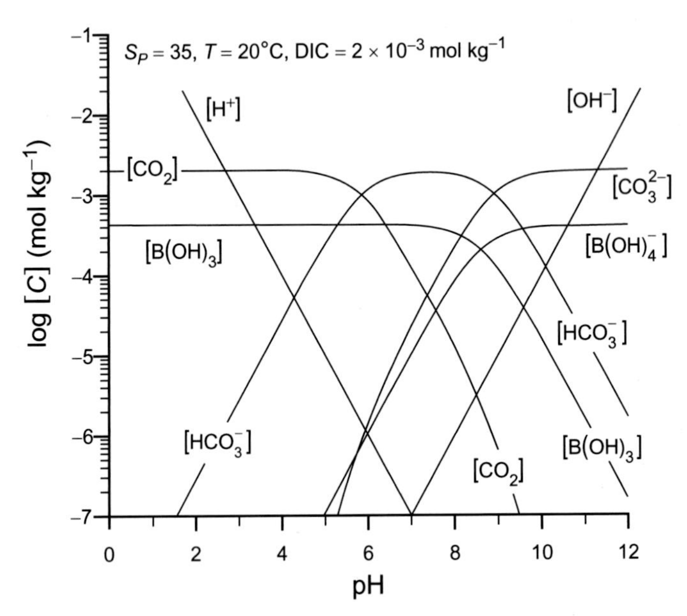

Figure 5.3 As in Fig. 5.2 but with the addition of the borate species $B(OH)_3$ and $B(OH)_4^-$. $B_T = 4.16 \times 10^{-4}$ mol kg^{-1}. The concentrations of H^+ and OH^- are added for reference using pK'_W = 13.2. (pK' values determined using the equations in Appendix F.) Modified from Emerson and Hedges (2008).

5.1.4 The Alkalinity of Seawater

To finish our understanding of the pH of seawater, we must deal with an extremely important but somewhat troublesome measure of the carbonate system – alkalinity. So far, we have presented a system of equilibrium equations that includes the equilibrium constants and total concentrations (Eqs. 5.16, 5.18, and 5.22 for the carbonate species; Eqs. 5.26 and 5.27 for borate) and which describes the predominant acid-base species over the entire pH range of seawater. However, none of these constants and concentrations constrains the specific position on the plot in Fig. 5.3. The location may be anywhere along the pH scale. For example, in the case of the carbonate system there are five unknowns, not including T, S, and pressure, (DIC, $[HCO_3^-]$, $[CO_3^{2-}]$, $[CO_2]$, and pH) but only three equations. If the concentration of DIC is measured, we are still one equation shy of being able to solve the system of equations uniquely and exactly define the pH. A measurement and equation that solves this problem is the expression for total alkalinity, A_T, which represents the charge balance or buffer capacity of the mixed electrolyte system of seawater. Just as the charge balance had to be identified in order to determine the pH at equilibrium on the simple acid-base plot in Fig. 5.1, so must the charge balance be evaluated to determine the pH at equilibrium on the acid-base plot for seawater (Fig. 5.3). The practical advantage for this new constraint is that it is measurable, and it is a total quantity like DIC, independent of temperature and pressure.

We present three different descriptions of alkalinity below, all of which are valid. Traditionally, Alkalinity has units of equivalents per kg (eq kg^{-1}) because it is a measure of charge balance, not carbon balance (e.g., Stumm and Morgan, 1996). In chemical oceanography literature, one often sees alkalinity with units of mol kg^{-1} instead. The reason for this is probably that, when calculating the concentration of different carbonate species, DIC and alkalinity are often added or subtracted, which is uncomfortable for quantities with different units. There are good reasons for accepting either units of moles or equivalents, which will be clearer as we describe alkalinity. We are going to stick with eq kg^{-1} here, but, since it does not matter quantitatively, there is no reason to be paranoid about it.

One way of thinking about alkalinity is by separating the anions that exchange protons from those that do not. Alkalinity is the difference in concentration (or the net charge) of cations and anions that do not exchange protons. One can calculate the alkalinity of standard seawater using the concentrations of conservative ions at a salinity of 35 presented in Table 1.4 and repeated in Table 5.3. The SO_4^{2-} and F^- ions are included among the species that do not exchange protons because the protonated forms in the pH range of seawater are so small that they are conservative to the five decimal places presented in the table. By this definition, the numerical value for total alkalinity, A_T, is equal to

$$\begin{aligned} A_T &= \text{Cation charge} - \text{Anion charge} = 0.60567 - 0.60325\ \left(\text{eq kg}^{-1}\right) \\ &= 0.00242\left(\text{eq kg}^{-1}\right). \end{aligned}$$

Note that if one added the molar concentrations rather than the equivalent concentrations (i.e., if one did not account for the charge on each ion) the numerical answer would be different and wrong.

Acids and bases that make up the total alkalinity must protonate in solution in a way that achieves charge balance. For example, the difference in equivalents evaluated in Table 5.3 determines the relative abundances of $[HCO_3^-]$ and $[CO_3^{2-}]$ that are required for charge balance. As the difference between A_T and DIC increases at constant DIC, there must be a higher carbonate concentration to achieve charge balance because CO_3^{2-} carries two equivalents and HCO_3^- only one.

A second conception of alkalinity is as the buffer capacity of seawater. In chemistry, a buffer is a solution that resists pH change, because chemical compounds in the solution combine with or release H^+, limiting the pH change. The alkalinity in a mixed electrolyte solution like seawater is the excess of bases (proton acceptors) over acids (proton donors) in the solution, i.e., a measure of the ability of the solution to react with H^+ added to the system. The sum of the bases that react with H^+ near seawater with pH of 8.2 is presented in Table 5.4 for surface seawater, which has low nutrient concentrations. In regions of the ocean where silicic acid and phosphate concentrations are measurable, they must also be included in the definition of total alkalinity since they can also accept H^+:

Table 5.3 Concentrations of cation and anion species that do not significantly exchange protons in the pH range of seawater ($S_P = 35$) in eq kg^{-1}. (From the compilation in Table 1.4.)

Cation	eq kg^{-1}	Anion	eq kg^{-1}
Na^+	0.46906	Cl^-	0.54586
Mg^{2+}	0.10564	SO_4^{2-}	0.05648
Ca^{2+}	0.02056	Br^-	0.00084
K^+	0.01021	F^-	0.00007
Sr^{2+}	0.00018		
Li^+	0.00002		
Total	0.60567		0.60325

Total Cations – Total Anions = 0.60567 – 0.60325 = 0.00242 eq kg^{-1}

Table 5.4 The concentrations of the dominant species that make up the total alkalinity ($A_T = 2420$ μeq kg^{-1}) of surface seawater at pH ~ 8.2 ($T = 20°C$, $S_P = 35$). The contribution of silicic acid and phosphate are not included because these have low concentrations in surface seawater.

	Concentration		
Species	μmol kg^{-1}	μeq kg^{-1}	% of A_T
HCO_3^-	1796	1796	75
CO_3^{2-}	255	510	21
$B(OH)_4^-$	108	108	4
OH^-	6	6	0.2

$$A_T = [HCO_3^-] + 2 \cdot [CO_3^{2-}] + [B(OH)_4^-] + [H_3SiO_4^-] + [HPO_4^{2-}] + 2 \cdot [PO_4^{3-}] + [OH^-]. \tag{5.30}$$

Notice that the coefficients of the concentrations on the right-hand side of Eq. 5.30 are equal to the number of protons that these species will accept. For example, CO_3^{2-} can accept two protons to become HCO_3^- and then H_2CO_3. In most cases, the number of protons the anion will accept is equal to its charge, except in the cases of HPO_4^{2-} and PO_4^{3-}. These species will protonate to $H_2PO_4^-$ but not all the way to H_3PO_4 in the pH range of the alkalinity definition (see later).

A third way of thinking about alkalinity focuses on the analytical method used to determine it, which is titration by a strong acid with accurately known concentration while measuring the pH. Figure 5.4 illustrates that, at the beginning of an acid titration, the hydrogen ion concentration in seawater increases (pH decreases) much more slowly than it would in pure water, because many of the added hydrogen ions react with the excess bases (CO_3^{2-}, HCO_3^-, $B(OH)_4^-$, etc.). Seawater alkalinity is the number of moles of H^+ that must be added to a kg of seawater to reach the titration equivalence point for carbonic acid, which can be seen graphically as the point in the titration where pH changes most rapidly for each addition of acid. Quantitatively, the equivalence point is found at the pH where $[H^+] = [HCO_3^-] + 2 \cdot [CO_3^{2-}] + [B(OH)_4^-] \ldots$, very close to the point where the H^+ and HCO_3^- lines cross each other in Fig. 5.3 near a pH of 4.3. Two seawater cases are given in

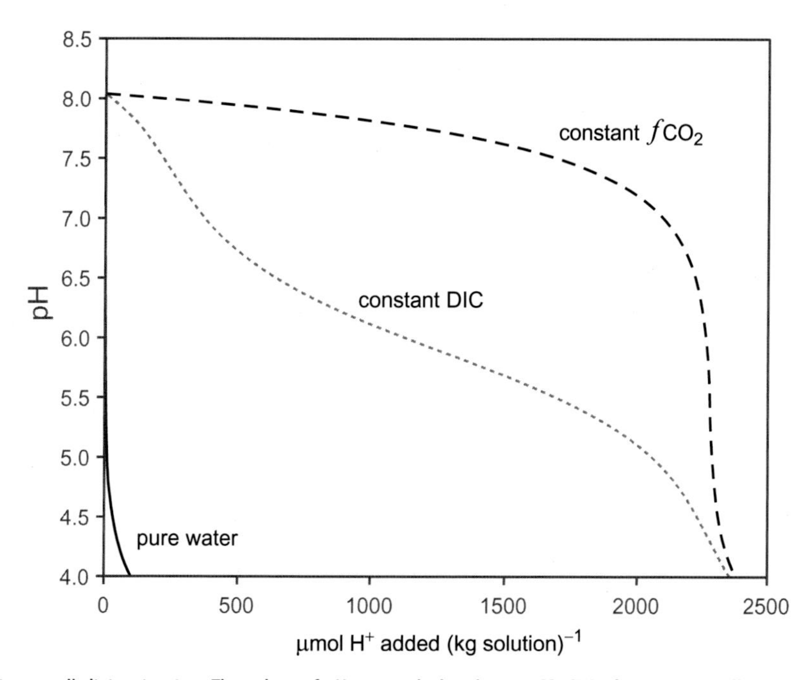

Figure 5.4 pH during an alkalinity titration. The values of pH were calculated using CO2SYS after incrementally adding H^+ to a surface seawater sample. Initial A_T is 2277 μeq kg^{-1} and DIC is 2024 μmol kg^{-1}. Three titration curves are shown: (1) pure water with no alkalinity, (2) seawater where the titration vessel is closed so DIC remains constant, and (3) seawater where the titration vessel is bubbled with air so the fCO_2^{sw} stays at 400 μatm.

Fig. 5.4. The curve labeled "constant DIC" is the pH pathway followed if the seawater titration is done in a closed titration cell so that CO_2 cannot escape during the titration. The curve labeled "constant fCO_2" is the pH pathway followed if the seawater titration is done in an open titration cell, while the solution is purged with air to force the excess CO_2 to escape and keep the dissolved CO_2 near equilibrium with atmospheric values. The pH change near the equivalence point in this case is much sharper, which is the reason that modern methods often suggest an open vessel titration.

Dickson (1981) defines alkalinity as

> The number of moles of hydrogen ion equivalent to the excess of proton acceptors (bases formed from weak acids with a dissociation constant $K \leq 10^{-4.5}$ at 25°C and zero ionic strength) over the proton donors (acids with $K > 10^{-4.5}$) in one kilogram of sample. (p. 611)

Proton acceptors with $K' \leq 10^{-4.5}(pK' \geq 4.5)$ in Table 5.2 include HCO_3^-, CO_3^{2-}, $B(OH)_4^-$, OH^-, $H_3SiO_4^-$, HPO_4^{2-}, and PO_4^{3-}, but not $H_2PO_4^-$, which means that in this definition HPO_4^{2-} and PO_4^{3-} will be titrated to $H_2PO_4^-$, but not to H_3PO_4. This is the reason that the stoichiometric coefficients of the phosphate species in Eq. 5.30 are one less than the charge.

To complete the precise definition of alkalinity, we subtract H^+ and the acids in Table 5.2 with $K' > 10^{-4.5}$, HSO_4, HF, and H_3PO_4:

$$\begin{aligned} A_T = {} & \left[HCO_3^-\right] + 2\cdot\left[CO_3^{2-}\right] + \left[B(OH)_4^-\right] + \left[H_3SiO_4^-\right] + \left[HPO_4^{2-}\right] + 2\cdot\left[PO_4^{3-}\right] \\ & + \left[OH^-\right] - \left[H^+\right] - \left[HSO_4^-\right] - [HF] - [H_3PO_4]. \end{aligned} \tag{5.31}$$

This rather long expression includes all known inorganic proton acceptors and donors in oxic seawater that follow Dickson's definition of the titration alkalinity. It includes two uncharged species at the very end, so it is not exactly consistent with the previous conception of alkalinity as a charge balance; however, in practice, the acidic species concentrations in seawater (H^+, HSO_4^-, HF, and H_3PO_4) are too low in the pH range of 7.0 to 8.0 to be significant and are frequently not included in the alkalinity definition. Including them here demonstrates the fate of protons during the course of acid addition to determine total alkalinity. (These species also play a more important role in more dilute environmental solutions like rainwater and in many freshwater lakes.)

The concentrations in Table 5.4 indicate that the ions of carbonate and borate define about 99 percent of the total alkalinity. Thus, calculations are sometimes made which include only these two species, and we define this as the carbonate and borate alkalinity, $A_{C\&B}$,

$$A_{C\&B} = \left[HCO_3^-\right] + 2\cdot\left[CO_3^{2-}\right] + \left[B(OH)_4^-\right]. \tag{5.32}$$

Another shortened form of alkalinity consists only of the carbonate species which make up about 96 percent of the total alkalinity (Table 5.4) and is termed the carbonate alkalinity, A_C,

$$A_C = \left[HCO_3^-\right] + 2\cdot\left[CO_3^{2-}\right]. \tag{5.33}$$

This definition is sometimes used for illustration purposes or back-of-the-envelope calculations, because of the simplicity of the calculations involved.

In anoxic waters, a whole new set of acids are created by the lower redox conditions. The most prevalent of these are hydrogen sulfide and ammonium (H_2S and NH_4^+, Table 5.2). Because of their pK', these species meet the criteria to be included in the titration alkalinity, and their concentrations can become as high as hundreds of μmol kg^{-1} in some highly reducing environments. For more common situations in which the water contains oxygen, these species are too low in concentration to be considered.

Discussion Box 5.1

- Based on Fig. 5.3, what are the approximate concentrations of the carbonate (CO_2, HCO_3^-, and CO_3^{2-}) and borate ($B(OH)_3$ and $B(OH)_4^-$) species in average surface seawater with a pH of 8.2? (Express your answers in μmol kg^{-1}.)
- If the surface ocean were to become more acidic, how would that impact atmospheric CO_2 levels based on this diagram?
- Imagine that you have a beaker of seawater at pH 8.2, to which you slowly add a strong acid (HCl for example). What happens to the protons as you add the acid? Based on Fig. 5.3, at what pH values will the pH change most quickly and at what values will it change most slowly as you add the acid?
- Compare these regions where you expect pH to change quickly or slowly to the titration curve in Fig. 5.4 for constant DIC. Is seawater well buffered over a broad range of pH values or only at narrow transition points?

5.2 Calculating Carbonate Equilibria and pH

We have now described the system of equations necessary for determining the pH and carbonate species concentrations in seawater. By including the definition and numerical value of alkalinity along with the system of equations used to determine the curves in Fig. 5.3, we have constrained the location on the plot to a single pH. The equations necessary to determine this location are summarized in Appendix 5A.1 for the progressively more complicated definitions of alkalinity: A_C, $A_{C\&B}$, and A_T.

In order to solve the equations to determine pH and the concentrations of the species that make up alkalinity, the apparent equilibrium constants, K', must be accurately known. These constants have been evaluated and reevaluated in seawater over the past 50 years. The pH scales and methods of measuring pH during these experiments have been different, which has complicated comparisons of the data until recently when many were converted to a common scale. Equations for carbonate system equilibrium constants as a function of temperature and salinity are presented by Dickson et al. (2007), Lueker et al. (2000), and Millero (1995) and summarized in Appendix F at the end of this book. CO2SYS, a

Table 5.5 Carbonate system parameters calculated for conditions in the surface and deep oceans at $S_p = 35$. Concentration units are μmol kg^{-1} unless otherwise specified. Calculated using CO2SYS with the K'_1 and K'_2 of Mehrbach et al. (1973) as refit by Lueker et al. (2000). The last three rows indicate the contribution of silica species, phosphate species, and OH^- to alkalinity.

Parameter	Surface Water	North Atlantic Deep Water	Antarctic Deep Water	North Pacific Deep Water
Measured Concentrations				
Depth (km)	0.0	4.0	4.0	4.0
T (°C)	20.0	2.0	2.0	2.0
A_T (μeq kg^{-1})	2 300	2 350	2 390	2 460
DIC (μmol kg^{-1})	1 950	2 190	2 280	2 370
[Si] (μmol kg^{-1})	0.0	60	130	160
[P] (μmol kg^{-1})	0.0	1.5	2.2	2.5
Calculated Carbonate Parameters				
pH	8.21	7.95	7.81	7.74
fCO_2^{sw} (μatm)	257	319	466	564
$[HCO_3^-]$	1 699	2 064	2 172	2 264
$[CO_3^{2-}]$	243	107	81	73
$[CO_2]$	8.3	19	27	33
$[B(OH)_4^-]$	110	68	51	45
A_{Si} (μeq kg^{-1})	0.0	1.3	2.0	2.2
A_P (μeq kg^{-1})	0.0	1.6	2.3	2.6
A_{OH} (μeq kg^{-1})	6.2	0.8	0.6	0.5

computer program to calculate carbonate species concentrations at equilibrium from any two measurements of the carbonate system, using the complete description of the alkalinity, A_T, and including the contributions from silicic acid and phosphate (Lewis and Wallace, 1998) is available for Excel, MATLAB, and Python online at the Ocean Carbon Data System website (www.ncei.noaa.gov/access/ocean-carbon-data-system/oceans/).

We demonstrate the changes in pH and carbonate species concentrations between waters from different locations in the ocean (Table 5.5) by calculating values using CO2SYS and typical A_T and DIC data from these locations. Both DIC and A_T increase from surface waters to the Deep Atlantic, Southern, and Pacific Oceans as one follows the route of the ocean "conveyor belt" (Fig. 1.16 and Discussion Box 1.2). Along this transect, pH changes from about 8.2 in surface waters to 7.7 in the deep Pacific Ocean, and CO_3^{2-} decreases from nearly 250 μeq kg^{-1} to less than a third of this value, ~75 μeq kg^{-1}. As discussed in the next section of this chapter, the change is driven by the addition of A_T and DIC to the deep waters by respiration and $CaCO_3$ dissolution. Notice that the contribution of the nutrients Si and P to the total alkalinity is only 0–3 μeq kg^{-1} or at most 0.3 percent of the total alkalinity. Although Si concentrations are much greater than those of P, the two nutrients have nearly equal contributions to the alkalinity (Table 5.5), because the pK' for the two phosphate reactions are closer to the pH of seawater than the pK' for silicic acid (see Table 5.2).

Table 5.6 Estimates of the errors in measurement and calculation of the carbonate system parameters. Measurement errors in the first row are from an assessment of ocean carbonate system best practices (Dickson, 2010) for an experienced laboratory using state-of-the art reference materials. Calculations were performed using CO2SYS at 20°C and 1 atmosphere pressure and the measured pair of species in column 1. Mean values for A_T, DIC, fCO_2, and pH were 2300 μeq kg^{-1}, 2042 μmol kg^{-1}, 400 (μatm), and 8.05. Deviations from the mean values were assigned given the measurement errors in row 1. The resulting calculation errors are the maximum error estimates assuming perfectly known values for the equilibrium constants.

Parameters	DIC (μmol kg^{-1})	A_T (μeq kg^{-1})	pH	fCO_2^{sw} (μatm)
Measurement error	2.5	2.5	0.005	2.0
Calculation error				
DIC, A_T			0.009	9.6
DIC, pH		5.6		5.5
DIC, fCO_2^{sw}		4.2	0.002	
A_T, pH	5.0			6.0
A_T, fCO_2^{sw}	3.1		0.002	
pH, fCO_2^{sw}	36	42		

The ability to accurately calculate carbonate system concentrations varies depending on which parameters are measured. While robust analytical methods currently exist for four of the carbonate system parameters (alkalinity, DIC, fCO_2^{sw}, and pH), the accuracy of calculating the rest of the carbonate ion species depends on the two parameters measured. This is demonstrated in Table 5.6, where carbonate system parameters are calculated from measured parameters varied through their maximum analytical error range. It is assumed in this exercise that the equilibrium constants K_1', K_2', and K_{H,CO_2} are known perfectly. The results demonstrate that measurement and calculation errors are of the same magnitude for all species, with the dramatic exception that A_T and DIC can be measured far more accurately than calculated by the fCO_2–pH pair (a difference of a factor of 10!). While fCO_2^{sw} and pH are currently the carbonate system measurements most readily determined by unmanned moorings and drifters, they are poor for defining the rest of the carbonate system because of the large errors in calculating A_T and DIC from this analytical pair (the last row in Table 5.6). Changes in fCO_2^{sw} and pH behave in almost compensating ways (an order of magnitude increase in CO_2 is accompanied by an order of magnitude increase in H^+, Fig. 5.3). This makes changes in other carbonate species concentrations relatively insensitive to changes in this concentration pair.

The error analysis in Table 5.6 is not the whole story, because it does not address the possibility of systematic errors in the equilibrium constants. This was assessed by Lueker et al. (2000), who compared the fCO_2^{sw} measured in seawater solutions at equilibrium with standard gases with fCO_2^{sw} calculated from A_T and DIC. They found that, for fCO_2^{sw} values less than 500 μatm, the equilibrium constants of Mehrbach et al. (1973), reinterpreted to the "total hydrogen ion concentration" scale, were the most accurate among all previous studies of laboratory-determined constants. The fCO_2^{sw} calculated from A_T and

DIC, with accuracies of ±1 μmol kg^{-1} and ±2 μeq kg^{-1}, respectively (about 0.05 percent and 0.1 percent errors), agreed with measured fCO_2^{sw} values to within ±3 μatm. This means that the equilibrium constants are accurate enough to calculate fCO_2^{sw} from DIC and A_T about as well as it can be directly measured. This is good news for determining air–sea CO_2 exchange; however, these researchers also found that the ability to verify the accuracy of equilibrium constants by comparing measured and calculated values deteriorates as fCO_2^{sw} increases above 500 μatm. The reason for the lack of agreement at fCO_2^{sw} values > 500 ppm may be that there are unknown organic acids and bases in the dissolved organic matter of seawater that alter its acid-base behavior at higher values of fCO_2^{sw}, but this has not been experimentally demonstrated in the open ocean.

While great advances in our understanding of the carbonate system have occurred in the last 20 years, it is also true that a version of the carbonate equilibrium constants determined more than 40 years ago (Mehrbach et al., 1973) are still preferred. As the fCO_2^{sw} in the surface ocean rapidly increases (it passed 400 ppm in 2017), it is essential that the next level of carbonate system accuracy be achieved so that internal consistency is possible in the future.

5.3 Processes that Control Alkalinity and DIC of Seawater

5.3.1 Terrestrial Weathering and River Inflow

On global spatial scales and over time periods comparable to, and longer than, the residence time of bicarbonate in the sea (~100 ky), the alkalinity and DIC of seawater are controlled by the species composition of rivers, which are determined by chemical weathering. In the generalized weathering reaction (see Chapter 2), CO_2 from the atmosphere reacts with water to release hydrogen ions that then react with rocks. Combining these two reactions for silicate rocks results in production of cations, bicarbonate, and silicic acid.

$$\frac{\begin{array}{c} Rock + H^+ + H_2O \rightarrow cations + clay + H_4SiO_4 \\ CO_2 + H_2O \rightarrow HCO_3^- + H^+ \end{array}}{Rock + CO_2 + H_2O \rightarrow cations + clay + HCO_3^- + H_4SiO_4}. \tag{5.34}$$

Bicarbonate is the main anion in river water because of the reaction of CO_2-rich soil water with both calcium carbonate and silicate rocks. Thus, neutralization of acid in reactions with more basic rocks during chemical weathering creates cations that are balanced by anions of carbonic acid. In this sense, the composition of rocks on land and the fact that CO_2 is the most important acid in the atmosphere has determined the overall alkalinity of the ocean.

Seawater has nearly equal amounts of alkalinity and DIC because the main source of these properties is riverine bicarbonate ion, which makes equal contributions to both parameters. The processes of $CaCO_3$ precipitation, hydrothermal circulation, and reverse weathering in sediments remove alkalinity and DIC from seawater and maintain present concentrations at about 2000 μmol (μeq) kg^{-1}. Reconciling the balance between river inflow and alkalinity removal from the ocean is still not well understood, as discussed in much greater detail in Chapter 2.

5.3.2 Alkalinity and DIC Changes within the Ocean

On timescales of oceanic circulation (1000 y and less), the internal distribution of carbonate system parameters is modified primarily by biological processes. Cross sections of the distribution of A_T and DIC in the world's oceans (Fig. 5.5) indicate that concentrations are higher in the deep ocean than the surface and that they become more concentrated in deep waters (1–4 km depth) from the North Atlantic to the Antarctic and into the Indian and Pacific oceans following the "conveyer belt" circulation (Fig. 1.16 and Discussion Box 1.2). Respiration of organic matter (OM) and dissolution of $CaCO_3$ cause these increases, so the chemical character of the particulate material that degrades and dissolves determines the relative changes in DIC and A_T. Note that the values in Fig. 5.5 also show some variations with salinity, for example higher values at the surface of the subtropics where evaporation concentrates DIC and A_T.

The stoichiometry of phosphorous, nitrogen, and carbon in OM that is respired in the ocean (see Chapter 3) is about

$$P : N : C = 1 : 16 : 106. \quad (5.35)$$

Organic carbon respiration and oxidation create CO_2, which dissolves in seawater. This increases DIC but does not change the alkalinity of the water because CO_2 is uncharged. Even though the CO_2 will dissociate to HCO_3^- and H^+, these species have equal and opposite impacts on alkalinity (see Eq. 5.31). The case for the nitrogen component in organic matter is not so simple, because respiration of organic N releases ammonia, which is then oxidized to dissolved NO_3^- (called nitrification, see Section 4.1.3). These reactions influence alkalinity, because nitrification is a redox reaction that involves the transfer of hydrogen ions into solution and ammonia is part of an acid-base pair. The net reaction and alkalinity change is:

$$\mathrm{OrgN} + 2O_2 \rightarrow NO_3^- + H^+ + H_2O \quad \textit{OM respiration + nitrification} \quad \Delta A_T = -1 \quad (5.36)$$

Since one proton is ultimately released into solution for every organic N (OrgN) reacted, the net alkalinity change is a decrease (see Eq. 5.31). Thus, when a mole of organic carbon as OM is respired, it causes the DIC to increase by one mole and the alkalinity to decrease by 16/106 (N/C) = 0.15 eq:

$$\Delta DIC_{OMresp} = 1.0; \; \Delta A_{T,OMresp} = -0.15. \quad (5.37)$$

Photosynthesis using nitrate as the source of bioavailable nitrogen has the exact opposite effect, lowering DIC by 1 and increasing alkalinity by 0.15 for each mole of organic carbon produced $\left(\Delta DIC_{OMphoto} = -1.0; \; \Delta A_{T,OMphoto} = 0.15\right)$.

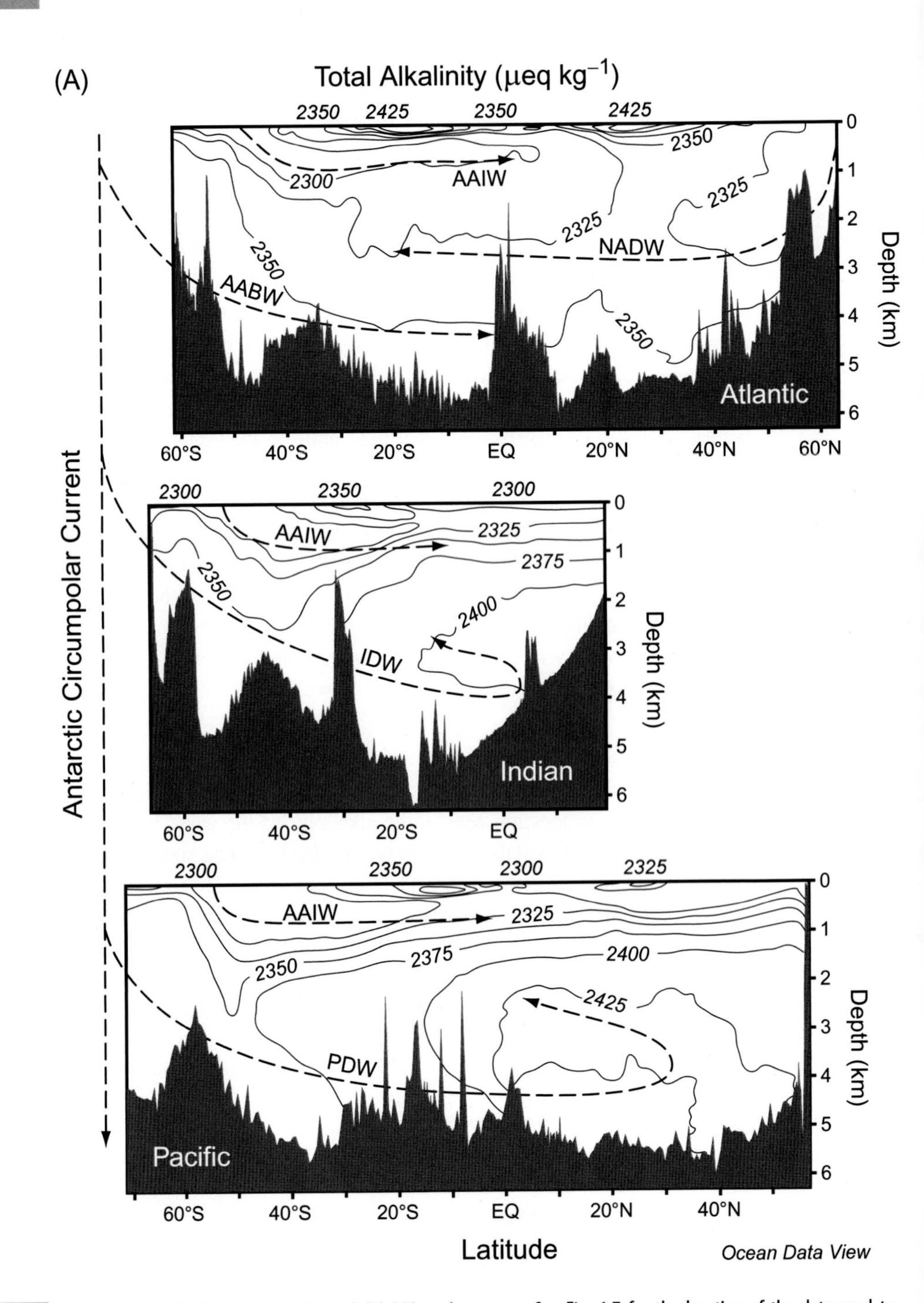

Figure 5.5 Cross sections of (A) total alkalinity and (B) DIC in the oceans. See Fig. 1.7 for the location of the data used to make the figure. Data are from Climate Variability and Predictability (CLIVAR) Repeat sections P16N & P16S for the Pacific, A16N & A16S for the Atlantic, and I09N and I08S for the Indian Ocean. (Data are compiled on the Ocean Carbon Data System, OCADS website: www.ncei.noaa.gov/products/ocean-carbon-data-system/)

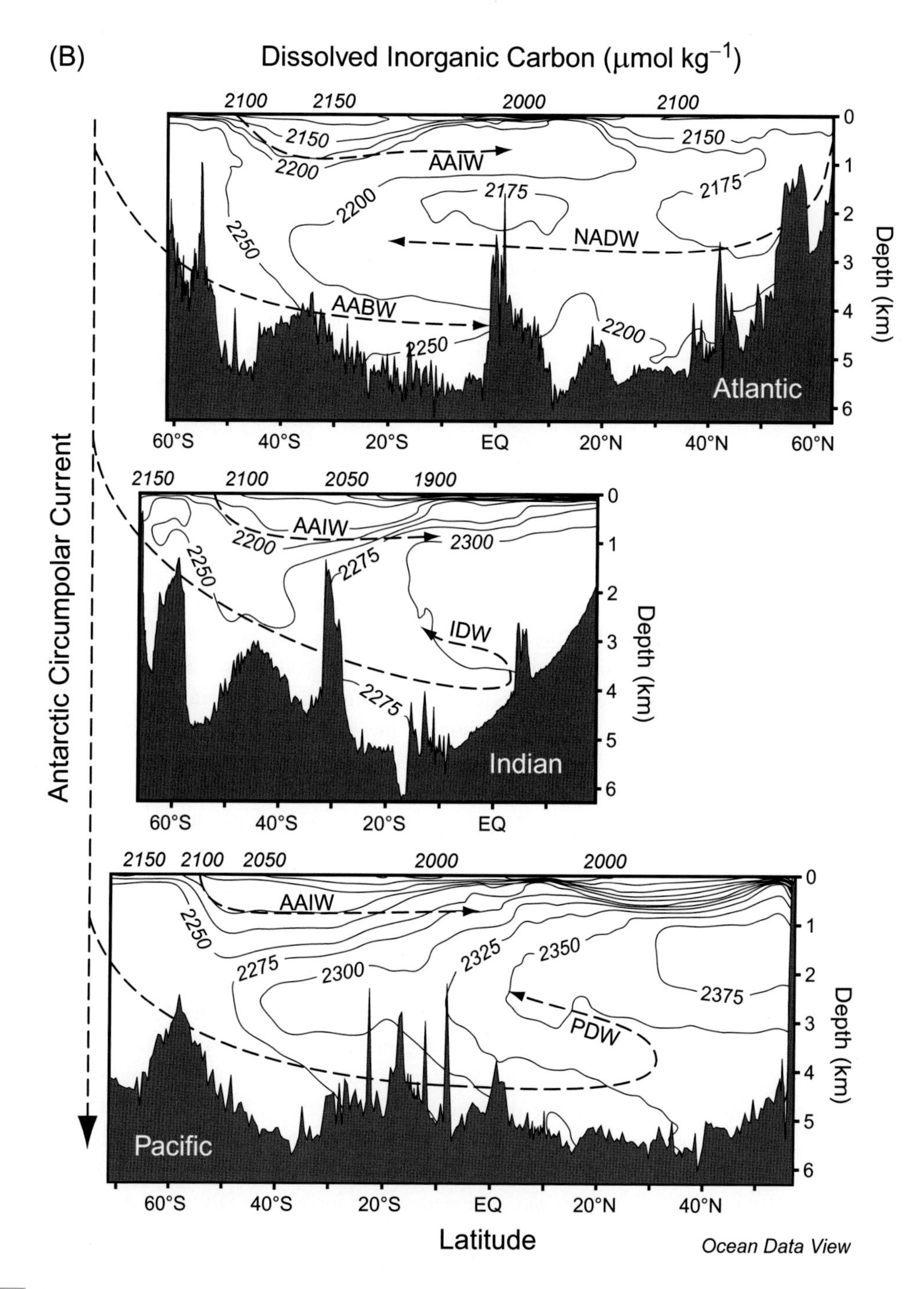

Figure 5.5 (*cont.*)

The change in DIC and A_T during $CaCO_3$ dissolution is very different from that resulting from organic matter respiration. One mole of calcium carbonate dissolution,

$$CaCO_3(s) \rightleftarrows Ca^{2+} + CO_3^{2-}, \tag{5.38}$$

causes an increase in alkalinity that is twice that of DIC because CO_3^{2-} introduces two charge equivalents for each mole of carbon change in solution. Thus:

$$\Delta DIC_{CaCO_3 diss} = 1.0;\ \Delta A_{T,CaCO_3 diss} = 2.0. \tag{5.39}$$

Here too, the impact of $CaCO_3$ formation on DIC and A_T is exactly opposite that of dissolution $\left(\Delta DIC_{CaCO_3 form} = -1.0, \Delta A_{T,CaCO_3 form} = -2.0\right)$. These relationships demonstrate that the change in alkalinity and DIC in seawater during respiration and dissolution of algae created in the surface ocean during photosynthesis depends greatly on the chemical character of the particulate material, i.e., the proportions of organic carbon to calcium carbonate. The ecology of the ocean's euphotic zone greatly influences the chemical changes observed everywhere in the sea.

Figure 5.6 is a plot of the salinity-normalized alkalinity, $A_{T,N}$, versus salinity-normalized dissolved inorganic carbon, DIC_N, for North Atlantic surface water $(100 - 1000\ m)$ and then along the deep-water conveyor belt (thermohaline) circulation from the North Atlantic to the South Atlantic, Southern, Indian, South Pacific, and North Pacific (deeper than 2500 m) Oceans. Salinity normalization (e.g., $DIC_N = DIC/S_P \times 35$) removes the impact of evaporation and precipitation on the concentrations, better highlighting the impact of biogeochemical processes. The lines in the figure illustrate that the relative changes of DIC_N and $A_{T,N}$ are not constant as it moves through the ocean and accumulates the products of respiration and shell dissolution. Between the surface Atlantic and the base of the thermocline the change in DIC_N: $A_{T,N}$ is about 10:1, whereas deeper than 2500 m, from the deep N. Atlantic to deep Indian and Pacific Oceans, the ratio is between 1:1 and 2:1.

The change in the ΔDIC_N:$\Delta Alk_{T,N}$ from the surface to the deep is due to the high OM: $CaCO_3$ ratio in particles and dissolved organic matter that exits the euphotic zone and the more rapid respiration of organic matter than dissolution of $CaCO_3$ as particles fall through the water. Recall from Section 3.2.2 that $CaCO_3$ is much denser than organic matter and sinks faster. This results in more organic matter being respired than $CaCO_3$ dissolved in the upper portion of the ocean. In deeper waters, the ΔDIC_N:$\Delta A_{T,N}$ ratio is close to one except in the Antarctic where the trend is relatively richer in DIC. Mineral secreting plankton in the Southern Ocean are dominated by diatoms, which form opal rather than $CaCO_3$ shells. Thus, particle dissolution at depth in the Southern Ocean releases DIC and H_4SiO_4 to the water but little alkalinity. The general 1:1 increase in DIC_N and $A_{T,N}$ in ocean deep waters outside of the Antarctic is probably strongly influenced by reactions at the sediment–water interface (see Sections 4.2.5 and 5.4.2 and Jahnke and Jackson (1987)). In carbonate-rich sediments, a large percentage of the CO_2 produced by organic matter respiration reacts with $CaCO_3$ to produce HCO_3^-, which ultimately creates an equal increase in DIC and A_T in solution (neglecting OrgN for simplicity here):

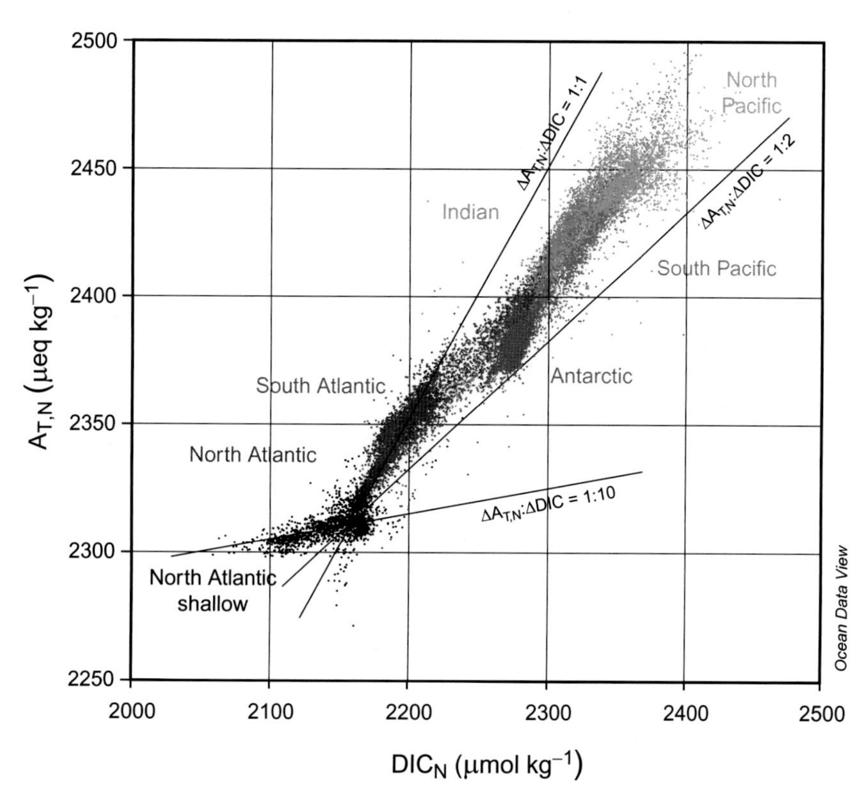

Figure 5.6 Salinity normalized ($S_P = 35$) total alkalinity, $A_{T,N}$, versus salinity normalized dissolved inorganic carbon, DIC_N, for the world's oceans. Data are for the deep ocean at depths >2500 m except for the points labeled "North Atlantic shallow," which come from 100–1000 m in the North Atlantic Ocean. Lines indicate different $\Delta A_{T,N} : \Delta DIC_N$ ratios. Data are from eWOCE (www.ewoce.org/data/#WHP_Bottle_Data). A black and white version of this figure will appear in some formats. For the color version, refer to the plate section.

$$\begin{array}{ll} & CH_2O(OM) + O_2 \rightarrow CO_2 + H_2O \\ & CaCO_3(s) + CO_2 + H_2O \rightarrow 2HCO_3^- + Ca^{2+} \\ \hline sum & CH_2O(OM) + CaCO_3(s) + O_2 \rightarrow 2HCO_3^- + Ca^{2+} \end{array} . \tag{5.40}$$

The relationship between the relative changes of DIC and A_T in seawater and the organic carbon respiration to $CaCO_3$ dissolution ratio in particulate matter is illustrated in Table 5.7. The $\Delta DIC:\Delta A_T$ ratio to be expected is calculated assuming one part $CaCO_3$ dissolution and progressively greater parts of organic carbon respiration using the stoichiometry in Eqs. 5.37 and 5.39. Solid phase OrgC:$CaCO_3$ ratios necessary to create the observed $\Delta DIC:\Delta A_T$ ratio in the shallower waters of the North Atlantic are between 6 and 8 – many more moles of carbon are released to solution from organic matter than from $CaCO_3$ at shallow depths. These ratios are still less than determined from very shallow depth gradients of A_T and DIC, which suggest an export flux capable of creating a trend of $\Delta DIC:\Delta A_T$ ~15 (Sarmiento et al. (2002). The reason that the $\Delta DIC:\Delta A_T$ ratios in Fig. 5.6

Table 5.7 Relative changes in DIC and A_T in seawater caused by dissolution of 1 μmol kg^{-1} of $CaCO_3$ along with respiration of 0–8 μmol kg^{-1} of organic carbon with RKR stoichiometry

OrgC respired (μmol kg^{-1})	$CaCO_3$ dissolved (μmol kg^{-1})	ΔDIC (μmol kg^{-1}) from OM	ΔDIC from $CaCO_3$	ΔDIC sum	ΔA_T (μeq kg^{-1}) from OM	ΔA_T from $CaCO_3$	ΔA_T sum	ΔDIC: ΔA_T
0	1	0	1	1	0	2	2	0.5
1	1	1	1	2	–0.15	2	1.85	1.1
2	1	2	1	3	–0.3	2	1.7	1.8
4	1	4	1	5	–0.6	2	1.4	3.6
6	1	6	1	7	–0.9	2	1.1	6.4
8	1	8	1	9	–1.2	2	0.8	11.2

are lower is presumably because much of the organic matter is respired in the top few hundred meters below the euphotic zone and the data along the 10:1 line in Fig. 5.6 are from a much greater depth range (100 – 1000 m). In deeper waters (>2500 m) where the $\Delta DIC:\Delta A_T$ ratio is about one, the organic carbon respiration to $CaCO_3$ dissolution ratio is also close to 1 as described in the previous paragraph. The deep $\Delta DIC:\Delta A_T$ ratio of one is less than the value of 4 that derives from box models (see Section 3.2.1 and Broecker and Peng (1982)) where the entire deep ocean is a weighted average of the thermocline and true deep waters.

It is helpful to have an intuitive understanding of how processes like photosynthesis or organic matter respiration, $CaCO_3$ formation or dissolution, and air–sea CO_2 exchange will affect pH and the relative proportions of the different carbonate species without having to make the full calculation using CO2SYS. This can be done either graphically or by determining the direction of the CO_3^{2-} ion change and then considering the implications of that change for pH and other properties. The first step for either method is to determine the changes in the total quantities DIC and A_T (or A_C), which can be exactly predicted for the different processes. The direction of the change in DIC and A_T by each process is presented as arrows in Fig. 5.7. Gas exchange affects only DIC; organic matter production and respiration affect DIC much more strongly than alkalinity; and $CaCO_3$ production and dissolution affect alkalinity more strongly than DIC. Processes that increase DIC strongly by adding CO_2 without having much impact on Alkalinity, namely organic matter respiration and CO_2 invasion, both cause pH and $[CO_3^{2-}]$ to decrease, since CO_2 is an acid. Notice that $CaCO_3$ formation, which lowers alkalinity more than DIC, also causes pH and $[CO_3^{2-}]$ to decrease. Formation of $CaCO_3$ by coccolithophorids and other organisms causes the surface ocean to become slightly more acidic, because CO_3^{2-} ions are removed from solution.

The impact of these processes on the carbonate system can be determined even without access to a graphical representation like Fig. 5.7. Again, we begin by determining the individual changes in A_T and DIC. Then the change in CO_3^{2-} ion concentration is estimated from the difference between A_T and DIC. The reason for determining this difference is

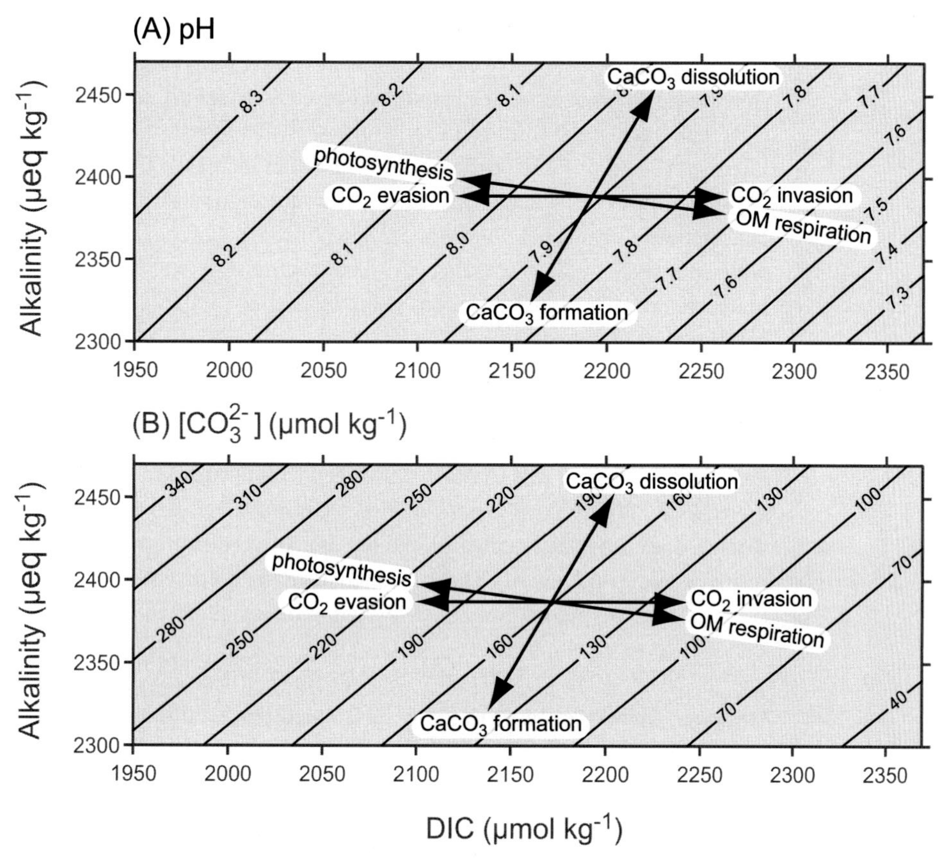

Figure 5.7 Impact of different processes on carbonate system parameters. In both panels, arrows show the direction that individual labeled processes cause total alkalinity and DIC to change. (A) Contours display pH values for different total alkalinity and DIC values. (B) Contours display carbonate ion concentrations $\left[CO_3^{2-}\right]$. Calculated using CO2SYS, assuming $T = 20°C$, $S_P = 35$, and negligible concentrations of phosphate and silicic acid.

illustrated using carbonate alkalinity, A_C, for simplicity. Subtracting the equation for DIC from that for carbonate alkalinity gives

$$\begin{aligned} A_C - DIC &= \left(\left[HCO_3^-\right] + 2 \cdot \left[CO_3^{2-}\right]\right) - \left(\left[HCO_3^-\right] + \left[CO_3^{2-}\right] + [CO_2]\right) \\ &= \left[CO_3^{2-}\right] - [CO_2] \cong \left[CO_3^{2-}\right]. \end{aligned} \tag{5.41}$$

(This is an example of why assigning units of eq kg^{-1} to alkalinity instead of mol kg^{-1} can be uncomfortable – we are subtracting numbers with different units.) Note that the approximation in the last step of Eq. 5.41 is most accurate in ocean waters with higher pH where CO_2 concentrations are negligible compared to CO_3^{2-}. See, for example, Table 5.5 where CO_3^{2-} and CO_2 concentrations become comparable as the pH decreases. The above equations indicate that the addition of more DIC than A_C to the water or the removal of more A_C than DIC results in a decrease in carbonate ion concentration $(\Delta A_C - \Delta DIC = \Delta CO_3^{2-})$. In that case, the process changing the carbonate chemistry (for example, organic matter respiration or CO_2 invasion) adds more acid in the form of CO_2 than base in the form of CO_3^{2-} to the seawater solution. Conversely, any process that increases the Alk_C – DIC difference results in an increased CO_3^{2-} ion concentration. Once the direction of the change in CO_3^{2-} ion concentration is known, the carbonate species concentration versus pH plot (Fig. 5.2) demonstrates that decreases in CO_3^{2-} are associated with decreasing pH, increasing CO_2 concentrations, and therefore increasing fCO_2^{sw}. Increases in CO_3^{2-} are associated with increasing pH, decreasing CO_2 concentrations, and therefore decreasing fCO_2^{sw}.

Two examples of calculations from these approximations are presented in Table 5.8. We begin with surface seawater and add 20 μmol kg^{-1} of CO_2 only, via gas exchange for example. The carbonate species concentrations for the initial solution, before CO_2 is added, are shown in the first row of Table 5.8. When CO_2 is added to the water, DIC in the solution increases by 20 μmol kg^{-1}, but A_T does not change, and we assume for our approximation that A_C does not change either (Table 5.8 2a). To predict the change in HCO_3^- and CO_3^{2-} in response to the addition of CO_2, we could take two different routes. First, by the laws of mass action, we would predict from the CO_2 hydration equation that bicarbonate would be formed

$$CO_2 + H_2O \rightleftarrows HCO_3^- + H^+. \tag{5.18}$$

However, there is no way to know how much this would affect the CO_3^{2-} concentration formed by the second carbonate dissociation reaction, because the first reaction produced both HCO_3^- and H^+, which appear on opposite sides of the second equilibrium reaction

Table 5.8 Demonstration of calculations using the $\Delta A_C - \Delta DIC$ approximation and the error involved (Eq. 5.41). (1) Typical distribution of carbonate species in surface seawater at chemical equilibrium ($T = 25°C$, $S_P = 35$). (2) Distribution of carbonate species after the addition of 20 μmol kg^{-1} of CO_2: (a) the $\Delta A_C - \Delta DIC$ approximation, (b) using the full carbonate equilibrium equations (CO2SYS) with ΔA_T. Note differences in the changes in HCO_3^- and CO_3^{2-}. (3) The same as (2) except for dissolution of the equivalent of 20 μmol kg^{-1} $CaCO_3$ instead. All concentrations are in μmol kg^{-1} except A_T and A_C, which are μeq kg^{-1}.

	A_T	A_C	DIC	HCO_3^-	CO_3^{2-}	CO_2
(1) Surf SW	2 300	2 187	1 950	1 697	244	9
(2) + 20 μmol kg^{-1} of CO_2						
(a) $\Delta A_C - \Delta DIC$ approximation		2 187	1 970	1 737	224	9
(1) – (2a) diff		**0**	**+20**	**+40**	**–20**	**0**
(b) A_T, DIC	2 300	2 193	1 970	1 729	231	10
(1) – (2b) diff	**0**	**+6**	**+20**	**+32**	**–13**	**+1**
(3) + 20 μmol kg^{-1} of CO_3^{2-}						
(a) $\Delta A_C - \Delta DIC$ approximation		2 227	1 970	1 697	264	9
(1) – (3a) diff		**+40**	**+20**	**0**	**+20**	**0**
(b) A_T, DIC	2 340	2 222	1 970	1 702	259	9
(1) – (3a) diff	**+40**	**+35**	**+20**	**+5**	**+15**	**0**

$$HCO_3^- \rightleftarrows CO_3^{2-} + H^+. \tag{5.16}$$

Is the increase in HCO_3^- or in H^+ more important? We are stuck unless we do the entire equilibrium/mass balance calculation.

The A_C – DIC approximation provides a way around this problem of two simultaneous interacting equilibria. By subtracting the change in dissolved inorganic carbon, ΔDIC, from the change in carbonate alkalinity, ΔA_C, the CO_3^{2-} concentration must decrease by about 20 μmol kg^{-1} (Eq. 5.41):

$$\Delta A_C - \Delta DIC = \Delta\left[CO_3^{2-}\right] - \Delta[CO_2] \cong \Delta\left[CO_3^{2-}\right] \cong -20\ \mu mol\ kg^{-1}.$$

Since the only carbonate species added was CO_2, it is reasonable to assume A_C cannot have changed. (We will check this below.) Thus, any change in CO_3^{2-} will require an opposite change in HCO_3^- of twice the magnitude to maintain constant carbonate alkalinity and a neutral solution. The only way both of these can happen is if HCO_3^- increases by 40 μmol kg^{-1} while CO_3^{2-} decreases by 20 μmol kg^{-1} (Table 5.8 line 2a). Comparing these calculations to Fig. 5.2 shows that the system is moving to the left on the plot, toward lower pH, higher CO_2, and higher fCO_2^{sw}.

We can evaluate the error in our approximation by exactly calculating the changes using CO2SYS and remembering that it is total alkalinity, A_T, that stays constant, not the carbonate alkalinity, A_C (Table 5.8 line 2b). The simple ΔA_C – ΔDIC approximation overestimates carbonate and bicarbonate changes by 25 and 50 percent, respectively, but the direction of the changes is correctly predicted by the approximation.

We can also demonstrate our approximation calculation and its error by imagining the addition of 20 μmol kg^{-1} of CO_3^{2-}, from the dissolution of $CaCO_3$, in the same surface water (Table 5.8 line 3). In this case, the CO_3^{2-} ion added to solution increases A_C by twice the number of moles of $CaCO_3$ added, 40 μeq kg^{-1}, and the DIC increases by the same number of moles of $CaCO_3$ added, 20 μmol kg^{-1}. Using the ΔA_C – ΔDIC approximation, we estimate an increase in CO_3^{2-} of 20 μmol kg^{-1} ($\Delta A_C - \Delta DIC = \Delta CO_3^{2-} = +20$ μmol kg^{-1}, Table 5.8 3a). In this specific case where the change in A_C is double that in CO_3^{2-}, there must be virtually no change in HCO_3^- to maintain charge balance. Comparing these calculations to Fig. 5.2 shows that the system is moving to the right on the plot, toward higher pH, lower CO_2, and lower fCO_2^{sw}. A check on the results of the approximation (Table 5.8 3b) reveals errors that are of the same magnitude as for the CO_2 addition.

Biogeochemical processes involving organic matter production and respiration, $CaCO_3$ formation and dissolution, and air–sea gas exchange result not only in easy to determine changes in DIC and alkalinity, but also in more difficult to determine pH changes and shifts in the proportions of the different carbonate species relative to each other. The complications of two interacting equilibria can make these shifts seem challenging to predict. The exact changes in all the species can be solved for using multiple simultaneous equations (Appendix 5A.1) or through computer programs like CO2SYS that encode the solution to these equations. The real worth of the A_C – DIC approximation is that it can be used in combination with Fig. 5.2 to provide a quick and intuitive means of determining the direction of change in each of the carbonate species and pH.

Discussion Box 5.2

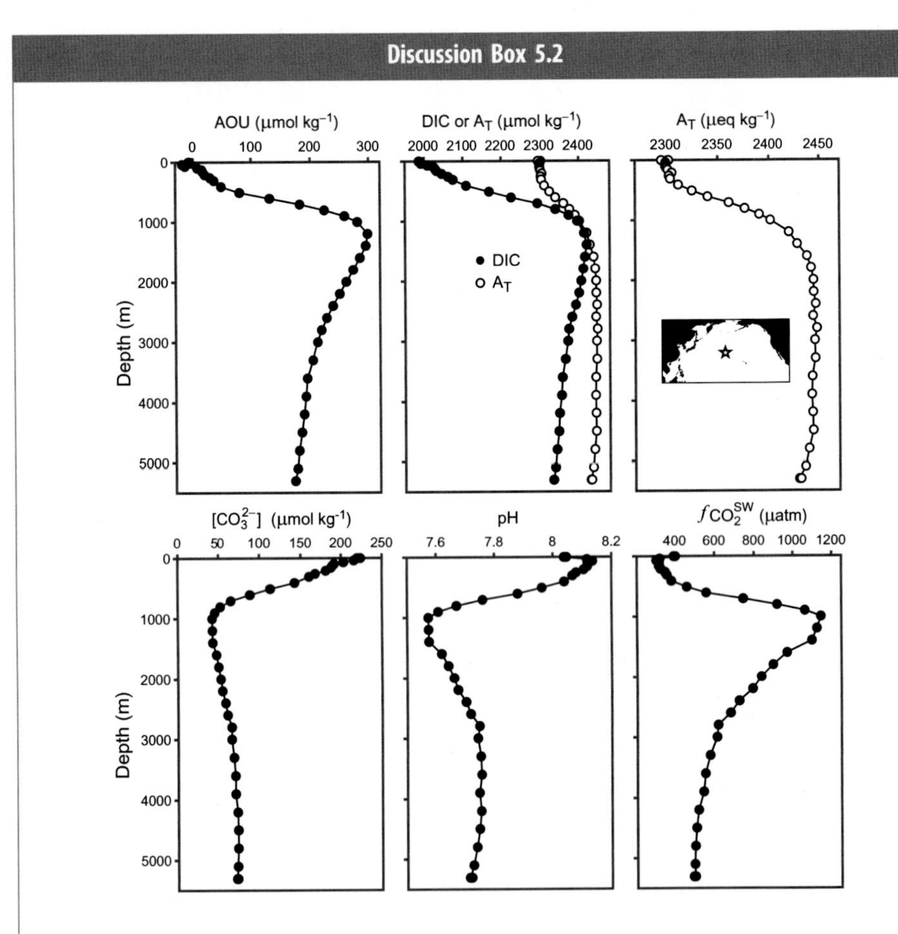

The biogeochemical processes we've discussed so far strongly impact the depth profiles of all the carbonate parameters. These figures show depth profiles of AOU, salinity-normalized DIC and total alkalinity, salinity-normalized total alkalinity on an expanded scale, carbonate ion concentration, pH, and fCO_2^{sw} at 31°N, 179°E in the North Pacific (the same North Pacific location we examined in Discussion Box 4.2).

- What creates the low values of DIC and alkalinity at the surface?
- What causes alkalinity to increase with depth? Why is the alkalinity maximum so much broader and deeper than the AOU maximum?
- What causes DIC to increase with depth? What causes the mid-depth maximum in DIC? Why is the maximum broader and deeper than the AOU maximum but shallower than the alkalinity maximum?
- Draw an approximate profile of total alkalinity minus DIC. Based on this, how do the DIC and total alkalinity profiles relate to the profiles of $[CO_3^{2-}]$, pH, and fCO_2^{sw}? How do the biogeochemical processes at work cause the carbonate parameter changes seen in the lower panels?

5.4 Mechanisms of Calcium Carbonate Dissolution and Burial

Between 20 and 30 percent of the calcium carbonate produced in the surface ocean escapes dissolution to be preserved in marine sediments. Given the profound impact of $CaCO_3$ dissolution and formation on the alkalinity and DIC of seawater, the fraction of $CaCO_3$ produced that is buried plays a very important role in the marine carbon cycle and the partial pressure of carbon dioxide in the atmosphere (see Chapter 8). Oceanographers who study the chemistry and mineralogy of ocean sediments have observed that the $CaCO_3$ content of marine sediments has changed with time in concert with glacial–interglacial periods. Understanding the mechanisms that control $CaCO_3$ preservation will help to clarify past changes in ocean chemistry and predict its future trajectory.

Sedimentary calcium carbonates are formed from the shells of marine plants and animals that consist primarily of two minerals – aragonite or calcite. Shallow water carbonates, mostly corals and shells of benthic algae (e.g., *Halimeda*), are heterogeneous in their mineralogy and chemical composition but are composed mainly of aragonite and magnesium-rich calcite (see Morse and Mackenzie, 1990 for a discussion). Carbonate shells (tests) of microscopic plants and animals, most of which live in the surface ocean, are primarily made of the mineral calcite, and are the main constituents of $CaCO_3$ in deep ocean sediments. The geographic distribution of $CaCO_3$ in ocean sediments (Fig. 5.8) indicates two distinct trends: topographic highs, like the mid-ocean ridges, are $CaCO_3$-rich while the deep abyssal planes are barren of this mineral, and there is relatively little $CaCO_3$ in the sediments of the North Pacific. Both of these observations are controlled by the

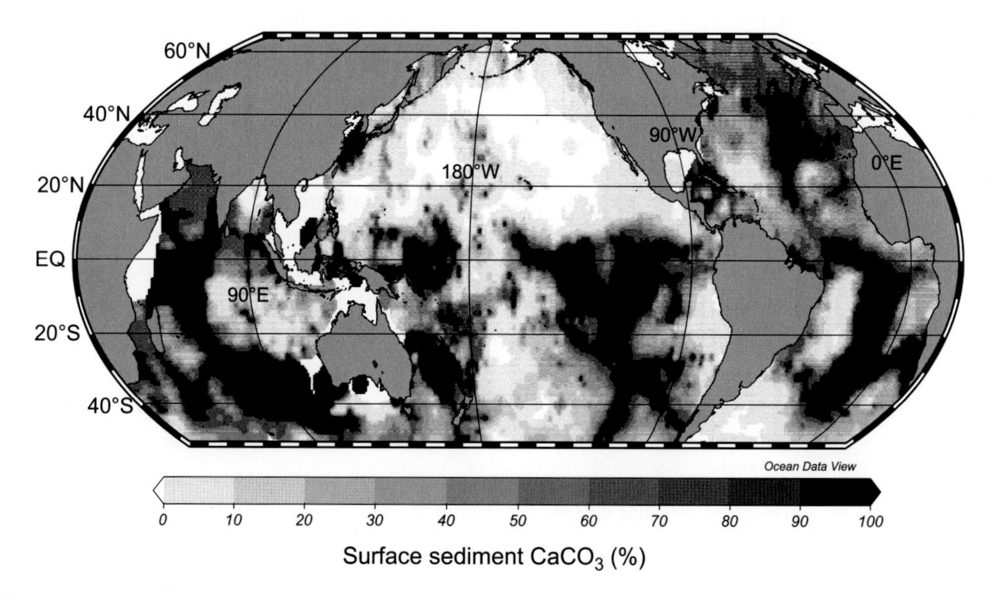

Figure 5.8 Global distribution of the weight percent $CaCO_3$ in surface sediments of the ocean deeper than 1000 m and located between 50°S and 60°N. White regions near the coast indicate no data. Redrawn from Archer (1996). A black and white version of this figure will appear in some formats. For the color version, refer to the plate section.

thermodynamics of $CaCO_3$ solubility in seawater and the carbonate chemistry of deep waters that bathe the sediments. The rest of this chapter focuses on the processes that control these distributions.

5.4.1 Thermodynamic Equilibrium

The solubility of $CaCO_3$ in seawater has been studied extensively as a mechanism to explain the observed $CaCO_3$ distribution in marine sediments (Fig. 5.8). The equation for dissolution of pure calcium carbonate:

$$CaCO_3(s) \rightleftarrows Ca^{2+} + CO_3^{2-}$$

has the following *apparent* (see Section 5.1.1) solubility product (equilibrium constant) in seawater:

$$K'_{SP} = \left[Ca^{2+}\right]\left[CO_3^{2-}\right] \tag{5.42}$$

because the activity of the pure solid $CaCO_3$ is assumed to be one by definition. The apparent solubility products of calcite and aragonite have been determined repeatedly in seawater solutions. We adopt the values of Mucci (1983) – K'_{SP} for calcite $= 4.30\ (\pm 0.20) \times 10^{-7}$ and K'_{SP} for aragonite $= 6.65(\pm 0.12) \times 10^{-7}$ mol^2 kg^{-2} at 25°C, $S_P = 35$, and one atmosphere of pressure. These data agree within error to previous measurements and represent many repetitions to give a clear estimate of the reproducibility $\left(\sim \pm 5\%\right)$.

Using the terminology discussed in Section 5.1.1 and in Table 5.1, the saturation state of seawater with respect to the solid, Ω, is the quotient of the ion concentration product (ICP $= [Ca^{2+}]\ [CO_3{}^{2-}]$) and the apparent solubility product, K'_{SP}, for the solid.

$$\Omega_{CaCO_3} = \frac{\left[Ca^{2+}\right]\left[CO_3^{2-}\right]}{K'_{SP}}. \tag{5.43}$$

If $\Omega = 1$, then the system is in equilibrium and should be stable; if $\Omega > 1$, the waters are supersaturated, and the laws of thermodynamics would predict that the mineral should precipitate removing ions from solution until Ω returns to one; and, if $\Omega < 1$, the waters are undersaturated and solid $CaCO_3$ should dissolve until the solution concentrations increase to the point where $\Omega = 1$ (Table 5.1). In practice, inorganic precipitation of $CaCO_3$ from supersaturated open ocean waters is rare, probably because the presence of high concentrations of Mg^{2+} in seawater blocks nucleation sites on the surface of the mineral. Supersaturated conditions thus tend to persist. Dissolution of $CaCO_3$, however, does occur when $\Omega < 1$. Since calcium concentrations are nearly conservative in the ocean, varying by a few percent, it is the apparent solubility product, K'_{SP}, and the carbonate ion concentration that largely determine the saturation state of carbonate minerals.

Equilibrium constants are strongly temperature and pressure dependent, but, since the temperature of seawater below 1000 meters varies little (~0–2°C), it is the pressure dependence of K'_{SP} that plays the main role in determining the solubility product in waters that bathe deep ocean sediments. The pressure dependence of the solubility product is related to the difference in volume, ΔV, occupied by the ions of Ca^{2+} and CO_3^{2-} in solution compared to the solid phase. The volume difference between the dissolved and solid phases is called the partial molal volume change, ΔV:

$$\Delta V = V_{Ca} + V_{CO_3} - V_{CaCO_3}. \tag{5.44}$$

The change in partial molal volume for calcite dissolution is negative, meaning that the volume occupied by solid $CaCO_3$ is greater than the combined volume of the component Ca^{2+} and CO_3^{2-} ions in solution. Values of the partial molal volume change determined by laboratory experiments and in situ measurements result in a range of – (35–45) $cm^3\ mol^{-1}$ (Sayles, 1980). Increasing pressure shifts the equilibrium toward the phase occupying the least volume, causing $CaCO_3$ to become more soluble with pressure (depth) in the ocean by a factor of about two for a depth increase of 4 km.

The pressure dependence of the solubility product controls the depth dependence of Ω_{CaCO_3} because $[CO_3^{2-}]$ is not strongly depth dependent below 1000m. This is the reason for $CaCO_3$ preservation on topographic highs. On the other hand, the variation in deep concentration of carbonate ion, $[CO_3^{2-}]$, is the reason for the Atlantic to North Pacific geographic trends in $CaCO_3$ preservation (Fig. 5.8). The high ratio of organic carbon to carbonate carbon in the particulate and dissolved organic material degrading and dissolving in the deep sea causes the deep waters to become more acidic and carbonate poor as they progress along the conveyer belt circulation from the deep North Atlantic to Indian and North Pacific Oceans. Carbonate ion concentrations change from ~250 $\mu mol\ kg^{-1}$ in surface waters to mean values in the deep waters of 113 $\mu mol\ kg^{-1}$ in the Atlantic, 83 $\mu mol\ kg^{-1}$ in the Indian, and 70 $\mu mol\ kg^{-1}$ in the deep North Pacific Oceans (Fig. 5.9), with little depth dependence of these values below about 1000 m. Thus, the tendency for $CaCO_3$ minerals to be preserved in deep sea sediments decreases "downstream" along the thermohaline circulation from the Atlantic to Indian to Pacific Oceans.

Depth profiles of CO_3^{2-} concentration from locations near the Equator of each ocean are plotted along with the CO_3^{2-} concentration required at solubility equilibrium with calcite and aragonite in Fig. 5.10. The depth where the measured dissolved concentration and the calculated saturation concentration, $[CO_3^{2-}]_{sat} = K'_{SP}/[Ca^{2+}]$, cross illustrates the depth where $\Omega = 1$ for these minerals. This depth is referred to as the saturation horizon for these minerals. Where $[CO_3^{2-}] > [CO_3^{2-}]_{sat}$, the minerals are supersaturated and tend to be preserved in sediments; where $[CO_3^{2-}] < [CO_3^{2-}]_{sat}$, the minerals are undersaturated and should dissolve, because they are thermodynamically unstable at these pressures and concentrations. Surface waters are greatly supersaturated in all oceans, but deeper waters are closer to thermodynamic saturation. The mean saturation horizon for calcite shoals from a depth of about 5 km in the equatorial Atlantic to 3.5–4.0 km in the Indian Ocean to less than 3.5 km in the equatorial Pacific. The saturation horizon for aragonite, the more soluble form of $CaCO_3$, is much shallower varying from 2.75 in the Atlantic Ocean to about 0.5 km in the equatorial Pacific.

There have been many attempts to correlate the presence of calcite in marine sediments with the degree of saturation in the overlying bottom waters. The terminology for the

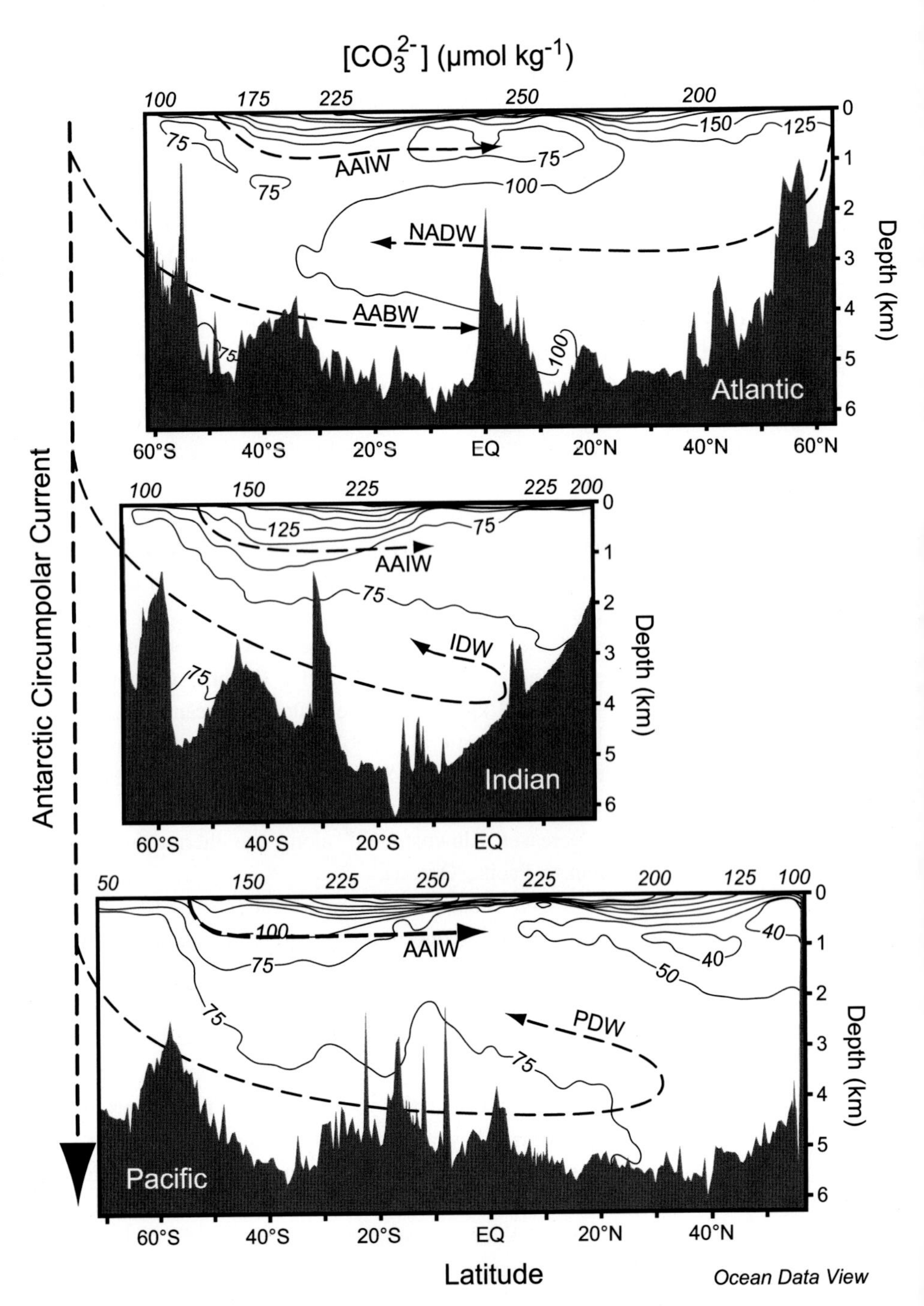

Figure 5.9 Cross sections of the carbonate ion concentration, $[CO_3^{2-}]$, in the oceans from the same locations and data source as in Fig. 5.5.

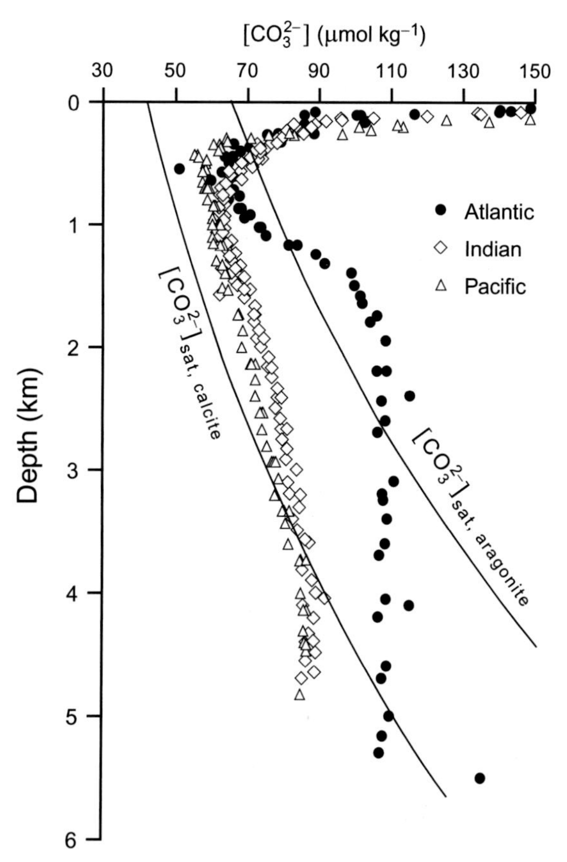

Figure 5.10 Profiles of $[CO_3^{2-}]$ as a function of depth for the equatorial Atlantic (closed symbols), Indian, and Pacific (open symbols) Oceans. $[CO_3^{2-}]$ concentration was calculated using CO2SYS from near equatorial profiles of total alkalinity and DIC from the database referenced in Fig. 5.5. $[CO_3^{2-}]$ data are compared to the carbonate concentrations at equilibrium with calcite, $[CO_3^{2-}]_{sat,\,calcite}$, and with aragonite, $[CO_3^{2-}]_{sat,\,aragonite}$. ($[CO_3^{2-}]_{sat} = K'_{SP}/[Ca^{2+}]$ where the K'_{SP} values are determined for 2°C and various pressures from Appendix G, and $[Ca^{2+}] = 10.28$ mmol kg^{-1}.)

presence of $CaCO_3$ in deep-sea marine sediments as a function of depth is a little esoteric (Fig. 5.11). The word "lysocline" means the depth in the ocean where the $CaCO_3$ content in sediments first begins to decrease with depth. The "carbonate compensation depth" (CCD) is the depth where the $CaCO_3$ content approaches zero percent. At the CCD, the deposition of calcium carbonate to the sea floor is exactly "compensated" by the rate of dissolution of $CaCO_3$ in sediments. The depth difference between these horizons varies from place to place, but is commonly as great as 1 km (Archer, 1996; Biscaye et al., 1976).

It seems as though the importance of thermodynamics versus kinetics in determining calcite preservation could be determined by comparing measurements of lysocline and CCD depths with the saturation horizon (SH). If thermodynamics were controlling the

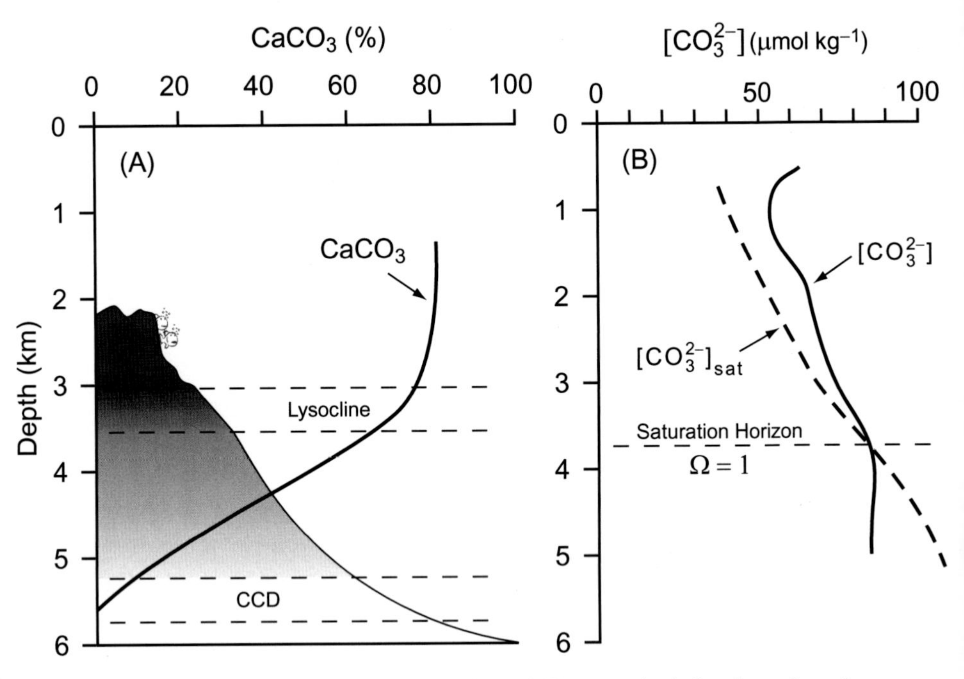

Figure 5.11 (A) A schematic representation of the lysocline (where the % $CaCO_3$ versus depth first shows signs of dissolution) and the carbonate compensation depth (CCD, where the % $CaCO_3$ becomes zero). (B) A similar sketch to Fig. 5.10 showing idealized depth profiles of observed $[CO_3^{2-}]$ and $[CO_3^{2-}]_{sat,\,calcite}$. The Saturation Horizon (SH) is where these two lines cross.

presence or absence of $CaCO_3$ in sediments then the lysocline–CCD transition should take place over a small depth interval near the saturation horizon. There are two main problems with attempts at this direct observation. The first issue is the poor accuracy with which we know the degree of saturation in the ocean because of the error on the value of K'_{SP}. An uncertainty of ± 5 percent in K'_{SP} at one atmosphere and ± 10 percent in the pressure effect compounds to an uncertainty in the saturation horizon depth of 0.5–1.0 km (Emerson and Hedges, 2003). The second issue is that the onset of $CaCO_3$ dissolution within sediments cannot be determined from measurements of the amount of $CaCO_3$ observed there. For example, if 90 percent of the particles sinking are $CaCO_3$, a 50 percent decrease in the $CaCO_3$ burial in the sediments would result in a change in the sediment composition from 90 to 82 percent $CaCO_3$. Although easy to measure, percentage of $CaCO_3$ is a blunt instrument for determining the onset of carbonate dissolution. Errors in evaluating the depths of both the saturation horizon and the onset of $CaCO_3$ dissolution complicate "field" tests of the mechanisms controlling calcite preservation. However, the simple observation that the transition from $CaCO_3$-rich sediments to no $CaCO_3$ takes place over depth ranges of up to 1 km suggests that calcite preservation cannot be controlled by thermodynamics alone.

5.4.2 Kinetics of $CaCO_3$ Dissolution

The dissolution rates of calcium carbonate minerals have been shown in laboratory experiments to follow the rate law

$$R_{CaCO_3} = k_{CaCO_3}(1 - \Omega)^n, \tag{5.45}$$

where k_{CaCO_3} is the dissolution rate constant, which has units of the amount of solid dissolved per gram of solid in the experiment or environment per time ($g_{dissolved}\ g_{solid}^{-1}\ d^{-1}$), and is often shortened to simply d^{-1} or multiplied by 100 and called % d^{-1}. Theoretically, the dissolution rate should be normalized to surface area of the material dissolving ($g_{dissolved}\ cm_{solid\ area}^{-2}\ d^{-1}$), but in practice the area term is difficult to measure accurately. The exponent, n, is one for diffusion-controlled reactions and usually some higher number for surface-controlled reactions. It is very difficult to determine $CaCO_3$ dissolution rates appropriate for in situ ocean conditions because the degree of undersaturation in the ocean, the value of $(1 - \Omega)$, is normally very small (0–0.2, Fig. 5.10). At such low degrees of undersaturation, the concentration and surface area of the solid changes slowly so that it is difficult to observe measurable changes in laboratory experiments. If the dissolution exponent (n) is 1, then rates measured far from equilibrium will be the same closer to equilibrium. If $n > 1$, however, measurements have to be made at environmentally relevant undersaturations.

Subhas et al. (2015) addressed this problem by devising a novel experiment to measure dissolution rates near saturation. They dissolved ^{13}C-labeled $CaCO_3$ crystals in seawater solutions with very carefully determined, environmentally relevant degrees of calcite undersaturation. By measuring the $\delta^{13}C$ change in the experimental seawater solution, very small amounts of $CaCO_3$ dissolution could be quantified over periods of time short enough to avoid significant changes in the surface area of the solid. Results of these experiments (Fig. 5.12) confirm the high order of calcite dissolution, resulting in a rate law described by

$$R_{CaCO_3} = 8.5\ (\%d^{-1})(1 - \Omega)^{3.85}. \tag{5.46}$$

Changes in the mechanism of dissolution as a function of undersaturation presumably cause the high order rate ($n = 3.85$), which has important ramifications for the interpretation of the mechanism of dissolution in nature.

The relatively slow rate of calcite dissolution implied by the high order rate law is consistent with the rather large depth difference between the lysocline and CCD (Fig. 5.11). However, the relationship between the saturation horizon and lycocline–CCD difference depends on another important issue that we have not yet considered – the role of organic matter respiration in sediments in promoting in situ calcite dissolution.

It was long suspected that organic matter respiration within the sediments would promote $CaCO_3$ dissolution because sediment pore waters should be more corrosive to $CaCO_3$ than the overlying waters, but this was not quantified until the 1980s and

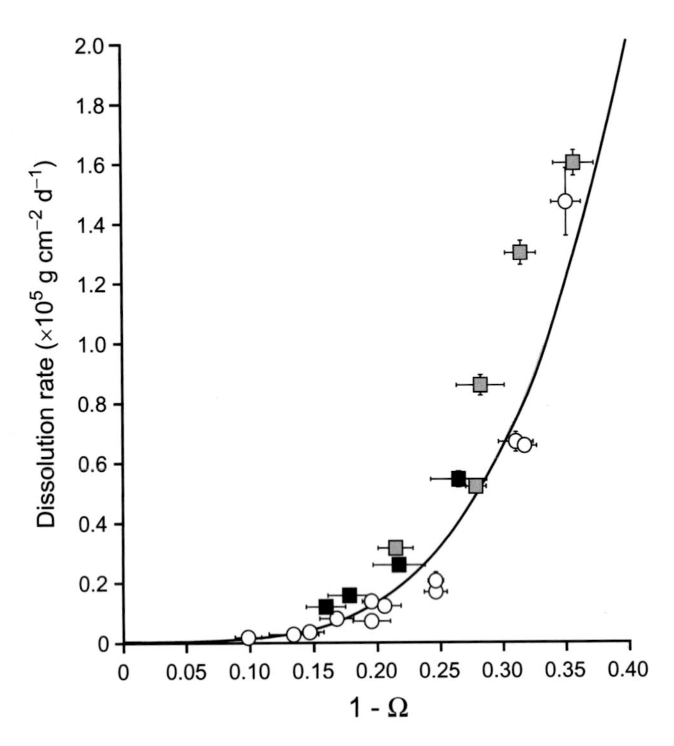

Figure 5.12 $CaCO_3$ dissolution rate normalized to surface area of the solids as a function of the degree of undersaturation $(1 - \Omega)$. The data are from the dissolution rate experiments of Subhas et al. (2015) in which open circles are 70–100 μm Aldrich calcite, black squares are 300–500 μm homegrown calcite, and gray squares are 500–700 μm homegrown calcite. Redrawn from Subhas et al. (2015).

1990s. Two factors led to the realization that organic matter respiration may dramatically alter the chemistry of sediments and therefore might significantly impact $CaCO_3$ burial in the ocean. First, sediment trap observations of the carbon content of particles reaching the ocean floor suggest that the molar ratio of organic carbon to $CaCO_3$ is about one. Surface sediment organic carbon measurements, however, indicate that this ratio is closer to 0.1, indicating that ~90 percent of the organic carbon reaching the sediment surface is respired rather than buried. Second, sediment pore water studies show strong decreases of oxygen in the pore waters of the top few centimeters of sediment. Pore water flux calculations require that most respiration of organic matter that reaches the sediments takes place within the sediments. Thus, most particles that reach the sea floor are stirred into the sediments by worms and other organisms before they have a chance to be respired while sitting on the surface. If this were not the case, and the particles were respired entirely at the sediment surface, there would be little oxygen depletion within the sediments.

Organic matter respiration within the sediments creates a microenvironment that is corrosive to $CaCO_3$, even if the bottom waters are not, because respiration adds DIC and

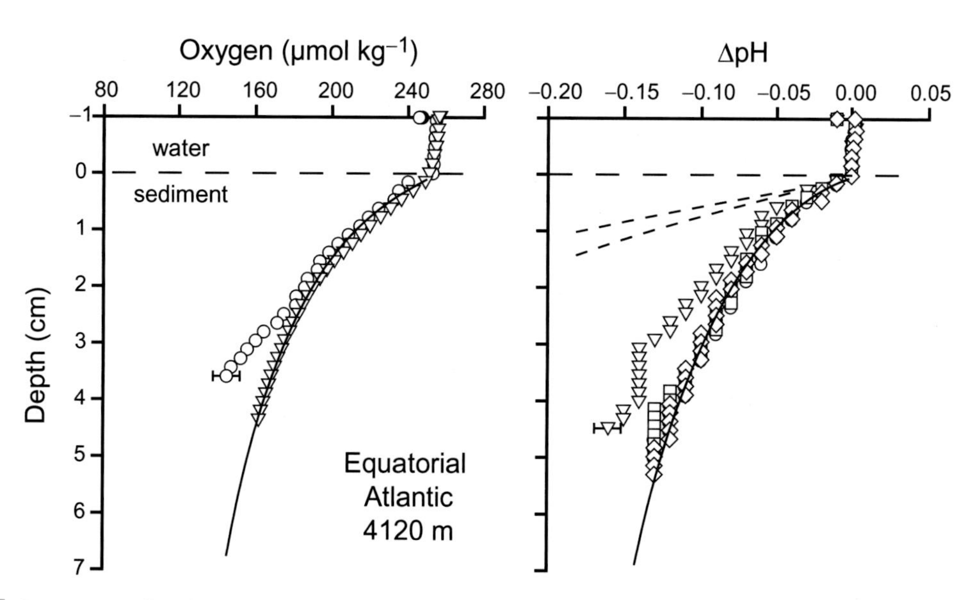

Figure 5.13 Pore water profiles of oxygen and ΔpH (the pH in the pore water minus that in the bottom water). The data were determined in situ at 4120 m depth using microelectrodes on a profiler at the Ceara Rise in the equatorial Atlantic where bottom waters are saturated or slightly supersaturated with respect to calcite. Points are individual measurements from different electrodes (different symbols). Slightly offset points in the overlying water at −1 cm are measurements made in the bottom water after the pore water profile. Dashed lines indicate the predicted ΔpH in the absence of $CaCO_3$ dissolution caused by organic matter respiration. Solid lines indicate the predicted ΔpH including $CaCO_3$ dissolution. Redrafted from Hales and Emerson (1997).

slightly reduces A_T in the pore water, lowering pH and carbonate ion concentrations. Using a simple analytical model and first-order dissolution rate kinetics, Emerson and Bender (1981) predicted that this effect should result in up to 50 percent of the $CaCO_3$ that rains to the sea floor at the saturation horizon being dissolved, even though the bottom waters are saturated with respect to calcite. Because organic matter respiration promotes $CaCO_3$ dissolution even at depths where bottom waters are saturated or supersaturated, the water column saturation horizon should be deeper than the depth where sediment dissolution begins (the lysocline). However, organic matter respiration should have little effect on the depth difference between the lysocline and CCD, because the respiration-driven $CaCO_3$ dissolution should affect all carbonate containing sediments equally.

The effect of organic respiration on $CaCO_3$ dissolution in sediments has been tested by determining gradients of oxygen and pH in sediment pore waters. These measurements had to be done on a very fine (millimeter) scale because the important region for the reaction is near the sediment–water interface. The test also required in situ measurements because the pH values of pore waters change when they are depressurized. These measurements (Archer et al., 1989; Hales and Emerson, 1996, 1997), some of which are reproduced in Fig. 5.13, showed that organic matter respiration causes significant $CaCO_3$ dissolution in marine sediments. In each case, the observed pH change was much smaller than it would have been due to organic matter respiration alone, proving that some $CaCO_3$ dissolved, raising alkalinity and pH levels. These conclusions were later confirmed by millimeter-scale measurements of pCO_2 and Ca^{2+} concentration in pore waters (e.g., Wenzhofer et al., 2001). Dissolution of $CaCO_3$ in response to organic matter respiration makes the burial of $CaCO_3$ in the ocean dependent not only on the thermodynamics and kinetics of $CaCO_3$ dissolution, but also on the organic carbon to calcium carbonate ratio in sinking particles.

5.5 Anthropogenic Influences

As of the writing of this textbook, the fCO_2 of the atmosphere has increased by about 40 percent above its preindustrial value due to burning of fossil fuels, and a third of this increase has taken place since the year 2000. The ocean has adsorbed about 30 percent of the total anthropogenic CO_2 introduced to the atmosphere, substantially moderating the atmospheric increase. Measurements used to determine the ocean's role in the global carbon cycle are the subject of Chapter 8. Here, we describe an important side effect of this ocean uptake of anthropogenic CO_2 pollution, namely ocean acidification.

The uptake of CO_2 from the atmosphere by the surface ocean causes an increase in DIC but no change in alkalinity, and we have shown in Section 5.3.2 and Fig. 5.7 that this will lead to a decrease in CO_3^{2-} concentration and a lower pH. We can quantify the expected change in surface ocean pH by using CO2SYS to solve for surface water pH based on the increasing fCO_2 of the atmosphere and a constant total alkalinity (Fig. 5.14). The pH of the surface ocean at equilibrium with atmospheric fCO_2 and surface ocean total alkalinity has decreased by about 0.13 pH units, which corresponds to an increase in the hydrogen ion concentration of about 35 percent (because pH is expressed on a log scale) and a decrease in carbonate ion concentration of about 21 percent. Measurements at the ocean time series sites at Hawaii and Bermuda since 1988 confirm these calculations (Fig. 5.15). The global pH change in response to ocean uptake of anthropogenic CO_2 has been documented in observations (Feely et al., 2004) and predicted by ocean circulation models (Orr et al., 2005).

The effect of lower pH on marine life, especially those organisms that make $CaCO_3$ shells, is a subject of intense research. It is not a purely inorganic question because organisms themselves have a profound effect on the formation of carbonate minerals. Shell formation occurs in a protected calcifying fluid separated from bulk seawater by a layer of tissue. Ion pumping enzymes in the membranes act to concentrate the Ca^{2+} and increase the pH of the calcifying fluid, thereby substantially raising Ω at the site of shell formation. Mussels have been discovered producing $CaCO_3$ shells near deep sea vents in

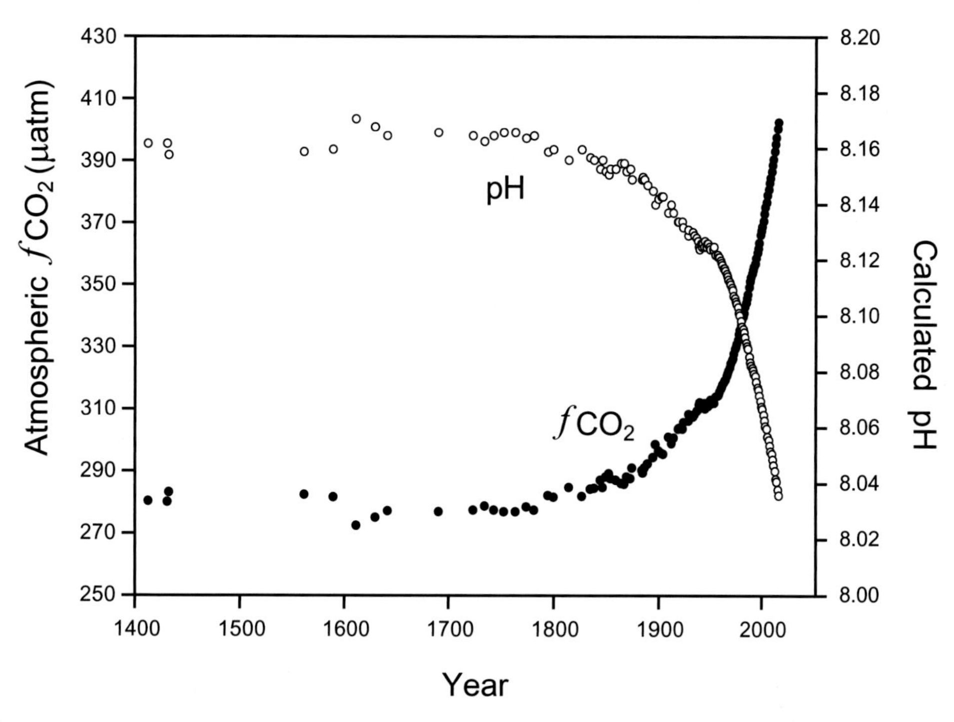

Figure 5.14 A calculation of mean ocean surface water pH as a function of time assuming equilibrium with atmospheric fCO_2 measurements. The fCO_2 data are from https://scrippsco2.ucsd.edu, a website maintained by Ralph Keeling and Stephen Piper at Scripps Institution of Oceanography, and represent the annual average of measurements from Hawaii and the South Pole merged with published ice core data. pH was calculated using CO2SYS with a mean ocean surface total alkalinity (A_T) of 2277 μeq kg^{-1}, temperature of 25°C, and salinity of 35.

ambient waters with pH less than 5.5 and Ω near 0.01, a testament to the ability of some biological organisms to evolve calcification adaptations to extreme conditions (Rossi and Tunnicliffe, 2017)! When these mussels die, their shells become exposed to ambient seawater and rapidly dissolve.

Larvae just beginning to produce their first shells are often found to be most sensitive to the Ω of the surrounding seawater, perhaps because their ion pumping machinery is not yet fully developed. One of the most complete studies of the effect of decreasing pH on aragonite secreting bivalves was carried out by a marine biologist (George Waldbusser) and chemical oceanographer (Burke Hales) working together at Oregon State University (Waldbusser et al., 2015). These researchers grew a Pacific oyster and a Mediterranean mussel in seawater solutions that were manipulated with acid and base to achieve A_T and DIC levels that varied in pH or fCO_2^{sw} or CO_3^{2-} (and hence $\Omega_{aragonite}$) while the other carbonate species remained relatively constant. Their work showed that what matters to the early development of the bivalve larvae (hours to days after fertilization) is the degree of aragonite saturation, $\Omega_{aragonite}$, not the pH itself or fCO_2^{sw} (Fig. 5.16). Notice that the larvae grow abnormally in these experiments at

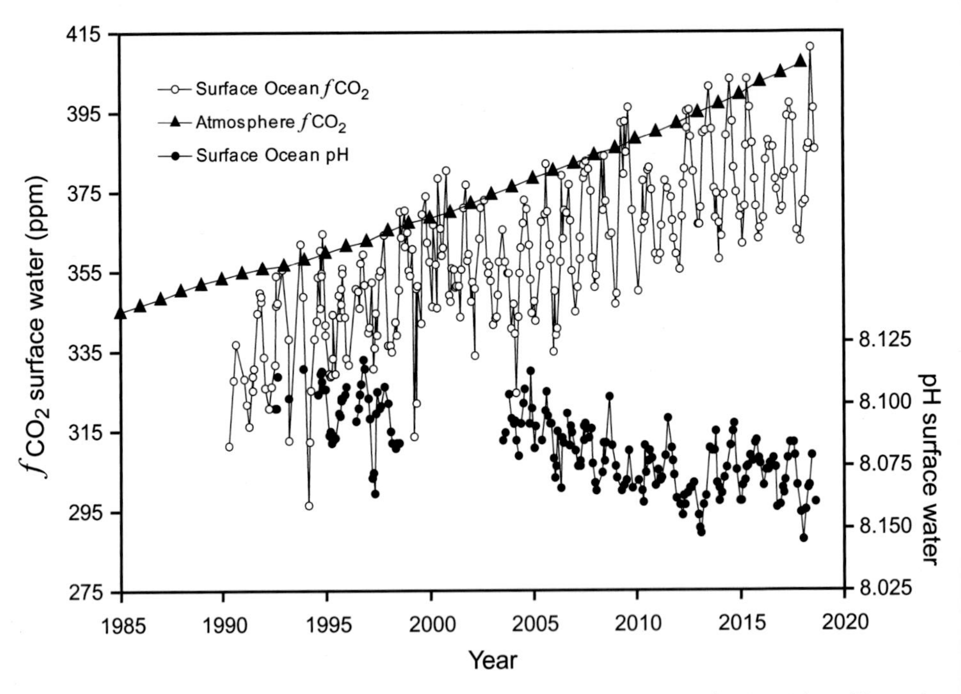

Figure 5.15 Atmospheric fCO_2 plotted along with surface ocean measurements of fCO_2 and pH at the Hawaii Ocean Time-series (HOT) as a function of time. The surface water data are from the Hawaii Ocean Time-series Data Organization & Graphical System (HOT-DOGS, https://hahana.soest.hawaii.edu/hot/hot-dogs/). The atmospheric data are from the same source referenced in the Fig. 5.14 caption.

$\Omega_{aragonite}$ of 2 or lower. Thus, the degree of saturation is important to the larvae creating the shell at values far above saturation equilibrium. The probable reason for this is that the cell maintains the region where aragonite precipitates (the calcification fluid) at a carbonate concentration higher than ambient to facilitate rapid aragonite precipitation. It requires energy to maintain such high internal carbonate ion concentrations, so the further away the environmental carbonate concentration is from the optimum intracellular carbonate concentration the more energy that must be used to maintain aragonite precipitation. At some point, the organism runs out of energy before the shell is formed.

It may be possible to manipulate seawater carbonate concentrations $[CO_3^{2-}]$ to high enough levels in shellfish hatcheries to maintain a viable industry in spite of ocean acidification. However, this intervention will not be possible for the world's coral reefs, pelagic pteropods, and coccolithophorids as the ocean grows more acidic. Which species will be most impacted and to what extent some organisms may evolve to adapt to these changing conditions is uncertain.

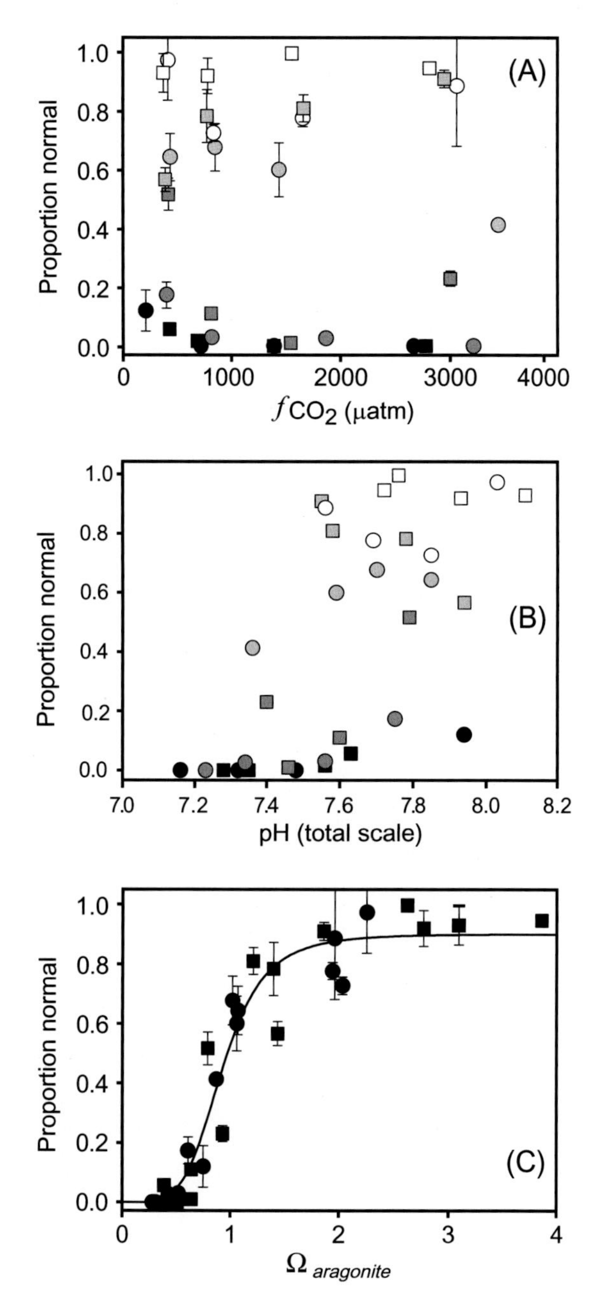

Figure 5.16 Results of mussel larvae incubations in seawater with different carbonate chemistry. The three graphs indicate the proportion of mussel larvae exhibiting normal shell development as a function of: (A) fCO_2, (B) pH, and (C) the degree of saturation with respect to aragonite, $\Omega_{aragonite}$. Normal shell development was assessed ~48 hours after fertilization using microscopic examination of 100–200 larvae sampled in triplicate from three separate incubations with the same chemistry in two separate experiments (circles and squares). In panels (A) and (B), symbol fill color indicates $\Omega_{aragonite}$, with black marking the lowest saturation states and white the highest. Redrafted from Waldbusser et al. (2015).

Discussion Box 5.3

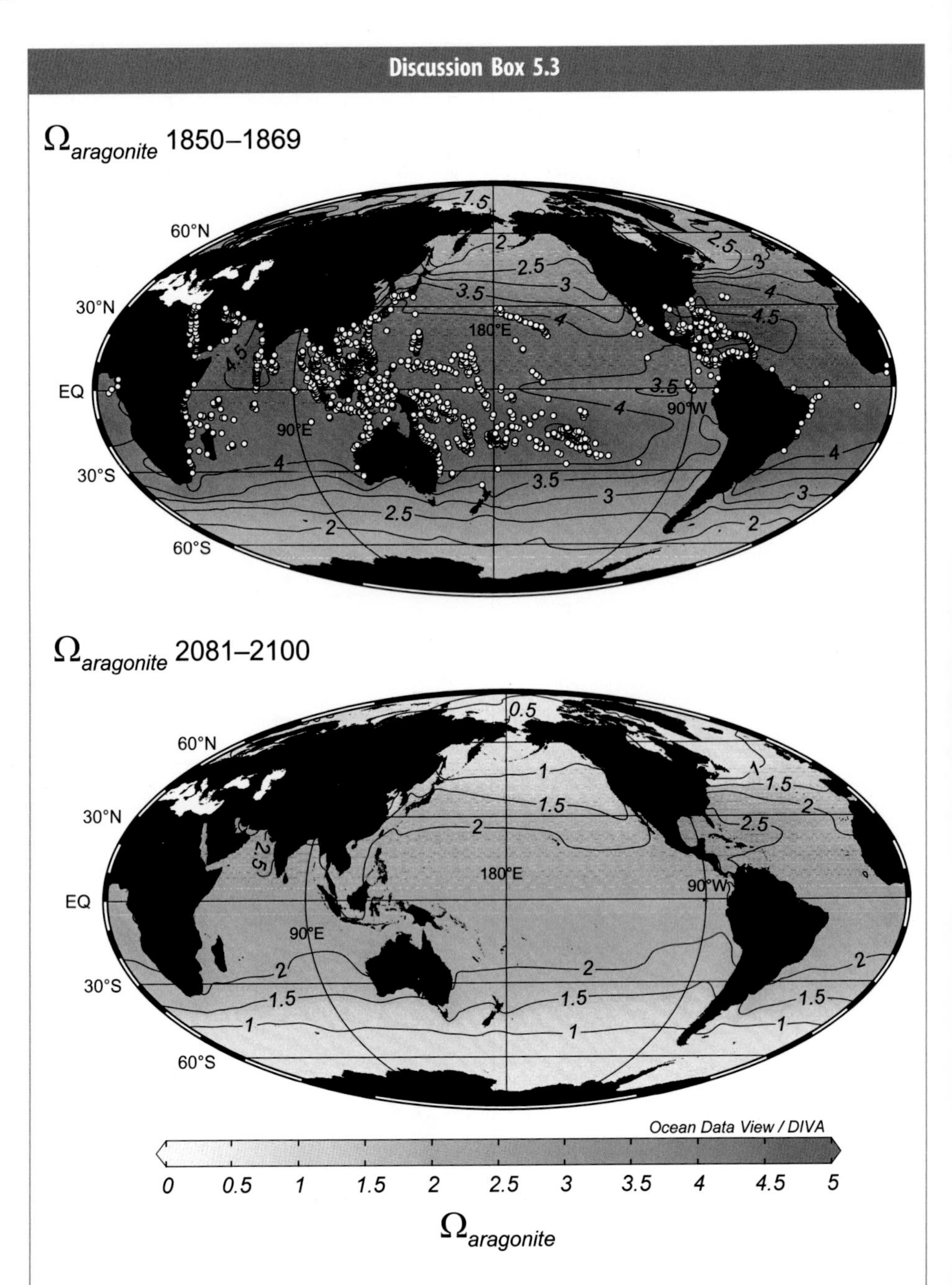

The aragonite and calcite saturation state of the ocean is rapidly declining, because of ocean uptake of anthropogenic CO_2. These figures show maps of estimated surface ocean aragonite saturation state ($\Omega_{aragonite}$) from model output for the decades 1850–1869 (upper panel) and projected for the decades 2081–2100 (lower panel). White circles in the upper panel indicate present-day locations of shallow coral reefs

Discussion Box 5.3 *(cont.)*

from ReefBase (www.reefbase.org). Model estimates, provided by Jim Christian of Fisheries and Oceans Canada (personal communication), are the mean values of seven Earth System model results. Future projections are for an IPCC high-emission scenario (ssp585).

- Which locations had naturally lower $\Omega_{aragonite}$ even before the industrial revolution? What likely causes the $\Omega_{aragonite}$ in these regions to be low?
- What $\Omega_{aragonite}$ levels were typical in the pre-industrial era in regions where corals typically grow? How does that compare with projected future levels for the same regions?
- Where are $\Omega_{aragonite}$ levels projected to fall below 1 by the end of this century? What about these regions makes them prone to undersaturated conditions?
- Where are organisms likely to be impacted by declining $CaCO_3$ saturation levels? How do you expect less formation of $CaCO_3$ in surface waters to affect the chemistry and ecology of the oceans?

5A.1 Appendix 5A.1 Carbonate System Equilibrium Equations in Seawater

Here we present the equations necessary for determining the concentrations of carbonate species in seawater for the three different definitions of alkalinity given in the text.

(a) Using Carbonate Alkalinity, A_C

(five equations, seven unknown chemical concentrations: A_C, DIC, HCO_3^-, CO_3^{2-}, CO_2, H^+, fCO_2^a)

$$A_C = [HCO_3^-] + 2[CO_3^{2-}], \tag{5.33}$$

$$DIC = [HCO_3^-] + [CO_3^{2-}] + [CO_2], \tag{5.22}$$

$$K_1' = \frac{[HCO_3^-] \times [H^+]}{[CO_2]}, \tag{5.18}$$

$$K_2' = \frac{[CO_3^{2-}] \times [H^+]}{[HCO_3^-]}, \tag{5.16}$$

$$K_{H,CO_2} = \frac{[CO_2]}{f_{CO_2}^a}. \tag{5.21}$$

(b) Using Carbonate and Borate Alkalinity, $A_{C\&B}$

(seven equations and 10 unknown chemical concentrations)

New unknown concentrations: $B(OH)_4^-$, $B(OH)_3$, and B_T, the last of which can be estimated from salinity.

substitute Eq. 5.32 for Eq. 5.33:

$$A_{C\&B} = [HCO_3^-] + 2[CO_3^{2-}] + [B(OH)_4^-]. \tag{5.32}$$

Include borate-related equations:

$$B_T = [B(OH)_3] + [B(OH)_4^-], \tag{5.27}$$

$$K_B' = \frac{[B(OH)_4^-] \times [H^+]}{[B(OH)_3]}. \tag{5.26}$$

(c) Using the Total Alkalinity, A_T, but No Anoxic Species

(18 equations and 25 unknown concentrations)

New unknown concentrations Si_T, $H_3SiO_4^-$, H_4SiO_4, P_T, PO_4^{3-}, HPO_4^{2-}, $H_2PO_4^-$, H_3PO_4, OH^-, $SO_{4,T}$, SO_4^{2-}, HSO_4^-, F_T, HF, F^-

substitute Eq. 5.31 for Eq. 5.32

$$A_T = [HCO_3^-] + 2[CO_3^{2-}] + [B(OH)_4^-] + [H_3SiO_4^-] + [HPO_4^{2-}] + 2[PO_4^{3-}] + [OH^-] - [H^+] - [HSO_4^-] - [HF] - [H_3PO_4]. \tag{5.31}$$

Include new species related mass balance and equilibrium equations:

$$Si_T = [H_4SiO_4] + [H_3SiO_4^-], \tag{5A.1.1}$$

$$P_T = [HPO_4^{2-}] + [PO_4^{3-}] + [H_2PO_4^-] + [H_3PO_4], \tag{5A.1.2}$$

$$SO_{4,T} = [SO_4^{2-}] + [HSO_4^-], \tag{5A.1.3}$$

$$F_T = [F^-] + [HF], \tag{5A.1.4}$$

$$K'_{Si} = \frac{[H_3SiO_4^-] \times [H^+]}{[H_4SiO_4]}, \tag{5A.1.5}$$

$$K'_{1P} = \frac{[H_2PO_4^-] \times [H^+]}{[H_3PO_4]}, \tag{5A.1.6}$$

$$K'_{2P} = \frac{[HPO_4^{2-}] \times [H^+]}{[H_2PO_4^-]}, \tag{5A.1.7}$$

$$K'_{3P} = \frac{[PO_4^{3-}] \times [H^+]}{[HPO_4^{2-}]}, \tag{5A.1.8}$$

$$K'_W = [OH^-][H^+], \tag{5A.1.9}$$

$$K'_S = \frac{[SO_4^{2-}][H^+]}{[HSO_4^-]}, \tag{5A.1.10}$$

$$K'_F = \frac{[F^-][H^+]}{HF}. \tag{5A.1.11}$$

References

Archer, D. E. (1996) An atlas of the distribution of calcium carbonate in sediments of the deep sea, *Global Biogeochemical Cycles*, **10**, 159–174.

Archer, D., S. Emerson, and C. R. Smith (1989) Direct measurement of the diffusive sublayer at the deep sea floor using oxygen microelectrodes, *Nature*, **340**, 623–626.

Biscaye, P. E., V. Kolla, and K. K. Turekian (1976) Distribution of calcium carbonate in surface sediments of the Atlantic Ocean, *Journal of Geophysical Research - Oceans*, **81**, 2595–2603.

Broecker, W. S., and T.-H. Peng (1982) *Tracers in the Sea*, 690 pp. Palisades: Eldigio Press.

Dickson, A. G. (1981) An exact definition of total alkalinity and a procedure for the estimation of alkalinity and total inorganic carbon from titration data, *Deep-Sea Research A*, **28**, 609–623.

Dickson, A. G. (2010) The carbon dioxide system in seawater: equilibrium chemistry and measurements, in *Guide to Best Practices for Ocean Acidification Research and Data Reporting* (eds. U. Riebesell, V. J. Fabry, L. Hansson, and J.-P. Gattuso), pp. 17–52, Luxembourg: IAEA Publications Office of the European Union.

Dickson, A. G., C. L. Sabine, and J. R. Christian (eds.) (2007) *Guide to Best Practices for Ocean CO_2 Measurements*, PICES Special Publication 3, 191 pp. Sidney: North Pacific Marine Science Organization.

Emerson, S. R., and M. I. Bender (1981) Carbon fluxes at the sediment-water interface of the deep-sea: Calcium carbonate preservation, *Journal of Marine Research*, **39**, 139–162.

Emerson, S., and J. Hedges (2003) Sediment diagenesis and benthic flux, in *The Oceans and Marine Geochemistry*, Vol. 6, *Treatise on Geochemistry* (ed. H. Elderfield), pp 293–319. Oxford: Elsevier-Pergamon.

Emerson, S. R., and J. I. Hedges (2008) *Chemical Oceanography and the Marine Carbon Cycle*, 453 pp. Cambridge: Cambridge University Press.

Feely, R. A., C. L. Sabine, K. Lee, et al. (2004) Impact of anthropogenic CO_2 on the $CaCO_3$ system in the oceans, *Science*, **305**, 362–366, doi:10.1126/science.1097329.

Hales, B., and S. R. Emerson (1996) Calcite dissolution in sediments of the Ontong-Java Plateau: in situ measurements of porewater O_2 and pH, *Global Biogeochemical Cycles*, **10**, 527–541.

Hales, B., and S. Emerson (1997) Calcite dissolution in sediments of the Ceara Rise: in situ measurements of porewater O_2, pH, and CO_2 (aq), *Geochimica et Cosmochimica Acta*, **61**, 501–514.

Jahnke, R. A., and G. A. Jackson (1987) Role of sea floor organisms in oxygen consumption in the deep North Pacific Ocean, *Nature*, **329**, 621–623.

Lewis, E. R., and D.W. R. Wallace (1998) Program developed for CO2 system calculations. ORNL/CDIAC – 105. Carbon Dioxide Information Analysis Center, Oak Ridge National Laboratory, U.S. Department of Energy, Oak Ridge, Tenn.

Lueker, T. J., A. G. Dickson, and C. D. Keeling (2000) Ocean pCO_2 calculated from dissolved inorganic carbon, alkalinity, and equations for K_1 and K_2: validation based on laboratory measurements of CO_2 in gas and seawater at equilibrium, *Marine Chemistry*, **70**, 105–119.

Mehrbach, C., C. H. Culberson, J. E. Hawley, and R. M. Pytkowicz (1973) Measurement of the apparent dissociation constants of carbonic acid in seawater at atmospheric pressure, *Limnology and Oceanography*, **18**, 897–907.

Millero, F. J. (1995) Thermodynamics of the carbon dioxide system in the oceans, *Geochimica et Cosmochimica Acta*, **59**, 661–677.

Morse, J. W., and F. T. Mackenzie (1990) *Geochemistry of Sedimentary Carbonates*, 706 pp. Oxford: Elsevier Science.

Mucci, A. (1983) The solubility of calcite and aragonite in seawater at various salinities, temperatures, and one atmosphere total pressure, *American Journal of Science*, **283**, 780–799.

Orr, J. C., V. J. Fabry, O. Aumont, et al. (2005) Anthropogenic ocean acidification over the twenty-first century and its impact on calcifying organisms, *Nature*, **437**, 681–686, doi:10.1038/nature04095.

Rossi, G. S., and V. Tunnicliffe (2017) Trade-offs in a high CO_2 habitat on a subsea volcano: condition and reproductive features of a bathymodioline mussel, *Marine Ecology Progress Series*, **574**, 49–64.

Sarmiento, J. L., J. Dunne, A. Gnanadesikan, et al. (2002) A new estimate of the $CaCO_3$ to organic carbon export ratio, *Global Biogeochemical Cycles*, **16**, 1107, doi:10.1029/2002GB001919.

Sayles, F. L. (1980) The solubility of $CaCO_3$ in seawater at 2°C based upon in-situ sampled pore water composition, *Marine Chemistry*, **9**, 223–235.

Stumm, W., and J. J. Morgan (1996) *Aquatic Chemistry*, 780 pp. New York: Wiley Interscience.

Subhas, A. V., N. E. Rollins, W. M. Berelson, et al. (2015) A novel determination of calcite dissolution kinetics in seawater, *Geochimica et Cosmochimica Acta*, **170**, 51–68, doi:10.1016/j.gca.2015.08.011.

Waldbusser, G. G., B. Hales, C. J. Langdon, et al. (2015) Saturation-state sensitivity of marine bivalve larvae to ocean acidification, *Nature Climate Change*, **5**, 273–280.

Weiss R. F. (1974) Carbon dioxide in water and seawater: the solubility of a non-ideal gas, *Marine Chemistry*, **2**, 203–215.

Wenzhöfer, F., M. Adler, O. Kohls, et al. (2001) Calcite dissolution driven by benthic mineralization in the deep sea: In situ measurements of Ca^{2+}, pH, pCO_2 and O_2, *Geochimica et Cosmochimica Acta*, **65**, 2677–2690.

Problems for Chapter 5

5.1. In problem 1.2 of Chapter 1, you were asked to make a recipe for artificial seawater from pure chemicals. The seawater was made up in a N_2-filled glove bag, so it contained no atmospheric CO_2. Now, consider what happens when this solution is removed from the globe bag and allowed to absorb CO_2 from the atmosphere.

a. What is the numerical value of the total alkalinity (A_T) while the solution was in the N_2 atmosphere and after it equilibrated with atmospheric CO_2?

b. What chemical species make up the alkalinity before and after?

5.2. The chemical oceanography literature on the nitrogen cycle tends to use ammonia, NH_3, and ammonium ion, NH_4^+, interchangeably, whereas only one species actually dominates at typical seawater pH.

a. Derive an expression for the concentration of NH_4^+ in terms of the total ammonia concentration, $(NH_4^+ + NH_3)$, hydrogen ion concentration, and apparent equilibrium constant.

b. Calculate the proportion of total ammonia $(NH_4^+ + NH_3)$ present as NH_4^+ for surface seawater with pH = 8.2 and for deep anoxic water with pH = 7.4. Assume temperature = 25°C and salinity = 35 for both calculations.

5.3. A surface water sample from the equatorial Atlantic has a DIC of 1780 μmol kg^{-1} and a pH of 8.12. The temperature of the water was 20°C at the time of collection and the salinity was 35.

a. Calculate the concentration of the various carbon species (CO_2, HCO_3^-, CO_3^{2-}, fCO_2^{SW}) using simultaneous equations and the information given above (i.e., without using CO2SYS or similar programs).

b. What is the carbonate alkalinity of the sample (A_C)?

c. If the total borate in the sample (B_T) is 416 μmol kg^{-1}, calculate the concentration of $B(OH)_4^-$ and determine the carbonate plus borate alkalinity ($A_{C\&B}$).

d. Use CO2SYS to check your values in part a.

5.4. The Mackenzie and Garrels mass balance (Chapter 2, Section 2.1.3) identified larger sources than sinks of HCO_3^- to the ocean. Recall that this excess HCO_3^- flux results from the weathering of silicate rocks, which transforms CO_2 to HCO_3^-. Such a process does not impact the total carbon in the ocean-atmosphere system but does increase the alkalinity of the ocean.

a. Use CO2SYS to calculate what average surface ocean DIC would have been in equilibrium with a pre-industrial atmospheric fCO_2 of 280 μatm, total alkalinity of 2300 μeq kg^{-1}, temperature of 18°C, and salinity of 35.

b. Based on the Mackenzie and Garrels mass balance, how much HCO_3^- is supplied to the ocean by rivers that is not removed by $CaCO_3$ precipitation each year?

c. Calculate the new ocean alkalinity value and the atmospheric fCO_2 in equilibrium with that ocean if this excess HCO_3^- flux persisted for 10 000 y, for 20 000 y, etc. up to 100 000 y. Make a plot of the atmospheric fCO_2 over time.

d. Why does atmospheric fCO_2 change when the total amount of carbon in the system does not?

5.5. This problem explores changes in ocean chemistry as water moves through the deep ocean following the thermohaline circulation.

a. Using Fig. 5.5, estimate the DIC and alkalinity concentrations at 3000 m and 40°N in the Atlantic and Pacific.

b. Calculate the change in DIC and change in alkalinity as deep-water flows from this point in the North Atlantic to this point in the North Pacific. (Note that this neglects the potential impact of Southern Ocean source waters here and assumes a direct flow path.)

(Problem continues on next page.)

c. Estimate the change in CO_3^{2-} concentration as deep-water flows from this point in the North Atlantic to this point in the North Pacific from the (Alk – DIC) approximation. Estimate the change in HCO_3^- along the same flow path.
d. Calculate the change in carbonate ion concentration between the deep North Pacific and the deep North Atlantic using CO2SYS. (Temperature and salinities for these sections appear in Fig. 1.7.)
e. Compare your answers from parts c. and d. with the change seen in Fig. 5.9. Which technique is more accurate? Why?
f. Is calcite more likely to dissolve in the deep North Atlantic or the deep North Pacific (at equal depths)? Why?
g. Using the DIC and alkalinity changes between the deep North Pacific and the deep North Atlantic that you calculated for part b, calculate the average contributions of organic carbon respiration and $CaCO_3$ dissolution that create these DIC changes.

5.6. Make a table showing whether you expect the carbonate system parameters listed in the top row to increase (↑), decrease (↓), or stay the same (→) as you subject your seawater to the processes listed on the left. Explain your reasoning.

Process	DIC	Alkalinity	$[CO_3^{2-}]$	$[HCO_3^-]$	pH	$[CO_2]$
Bubble with N_2 gas						
Respire OM						
Dissolve $CaCO_3(s)$						

5.7 This problem compares the pH of sediment pore waters in regions where diatoms dominate in surface waters to those in regions where coccolithophores dominate. Two theoretical locations have the same bottom water conditions of $O_2 = 200$ μmol kg^{-1}, DIC = 2090 μmol kg^{-1}, total alkalinity = 2230 μeq kg^{-1}, temperature = 2°C, salinity = 34.5, and pressure = 4000 dbar. Calcite K'_{SP} for these conditions is 9.36×10^{-7} mol^2 kg^{-2}.

a. Calculate the pH and CO_3^{2-} ion concentration of the bottom water using CO2SYS. Calculate the Ca^{2+} concentration based on the salinity. What is the calcite saturation state (Ω) for the bottom waters?
b. Suppose that oxygen in the pore waters is drawn down to 100 μmol kg^{-1} by organic matter respiration. For the location where diatoms dominate, respiration will be the only process affecting carbonate chemistry, because SiO_2 dissolution has no significant effect on the pH of seawater. Calculate the DIC, total alkalinity, pH, CO_3^{2-} ion concentration, and calcite saturation state of the pore waters.
c. Suppose the same amount of respiration affects the sediments in the location with calcite present in the sediments. The calcite present in the sediments will dissolve when pore waters are undersaturated until $[CO_3^{2-}]$ increases to saturation equilibrium with respect to calcite. Calculate the DIC, total alkalinity, and pH of the pore waters after the respiration and dissolution occurs.

6 Stable Isotope Tracers

6.1 Isotopes in the Environment 227
6.2 Analytical Methods and Terminology 229
6.3 Equilibrium Isotope Fractionation 231
6.3.1 Oxygen Isotopes, $\delta^{18}O$, in $CaCO_3$, a Tracer for Temperature Change 232
Discussion Box 6.1 236
6.3.2 Boron Isotopes, $\delta^{11}B$, a Tracer for pH 241
6.4 Kinetic Isotope Fractionation 244
Discussion Box 6.2 245
6.4.1 $\delta^{13}C$-DIC, a Tracer of Biological Processes 246
6.4.2 Triple Isotopes of Oxygen, a Tracer for Ocean Photosynthesis 249
6.4.3 $\delta^{18}O$ in Molecular Oxygen, a Tracer for Respiration 251
6.4.4 $\delta^{15}N$, a Tracer for the Marine Nitrogen Cycle 255
6.5 Rayleigh Fractionation 259
Discussion Box 6.3 263
6.6 Anthropogenic Influences 263
Appendix 6A.1 Relating the Stable Isotope Terms K, α, ε, and δ 266
Appendix 6A.2 Derivation of the Rayleigh Fractionation Equation 267
References 268
Problems for Chapter 6 271

Biological, physical, and geological processes all impact the distribution of isotopes within the ocean environment, making analyses of isotopic compositions a mainstay of the chemical perspective of oceanography. Interpretating these isotopic distributions has helped make chemical oceanography a strongly interdisciplinary science. Small contrasts in stable isotope compositions carry information for discriminating sources of an element, such as the fraction of atmospheric CO_2 that has originated from fossil fuel combustion, and evaluating the rates of processes like photosynthesis and respiration. Stable isotope signatures can persist over geologic time, even through severe changes in temperature and chemical composition, allowing isotopic distributions within fossils to provide information about the temperatures and partitioning of elements among different reservoirs in ancient environments.

This chapter presents a brief review of the principles, terminology, and application of stable isotope distributions in marine science. Using modern instruments, isotope ratios of almost all elements can be measured. However, the application of isotope ratios to environmental problems began with isotope ratios of light elements that could be introduced into mass spectrometers in the gaseous form (H, O, C, N, and S). These are also the elements that make up the bulk of most organic molecules and $CaCO_3$, so they have been extensively studied. Our discussion will mainly center on what has been learned about marine chemistry from the distribution of stable isotopes in these elements. After a brief introduction to isotopes and review of analytical methods and terminology, we focus on the two main classes of isotopic fractionation – equilibrium and kinetic. We present examples of how isotopic distributions that result from these mechanisms reveal information about past and present environments. A short discussion of Rayleigh distillation follows, because of its importance to environmental isotope distributions. We end with a few words about how anthropogenic activities have altered the carbon isotope distribution in the ocean and atmosphere.

6.1 Isotopes in the Environment

About 15 billion years ago, immediately following the Big Bang, the light elements of H ($\approx$99 percent), He ($\approx$1 percent), and trace amounts of Li formed as the universe cooled. Subsequent nuclear reactions during star formation and collapse created (and are still creating) all the remaining elements. On Earth today, there are 92 naturally occurring elements, each of which has a unique number of protons (atomic number, Z) in its nucleus; however, the number of accompanying neutrons (N) often varies. The masses of protons and neutrons are both very close to 1 atomic mass unit (1 g mol^{-1}), making their numeric sum equivalent to the atomic weight (A) of the atom. Atoms of the same element (same number of protons) with different numbers of neutrons, and hence different atomic weights, are referred to as *isotopes* of that element. The example of oxygen isotopes is provided in Table 6.1. Because different isotopes of the same element have the same number of electrons, they exhibit *almost* identical chemical properties. However, very small differences do occur, because heavier isotopes of an element typically form slightly stronger bonds to other atoms and because molecules containing heavier isotopes move somewhat more slowly at a given temperature due to their greater mass.

Isotopes can be categorized into stable and radioactive forms, based on whether they spontaneously convert into other nuclei at a discernable rate. For example, oxygen occurs in three stable isotope forms as ^{16}O, ^{17}O, and ^{18}O, with 8, 9, or 10 neutrons, respectively

Table 6.1 The isotopic composition of oxygen

Isotope	Protons (Z)	Neutrons (N)	Atomic wt. (A)	Atom (%)
^{16}O	8	8	16	99.8
^{17}O	8	9	17	0.04
^{18}O	8	10	18	0.2
$^{19}O^a$	8	11	19	0

^{a} $t_{½} = 26.5$ s

(Table 6.1). The ^{19}O isotope with 11 neutrons, however, is radioactive and has zero natural abundance because formation on Earth outside the laboratory does not occur, and its half-life (the time to decrease its concentration by a factor of two) of 26.5 s does not allow measurable quantities to persist. The listed percentages of the three stable oxygen isotopes in Table 6.1 are approximate, because isotopic compositions vary among many natural substances.

The "stair step" pattern in Fig. 6.1 illustrates the proton and neutron distributions among the elements. A wide step at a given atomic number (Z) indicates a variety of different isotopes for that element. Notice that the number of neutrons in naturally occurring isotopes tends to be equal to or greater than the number of protons. The excess of neutrons over protons increases at higher atomic numbers, causing the isotopic distribution in Fig. 6.1 to shift to the right of the 1:1 line. The number of stable isotopes also increases with atomic number. Finally, most nuclei beyond Hg, with Z in excess of 80 and A greater than 210, are unstable. Most naturally occurring nuclides in this super-heavy region are relatively short-lived daughters that ultimately cascade from decay of three extremely long-lived radioactive parents, ^{238}U, ^{235}U, and ^{232}Th. The utility of radioactive isotopes in chemical oceanography is the subject of the next chapter (Chapter 7).

Most elements have more than one stable isotope (Fig. 6.1). In general, the Earth's crust exhibits relatively homogeneous distributions among the isotopes of each naturally occurring element, with spatial and temporal variability typically on the order of one percent or less. Nevertheless, these small differences in relative abundance contain important information on material sources and processes. Although stable isotope studies are now being carried out for a wide range of elements because of continuous improvements in technology, this chapter will focus primarily on the most common applications for low mass ($Z < 20$) species, many of which occur in both organic and inorganic forms.

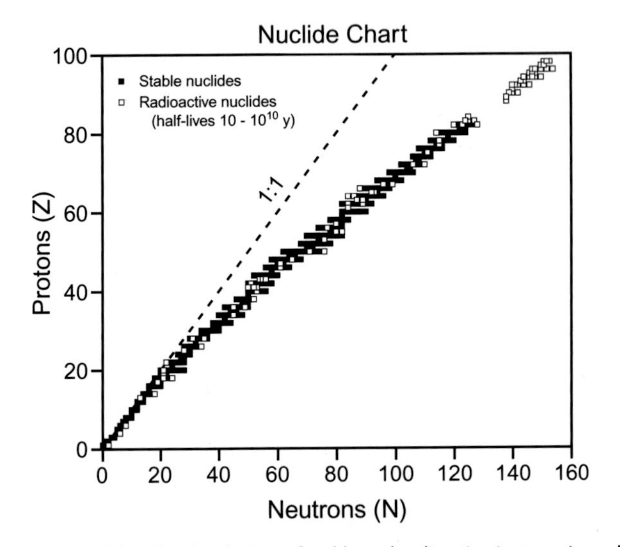

Figure 6.1 Chart of naturally occurring nuclides. The distribution of stable and radioactive isotopes in a plot of atomic number (Z = number of protons) versus number of neutrons (N). The horizontal portion of the "stair steps" formed by the stable nuclides (■) represents the number of stable isotopes for any given element.

6.2 Analytical Methods and Terminology

The keys to measuring small differences in isotope abundances are instrument sensitivity and stability. To obtain high stability, the relative amounts of at least two isotopes are measured simultaneously, repeatedly alternating between sample and standard materials. Isotopes of virtually any mass can now be measured using many different forms of sample introduction into modern mass spectrometers. Along with the expansion of isotope studies to nearly all elements of the periodic table, stable carbon, oxygen, nitrogen, and hydrogen isotopic measurements are now possible on nano to pico gram amounts of individual organic molecules separated by gas chromatography and analyzed "on the fly" by a downstream ratio mass spectrometer.

The lighter isotopes are usually measured in gaseous molecules that contain the target element as a major component. The analyte gas must not chemically react with metal or be prone to stick to surfaces (leaving out H_2O). Examples of the most commonly measured light isotopes and their corresponding analyte phase, abundances, and standard materials are presented in Table 6.2. Ideally, a standard material against which all sample isotope ratios are measured and expressed should be readily available in a homogeneous form. An ideal example is atmospheric N_2 for nitrogen isotopes. Another widely available reference material for H and O is Vienna Standard Mean Ocean Water (VSMOW), which is available for purchase. At the other availability extreme is the PeeDee Belemnite, or PDB standard, a small calcium carbonate fossil from the Carolina Peedee formation that was exhausted in the early days of stable isotope analysis. Although the

Table 6.2 Analyte standard materials for light isotope analysis

Element (analyte)	Stable isotopes	Atom (%)	Standard material
Hydrogen (H_2)	^{1}H	99.99	VSMOW[a]
	^{2}H	0.01	
Boron (BO_2^-, $Na_2BO_2^+$)	^{11}B	0.80	NIST SRM[b] Boric acid
	^{10}B	0.20	
Carbon (CO_2)	^{12}C	98.9	PDB $CaCO_3$
	^{13}C	1.1	
Nitrogen (N_2)	^{14}N	99.64	Air
	^{15}N	0.36	
Oxygen[c] (CO_2, O_2)	^{16}O	99.8	PDB $CaCO_3$; VSMOW[a]; Air
	^{17}O	0.04	
	^{18}O	0.2	
Sulfur (SO_2)	^{32}S	95.0	Canyon Diablo troilite (FeS)
	^{33}S	0.8	
	^{34}S	4.2	
	^{36}S	0.01	

[a] VSMOW = Vienna Standard Mean Ocean Water
[b] NIST SRM = National Institute of Standards and Technology Standard Reference Material
[c] Several standards are in regular use for oxygen

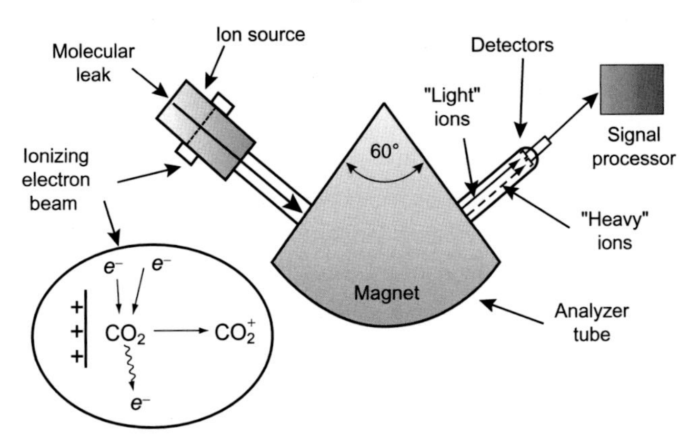

Figure 6.2 Schematic diagram of the main functions of a ratio mass spectrometer. The paths of charged molecules generated in the ion source are bent by the magnet and then sensed by physically spaced detectors at the end of the analyzer tube. The inset illustrates the reactions that occur during ionization of a CO_2 molecule as electrons from the electron beam collide with the molecule in the ion source. Modified from Emerson and Hedges (2008).

PDB-based scale is still used for reporting stable carbon isotope compositions, other secondary reference materials with known isotopic ratios relative to PDB now serve as working standards in stable isotope laboratories.

Stable isotope abundances are measured using ratio mass spectrometers (Fig. 6.2), which operate as continuous ion separators. In the example shown in Fig. 6.2, molecules of the analyte gas prepared from a sample are introduced into the ion source of the instrument through a non-fractionating molecular leak. The entry path to the mass spectrometer must be tiny to maintain a high vacuum within the mass analyzer. The incoming gas molecules are bombarded with electrons that are "boiled off" of a heated filament. Some of the high-energy electrons strike the neutral gas molecules and knock out an electron, producing singly charged positive ions (see inset in Fig. 6.2). These ions are repelled by positively charged plates in the ion source and are accelerated to constant velocity down the analyzer tube. The analyzer tube is evacuated so that the ions do not collide with other particles as they pass through a magnetic field. The path of a charged particle moving within a magnetic field is deflected in the direction of the field. The extent of deflection is greater for ions with a smaller mass to charge ratio (e.g., $^{12}CO_2^+$, mass/charge = 44) so that they are separated from their more massive counterparts (e.g., $^{13}CO_2^+$, mass/charge = 45) on the way down the analyzer tube. The process is similar to separating chaff from grain by lofting the mixture into a wind that carries the lighter chaff further away from its initial position. The net result in a mass spectrometer is that the initial mixture of ions is resolved spatially into separate beams containing ions of different mass/charge ratio. The "heavy" and "light" ions are focused onto different detectors that simultaneously create electrical currents proportional to the number of ions striking them. This procedure is carried out repeatedly for gases of the same type that are generated from sample and standard materials and alternately introduced into the ratio mass spectrometer.

The advantage of alternating analysis of sample and standard gases is that fluctuations in instrument response cancel over time.

The isotopic composition of the sample is presented as the relative deviation of its isotopic ratio from the ratio of the standard material measured during the analysis,

$$\delta H = \left[\frac{\left(H/L\right)_{sample}}{\left(H/L\right)_{standard}} - 1 \right] \times 1000 = \left[\frac{r_{sample}}{r_{standard}} - 1 \right] \times 1000. \qquad (6.1)$$

where H and L represent the "heavy" and "light" isotopes of the target element in the analyte, r is the H/L ratio, and most marine chemists pronounce this quantity as "del." In this formulation, the sample composition is normalized to that of the standard. The results are expressed in parts-per-thousand or per mil (‰) because the absolute relative differences between sample and standard are typically small. In the early days of stable isotope analysis, H/L ratios were calculated as absolute values (e.g., 0.0110 for $^{13}C/^{12}C$), a format that proved to be cumbersome.

"Isotopic fractionation" occurs when a reaction or process favors one isotope relative to another, leading to products and reactants with different stable isotope composition. Isotopic fractionation commonly occurs when molecules or atoms with the same elements, but different numbers of neutrons, are involved in spontaneous equilibrium reactions. Equilibrium reactions involving the same chemical species but with different numbers of neutrons have slightly different equilibrium constants. Such systems will exhibit small differences in isotopic abundance among their component elements at thermodynamic equilibrium. This phenomenon is called an "equilibrium isotope effect."

Isotopic fractionation also commonly occurs due to differential bond breaking or diffusion rates for elements or molecules with the same proton number but different atomic weights. Products containing the lighter isotope will typically form or diffuse faster, leaving behind a pool of reactants that is enriched in the heavier isotope. This phenomenon is called a "kinetic isotope effect." Fractionation caused by bond breaking occurs only for atoms directly involved in the bond that is being broken (or formed). This is particularly relevant for large organic molecules, where a pronounced bulk isotope fractionation occurs only when atoms at the cleavage point represent a large fraction of the total element in the parent ion or molecule.

We categorize our discussion of the utility of stable isotopes in chemical oceanography into equilibrium and kinetic isotope effects. Photosynthesis and respiration (Chapters 3 and 4) are non-equilibrium biological processes and thus dominated by kinetic isotope effects. The carbonate system (Chapter 5) involves inorganic reactions that are at or near equilibrium in seawater and sediments and is thus dominated by equilibrium isotope effects.

6.3 Equilibrium Isotope Fractionation

Equilibrium isotope effects are most likely to be exhibited by inorganic chemical species that rapidly interconvert between forms containing the same elements. Organic molecules almost never directly exhibit equilibrium isotope effects because the covalent bonds linking the component atoms do not continuously break and reform. We begin this section on equilibrium fractionation with a discussion of how temperature affects the stable isotope ratios in the most ubiquitous authigenic mineral – calcium carbonate.

6.3.1 Oxygen Isotopes, $\delta^{18}O$, in $CaCO_3$, a Tracer for Temperature Change

The carbonate buffer system, involving water (H_2O), the dissolved carbonate species (CO_2, HCO_3^-, and CO_3^{2-}), and solid calcium carbonate ($CaCO_3$) is an oceanographically important example of chemical reactions that exhibit equilibrium isotope effects for both carbon and oxygen. In such systems, chemical interconversion and associated isotopic exchange continue even when the system is at chemical and isotopic equilibrium, although under this state no net changes in chemical concentration or isotopic distribution are observed. It is also possible for a system to be at chemical equilibrium but not isotopic equilibrium. For example, if one instantaneously replaced $C^{16}O_2$ with $C^{16}O^{18}O$ in a seawater sample, it would take several minutes for the heaver isotope to distribute itself among all the inorganic carbonate species.

The dissolved reactions among water, CO_2, HCO_3^-, and CO_3^{2-}; and the heterogeneous exchange between CO_3^{2-} and $CaCO_3(s)$ are respectively (see Chapter 5):

$$CO_2 + H_2O \leftrightarrow HCO_3^- + H^+ \qquad K_1' = \frac{[HCO_3^-][H^+]}{[CO_2]}, \tag{6.2}$$

$$HCO_3^- \leftrightarrow CO_3^{2-} + H^+ \qquad K_2' = \frac{[CO_3^{2-}][H^+]}{[HCO_3^-]}, \tag{6.3}$$

$$CaCO_3(s) \leftrightarrow CO_3^{2-} + Ca^{2+} \qquad K_{SP}' = \frac{[Ca^{2+}][CO_3^{2-}]}{[CaCO_3]}. \tag{6.4}$$

Combining these reactions results in an expression relating CO_2, H_2O and $CaCO_3(s)$:

$$CO_2 + H_2O + Ca^{2+} \leftrightarrow CaCO_3(s) + 2H^+$$

$$K^{16\prime} = K_1'K_2' / K_{SP}' = \frac{[H^+]^2[CaCO_3]}{[Ca^{2+}][H_2O][CO_2]}. \tag{6.5}$$

Normally, the concentrations of H_2O and $CaCO_3$ would not appear in the equilibrium constant expressions, because by thermodynamic convention the activities of the nearly pure phases water and calcium carbonate are one. However, in this case, we retain the symbols for these concentrations because we seek an expression for the relative isotope ratio between H_2O and $CaCO_3$. Reaction 6.5 can be written for both ^{18}O and ^{16}O, and, for simplicity, we treat Eq. 6.5 as the ^{16}O reaction. The version for ^{18}O is

$$CO_2 + H_2{}^{18}O + Ca^{2+} \leftrightarrow CaCO_2{}^{18}O(s) + 2H^+,$$

$$K^{18\prime} = \frac{[H^+]^2[CaCO_2{}^{18}O]}{[Ca^{2+}][H_2{}^{18}O][CO_2]}. \tag{6.6}$$

Rearranging the ^{16}O equation and combining it with Eq. 6.6 results in an exchange reaction of the oxygen isotopes in H_2O with those in $CaCO_3$:

$$CaCO_3 + CO_2 + H_2{}^{18}O \leftrightarrow CaCO_2{}^{18}O + CO_2 + H_2O,$$

$$K' = K^{18\prime}/K^{16\prime} = \frac{[CaCO_2{}^{18}O][CO_2][H_2O]}{[CO_2][H_2{}^{18}O][CaCO_3]} = \frac{[CaCO_2{}^{18}O][H_2O]}{[H_2{}^{18}O][CaCO_3]}. \quad (6.7)$$

An exchange reaction has the same chemical species on both sides of the equilibrium equation but indicates how the isotopes are transferred from the species on the left side to those on the right. It is not possible from this equation to determine whether the ^{16}O in $CaCO_3$ exchanges with the ^{18}O in CO_2 or H_2O. However, in this case, it does not matter because the isotopes of water and CO_2 continuously exchange (via Eq. 6.2), and there are vastly more oxygen atoms in H_2O than in CO_2 (the CO_2 concentration in surface ocean water is about 10 μmol kg^{-1} while H_2O is 55.6 mol kg^{-1}.) The equilibrium constant in Eq. 6.7, like all thermodynamic equilibrium constants, is a function of temperature and pressure. As is almost always observed for equilibrium isotope effects, the extent of isotopic fractionation becomes less as the temperature of the system increases.

By convention, the magnitude of the isotope effect is expressed as a fractionation factor (α) which is the heavy over light ratio (*H*/*L*) of the product divided by that of the reactant. If the product is enriched in the heavy isotope relative to the reactant, α is greater than one; if the product is relatively depleted in the heavy isotope, α is less than one. For reaction 6.7, α has the form:

$$\alpha_{CaCO_3-H_2O} = \frac{\left(H/L\right)_{product}}{\left(H/L\right)_{reactant}} = \frac{[CaCO_2{}^{18}O]/[CaCO_3]}{[H_2{}^{18}O]/[H_2O]}. \quad (6.8)$$

Note that it is important to define which species is considered the product and which is the reactant in an equilibrium reaction, so the fractionation factor is often marked as $\alpha_{product\text{-}reactant}$. For this example, where all stoichiometric coefficients in the balanced equation are equal to one, α has the same value as K' in Eq. 6.7. This will not be the case for more complicated isotope equations with larger stoichiometric coefficients, because K' will have different exponents.

The fractionation factor is often written in the alternate form ε:

$$\varepsilon = (\alpha - 1) \times 1000. \quad (6.9)$$

This form of the fractionation factor is closely related to the difference in isotopic composition between the product and the reactant:

$$\varepsilon \approx \delta_{product} - \delta_{reactant}. \quad (6.10)$$

A derivation of the numeric relationships between *K*, α, ε, and the difference in the isotopic composition between the product and reactant for the oxygen isotope exchange reaction is presented in Appendix 6A.1. Fractionation factors for many of the important reactions among molecules containing the elements H, B, C, N, and O are presented in Table 6.3.

The classic application of an equilibrium isotope effect to oceanographic research is the use of stable oxygen isotopes to estimate the temperature change in an ancient environment in which a carbonate shell formed. The reaction on which this application is based is Eq. 6.7, which has a temperature dependent equilibrium constant. When $CaCO_3$ precipitates, the carbonate ion is incorporated into the calcite or aragonite ($CaCO_3$) shell of a

Table 6.3 Fractionation factors, $\varepsilon = (\alpha\text{-}1) \times 1000$, for important equilibrium (equations with two-way arrows) and kinetic (one-way arrows) reactions among the elements H, B, C, O, and N. Equilibrium fractionation factors are for 20° C. Kinetic fractionation factors are approximate as they vary in the marine environment.

	Reaction	ε (‰)	Reference
H	Evaporation / precipitation		
	$H_2O(g) \rightleftarrows H_2O(l)$	+78	Dansgaard, 1965
B	Acid-base equilibria		
	$B(OH)_3 + H_2O \rightleftarrows B(OH)_4^- + H^+$	–27	Foster and Rae, 2016
C	CO_2 solubility		
	$CO_2(g) \rightleftarrows CO_2(aq)$	–1.1	Knox et al., 1992
	Carbonate equilibria	+8.4	
	$CO_2 \rightleftarrows DIC$	pH = 8.15	Zhang et al., 1995
	Photosynthesis		
	$CO_2 + H_2O \rightarrow CH_2O(OM) + O_2$	–14 to –19	O'Leary, 1981
	Respiration		
	$CH_2O(OM) + O_2 \rightarrow CO_2 + H_2O$	~0	
	$CaCO_3$ precipitation	+1	
	$Ca^{2+} + HCO_3^- \rightleftarrows CaCO_3(s) + H^+$	(Calcite)	Romanek et al., 1992
	Methane formation		
	$7H_2 + 2CO_2 + 2H^+ \rightarrow 2CH_4^+ + 4H_2O$	–40 to –90	Lansdown et al., 1992
O	Evaporation / precipitation		
	$H_2O(g) \rightleftarrows H_2O(l)$	+9	Dansgaard, 1965
	O_2 solubility		
	$O_2(g) \rightleftarrows O_2(aq)$	+0.7	Knox et al., 1992
	Marine photosynthesis		
	$CO_2 + H_2O \rightarrow CH_2O(OM) + O_2$	+4	Luz and Barkan, 2011
	Respiration		
	$CH_2O(OM) + O_2 \rightarrow CO_2 + H_2O$	–20	Kiddon et al., 1993
N	Solubility		
	$N_2(g) \rightleftarrows N_2(aq)$	+0.7	Knox et al., 1992
	Nitrogen fixation		
	$4H^+ + 6H_2O + 2N_2 \rightarrow 4NH_4^+ + 3O_2$	–2	Carpenter et al., 1997
	NO_3^- uptake during photosynthesis		
	$2H^+ + NO_3^- + H_2O \rightarrow NH_4^+ + 2O_2$	–5	Altabet and Francois, 1994
	Denitrification		
	$4NO_3^- + 5CH_2O(OM) + 4H^+ \rightarrow 2N_2 + 5CO_2 + 7H_2O$	–25	Granger et al., 2008

plant or animal growing in the water. Thus, changes in $\delta^{18}O$ of the $CaCO_3$ shell can be used as a record of the changes in temperature during shell formation. This ^{18}O paleotemperature method involves several assumptions. The first is that either the organism precipitated $CaCO_3(s)$ (calcite or aragonite) in isotopic equilibrium with dissolved CO_3^{2-} and H_2O, or that any non-equilibrium isotope (vital) effect created by the organism is known as a function of temperature. This can be evaluated empirically from either laboratory incubation experiments or by measuring field samples that are known to have grown in different temperature environments. The second assumption is that the $\delta^{18}O$ of the shell has remained unchanged since the time it was precipitated by the

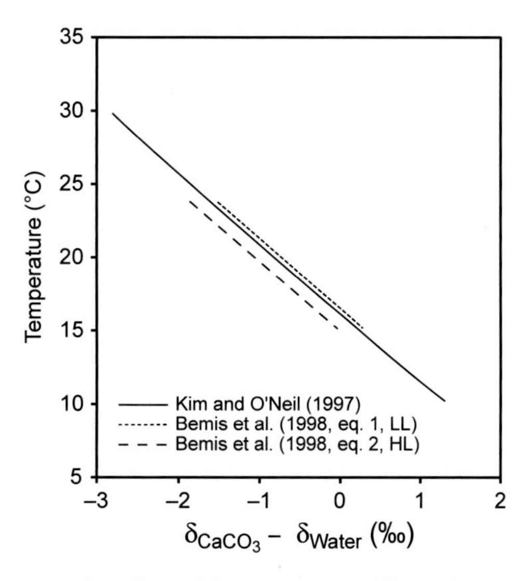

Figure 6.3 Laboratory-determined temperature dependence of the oxygen isotope difference between $CaCO_3$ (calcite), δ_{CaCO_3}, and the water, δ_{Water}, in which it formed. The experiments and the equations that define the relationships are: (solid line) inorganic calcite precipitated in the laboratory (Kim & O'Neil, 1997), $T(°C) = 16.1 - 4.64(\delta_{CaCO_3} - \delta_{Water}) + 0.09(\delta_{CaCO_3} - \delta_{Water})^2(‰)$; (dotted line) calcite precipitated by the planktonic foraminifer *Orbulina universa* under low light (LL) conditions (Bemis et al. 1998), $T(°C) = 16.5(\pm 0.2) - 4.80(\pm 0.16)(\delta_{CaCO_3} - \delta_{Water})(‰)$; and (dashed line) calcite precipitated by the planktonic foraminifer *O. universa* under high light (HL) conditions (Bemis et al. 1998), $T(°C) = 14.9(\pm 0.1) - 4.80(\pm 0.08)(\delta_{CaCO_3} - \delta_{Water})(‰)$. There are many different versions of these curves for both inorganically and organically produced $CaCO_3$.

organism. Finally, it is assumed that the $\delta^{18}O$ of the original water is known or has not changed. This final assumption is the most challenging to determine independently and often poses the largest uncertainty in the application of isotopes to reconstruct temperature. One way to address this uncertainty uses "clumped isotopes," which have a temperature effect independent of the $\delta^{18}O$ of the water in which the fossil grew. We discuss this in more detail at the end of this section.

The $\delta^{18}O$ versus temperature curves obtained empirically by growing calcite and aragonite precipitating organisms in water of known $\delta^{18}O$ are illustrated in Fig. 6.3. Carbonates precipitated at different temperatures, in this case over 0–24°C, were individually isolated and analyzed for their $\delta^{18}O$ composition. The isotopic compositions of each carbonate mineral are displayed as the difference between the $\delta^{18}O$ of the carbonate, δ_{CaCO_3}, and the $\delta^{18}O$ of the ambient water, δ_{Water} in which it grew. This formulation allows temperature to be determined from δ_{CaCO_3} if the δ_{Water} is known. A number of empirical equations of this form with slightly different numerical coefficients have been derived for pure inorganically produced $CaCO_3$ and for biologically produced $CaCO_3$ from different organisms. Bemis et al. (1998) present a review of these expressions.

The stable oxygen isotope compositions of the carbonate samples and of the water are determined in a mass spectrometer using CO_2 gas that is derived from the samples during sample preparation. The $\delta^{18}O$ of $CaCO_3$ is determined on CO_2 released from the $CaCO_3$

sample with anhydrous phosphoric acid. The H_3PO_4 must not contain water, which would rapidly equilibrate with the generated CO_2 and change its $\delta^{18}O$. The value of δ_{Water} cannot be determined directly on H_2O vapor, because water is a "sticky" molecule that does not behave well in mass spectrometers. Instead, a small amount of CO_2 is equilibrated with water at a well-known temperature. Because CO_2 and H_2O come to equilibrium after a few minutes, the oxygen isotope ratio of the CO_2 in the incubation will reflect that of the water after the equilibrium fractionation between CO_2 and H_2O is subtracted.

Discussion Box 6.1

Upon first encountering it, some people find the δ and α notation rather cumbersome. Let's do some calculations to clarify what makes these such a useful way of expressing isotopic composition and fractionation.

- Imagine you've made a measurement of $\delta^{15}N$ of N_2 dissolved in seawater that is 0.7‰ relative to a standard of atmospheric air. Convert this measurement to express it as $r_{sample}/r_{standard}$.
- Reported isotopic abundances for ^{14}N and ^{15}N in air are 0.99634 and 0.00366, respectively. Calculate the isotopic abundances for ^{14}N and ^{15}N in the seawater measurement.
- Which of these three ways of expressing the isotopic composition of dissolved N_2 do you prefer? Why?
- Imagine you've made a measurement of $\delta^{13}C$ of organic carbon in a leaf that is –26‰ relative to the PDB standard. The $\delta^{13}C$ of atmospheric CO_2 that the leaf used during photosynthesis was –7‰. Calculate the fractionation factor expressed as an α value and an ε value.
- How similar is your fractionation factor as an ε value to the difference in isotopic composition (δ values) between the product and the reactant?

Application of the $\delta^{18}O$ Thermometer to Past Temperature Changes. The first application of the paleotemperature method was undertaken by Harold Urey (Urey et al., 1951), who analyzed incremental sections along the radius of the shell of a fossil belemnite (a type of squid-like cephalopod). This bullet-shaped shell, from the Cretaceous Peedee formation of South Carolina, was the original standard material for $\delta^{13}C$ analysis (Table 6.2). The isotopic "diary" of the Peedee belemnite (Fig. 6.4) records four cool extremes separated by three warm periods, in what appears to be a 3.5-y life history laid down approximately sixty million years ago. The absolute temperature scale is uncertain, because the $\delta^{18}O$ of water in the ancient sea where the belemnite lived is unknown; however, the ability to detect temperature changes amounted to a remarkable discovery at that time.

A more sweeping application of the paleotemperature method has been developed by paleoceanographers who measure the $\delta^{18}O$ of foraminifera (millimeter-sized carbonate secreting animals that grow in either surface waters or on the sea floor) from long sediment cores representing several million years of Earth history. The last 700 000 y record

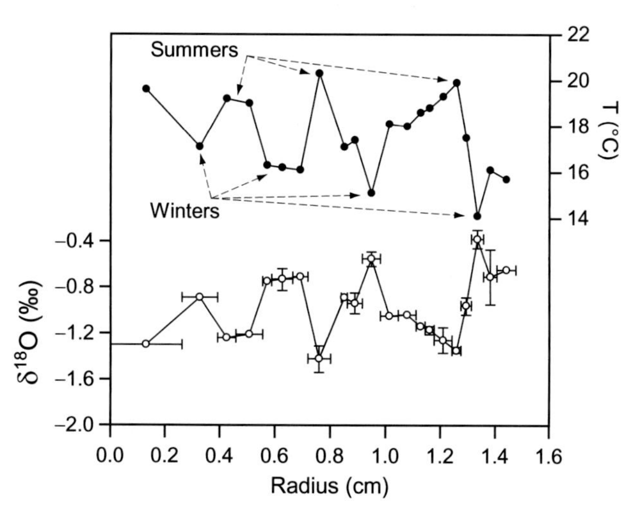

Figure 6.4 The original application of oxygen isotope thermometry using $\delta^{18}O$ of $CaCO_3$ in a fossil shell. The thermal history (*top*) of the surroundings in South Carolina, USA, during the Cretaceous period as recorded by oxygen isotope changes (*bottom*) as a function of shell radius in the Peedee belemnite. Error bars on $\delta^{18}O$ are derived from replicates, where available. Redrawn from Urey et al. (1951).

(Fig. 6.5) indicates approximately eight glacial/interglacial cycles, with total $\delta^{18}O$ offsets in the planktonic foraminifera shells of approximately 1.8‰. If this offset was caused by temperature change alone, then the temperature of the surface ocean would have had to change by about 5°C from glacial to interglacial times. The main difficulty in interpreting such fluctuations in terms of paleotemperatures is that the more positive $\delta^{18}O$ values in carbonates precipitated during glacial times partially result from a change in the $\delta^{18}O$ of seawater during ice ages.

As will be discussed later in this chapter, polar ice has a much lighter oxygen isotope ratio $\left(\delta^{18}O \approx -35‰\right)$ compared to seawater, which is the standard and therefore has a $\delta^{18}O = 0‰$ by definition. Fractionation during evaporation and precipitation and the net transfer of water from the ocean to the atmosphere to polar ice sheets discriminates against isotopically heavier water molecules, leaving water that arrives at the poles as snow depleted in ^{18}O. Since ice sheets grew larger during glacial periods, at those times there would be more polar ice with depleted ^{18}O, which requires ocean water during the same periods to have higher ^{18}O. Such a change is in the same direction of the observed changes (Fig. 6.5).

It is possible to obtain a first-order estimate of how much ice expansion would change the $\delta^{18}O$ of seawater using a simple mass balance calculation for oxygen isotopes in water. A first-order global hydrologic balance states that the mass of water on the surface of the Earth, M_T, is equal to the mass in the oceans (subscript $_O$) and the mass in polar ice caps (subscript $_I$):

$$M_T = M_O + M_I. \tag{6.11}$$

This equation also represents the ^{16}O mass balance because the light isotope is by far the most abundant. A similar mass balance for ^{18}O in H_2O yields

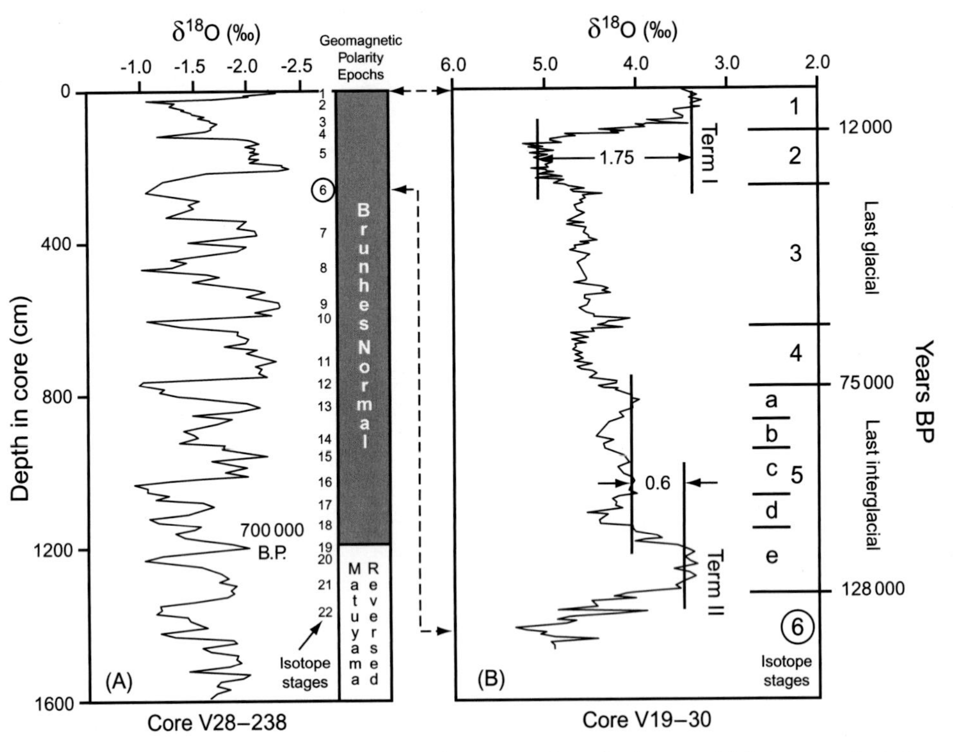

Figure 6.5 The oxygen isotope record in foraminifera tests in sediment cores from the deep sea. (A) is data over the last million years and (B) zooms in on the last 130 000 years. The $\delta^{18}O$ scales are different in (A) and (B) because the isotope record on the left is for planktonic foraminifera and the one on the right is for benthic foraminifera. The present planktonic-benthic difference is ~5‰. The core on the left is V28–238 from the equatorial Pacific (Shackleton and Opdyke, 1973). The core on the right is also from the equatorial Pacific, V19–30 (Chappell and Shackleton, 1986). Redrawn from Crowley (1983) and Broecker (2002).

$$M^{18}{}_T = M^{18}{}_O + M^{18}{}_I. \tag{6.12}$$

The ^{18}O mass balance can be written in terms of the measured ratio of ^{18}O to $^{16}O(r = {}^{18}O/{}^{16}O)$ and the ^{16}O mass balance in Eq. 6.11:

$$M_T r_T = M_O r_O + M_I r_I. \tag{6.13}$$

If we switch the terminology in this equation from mass, M, to the depth of water in the ocean, H, we can use independent estimates of glacial–interglacial sea level changes. The mass of water in the ocean is equal to seawater density times ocean area times ocean water depth, $M = \rho A H (\mathrm{kg\ m^{-3}(m^2)\ m})$. Substituting this equation into Eq. 6.13 and dividing both sides by ocean area, A, and density, ρ, because they do not appreciably change between glacial and interglacial times leaves

$$H_T r_T = H_O r_O + H_I r_I. \tag{6.14}$$

Today, the mean ocean depth, $H_O{}^t$, is 3682 m (the superscript, *t*, means today), while the amount of water in today's polar ice caps, $H_I{}^t$, is equivalent to ~80 m of ocean depth. (Sea level will rise about 80 m when all present polar ice has melted.)

One can write two equations like 6.14, one for today and one for glacial times (superscript ^{g}). Since the totals on the left side of these two equations are the same ($H_T{}^t = H_T{}^g$, because water moves from ocean to ice, but the total does not change), they can be combined to give

$$H_O{}^t r_O{}^t + H_I{}^t r_I{}^t = H_O{}^g r_O{}^g + H_I{}^g r_I{}^g. \tag{6.15}$$

Dividing the equation by $r_O{}^t$ puts all of the isotope ratios relative to today's ratio in seawater, which is the standard so the equation can be written in terms of δ values using Eq 6.1.

$$H_O{}^t + H_I{}^t(r_I{}^t/r_O{}^t) = H_O{}^g(r_O{}^g/r_O{}^t) + H_I{}^g(r_I{}^g/r_O{}^t). \tag{6.16}$$

Where,

$$(r/r_O{}^t) = (\delta/1000 + 1). \tag{6.17}$$

These last two equations can be solved for the δ values of seawater in glacial times, $\delta_O{}^g$, using what we know about the changes in sea level and the $\delta^{18}O$ of polar ice between glacial and interglacial periods. During the last glacial maximum sea level was about 100 meters lower than today, thus $H_O{}^g = 3582$ m and $H_I{}^g = 180$ m. The $\delta^{18}O$ of polar ice today is ~ –35‰ and that measured during the last glacial maximum was ~ –41‰ (Grootes et al., 1993). Using these values, we can calculate that the average glacial seawater isotopic composition, $\delta^{18}O_O{}^g$, was +1.2‰, about two-thirds of the measured $\delta^{18}O$ change in foraminifera calcite!

There are several uncertainties in this calculation. For example, the distribution of $\delta^{18}O$-H_2O within polar ice sheets during the last glacial period is not well known. That makes our estimate of $\delta^{18}O_I{}^g$ and therefore $\delta^{18}O_O{}^g$ only an approximation; however, the calculation shows that the change in the isotope ratio of the ocean source cannot be overlooked when interpreting the variations shown in Fig. 6.5.

In the early 1970s, Nick Shackleton from Cambridge University determined the $\delta^{18}O$ in the shells of both planktonic and benthic foraminifera in a sediment core from the equatorial Pacific (Shackleton and Opdyke, 1973). Most analyses before then had been done on planktonic foraminifera because they are far more abundant in sediments. The new $\delta^{18}O$ analysis showed that the changes in oxygen isotopes of both planktonic and benthic foraminifera were about the same. Deep ocean temperature today is about 2°C, and the freezing point is about –2°C, so temperature change cannot be responsible for all of the signal in the benthic foraminifera proving that at least some of the $\delta^{18}O$ change must be due to the source water itself.

Almost 20 years after Shackleton's foraminifera analysis, the mechanisms behind the glacial–interglacial $\delta^{18}O$ changes were resolved by actually measuring the $\delta^{18}O$ in the pore waters of marine sediments in very long (~100 m) sediment cores raised by the Ocean Drilling Project (ODP) (Adkins and Schrag, 2001). As sediments accumulate on the sea floor, the ambient water is also incorporated among the sediment grains. The pore water exchanges by molecular diffusion with overlying bottom water as the sediments accumulate, but, since molecular diffusion is a relatively slow process, the signature of the $\delta^{18}O$–H_2O of previous times is not totally erased. The observed pore water changes are much smaller than the true glacial–interglacial change because of diffusive smoothing, but this process can be accurately modeled because we know molecular diffusion rates very well. The observed change between today's bottom waters and the pore waters from the last glacial maximum at about 25 m depth in the sediment core is about 0.3‰, which, when fit with a one-dimensional molecular diffusion model, implies a bottom water $\delta^{18}O$ change of 0.95‰. Since the mean change in foraminiferal $\delta^{18}O$ between Holocene and glacial times from 20 globally distributed cores is 1.75‰ (Broecker, 2002), about half of this observed change in foraminifera calcite is caused by seawater becoming heavier during the glacial period and about half by temperature change. This implies that the glacial–interglacial change in the temperature of surface waters was 2–3°C, not 5–6°C as originally suggested, and that deep ocean water was about 3°C colder during glacial times – perilously close to freezing!

Clumped Isotopes. A relatively new development in the use of oxygen isotopes as tracers of temperature change is called "Clumped Isotope Thermometry," in which isotope ratio changes are dependent on temperature change but independent of the $\delta^{18}O$ of the water in which the shell grew. In the case of $CaCO_3$, rare, heavy isotopes of both carbon, ^{13}C, and oxygen, ^{18}O, are found together or "clumped" as $Ca^{13}C^{18}O^{16}O_2$, in a multiply substituted isotopologue. ("Isotopologues" are different isotopes and configurations of isotopes in a chemical compound. "Multiply substituted isotopologues" are compounds that contain more than one rare isotope.) The presence of this species is a result of the exchange reaction between carbonate ions that is then preserved in the $CaCO_3$ mineral (Eiler, 2011; see Fig. 6.6):

$$^{12}C^{16}O_2{}^{18}O^{2-} + {}^{13}C^{16}O_3^{2-} \rightleftarrows {}^{13}C^{16}O_2{}^{18}O^{2-} + {}^{12}C^{16}O_3^{2-} \tag{6.18}$$

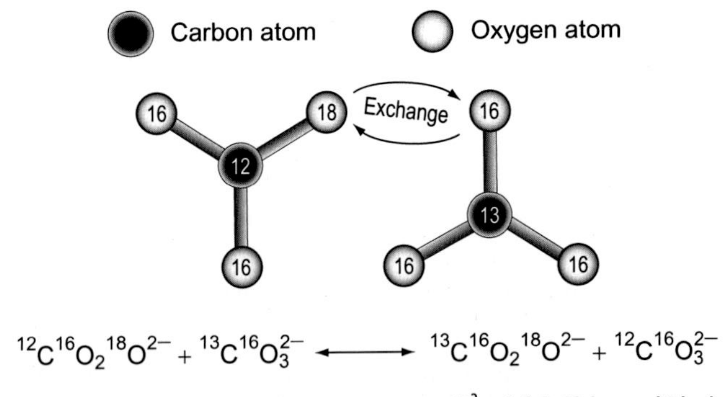

Figure 6.6 Schematic illustration of a clumped isotope exchange reaction in CO_3^{2-} (of $CaCO_3$) on which clumped isotope thermometry is based. A single ^{18}O atom in $^{12}C^{16}O_2{}^{18}O^{2-}$ exchanges with an ^{16}O atom in an adjacent $^{13}C^{16}O_3^{2-}$ creating CO_3^{2-} with the rare isotopes of both C and O. Modified from Eiler (2011).

This is a homogeneous reaction, because it is written as though it takes place entirely within a single solid phase, where the carbonate ions in reaction, 6.18 are part of the $CaCO_3$ solid. The reaction is an equilibrium reaction, and therefore the distribution of reactants and products is temperature dependent. Phase homogeneity distinguishes this isotope temperature gauge from the traditional one indicated in Eqs. 6.2, 6.3, and 6.4 where dissolved ions, water, and the solid phase are all involved. The actual mechanism that controls the distribution of the isotopes during solid $CaCO_3$ formation is uncertain, but observations indicate that the clumped isotope distribution can be interpreted as a homogeneous solid phase reaction during $CaCO_3$ formation. Under normal circumstances, the distribution of isotopes is set at the time of shell formation, and isotopes do not redistribute in a temperature sensitive way after the solid forms.

The chemical bonds of the multi-substituted species containing both heavy isotopes have lower vibrational energy than the other isotopologues and are thus more thermodynamically stable (Eiler, 2011). This results in the equilibrium abundance of this species being greater than predicted if the isotopes were randomly distributed. As temperature increases, the differences between chemical equilibrium and a random distribution diminish, which is why determining the abundance of the multi-substituted isotopologue in $CaCO_3$ is a tracer of the temperature when the solid precipitated and is independent of the isotope ratios in the water.

The temperature dependence of clumped isotopes in marine foraminifera and corals has been determined and compared with observations in inorganic laboratory experiments. At the time of writing this book, this method is just beginning to be applied to distinguishing temperature and ice volume changes during Pleistocene glacial–interglacial cycles (e.g., Grauel et al., 2013). Undoubtedly complications with the method will arise, as they have in the development of other environmental proxies. However, the payoff is great because, by combining traditional oxygen isotope ratios of $CaCO_3$ with "clumped isotope" determinations, both the temperature of the environment and the $\delta^{18}O$ of the water in which the carbonate secreting species grew can be determined.

6.3.2 Boron Isotopes, $\delta^{11}B$, a Tracer for pH

Boron isotopes have received a lot of attention because their equilibrium fractionation has proven to be a reliable tracer for ocean pH, and therefore aspects of the ocean carbon cycle. Boron is conservative in seawater and the second most important acid-base pair in total alkalinity, with a proton exchange between borate, $B(OH)_4^-$, and boric acid, $B(OH)_3$ (Chapter 5, Eqs. 5.26–5.29). There are two main isotopes of boron, ^{11}B (80.1%) and ^{10}B (19.9%), and there is a 27‰ fractionation between the two species in seawater, such that $B(OH)_3$ is 27‰ heavier than $B(OH)_4^-$. We will show how this makes the isotopic composition of both species a function of pH. The utility of this method for paleo-pH applications stems from the observation that trace amounts of $B(OH)_4^-$ (but no $B(OH)_3$) are incorporated within calcium carbonate during the formation of plant and animal shells.

Since the borate system is at equilibrium in seawater, it is possible to predict exactly how the isotope ratio of the two species in the acid-base pair vary as a function of pH. The distribution of $B(OH)_3$ and $B(OH)_4^-$ concentrations as a function of pH (Fig. 6.7A) is

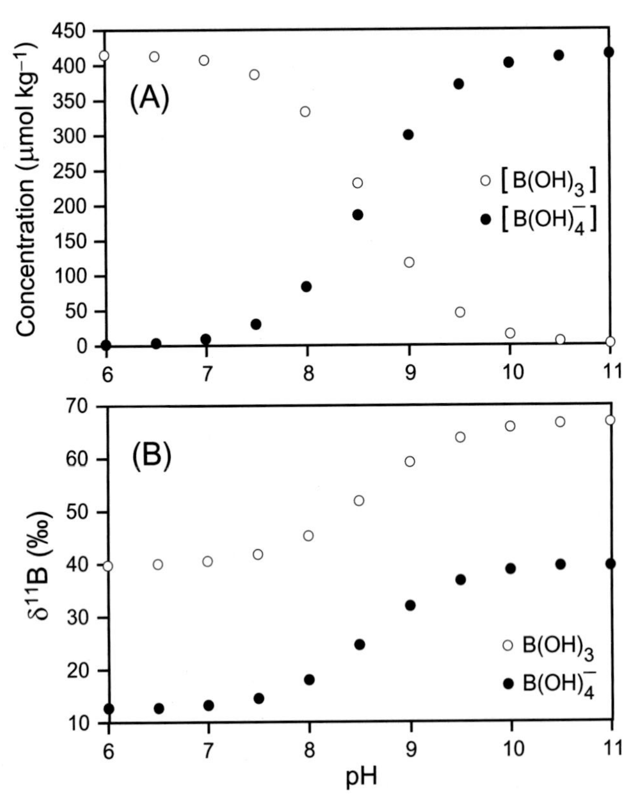

Figure 6.7 The basis for the boron isotope pH tracer. (A) The change in concentrations of $B(OH)_3$ and $B(OH)_4^-$ as a function of pH in seawater using the equilibrium constant for borate given in Appendix F (25°C, $S_P = 35$) and the concentration of total boron in seawater (416×10^{-6} mol kg^{-1}). (B) Changes in the $\delta^{11}B$ (‰) of both boron species, $B(OH)_3$ and $B(OH)_4^-$, as a function of pH using the isotope mass balance for the two species, a 27.1‰ fractionation between $B(OH)_3$ and $B(OH)_4^-$, and a $\delta^{11}B$ of 39.6‰ for total dissolved boron in the ocean.

determined from the equilibria in Eqs. 5.26–5.29, the conservative concentration of total boron, B_T, in seawater (416 μmol kg^{-1}, $S_P = 35$), and the equilibrium constant $K_B{}' = 2.53 \times 10^{-9} (T = 25°C,\ S_P = 35)$ for the reaction. The value of the equilibrium constant as a function of temperature and salinity can be determined from the equations in Appendix F.

The $\delta^{11}B$ of $B(OH)_4^-$ and $B(OH)_3$ (Fig. 6.7B) as a function of pH is derived following a similar mass balance procedure to that used previously for the difference in $\delta^{18}O$ of seawater between glacial and interglacial times (Eqs. 6.11–6.17). The mass balance equation for ^{10}B is

$$^{10}B_T = \left[^{10}B(OH)_3\right] + \left[^{10}B(OH)_4^-\right] \tag{6.19}$$

and for ^{11}B is

$$^{11}B_T = \left[^{11}B(OH)_3\right] + \left[^{11}B(OH)_4^-\right]. \tag{6.20}$$

The ^{11}B equation can be recast as the concentration of ^{10}B times the ratio of ^{11}B to ^{10}B, $r = (^{11}B/^{10}B)$:

$$^{10}B_T r_T = \left[^{10}B(OH)_3\right] r_{B(OH)_3} + \left[^{10}B(OH)_4^-\right] r_{B(OH)_4^-}. \tag{6.21}$$

From Eq. 6.1:

$$r = (\delta/1000 + 1) r_{std}. \tag{6.22}$$

There are three isotope ratios in the marine borate system: the isotope ratio of the total boron, B_T, which is conservative in the ocean at $\delta^{11}B\text{-}B_T = 39.6‰$ (with respect to the NIST SRM boric acid standard), and the values for $\delta^{11}B\text{-}B(OH)_3$ and $\delta^{11}B\text{-}B(OH)_4^-$, which are separated by a difference between product and reactant of 27.2‰ as $\delta^{11}B\text{-}B(OH)_3 - \delta^{11}B\text{-}B(OH)_4^- = 27.2‰$. Substituting Eq. 6.22 into Eq. 6.21 and rearranging yields an equation for the isotope ratio of the $B(OH)_4^-$ species, which can then be combined with Eqs. 5.28 and 5.29 to yield the isotope ratio as a function of pH (Fig. 6.7B):

$$\delta^{11}B\text{-}B(OH)_4^- = 39.6 - 27.2 \frac{[H^+]}{[H^+] + K'_B} \tag{6.23}$$

Laboratory experiments of the $\delta^{11}B\text{–}CaCO_3$ isotope ratio in inorganic calcite and calcitic tests of cultured foraminifera grown as a function of seawater pH provide empirical evidence for the boron isotope ratio pH dependence (e.g., Foster and Rae, 2016). Observations do not follow exactly the predictions of Eq. 6.23, and different foraminifera species have different trends (vital effects), but there is clear evidence that the boron isotope ratio in $CaCO_3$ is a function of the pH of the water in which the mineral shell grew.

A striking paleoclimate application of the boron isotope pH tracer examines seawater pH change during a massive release of carbon to the atmosphere at the Paleocene-Eocene Thermal Maximum (PETM) (Penman et al., 2014). At this time, about 55 million years ago, previous observations had revealed a sudden decline in the $\delta^{13}C$ in planktic foraminifera of about 3‰ (Fig. 6.8B, Zachos et al., 2003). This observation has been interpreted as a very rapid release of carbon to the atmosphere as either CO_2 or CH_4, both of which are isotopically light (–21‰ and <–40‰, respectively) relative to ocean bicarbonate, which is near zero. Since CH_4 oxidizes to CO_2 in the atmosphere on a timescale of about one year, they would both cause a massive increase in CO_2. However, the amount of carbon required to explain the lower $\delta^{13}C$ depends on the $\delta^{13}C$ of the carbon source, which is uncertain. Instead, the surface-ocean pH decrease resulting from the atmospheric CO_2 increase can be used to estimate the magnitude of the fCO_2 change in the atmosphere. Observations of the $\delta^{11}B$ in planktonic foraminifera indicate a decrease of about 0.8‰ occurring at the same time as the decrease in $\delta^{13}C$ (Fig. 6.8). The amount of added carbon required to cause such an ocean pH change is on the order of 5000 Pg-C, which is of the same magnitude as the increase expected if humans burn all of the available fossil fuels. The ancient record should help us understand consequences of the anthropogenic CO_2 increase that we will experience if we continue with the present energy sources. The PETM does not represent a perfect analog because the CO_2 increase at that time occurred on a timescale of about 70 ky, which is more than 10 times slower than the fossil fuel–induced changes that are currently underway.

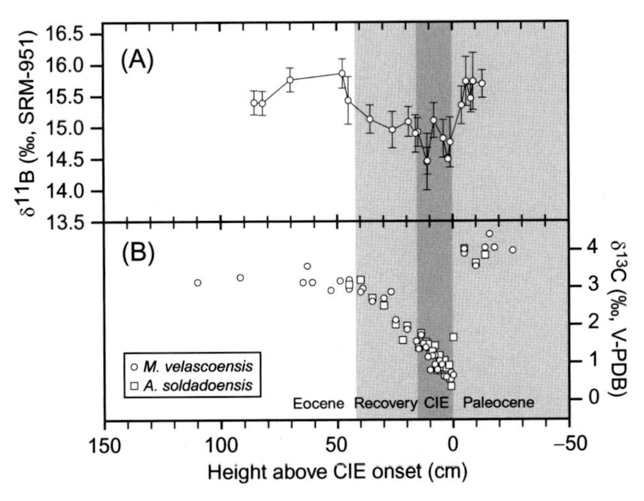

Figure 6.8 The change in the boron (A) and carbon (B) isotopic composition of foraminifera from a sediment core that spans the Paleocene-Eocene Thermal Maximum (PETM). The x-axis indicates the sediment core distance (cm) above the Carbon Isotope Excursion (CIE) event. The legend specifies the foraminifera species (*Morozovella velascoensis* and *Acarinina soldadoensis*). The boron isotope data are from Penman et al. (2014) and the $\delta^{13}C$ data are from Zachos et al. (2003). Modified from Penman et al. (2014).

6.4 Kinetic Isotope Fractionation

Kinetic isotope effects take place under non-equilibrium physical or chemical conditions. Physically based isotope effects often occur because molecules of the same compound, but containing different stable isotopes, move at different rates (due to unequal masses). For example, carbon dioxide occurs as molecules of $^{12}CO_2$ (molecular weight, $m_{wt} = 44$) and $^{13}CO_2(m_{wt} = 45)$, which must have the same kinetic energy when both are in the same environment ($E_k = ½mv^2$, where m is mass and v is velocity). At the same temperature, the relationship between velocity and mass is

$$½ \times M_{44} \times v_{44}^2 = ½ \times M_{45} \times v_{45}^2 \tag{6.24}$$

$$\frac{v_{44}}{v_{45}} = \frac{\sqrt{45}}{\sqrt{44}} = 1.011. \tag{6.25}$$

Thus, $^{12}CO_2$ molecules diffusing in pure CO_2 gas travel 1.1% (11‰) faster than $^{13}CO_2$ molecules, with the net result that $^{13}CO_2$ will trail behind in a "pack" of diffusing carbon dioxide molecules. Calculating the isotope effect of gas diffusion in air requires accounting for the molecular weight of both the diffusing gas and the medium in which it diffuses. Molecules of the same gas containing more massive isotopes also diffuse more slowly in water, but the isotope effect is much smaller than in the gas phase based on experimental evidence for noble gases (Tempest and Emerson, 2013) and on theoretical predictions (Bourg and Sposito, 2008).

Chemically based kinetic isotope effects occur because molecules of the same compound, but containing different isotopes, react at different rates. In general, heavier isotopes of the same element form stronger chemical bonds that are less likely to break during chemical reactions. Overall, molecules of the same compound containing lighter isotopes will move and react faster than molecules containing heavier isotopes. In this case, the kinetic isotope fractionation factor, α, is defined as the ratio of transfer or reactions rates for the heavy and light isotopologues, $\alpha = k'_H / k'_L$, rather than the ratio of equilibrium constants, as in equilibrium isotope effects.

Discussion Box 6.2

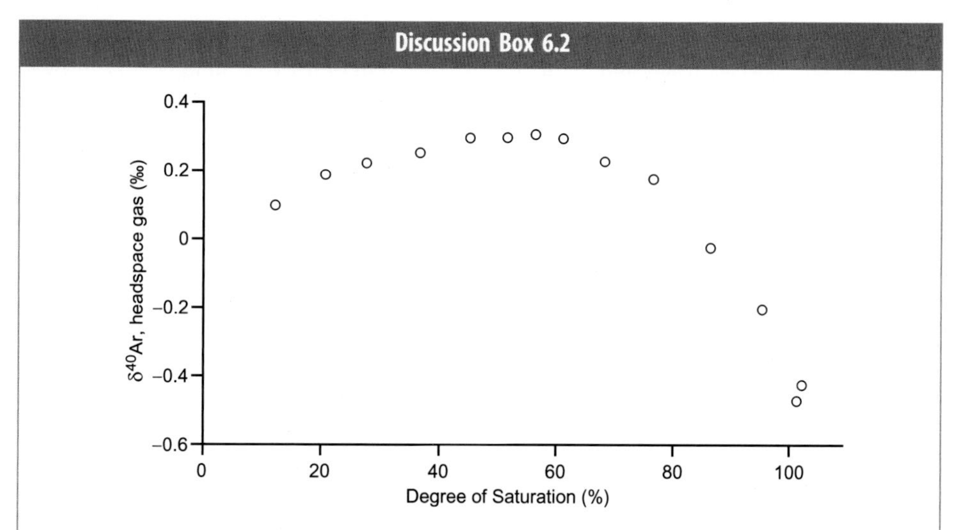

This figure shows the results of an experiment reported in Tempest and Emerson (2013) in which water containing no dissolved gas is exposed to a headspace of pure argon gas. The plot shows the δ^{40}Ar of the argon gas in the headspace over the course of the experiment versus the percent saturation in the water. The standard used for the δ^{40}Ar–Ar measurements is the initial headspace gas at the start of the experiment. The Ar saturation is the measured Ar concentration divided by that expected at saturation equilibrium times 100 to give it units of percent ($Ar_{meas}/Ar_{eq} \times 100$).

- By what process does argon gas move from the headspace to the water? What would you expect a graph of Ar saturation versus time during the experiment to look like?
- What type of fractionation dominates in the early part of the experiment? Why does the δ^{40}Ar–Ar of the headspace gases initially increase?
- What type of fractionation dominates at the end of the experiment? Why does the δ^{40}Ar–Ar of the headspace gases end at a lower value than at the beginning?
- From this data, how would you go about determining the fractionation factor associated with the initial period of the experiment?
- What additional information would you need to know to calculate the fractionation factor associated with the end-of-experiment fractionation?

6.4.1 δ^{13}C-DIC, a Tracer of Biological Processes

Both equilibrium and kinetic fractionation factors among CO_2, HCO_3^-, and organic matter control the carbon isotope ratios in the marine, atmospheric, and terrestrial reservoirs (Fig. 6.9). Essentially, all isotope effects involved with the formation and destruction of organic matter are kinetic, because these enzyme-catalyzed reactions typically occur in only one direction and do not come to an equilibrium. Most terrestrial land plants and marine phytoplankton exhibit an enzymatic isotope effect leading to fractionation of carbon during photosynthesis. Enzymatic fractionation during photosynthesis results in a fractionation factor of about $\varepsilon = -30‰$ (O'Leary, 1981). This entire effect is rarely observed, because the isotopic ratio of the internal reservoir of CO_2 is controlled by the rate of CO_2 flux across a membrane that separates the reservoir from the atmosphere or surface water and by the rate of depletion of the CO_2 in the reservoir by carbon fixation. If carbon fixation is much faster than diffusion, every CO_2 molecule that crosses the membrane from the environment is consumed, and the isotope effect is only that due to diffusion across the membrane. If the flux across the membrane is faster than fixation, the

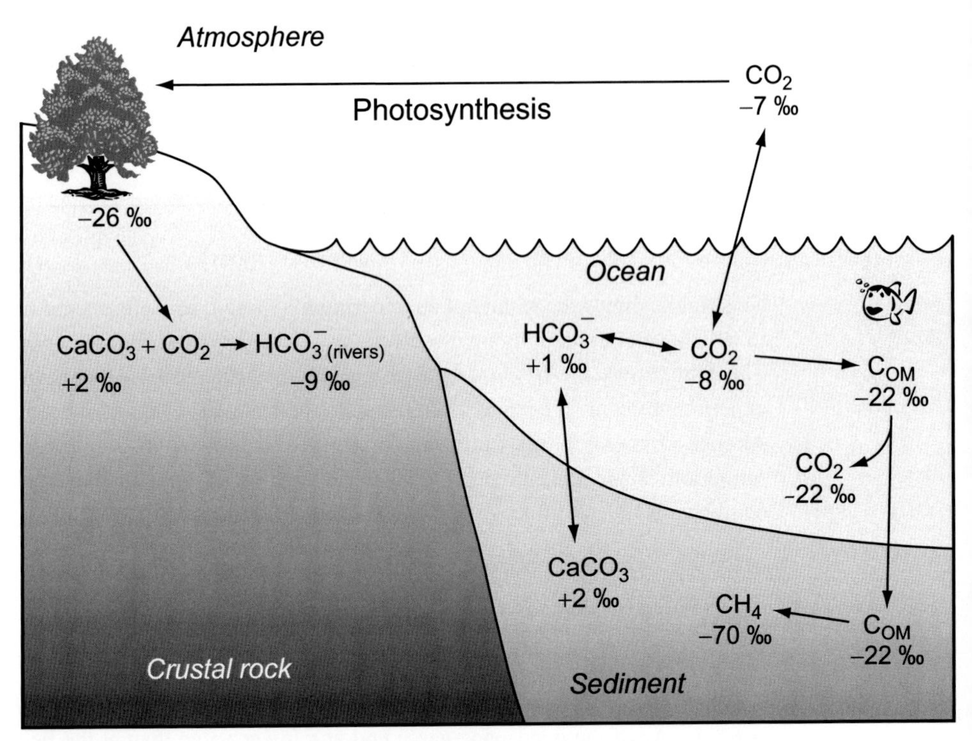

Figure 6.9 Schematic diagram of the stable carbon isotopic composition among the reservoirs of atmospheric CO_2, seawater dissolved HCO_3^-, CO_2, and CH_4, and the solids organic carbon (C_{OM}) and $CaCO_3$. Arrows indicate exchanges – double ended arrows are equilibrium fractionations and single ended are kinetic fractionations. Numbers under the chemical symbols represent approximate mean isotopic ratios in ‰, though these vary in the environment. See Table 6.3 for an estimate of the fractionation factors. Modified from Emerson and Hedges (2008).

internal reservoir and the external environment have similar fCO_2 and isotopic composition, and the fractionation factor is only that due to fixation.

During respiration, carbon isotopes are fractionated very little. Thus, for animals, the adage "you are what you eat" generally holds. The consequence of significant isotopic fractionation during photosynthesis, but little during respiration, is that surface waters become enriched in ^{13}C as ^{12}C is preferentially incorporated into organic carbon that is transported out of the euphotic zone in the form of dead plants and animals or dissolved organic matter (Fig. 6.10A). Organic matter that exits the euphotic zone degrades in the deep sea, slightly lowering the deep-water $\delta^{13}C$-DIC. The isotope effect is amplified in the surface waters, because the volume of the euphotic zone is small compared to the deep dark ocean. Because of these fractionations and dynamics, carbon isotopes in surface waters of the ocean are useful tracers of the photosynthetically driven flux of organic matter to the deep ocean, the biological pump.

Net photosynthesis, air–sea gas exchange, and water mass mixing create seasonal trends in the DIC and $DI^{13}C$ of the upper ocean (Fig. 6.10 (B) & (C)) because isotopic fractionation affects each process differently. This is the reason that equations describing the mass balance of DIC and $DI^{13}C$ are independent and can be used along with gradients like those in Fig. 6.10 to determine the rate of net community production (Gruber et al., 1998; Quay and Stutsman, 2003). The upper ocean mass balance for [DIC] (mol m^{-3}) in terms of the general upper ocean metabolic tracer equation presented in Chapter 3 (Eq. 3.21) equates the time rate of change in the upper ocean to physical and biological processes:

$$\frac{d}{dt}\int_{z=0}^{z=h} [\mathrm{DIC}]dz = F_{A-W} + F_z + F_{\mathrm{xy}} - \int_{z=0}^{z=h} J\, dz \quad \left(\mathrm{mol\ m^{-2}\ d^{-1}}\right). \tag{6.26}$$

The integration on the left side is over the winter mixed layer depth, h (m), for reasons described in Section 3.2. The dominant physical fluxes on the right-hand side are: atmosphere–ocean gas exchange, F_{A-W}, vertical exchange across the base of the winter mixed layer, F_z, and the horizontal advective fluxes, F_{xy}. For simplicity, the horizontal flux is indicated here as a flow in the dominant direction rather than separately in the x and y (latitude and longitude) directions. The biological flux, J, is the net community production (NCP) which removes DIC. If the DIC and organic carbon are at steady state over an annual cycle, the annual NCP is equal to the vertical organic carbon export at depth h (the biological pump).

Each of the terms describing physical processes in Eq. 6.26 are expanded as the product of a transport coefficient and concentration gradient. Because there are not enough ocean measurements of $\delta^{13}C$-DIC to determine horizontal gradients, Quay and Stutsman (2003) assumed that the most important physical flux is vertical. This is a necessary assumption because of insufficient data coverage, but probably restricts use of this method to regions like the subtropical gyres where there are not strong horizontal gradients. With this assumption and the expansion of the physical flux terms, the DIC equation becomes

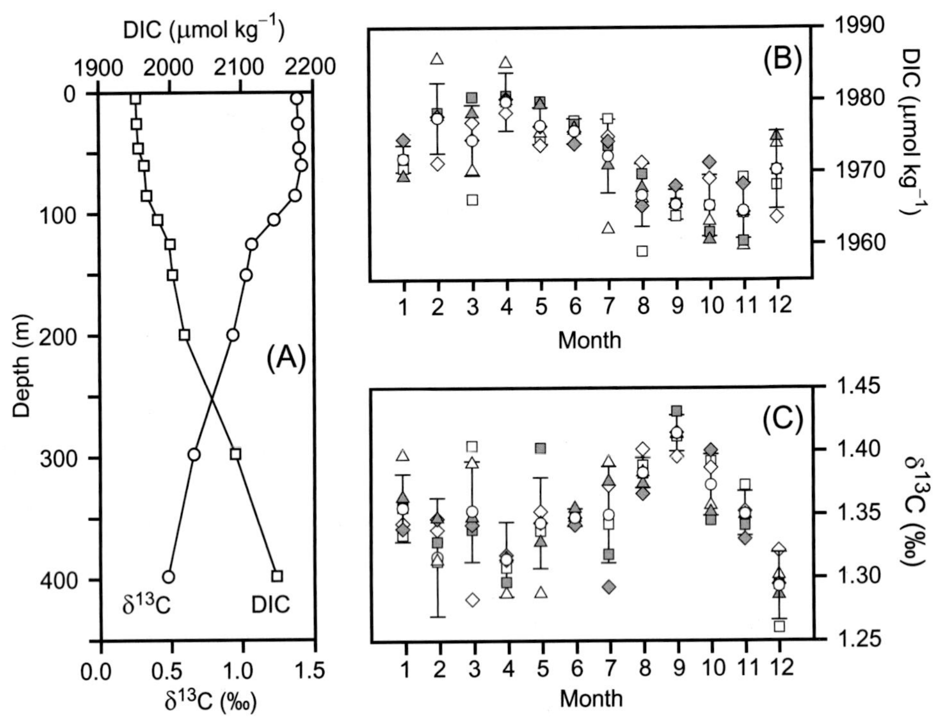

Figure 6.10 DIC and carbon isotope ratio of DIC, δ^{13}C-DIC, at the Hawaii Ocean Time-series (HOT) station ALOHA. (A) Depth profile of DIC (□) and δ^{13}C-DIC (o); (B) Monthly values of DIC in the mixed layer at HOT; and (C) Monthly values of δ^{13}C-DIC in the mixed layer at HOT. Symbols in B and C indicate different years between 1994 and 1999. Modified from Quay and Stutsman (2003).

$$\frac{d}{dt}\int_{z=0}^{z=h}[\mathrm{DIC}]dz = -k_{\mathrm{CO_2}}K_{H,\mathrm{CO_2}}\left(f\mathrm{CO}_2^{SW} - f\mathrm{CO}_2^{a}\right) + K_z\left(\frac{d[\mathrm{DIC}]}{dz}\right)_{z=h} - \int_{z=0}^{z=h} J\,dz. \tag{6.27}$$

The air–sea gas flux is the product of the gas exchange mass transfer coefficient ($k_{\mathrm{CO_2}}$, m d^{-1}), the Henry's law coefficient for CO_2 ($K_{H,\mathrm{CO_2}}$, mol m^{-3}atm^{-1}) and the air–sea difference in the fugacity of CO_2. (See Appendix A2.3 of Chapter 2.) Superscripts indicate surface water and atmosphere, ^{SW} and ^{a}, respectively. The vertical flux at the base of the winter mixed layer is parameterized as an eddy diffusion flux, in which the gradient at depth h is multiplied by an eddy diffusion coefficient, K_z (m^2 d^{-1}).

The upper ocean mass balance for DI^{13}C is determined, as before, by multiplying the concentrations in Eq. 6.27 by the isotope ratio, $r = {}^{13}\mathrm{C}/{}^{12}\mathrm{C}$, and by fractionation factors, α, for processes such as solubility and chemical equilibrium (equilibrium fractionation) and the kinetic fractionations of molecular diffusion and net biological production. The ratios, r, are introduced so that they can ultimately be replaced with the δ values using Eq. 6.1.

$$\frac{d}{dt}\int_{z=0}^{z=h}[\mathrm{DI}^{13}\mathrm{C}]\,dz = -k_{\mathrm{CO_2}}\alpha_{ge}K_{H,\mathrm{CO_2}}\alpha_{sol}\{f\mathrm{CO}_2^{SW}\alpha_{\mathrm{CO_2-DIC}}r_{\mathrm{DIC}}^{SW} - f\mathrm{CO}_2^{a}r^{a}\} + K_z\left(\frac{d[\mathrm{DIC}]r_{\mathrm{DIC}}}{dz}\right)_{z=h} - \int_{z=0}^{z=h} J\, r_{\mathrm{DIC}}^{SW}\alpha_P dz. \quad (6.28)$$

The r values represent the $^{13}C/^{12}C$ ratio in surface water DIC, r_{DIC}^{SW}, in atmospheric CO_2, r^a, and the gradient in $\mathrm{DI}^{13}\mathrm{C/DIC}$ at $z=h$, r_{DIC}. The fractionation factors are for the processes of: gas exchange, in which the ratio of the mass transfer coefficients, $\alpha_{ge} = {}^{13}k/{}^{12}k$, has been determined experimentally; solubility, in which the ratio of the Henry's law coefficients, $\alpha_{\mathrm{sol}} = {}^{13}K_{H,\mathrm{CO_2}}/{}^{12}K_{H,\mathrm{CO_2}}$, has also been measured in laboratory experiments; equilibrium between DIC and CO_2 in surface waters, $\alpha_{\mathrm{CO_2-DIC}}$; and photosynthetic production of organic matter, α_P. The reason for the complicated term expressing the $f^{13}CO_2$ of surface waters, $f\mathrm{CO}_2^{SW}\alpha_{\mathrm{CO_2-DIC}}r_{\mathrm{DIC}}^{SW}$, is that the isotope ratio is measured in DIC, not CO_2. $\alpha_{\mathrm{CO_2-DIC}} = r_{\mathrm{CO_2}}/r_{\mathrm{DIC}}$ so $r_{\mathrm{CO_2}}\left(\text{or } f^{13}\mathrm{CO_2}/f\mathrm{CO_2}\right)^{\mathrm{SW}} = \alpha_{\mathrm{CO_2-DIC}}\,r_{\mathrm{DIC}}^{SW}$. The value of the equilibrium CO_2-DIC fractionation factor is –8‰ ($\alpha_{\mathrm{CO_2-DIC}} = 0.992$, Fig. 6.9).

The great advantage of this mass balance approach is that the two independent equations make it possible to solve for two unknowns: the upward supply of DIC by vertical mixing, $K_z(d\mathrm{DIC}/dz)_{z=h}$, and J. An independent estimate of the eddy diffusion coefficient, K_z, which is very hard to quantify, is not necessary to calculate the carbon export from the surface ocean by this method. Comparisons of Annual Net Community Production (ANCP) determined by the ^{13}C method and oxygen mass balance yield very similar results (Emerson et al., 1997, Yang et al., 2019).

6.4.2 Triple Isotopes of Oxygen, a Tracer for Ocean Photosynthesis

The isotopes of oxygen are another tracer for photosynthesis, however the procedure is not based on the simple production of O_2 from H_2O and the attending change in the $\delta^{18}O$-O_2 in the surface ocean, as described above for $\delta^{13}C$-DIC. Because there is little fractionation of oxygen during photosynthesis (Table 6.3), surface water $\delta^{18}O$-O_2 is only slightly affected by biological production. An entirely different process involving the third and least abundant isotope of oxygen, ^{17}O, has been used to measure gross photosynthesis rates. We would like to emphasize, before continuing, that this method estimates "gross" production rather than net community production as is the case for the $\delta^{13}C$-DIC method. Gross production is expected to be much higher than ANCP (see Chapter 3).

The triple isotope tracer of photosynthesis is based on an unusual property of the isotopes of O_2. Normally, the difference in the reaction rates of heavy and light isotopes is proportional to the difference in the masses. For oxygen isotopes the difference in mass between ^{17}O and ^{16}O is half the difference between ^{18}O and ^{16}O, thus the fractionation (discrimination) of ^{17}O relative to ^{16}O for most processes is half that of the discrimination between ^{18}O and ^{16}O. Because mass dependent fractionation is normal, most of

Earth's large reservoirs of oxygen, including seawater, have a $\delta^{17}O/\delta^{18}O$ ratio of about 0.5.

An exception to mass-dependent fractionation occurs in the stratosphere, where ultraviolet radiation induces reactions among O_2, O_3, and CO_2 that result in mass-independent fractionation. Because of these reactions and mixing between the troposphere and stratosphere, the lower atmosphere has a $\delta^{17}O$-O_2 that is 0.250‰ less than that predicted from its $\delta^{18}O$-O_2 if only the mass-dependent processes of photosynthesis and respiration were important.

The three main processes influencing the isotopic composition of O_2 in the ocean are photosynthesis, respiration, and gas exchange. Photosynthesis produces O_2 that has $\delta^{17}O$-O_2 and $\delta^{18}O$-O_2 values close to the isotopic composition of water because there is little fractionation and it is mass-dependent. Respiration removes O_2 with a relatively large fractionation (see next section), which is also mass dependent. Both of these processes influence the absolute values of the oxygen concentration and their $\delta^{17}O$-O_2 and $\delta^{18}O$-O_2 values but the relative ratio of the isotope values is not affected because the fractionations are mass dependent. Gas exchange, on the other hand, introduces O_2 to the surface ocean with anomalously low $\delta^{17}O$-O_2 due to the mass-independent fractionation in the stratosphere. The oxygen isotope ratios observed in the surface ocean are, thus, a mixture of the isotopic signatures of all three of these processes.

Luz and Barkan (2000) defined a new term combining the $\delta^{17}O$ and $\delta^{18}O$ of oxygen to quantify the ^{17}O excess:

$$\delta^{17}\Delta = 1000\left(\delta^{17}O\text{-}O_2 - 0.518\ \delta^{18}O\text{-}O_2\right), \tag{6.29}$$

where the standard used for the δ values is atmospheric O_2 and the factor 1000 transforms the $^{17}\Delta$ to units of per meg (1 per meg is 1 part in 10^6) from units of per mil (‰, 1 part in 10^3). The value of 0.518 is the best estimate of the mass dependence of the fractionations by respiration. (A more recent equation defining $^{17}\Delta$ is presented in Luz and Barkan, 2005.) Equation 6.29 defines $^{17}\Delta$ as a measure of the excess or deficit in $^{17}O/^{16}O$ relative to $^{18}O/^{16}O$ compared to the atmospheric value. Respiration has no effect on $^{17}\Delta$ because of the way it is defined; respiration increases $\delta^{18}O$-O_2 at 0.518 of the rate that it increases $\delta^{17}O$-O_2 such that the effects cancel in $^{17}\Delta$. Since the atmosphere is the standard, a $^{17}\Delta$ value of pure atmospheric gas is zero (both $\delta^{17}O$-O_2 and $\delta^{18}O$-O_2 are zero). The $^{17}\Delta$ of dissolved O_2 in equilibrium with atmospheric pO_2 is offset from zero by the solubility equilibrium fractionation, which is about +8 per meg. As the fraction of surface ocean oxygen that originates from photosynthesis becomes greater, the $^{17}\Delta$ value increases to a maximum possible value of 250 per meg, reflecting the isotopic composition of water. The only biological process responsible for influencing the $^{17}\Delta$ excess in the surface ocean O_2 is gross oxygen production (GOP), which is the oxygen equivalent of gross primary production (GPP) for carbon. The $^{17}\Delta$ value observed in the surface ocean is, thus, a mixture of the influence of gas exchange and photosynthesis.

Deriving GOP values from triple isotope measurements requires an upper ocean mass balance model (Juranek and Quay, 2013) that takes into account air–sea gas exchange and mixing from below the mixed layer of the type described for DIC above. Horizontal influences certainly could be important near ocean fronts, but there is little information

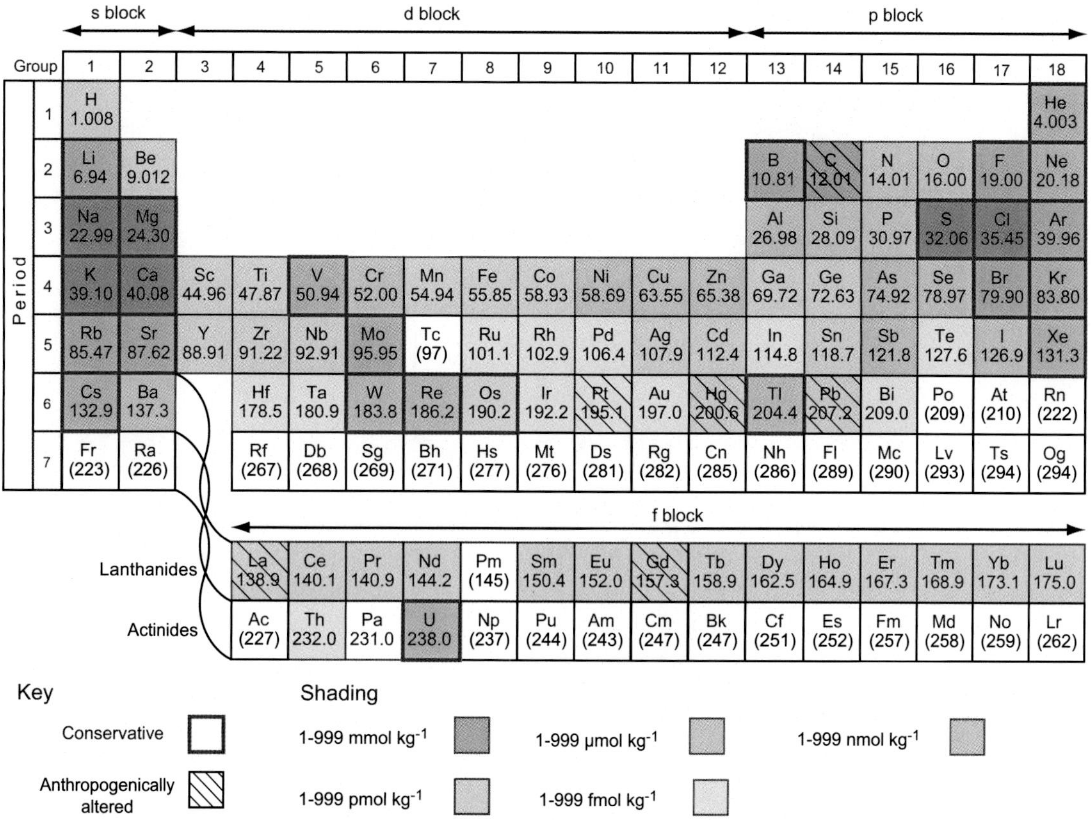

s block | d block | p block

Period / Group	1	2	3	4	5	6	7	8	9	10	11	12	13	14	15	16	17	18
1	H 1.008																	He 4.003
2	Li 6.94	Be 9.012											B 10.81	C 12.01	N 14.01	O 16.00	F 19.00	Ne 20.18
3	Na 22.99	Mg 24.30											Al 26.98	Si 28.09	P 30.97	S 32.06	Cl 35.45	Ar 39.96
4	K 39.10	Ca 40.08	Sc 44.96	Ti 47.87	V 50.94	Cr 52.00	Mn 54.94	Fe 55.85	Co 58.93	Ni 58.69	Cu 63.55	Zn 65.38	Ga 69.72	Ge 72.63	As 74.92	Se 78.97	Br 79.90	Kr 83.80
5	Rb 85.47	Sr 87.62	Y 88.91	Zr 91.22	Nb 92.91	Mo 95.95	Tc (97)	Ru 101.1	Rh 102.9	Pd 106.4	Ag 107.9	Cd 112.4	In 114.8	Sn 118.7	Sb 121.8	Te 127.6	I 126.9	Xe 131.3
6	Cs 132.9	Ba 137.3		Hf 178.5	Ta 180.9	W 183.8	Re 186.2	Os 190.2	Ir 192.2	Pt 195.1	Au 197.0	Hg 200.6	Tl 204.4	Pb 207.2	Bi 209.0	Po (209)	At (210)	Rn (222)
7	Fr (223)	Ra (226)		Rf (267)	Db (268)	Sg (269)	Bh (271)	Hs (277)	Mt (276)	Ds (281)	Rg (282)	Cn (285)	Nh (286)	Fl (289)	Mc (290)	Lv (293)	Ts (294)	Og (294)

f block

Lanthanides	La 138.9	Ce 140.1	Pr 140.9	Nd 144.2	Pm (145)	Sm 150.4	Eu 152.0	Gd 157.3	Tb 158.9	Dy 162.5	Ho 164.9	Er 167.3	Tm 168.9	Yb 173.1	Lu 175.0
Actinides	Ac (227)	Th 232.0	Pa 231.0	U 238.0	Np (237)	Pu (244)	Am (243)	Cm (247)	Bk (247)	Cf (251)	Es (252)	Fm (257)	Md (258)	No (259)	Lr (262)

Key

Conservative

Anthropogenically altered

Shading

1-999 mmol kg^{-1}

1-999 μmol kg^{-1}

1-999 nmol kg^{-1}

1-999 pmol kg^{-1}

1-999 fmol kg^{-1}

Figure 1.9 Periodic table of the elements indicating their atomic mass and seawater dissolved concentrations. Values in parentheses are the atomic mass of the longest-lived radioisotope of that element. Shading identifies the concentration range of the element dissolved in seawater. Elements with a thick border are conservative; elements that are hatched indicate measurable anthropogenic alteration (Hatje et al., 2018); and elements with no shading have uncertain or lower than femtomolar seawater concentrations.

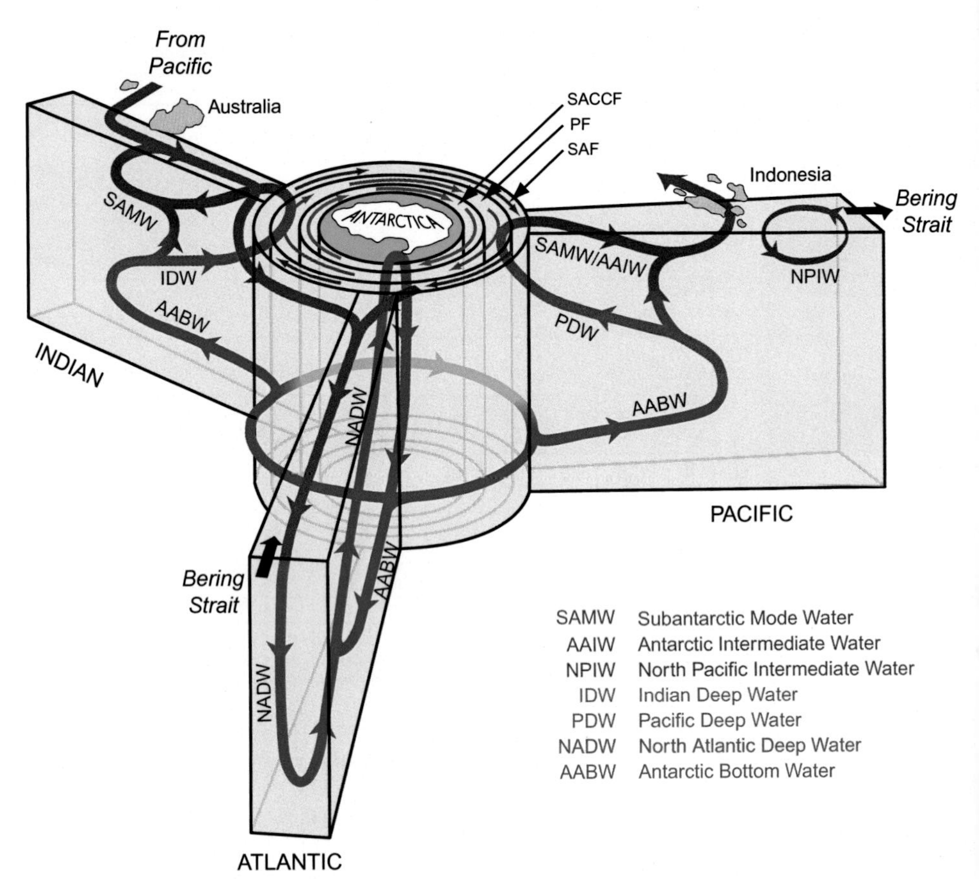

Figure 1.16 Antarctic-centric schematic of the overturning circulation indicting the major deep water flows. From bottom to top: the Antarctic Bottom Water (AABW); North Atlantic Deep Water (NADW), Pacific and Indian Deep Water (PDW, IDW), Subantarctic Mode Water and Antarctic Intermediate Water (SAMW/AAIW), and the surface return flows. Solid circles surrounding Antarctica indicate fronts in the Antarctic Circumpolar Current. They are the Subantarctic Front (SAC), Polar Front (PF), and the Southern Antarctic Circumpolar Current Front (SACCF). Redrafted from Talley et al. (2011).

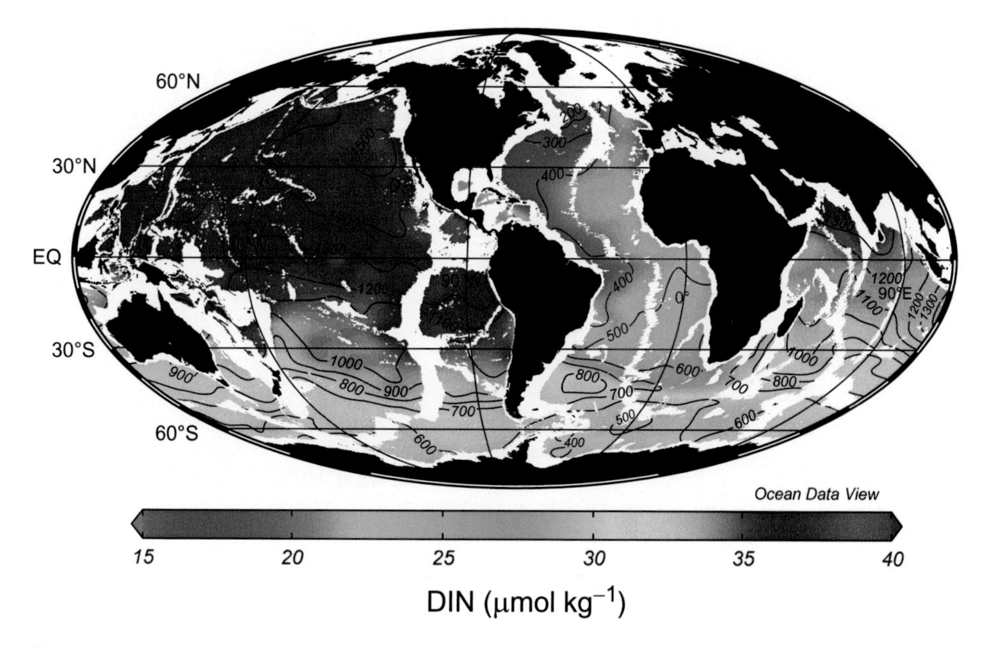

Discussion Box 1.2: A map of dissolved inorganic nitrogen (DIN, primarily nitrate concentrations) and ages at a depth of 3000 m in the ocean (redrafted from Key et al., 2004). The white areas indicate regions where the ocean is shallower than 3000 m. Nitrate concentrations are shown by the shading (in units of μmol kg^{-1}). The ages of the water in different locations (time since the water was last at the surface, derived from ^{14}C measurements) are shown by the contours (in units of years).

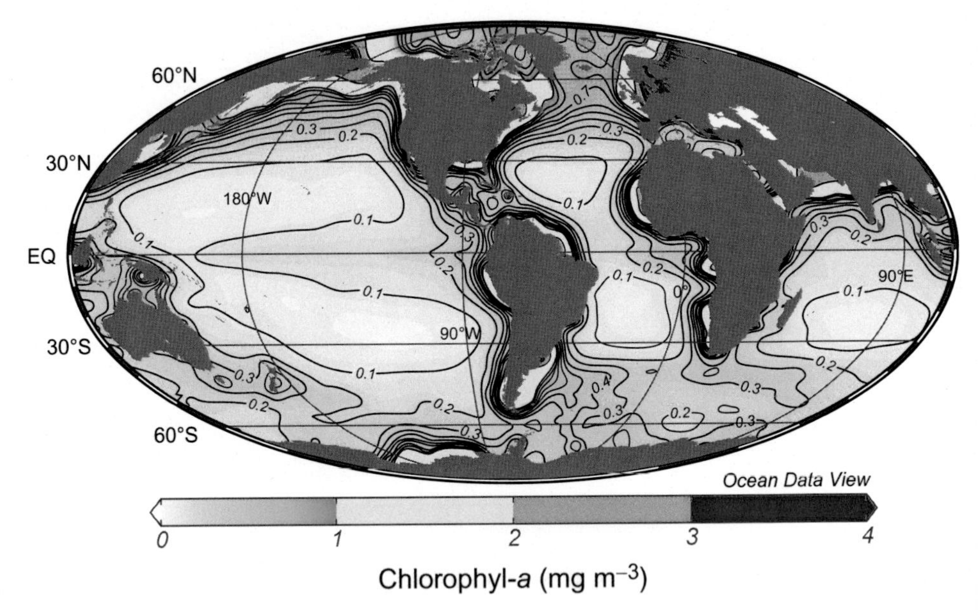

Figure 1.17 Global map of mean annual surface ocean chlorophyll-*a* concentration from MODIS satellite color imagery for the period 2009–2013. (https://oceancolor.gsfc.nasa.gov/data/aqua/).

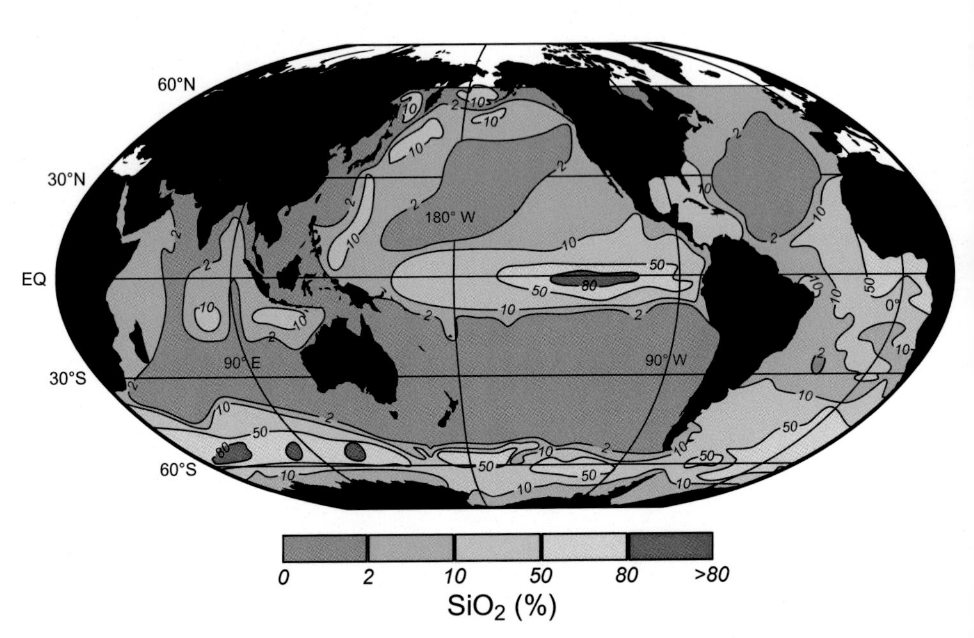

Figure 2.11 The global distribution of biogenic SiO_2 (in wt%) in marine sediments. Redrafted from Broecker and Peng (1982).

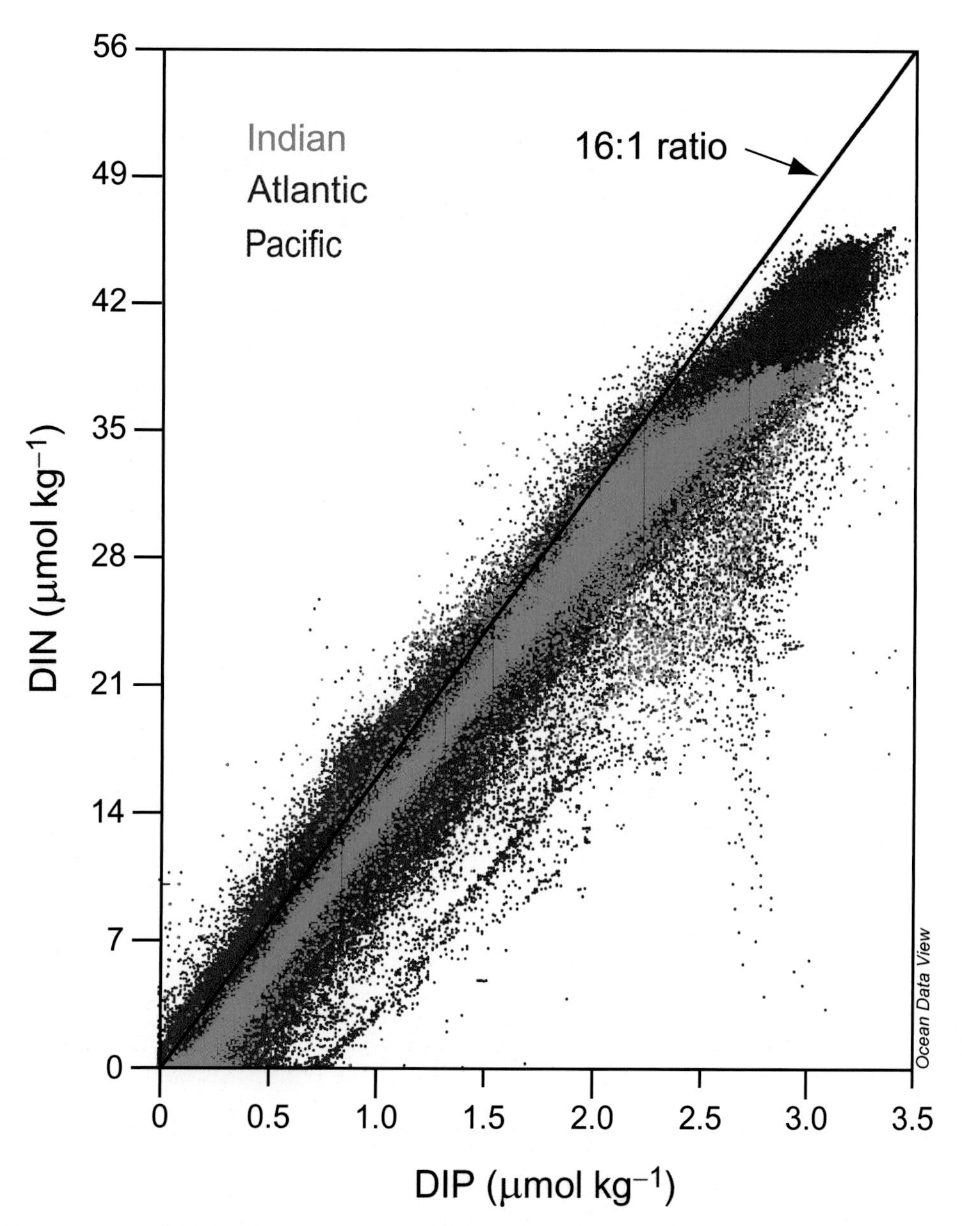

Figure 3.2 The concentration of dissolved inorganic nitrogen and phosphorus (DIN vs. DIP) in the world's oceans. Data are from the CLIVAR and Carbon Hydrographic Data Office (CCHDO, https://cchdo.ucsd.edu). The 16:1 ratio is presented for reference. Note that ratios are slightly higher in the Atlantic than in the Pacific and Indian Oceans because of denitrification reactions in the low-oxygen-containing waters and sediments of the Pacific and Indian Oceans (see Chapter 4).

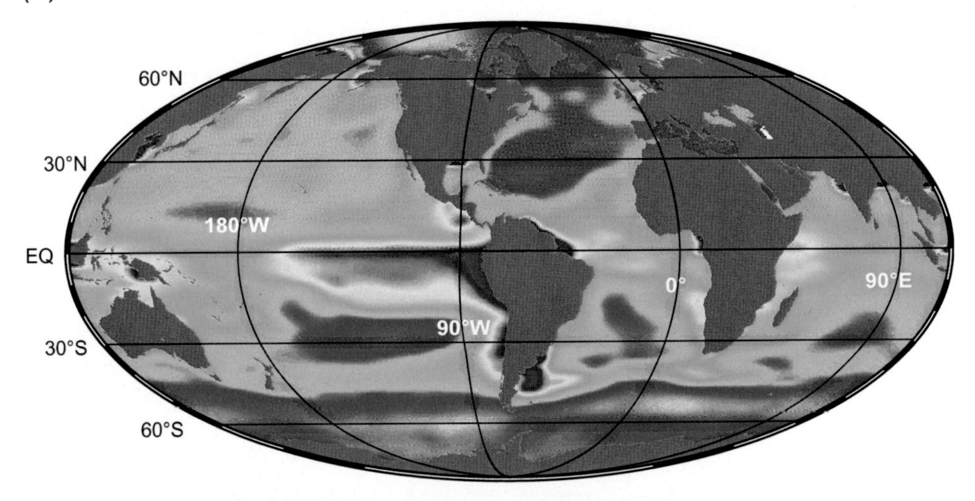

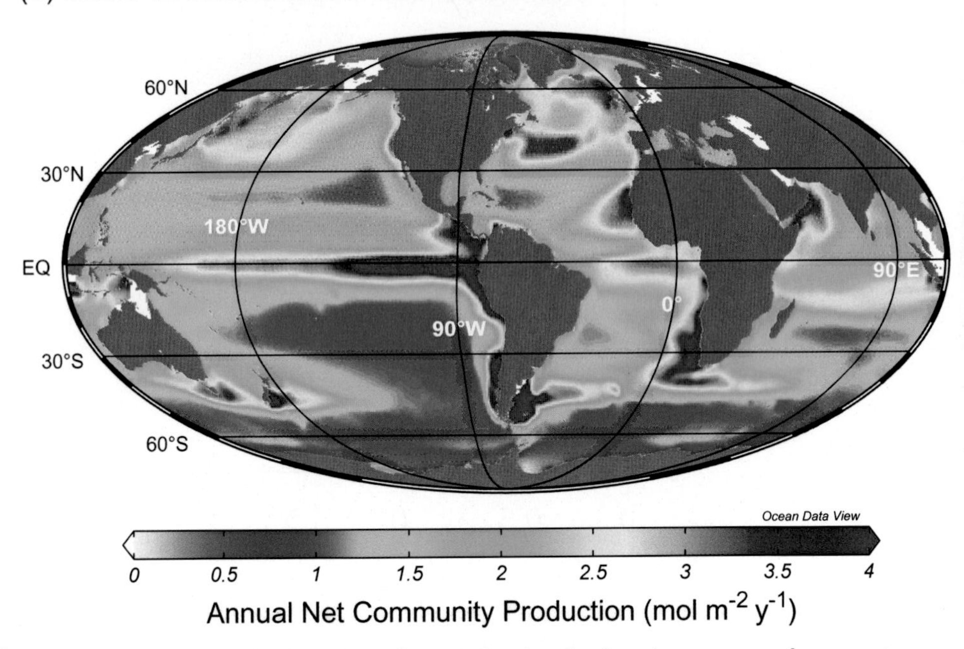

Figure 3.11 Global maps of the biological pump or particulate organic carbon flux from the upper ocean from separate sources. (A) Annual carbon export determined from the relationship C export = NPP (C export/NPP). NPP was determined from satellite-measured ocean color and optical back scatter from 2015 and the CbPM algorithm (Westberry et al., 2008). The export/NPP ratio was determined from the ecosystem model of Laws et al (2011). (B) Carbon export determined from an ocean GCM with the BEC ecosystem model (from Laufkötter et al., 2016).

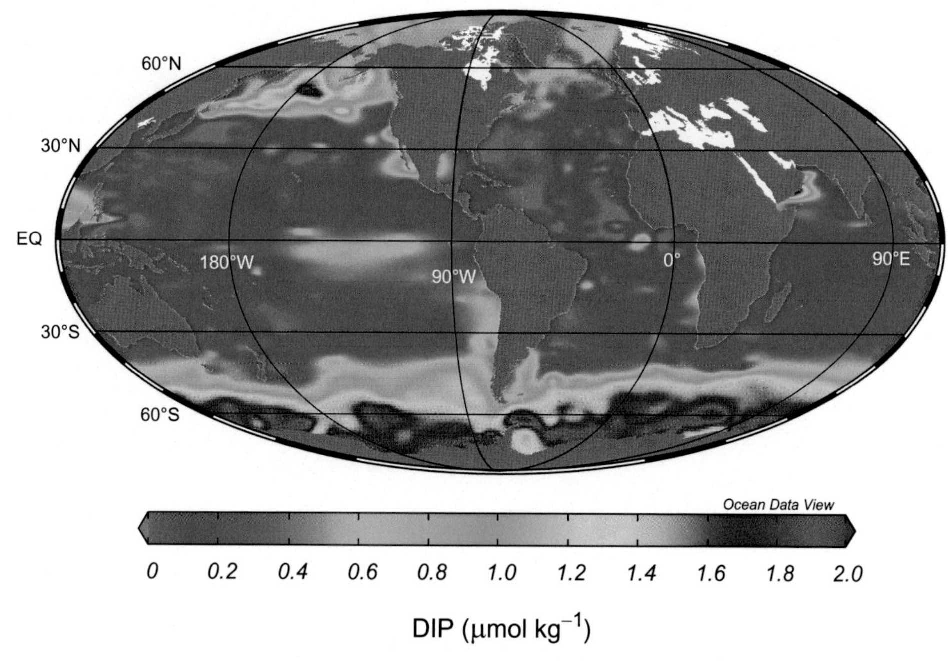

Figure 4.2 The concentration of dissolved inorganic phosphorus (DIP, $\mu mol\ kg^{-1}$) in surface waters of the ocean. Notice elevated concentrations at high latitudes and the Equator. Bottle data are from CCHDO (CLIVAR and Carbon Hydrographic Data Office, https://cchdo.ucsd.edu).

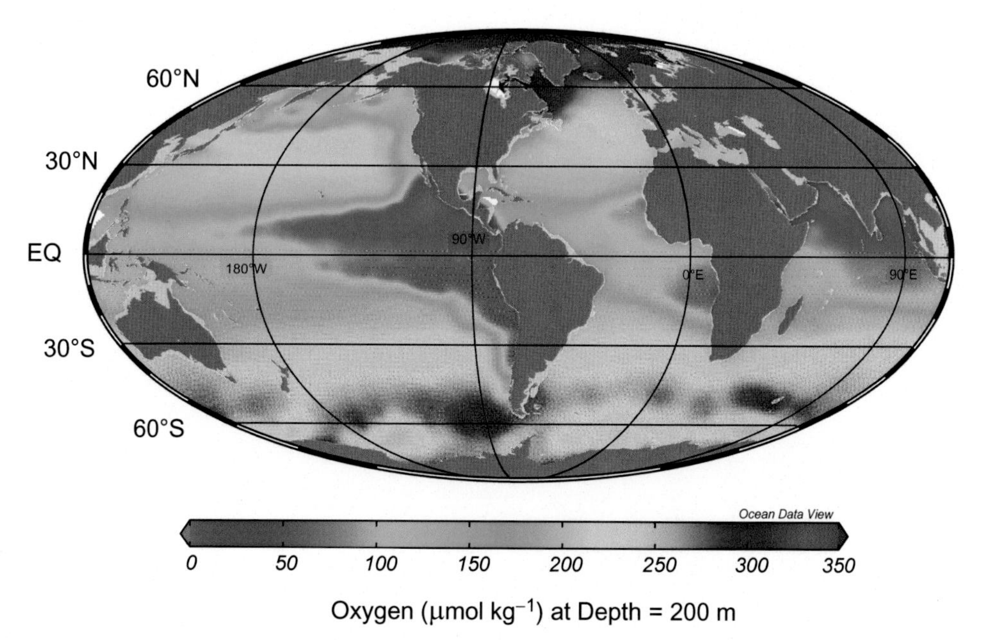

Figure 4.3 The oxygen concentration in the world's ocean at 200 meters depth. Data are from the Global Ocean Data Analysis Project version 2 (GLODAP v2) at the Ocean Data View (ODV) web site (https://odv.awi.de).

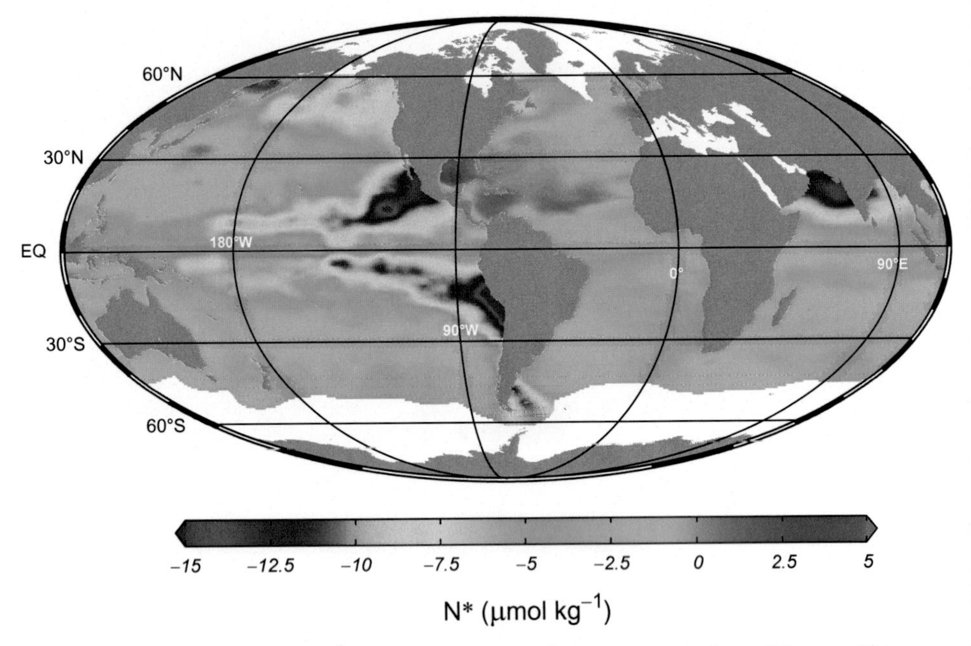

Figure 4.7 N* (DIN – 16*DIP in units of μmol kg^{-1}) on potential density surface $\sigma_\theta = 26.0$ in the world's ocean. White areas at high latitudes indicate regions without water of this density. Calculated from data in the World Ocean Atlas (WOA) 2013.

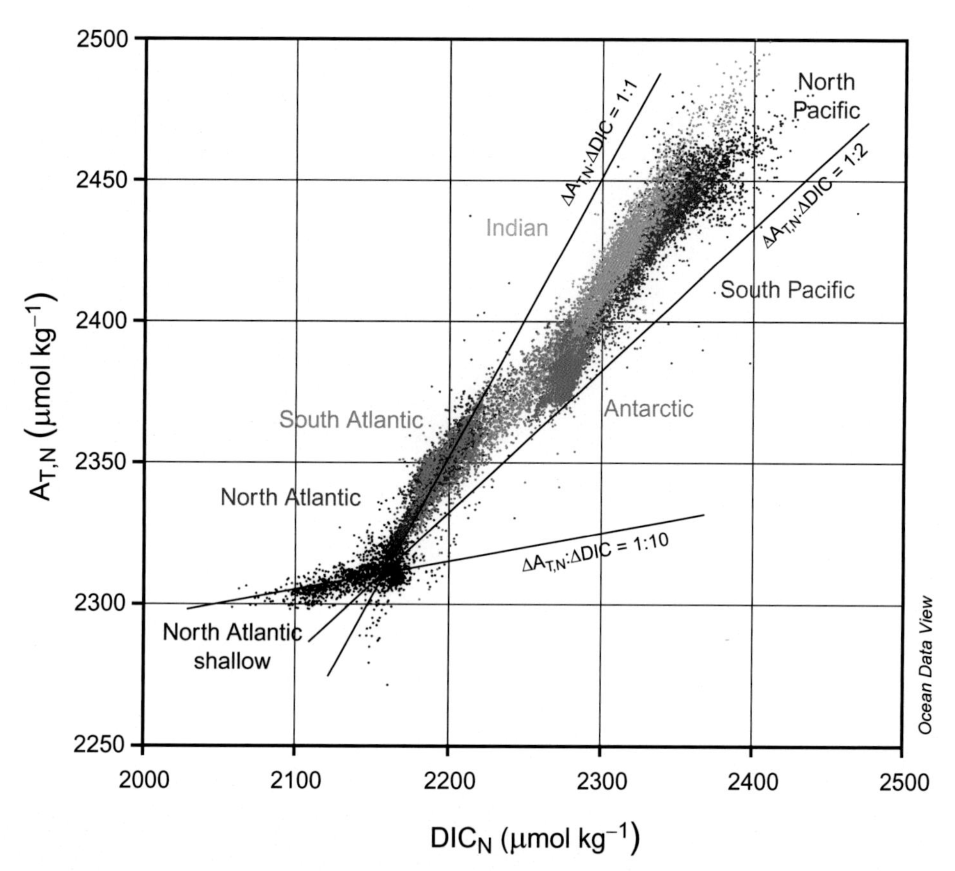

Figure 5.6 Salinity normalized ($S = 35$) total alkalinity, $A_{T,N}$, versus salinity normalized dissolved inorganic carbon, DIC_N, for the world's oceans. Data are for the deep ocean at depths >2500 m except for the points labeled "North Atlantic shallow," which come from 100–1000 m in the North Atlantic Ocean. Lines indicate different $\Delta A_{T,N} : \Delta DIC_N$ ratios. Data are from eWOCE (www.ewoce.org/data/#WHP_Bottle_Data).

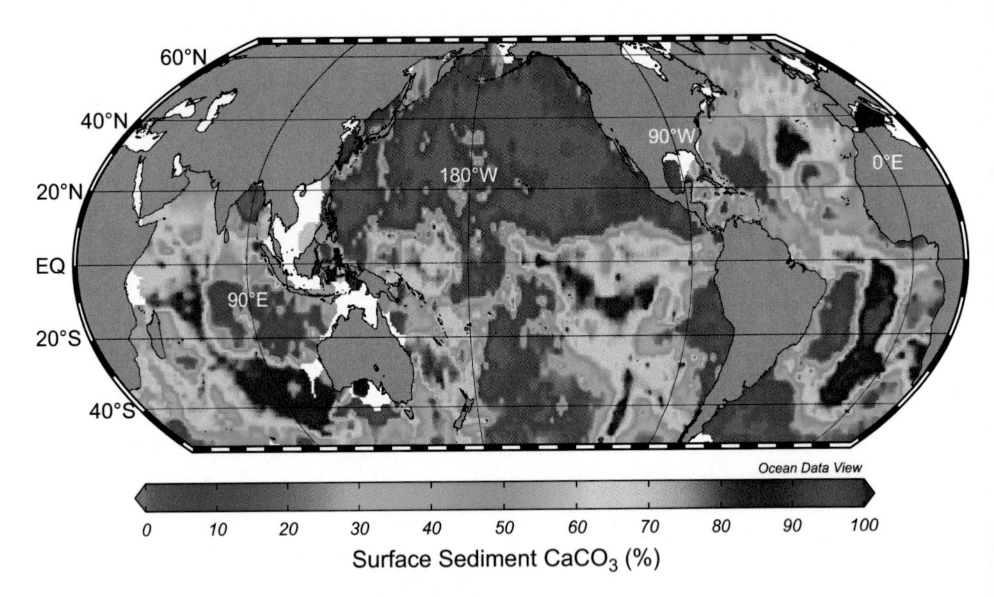

Figure 5.8 Global distribution of the weight percent $CaCO_3$ in surface sediments of the ocean deeper than 1000 m and located between 50°S and 60°N. White regions near the coast indicate no data. Redrawn from Archer (1996).

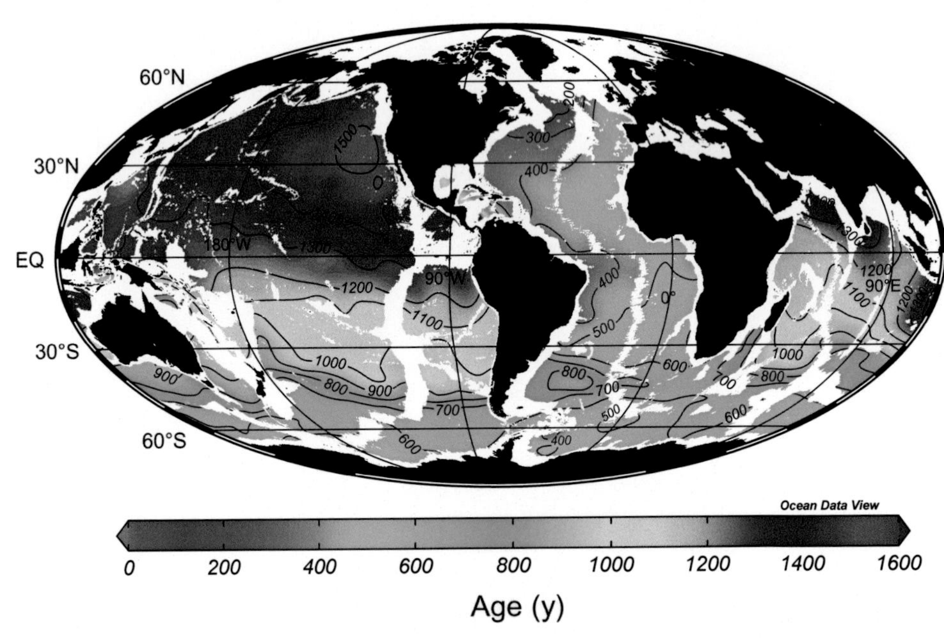

Figure 7.4 The ^{14}C age of DIC in the world's ocean at a depth of 3500 m based on radiocarbon observations compiled by the Global Ocean Data Analysis Project (GLODAP) and a matrix inversion method (from Gebbie and Huybers, 2012).

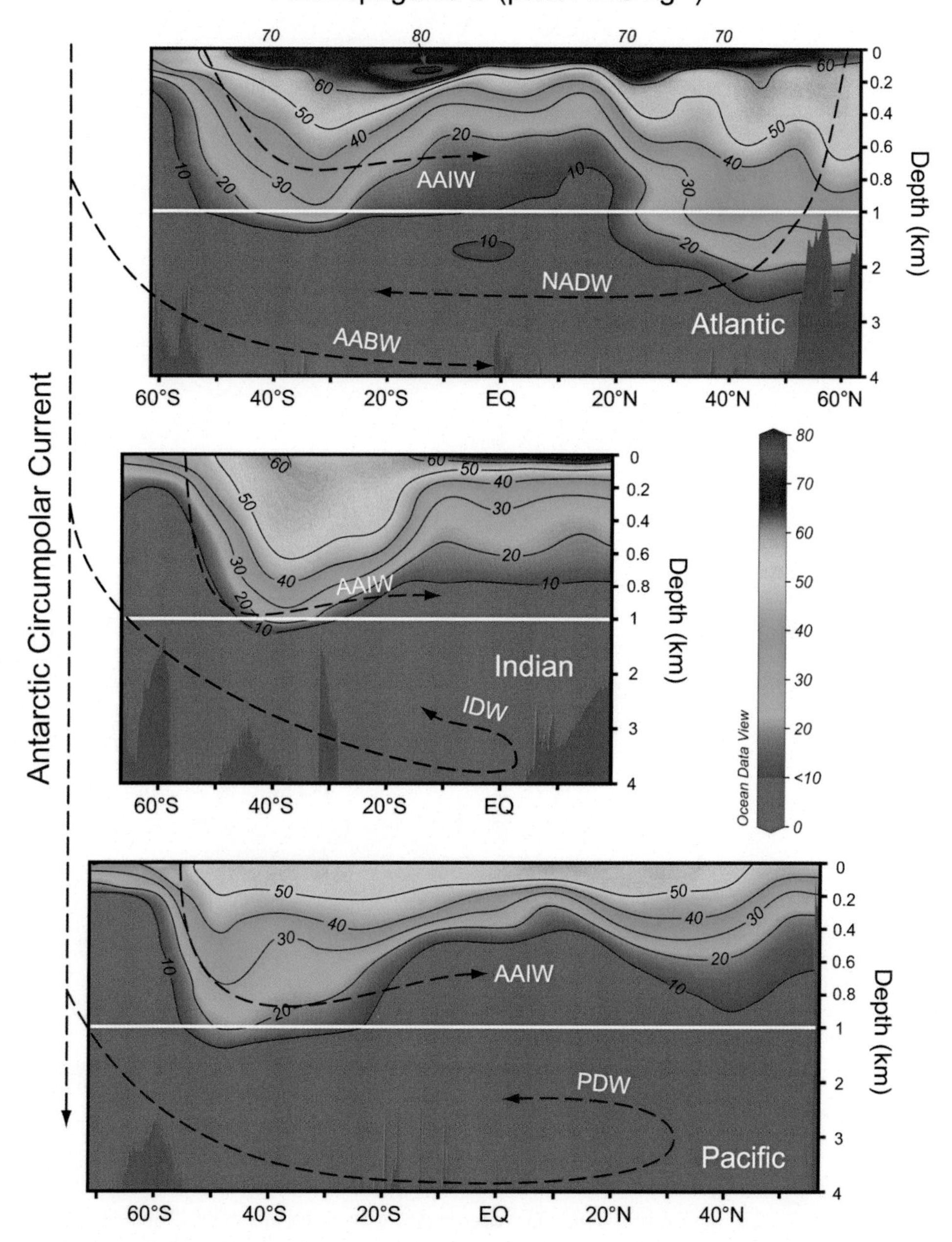

Figure 8.9 Vertical sections of the concentration of anthropogenic carbon in the ocean as of 2007. The contours are μmol kg^{-1} anthropogenic carbon in the measured DIC. These estimates are determined by a combination of multilinear regression (MLR) and DIC_{anth} techniques. Values are the sum of data from the preindustrial period to 1994 (Sabine et al., 2004, as gridded by Key et al., 2004) and the period 1994 to 2007 (Gruber et al., 2019).

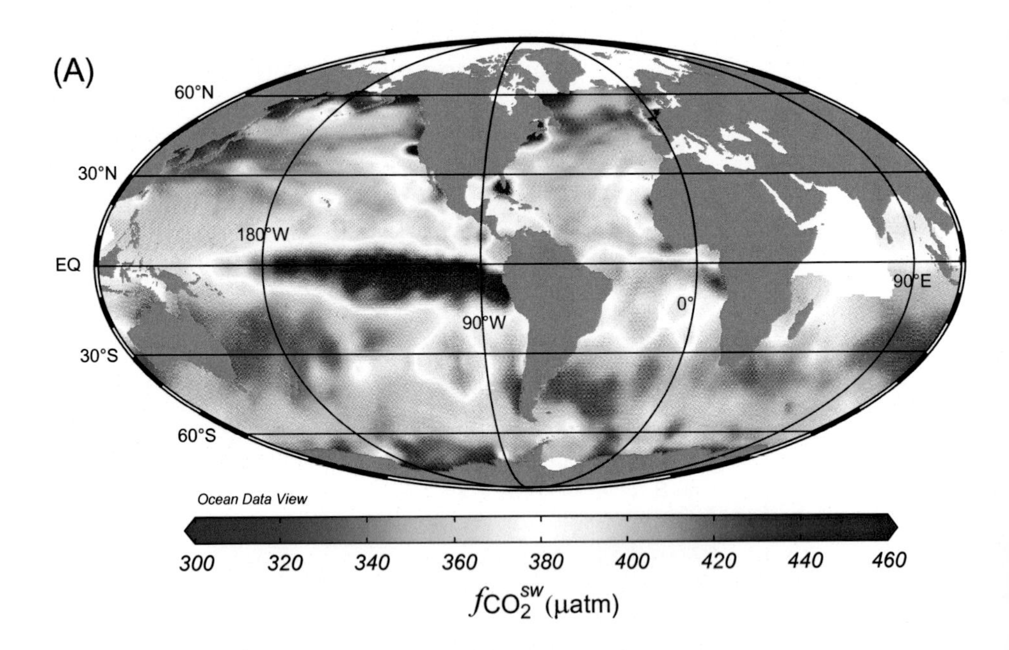

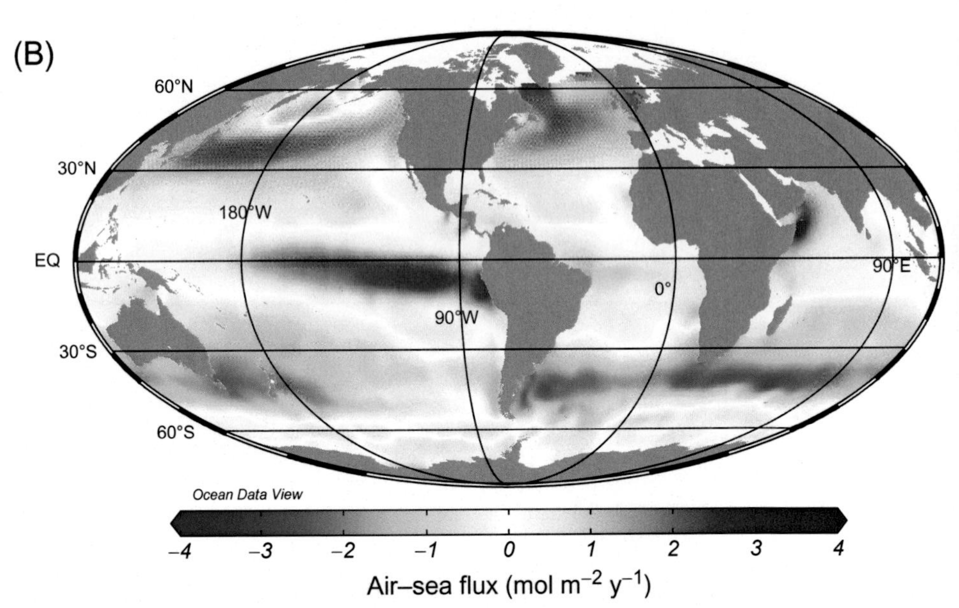

Figure 8.10 The global distribution of surface ocean fCO_2 (μatm) (A) and the air–sea CO_2 flux (mol-C m^{-2} y^{-1} where a positive flux is to the atmosphere) (B). Values in (A) are the spatial average of all available surface ocean fCO_2 data measured in the 2000–2010 decade and archived in the Surface Ocean CO_2 Atlas, SOCAT, www.socat.info/. Fluxes in (B) are determined from the SOCAT database available in 2020 (Landschützer, personal communication) using the neural network method described in Landschützer et al. (2014).

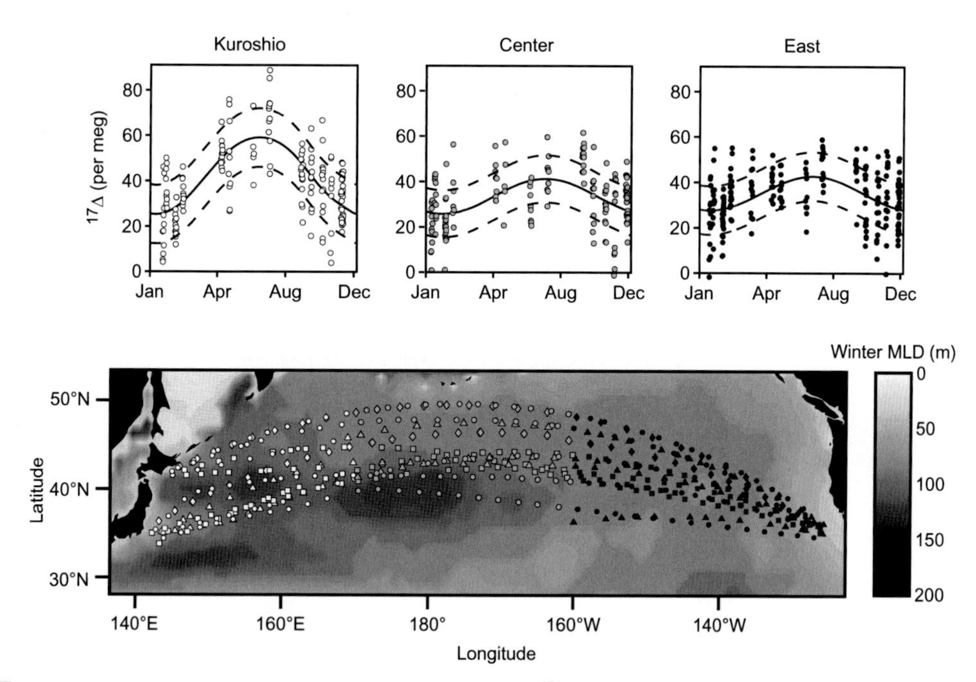

Figure 6.11 Seasonal variations in the triple isotope ratio anomaly of O_2, $\Delta^{17}O$ (per meg), in surface waters of the North Pacific Ocean (top three panels) determined by measurements at the locations indicated on the map. Seawater samples were measured in the engine room of container ships on 16 basin-wide transects between Asia and Canada at the location of the different symbols. Data in the top three panels indicate seasonal changes in the three regions indicated by the shade of symbols on the map: "Kuroshio," "Center," and "East." Gray scale on the map indicates the depth of the surface mixed layer. Redrafted from Palevsky et al. (2016).

about the horizontal variability of $^{17}\Delta$, so it is assumed to be negligible. Results from surface samples collected in the engine room of container ships that cross the North Pacific (Fig. 6.11, Palevsky et al., 2016) demonstrate the promise of this method for obtaining GOP rates. Values above $^{17}\Delta = 8$ per meg (solubility equilibrium with the atmosphere) indicate the influence of photosynthesis. Scatter in the data is rather large because the ^{17}O abundance is low, creating significant analytical error, and because biological rates are spatially variable. Both the magnitude and seasonality of $^{17}\Delta$ and thus gross productivity rates in the westernmost "Kuroshio" region are greater than in the Center and East Pacific regions.

6.4.3 $\delta^{18}O$ in Molecular Oxygen, a Tracer for Respiration

In the last section, we introduced the fact that the kinetic fractionation patterns for isotopes of molecular oxygen, O_2, are different than for carbon. It had long been assumed that there was little or no fractionation of oxygen produced during photosynthesis and a relatively large fractionation during respiration ($\varepsilon \approx -20‰$). More recent laboratory experiments

(Luz and Barkan, 2011; Table 6.3), however, indicate that during photosynthesis oxygen is produced that is somewhat heavier than the water in which the plants grew ($\varepsilon \approx +4‰$). Discrimination against the lighter isotope is opposite to that expected during a traditional kinetic fractionation.

Because O_2 is an insoluble gas, about 95% of it resides in the atmosphere (Chapter 2), and it has an isotope ratio of +23.5‰ with respect to the ocean water standard VSMOW (Table 6.2). A mass balance between the global processes of photosynthesis and respiration sets the oxygen isotope ratio of the atmosphere and is referred to as the Dole Effect after its discoverer, Malcolm Dole. Globally, we know that the rates of photosynthesis, P, and respiration, R, are about equal because the O_2 concentration of the atmosphere is relatively stable.

$$P_{O_2} = R_{O_2}. \tag{6.30}$$

The corresponding equation for ^{18}O written in terms of the ratios, $r = (^{18}O)/(^{16}O)$, and fractionations, α, is

$$P_{O_2} r_O \alpha_P = R_{O_2} r_A \alpha_R, \tag{6.31}$$

where the subscripts $_O$ and $_A$ represent ocean water, the source of molecular O_2 during photosynthesis, and the atmosphere, respectively. Converting the fractionation factors for photosynthesis and respiration to alpha values results in $\alpha_P = 1.004$ and $\alpha_R = 0.980$. Canceling P_{O_2} and R_{O_2} because they are equal and rearranging gives

$$r_A / r_O = \alpha_P / \alpha_R. \tag{6.32}$$

Because the isotope ratio in seawater is the $^{18}O/^{16}O$ standard, $r_A/r_O = r_A/r_{std}$, and combining with Eq. 6.1 yields

$$(\delta^{18}O\text{-}O_{2,A}/1000 + 1) = 1.004/0.980 = 1.024, \tag{6.33}$$

which results in a $\delta^{18}O\text{-}O_{2,A}$ of +24‰, similar to the measured value of 23.5‰.

The influence of land photosynthesis on the $\delta^{18}O$ of O_2 is more complicated because most photosynthetic oxygen is produced in leaves where evaporation leads to isotope enrichment of the leaf water. However, a modern analysis of the global sources contributing to the Dole effect (Luz and Barkan, 2011) argues that marine and terrestrial biological oxygen sources have roughly equal contributions.

Beneath the euphotic zone in the ocean, respiration dominates over photosynthesis. Because respiration fractionates O_2, the $\delta^{18}O$ of dissolved oxygen in the ocean becomes progressively heavier as the oxygen concentration decreases (Fig. 6.12). Factors that control the $\delta^{18}O$ of O_2 in the aphotic ocean are the isotope fractionation during respiration as well as the physical processes of mixing and advection. Lavine et al. (2009) used a hierarchy of models to show that it is difficult to interpret the data in Fig. 6.12 using a single respiration fractionation factor. Different lines in Fig. 6.12 indicate predictions using a simple endmember mixing model and a Rayleigh distillation model with two different fractionation factors. (The latter model is described near the end of this chapter.) If it were possible to preserve the $\delta^{18}O$ of O_2 in some mineral form that precipitates in ocean

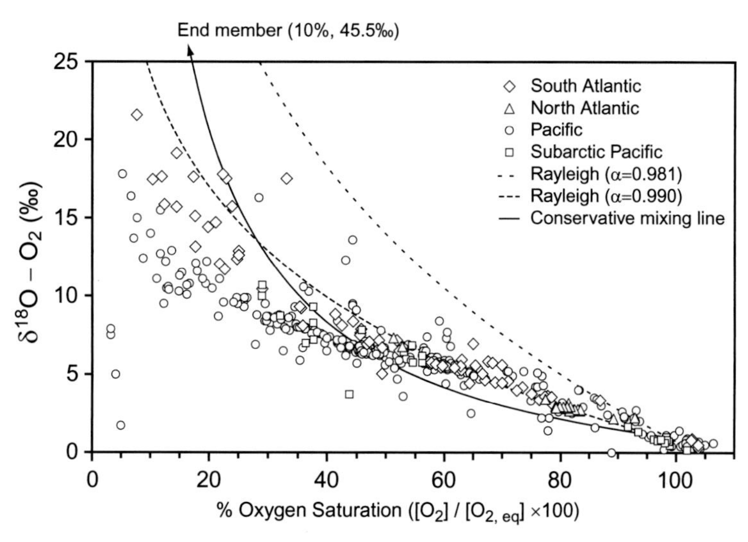

Figure 6.12 The relationships between the $\delta^{18}O$ of molecular O_2 (with respect to the $\delta^{18}O$ of atmospheric O_2) and the percent oxygen saturation (%) from the Atlantic and Pacific Oceans. Near-surface values are those around 100 % saturation and $\delta^{18}O = 0$‰. The solid line indicates endmember mixing between surface waters and the value indicated above the graph. Dashed lines indicate the Rayleigh fractionation model trends using respiration fractionation factors of −19 and −10‰, respectively. Modified from Lavine et al. (2009) who compiled this data.

sediments, say an oxide of Fe or Mn, it could be used as a paleoceanographic tracer for the extent of O_2 depletion in the deep ocean. This, however, appears to be wishful thinking, as no solid material that faithfully records the dissolved O_2 isotope ratio has yet been identified.

There is, however, a paleo record of the $\delta^{18}O$ of the atmosphere trapped in air bubbles of polar ice cores. The oxygen concentration of air in ice bubbles appears to be stable over many hundreds of thousands of years, presumably because of the very low temperatures and low numbers of bacteria in the ice. Sowers et al. (1993) showed that variations in $\delta^{18}O$-O_2 in ice core bubbles resembled the $\delta^{18}O$ of foraminiferal $CaCO_3$ from deep-sea sediments. If one assumes that fractionation factors have not changed, then the sole reason for changes in the $\delta^{18}O$-O_2 of the atmosphere trapped in ice bubbles is the variation in the isotope ratio of seawater, which we showed earlier in this chapter is determined by the volume of polar ice.

Interpreting $\delta^{18}O$-O_2 changes measured in the bubbles of ice cores in terms of glacial–interglacial changes in ocean water $\delta^{18}O$ made it possible to compare the timing of changes in atmospheric fCO_2 and CH_4, which are also measured in the atmosphere bubbles, with glacial–interglacial temperature changes. Before Sowers et al. (1993) made the $\delta^{18}O$-O_2 measurements on the atmosphere trapped in bubbles, the observed changes in fCO_2 and CH_4 from bubbles were compared with changes in the $\delta^{18}O$-H_2O of the ice itself, which records local temperature changes. However, there is an offset in timing between the age of the ice and the age of the atmosphere trapped in that same ice.

Because bubbles do not fully close until there is a substantial ice thickness, air age in bubbles is younger than the ice age at the same ice core depth. Models of this process indicate the age difference is dependent on many ice deposition factors that also change from glacial to interglacial times. The error in the ice age – gas age difference is several thousand years in old ice (>10 ky) and several hundreds of years for younger ice (<10 ky). By measuring changes in $\delta^{18}O$-O_2, CO_2, and CH_4 in the same air bubbles, the timing of the change in ice volume and that of atmospheric CO_2 and CH_4 chemistry can be directly compared. The only correction that must be made is the lag between changes in the $\delta^{18}O$-H_2O of seawater and the atmospheric $\delta^{18}O$-O_2 response, which is determined by the residence time, τ_{atm}, of O_2 in the atmosphere with respect to global photosynthetic O_2 production – about 2000 years.

The time course of changes in atmospheric fCO_2, CH_4, and $\delta^{18}O$-O_2 has been measured during the last four glacial–interglacial transitions in the Vostok ice core (Fig. 6.13). Changes in CO_2 are correlated to ocean circulation and biogeochemical process; CH_4 changes reflect the extent of wetlands on land; and, as discussed above, the $\delta^{18}O$ of O_2 is primarily a tracer of the $\delta^{18}O$ of seawater. With these three tracers, one can judge the timing of the physical processes that occurred during glacial–interglacial transitions. In each case, the vertical dashed line in the figure is located at precipitous jumps in CH_4 concentration. These jumps roughly coincide with the beginning of the increase in $\delta^{18}O$-O_2

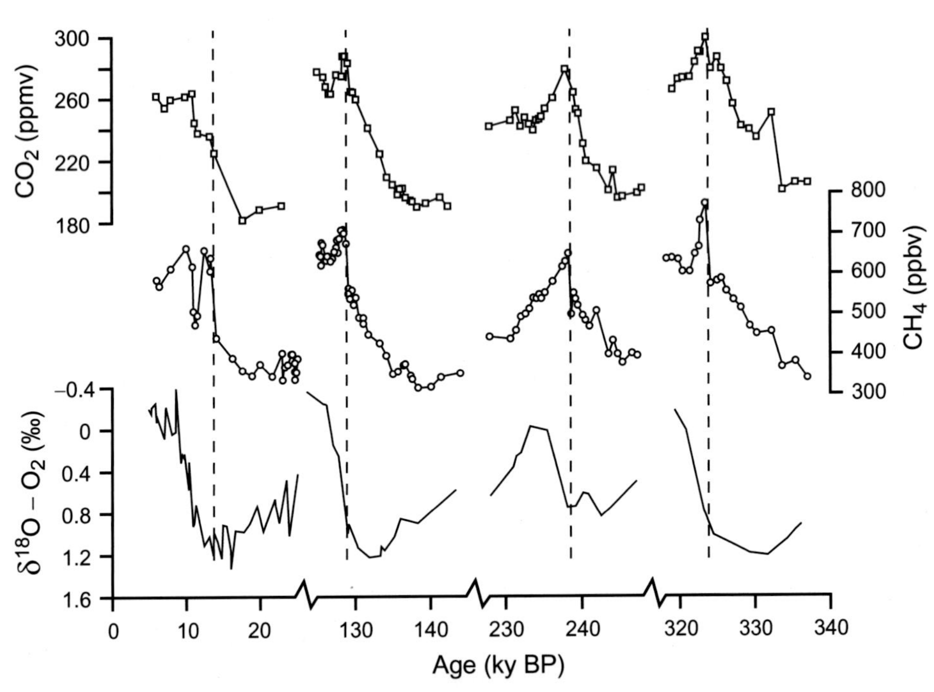

Figure 6.13 The relationship among the gases CO_2, CH_4, and the $\delta^{18}O$ of O_2 in air bubbles in the Vostok ice core from the Antarctic ice sheet spanning the last four glacial–interglacial transitions. Vertical dashed lines are drawn at the times of the "jumps" in the CH_4 concentrations. Redrawn from Petit et al. (1999).

as one moves from a glacial to an interglacial period, marking the beginning of substantial melting of the ice. Remember that the residence time for O_2 in the atmosphere is ~2 ky, so ice melting would have begun ~2 ky earlier than the atmospheric $\delta^{18}O$-O_2 change. Close inspection of the CO_2 and CH_4 data indicates that they both begin to rise, in each transition, 4–8 ky before the $\delta^{18}O$-O_2 begins to become lighter. Hence, the processes controlling atmospheric fCO_2 lead the change in ice volume during the glacial–interglacial transitions. One major theory for the cause of glacial to interglacial transitions is that increases in solar insolation initiate ice melting, which is then enhanced by lower albedo as glacial ice retreats (Ruddiman, 2001). However, the data from the ice core instead suggest that glacial–interglacial changes in ocean and in tropical land area processes occur before ice melting.

6.4.4 $\delta^{15}N$, a Tracer for the Marine Nitrogen Cycle

Stable nitrogen isotopes can be used to understand sources and sinks in the nitrogen cycle and how these cycles have changed in the past. We introduced the marine nitrogen cycle in Chapter 4 (Fig. 4.6), and the stable isotope ratios for the major nitrogen compounds are labeled in Fig. 6.14. The $\delta^{15}N$ of atmospheric N_2 is zero, because it is the standard. Dissolved NO_3^-, the most abundant form of fixed or bioavailable nitrogen (NO_3^-, NO_2^-, NH_4^+), is heavier than N_2 because of fractionation during denitrification which reduces NO_3^- to N_2 in oxygen deficient zones. Organic matter formed in the surface ocean has an isotope ratio that is the product of the nitrogen source – either N_2 from the atmosphere or NO_3^- from below the mixed layer, modified by the fractionation during photosynthesis ($\varepsilon \sim -5‰$).

The global marine nitrogen cycle can be categorized into two major subcycles for the purpose of describing the stable isotope transformations. The first includes the ocean-wide sources and sinks of fixed nitrogen, which are nitrogen fixation and denitrification plus anammox, respectively (see Section 4.1.3). These pathways are indicated by the dashed lines in Fig. 6.14. The other cycle is internal to the ocean and driven by the flux of organic N to the deep ocean balanced by the rate of transport of NO_3^- from the deeper ocean to the euphotic zone, indicated by the solid lines in Fig. 6.14. The residence times of these two pathways are not greatly different, with the internal pathway being limited by ocean mixing with a timescale of 500 to 1000 years and the global source and sink cycles being a few thousand years based on the NO_3^- inventory and current estimates of denitrification plus anammox (Altabet, 2007). We will first describe what controls the $\delta^{15}N$ of ocean NO_3^- based on the nitrogen fixation – denitrification plus anammox cycle and then discuss processes that control the $\delta^{15}N$ of organic matter produced during photosynthesis.

Mass Balance between Nitrogen Fixation and Denitrification. Processes that control the isotope ratio of NO_3^- in the deep ocean can be quantified by a mass balance of the sources and sinks. At steady state, the flux of fixed nitrogen to the ocean by nitrogen fixation, F_{fix}, is equal to the flux out of the ocean by denitrification plus anammox, F_{den}:

$$F_{fix} = F_{den}. \tag{6.34}$$

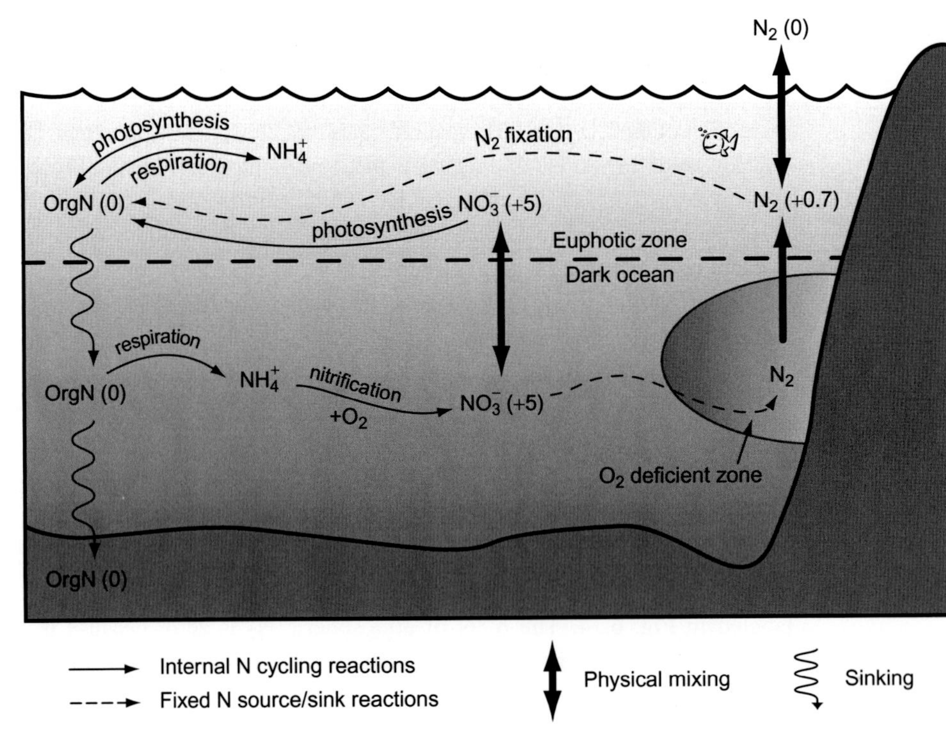

Figure 6.14 The $\delta^{15}N$ of the main nitrogen reservoirs in the ocean and atmosphere (values in parentheses) and exchanges among the reservoirs. Dashed lines indicate the nitrogen fixation and denitrification cycle, which are the main sources and sinks of fixed nitrogen (NO_3^-, NH_4^+, and organic N). Solid lines indicate fractionations during the photosynthesis and respiration cycle.

This equation indicates the fluxes of ^{14}N, which is the dominant isotope (Table 6.2). Note that the assumption of steady state for the mass balance of a reactive species like NO_3^- with a relatively short residence time is open to scrutiny, and we will return to this issue at the end of this section. A similar mass balance for ^{15}N can be constructed using the terms in the above equation along with the nitrogen isotope ratios, $r = {}^{15}N/{}^{14}N$, and the fractionation factors that accompany the reactions, α:

$$F_{fix} r_{N_2(aq)} \alpha_{fix} = F_{den} r_{NO_3^-} \alpha_{den}. \tag{6.35}$$

Here we assumed that the fractionation factors for denitrification and anammox are the same. Combining these two expressions for the steady-state situation leaves an equation with only ratios and fractionation factors:

$$r_{N_2(aq)} \alpha_{fix} = r_{NO_3^-} \alpha_{den}. \tag{6.36}$$

At this point, we need to make a short detour to the air–water interface, because atmospheric fN_2^a is the standard in the stable isotope notation (Eq. 6.1), so we can simplify our derivation by replacing the dissolved nitrogen $N_2(aq)$ with fN_2^a. The Henry's law equilibrium between atmospheric fN_2 and the dissolved concentration in the ocean is

$$K_{H,N_2} f N_2^a = N_2(aq). \tag{6.37a}$$

Using r to indicate the ratios $^{15}N/^{14}N$ for the atmosphere, $r_{N_2^a}$, and dissolved N_2, $r_{N_2(aq)}$, the solubility equation for ^{15}N becomes

$$^{15}K_{H,N_2} f N_2^a \, r_{N_2^a} = N_2(aq) \, r_{N_2(aq)}. \tag{6.37b}$$

Dividing Eq. 6.37b by Eq. 6.37a gives the relationship between the isotope ratio in air and that in water at solubility equilibrium:

$$r_{N_2(aq)} = \left(\frac{^{15}K_{H,N_2}}{K_{H,N2}} \right) r_{N_2^a} = \alpha_{sol} \, r_{N_2^a}, \tag{6.37c}$$

where the subscript, $_{sol}$, indicates the solubility fractionation factor, which is the ratio of the Henry's law constants. Substituting Eq. 6.37c into Eq. 6.36 yields the isotope ratio of NO_3^- in terms of fractionation factors:

$$\frac{r_{NO_3^-}}{r_{N_2^a}} = \alpha_{sol} \frac{\alpha_{fix}}{\alpha_{den}} = 1.0007 \times 0.998/0.975 = 1.024, \tag{6.38}$$

$$\delta^{15}N\text{-}NO_3^- = +24‰.$$

This calculated value for $\delta^{15}N\text{-}NO_3^-$ of +24‰ is much greater than that measured for NO_3^- in the deep ocean (~ +5‰) indicating a major problem with our calculation. The problem is that the laboratory measured fractionation factor for denitrification (Table 6.3) is much too high to be used in a whole-ocean model, because denitrification occurs both in oxygen deficient regions of the water column and in anoxic sediments. In a semi-closed system like sediment pore waters, the entire isotope effect is not observed because exchange with bottom waters is limited. In these circumstances, nearly every NO_3^- molecule diffusing into the sediments is denitrified; the discrimination between ^{15}N and ^{14}N does not matter if every molecule eventually undergoes the reaction. Measurements of the $\delta^{15}N$ of the NO_3^- flux into denitrifying sediments, using a chamber that isolates a portion of the sea floor from the surroundings, indicate that fractionation during denitrification in deep sea sediments is very small (~ –1.5‰, Brandes and Devol, 2002). If you combine these two locations of denitrification in our calculation – the water column with a denitrification fractionation of –25‰ and the sediments with no fractionation, you can calculate the portion of the global denitrification that occurs in marine sediments necessary to produce a $\delta^{15}N\text{-}NO_3^-$ of +5. This cycle has been represented in much more elaborate models of the ocean's nitrogen cycle (DeVries et al., 2012), and is the subject of problem 6.4 at the end of this chapter.

The $\delta^{15}N$ of Marine Organic Matter and Its Utility for Past N Cycle Changes. The $\delta^{15}N$ of organic matter in ocean surface waters is dependent on the isotope ratio of the nitrogen substrate and the fractionation factor during photosynthesis (~ –5‰). As in the denitrification example in the previous section, the amount of isotopic fractionation that occurs during photosynthesis is a function of nutrient availability versus uptake rates. In situations where surface ocean NO_3^- is drawn down to less than 1 μmol kg^{-1} (like the subtropical oceans), the $\delta^{15}N$ of organic nitrogen formed during photosynthesis will be equal to that of the nitrogen substrate – either NO_3^- that upwells to surface waters (global $\delta^{15}N \sim +5‰$) or N_2 gas in regions of nitrogen fixation ($\delta^{15}N \sim -1‰$). In this case, there is little fractionation during photosynthesis because the nitrogen source is consumed entirely.

In situations where the NO_3^- in the euphotic zone is not completely utilized by phytoplankton (like coastal waters, the Equator, or high latitudes), the $\delta^{15}N$ of the organic matter formed will be ~5‰ lighter than the substrate NO_3^-, due to the kinetic fractionation factor during photosynthesis. In this case, the $\delta^{15}N$ of organic matter exiting the surface waters and ultimately stored in sediments can indicate an increase or decrease in the drawdown of surface water NO_3^- due to a change in the ratio of the sources and sinks (for example, a change in the supply of NO_3^- to the euphotic zone). We present, in the following paragraphs, examples of both applications of the $\delta^{15}N$ of organic matter in dated sediments to determine changes (or stability) of the marine nitrogen cycle through time.

Because the residence time of NO_3^- in the ocean is so short (a few thousand years), the mean concentration in deep waters may be quite variable on this timescale. A change in mean ocean NO_3^- concentration would have affected the biological pump, which would in turn have important consequences for atmospheric fCO_2 and the concentration of oxygen in the deep ocean. With this question of the stability of the global NO_3^- inventory in mind, Altabet (2007) measured the $\delta^{15}N$ of organic matter in sediment cores from continental margins that were long enough and well dated enough to encompass a period of at least one residence time for NO_3^-. The concentration of NO_3^- in surface waters of the regions studied is low enough today that the $\delta^{15}N$ of organic matter formed in the surface ocean reflects the $\delta^{15}N$ of the NO_3^- substrate. These regions also have little nitrogen fixation today indicating that this process was likely negligible in the recent past. The history of the isotopic composition in the organic matter that escaped the surface ocean should therefore record changes in the $\delta^{15}N$ of NO_3^- in the deep water below the euphotic zone. Results from the Ecuadorian continental margin (Fig. 6.15) indicate no change in the organic matter $\delta^{15}N$ of +5.5‰ within the ±0.5‰ scatter of the data for the length of the core, arguing for a stable marine NO_3^- concentration in the deep ocean for at least the last 4000 years.

An entirely different interpretation of changes in the $\delta^{15}N$ of organic matter in ocean sediment cores results from measurements in locations where surface water NO_3^- concentrations are high. In such regions, $\delta^{15}N$ changes in sedimentary organic matter over glacial–interglacial timescales indicate a change in the utilization of the surface water NO_3^- pool. An increase in the $\delta^{15}N$ of organic matter indicates a stronger drawdown in NO_3^- to lower concentrations because of the kinetic isotope fractionation during photosynthesis. The cause of the NO_3^- drawdown could be either an increase in the rate of carbon export or a decrease in the flux of NO_3^- to the surface ocean – both of which would increase the $\delta^{15}N$ of surface water NO_3^-. This method was used to suggest a decrease in surface ocean nitrate in the ice age Antarctic ocean (Robinson and Sigman, 2008).

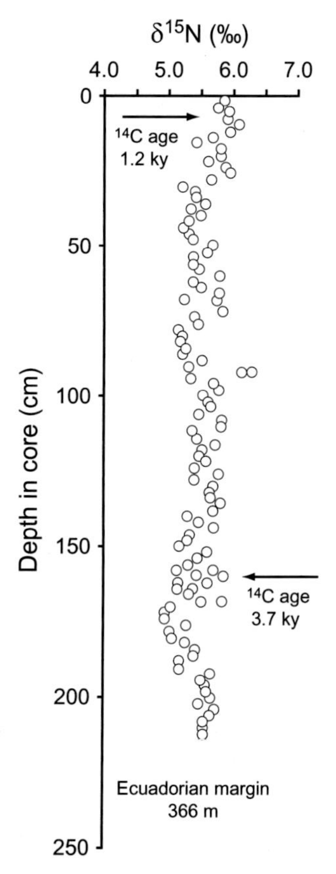

Figure 6.15 The $\delta^{15}N$ of organic matter in sediments from a core on the Ecuadorian continental margin in 366 m water depth. Age estimates of the organic matter are indicated by ^{14}C dates. Redrafted from Altabet (2007).

6.5 Rayleigh Fractionation

Many isotope fractionation reactions in the environment occur in situations where the reaction product does not interact further with the original reactants. A model for this type of system was developed to explain fractional distillation of mixed liquids before the dawn of the twentieth century by Lord Rayleigh (Rayleigh, 1896), and it bears his name, Rayleigh distillation. This model is used as a first approximation for interpreting many isotopic measurements in the environment. The classic environmental application is the change in $\delta^{18}O$ of rain over the lifetime of a cloud as it cools, and it has also been employed to explain the change in $\delta^{18}O$-O_2 during respiration in deep waters (Fig. 6.12).

The prerequisite for applying this model is that the change in concentration of the light, [*L*], and heavy, [*H*], isotopes can be approximated by first-order reactions, so that the rate of change is proportional to the concentration. This could happen in the case where the reactant and product are in equilibrium chemically and isotopically and the product is physically removed from the system (rain from a cloud, for instance), or in the case of an irreversible kinetic reaction like radioactive decay in which the product cannot reform the reactant:

$$d[L]\Big/dt = k_L[L], \tag{6.39}$$

$$d[H]\Big/dt = k_H[H], \tag{6.40}$$

where the k values are first-order rate constants. Dividing the heavy equation by the light yields

$$\frac{d[H]}{d[L]} = \frac{k_H}{k_L}\frac{[H]}{[L]} = \alpha\frac{[H]}{[L]} = \alpha\, r, \tag{6.41}$$

where the ratio of the rate constants is the isotope fractionation factor, α, and the isotope ratio is designated r. After a short series of mathematical manipulations (see Appendix 6A.2) one can represent the isotopic ratio of the remaining reactant at time t divided by the ratio at the beginning of the reaction ($t = 0$) in terms of the fraction of the remaining reactant, f, and the fractionation factor, α:

$$r_t/r_{t=0} = f^{(\alpha-1)}. \tag{6.42}$$

The Rayleigh fractionation equation can be transformed into isotopic δ notation using Eq. 6.1:

$$\delta_t = (\delta_{t=0} + 1000)\ f^{(\alpha-1)} - 1000. \tag{6.43}$$

Isotopic fractionation of H_2O during formation and destruction of clouds is an interesting combination of equilibrium and kinetic isotope effects accompanying water evaporation, transport, and precipitation. Both kinetic and equilibrium isotope effects are important during evaporation from the ocean surface depending on the humidity of the surrounding air, but only the equilibrium isotope effect is imparted when water molecules condense from vapor in a cloud back into liquid form and the liquid "rains out" of the cloud. While our discussion focuses on oxygen isotope fractionation, the same systematics apply to hydrogen, in which equilibrium isotope effects between liquid and gaseous water are about 10 times larger than those for oxygen.

At a temperature of 20°C, the $\delta^{18}O$ equilibrium isotope effect between water vapor and liquid water is about 9‰ (Fig. 6.16), which occurs because H_2O molecules containing ^{18}O have a slightly lower vapor pressure than those containing ^{16}O. When air masses containing water vapor evaporated from the ocean are cooled via transport to colder regions, the vapor transforms to liquid water because the vapor pressure of water progressively decreases with lower temperature. Because the $\delta^{18}O$ in the condensate (rain) is approximately 9‰ more enriched than the water vapor, by mass balance the water vapor remaining in the cloud must become progressively more depleted in ^{18}O. Thus, an ocean-derived cloud that cooled from 20 to 10°C would lose approximately half its initial water vapor content (Fig. 6.16), and, in the process, decrease its isotopic

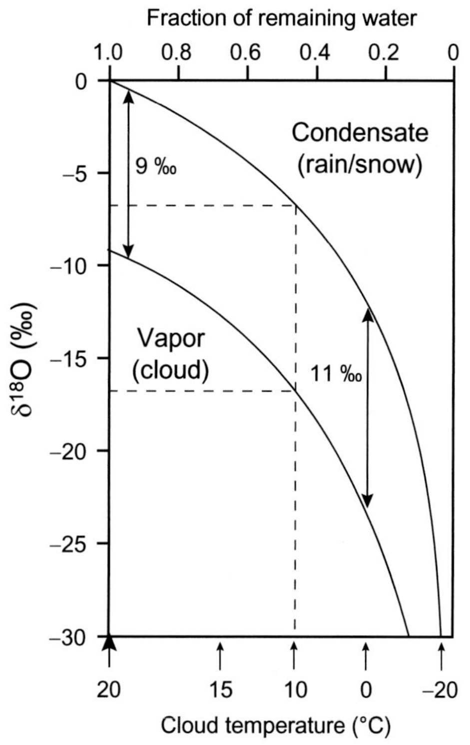

Figure 6.16 An idealized illustration of the differences between the $\delta^{18}O$ of water condensate and vapor as a function of the fraction of the remaining water during the Rayleigh fractionation process. Envision a cloud that forms at 20°C and remains a closed system except for water that rains out as it cools from 20°C to −20°C. The equilibrium fractionation factor is temperature dependent, 9‰ at 20°C and 11‰ at 0°C. Modified from Dansgaard (1965).

composition from −9 to −17‰. The rain falling from the cloud would follow this depletion in ^{18}O, and would have a $\delta^{18}O$ near 0‰ at 20°C but near −7‰ at 10°C. The change in $\delta^{18}O$ of the remaining cloud vapor in Fig. 6.16 is derived from the Rayleigh fractionation equation taking into account that there is a slight change in α with temperature. Note that if one were to somehow collect all the precipitation falling from the cloud over its lifetime, the $\delta^{18}O$ of liquid water collected would, by mass balance, equal the $\delta^{18}O$ of the original cloud vapor.

This "milking" process continues as long as the temperature decreases, which generally occurs as clouds move to higher latitudes or to greater elevations, for example moving inland over mountains (Fig. 6.17). The $\delta^{18}O$ of the precipitation falling at any location reflects both the ~ +9‰ equilibrium fractionation effect between the water vapor and condensate and the cumulative $\delta^{18}O$ depletion over time of the cloud's remnant vapor. The result of this Rayleigh fractionation behavior on a global scale is exemplified by the global distribution of the $\delta^{18}O$ of rainwater (Fig. 6.18). Values in low latitudes near the Equator are near zero, because this region is the source of most water vapor and the $\delta^{18}O$ resembles

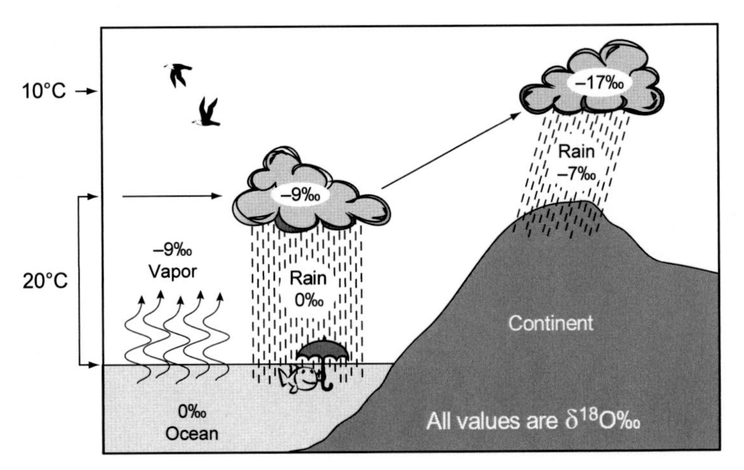

Figure 6.17 Schematic diagram of the oxygen isotopic fractionation among seawater, the atmosphere, and rain. Modified from Siegenthaler (1976).

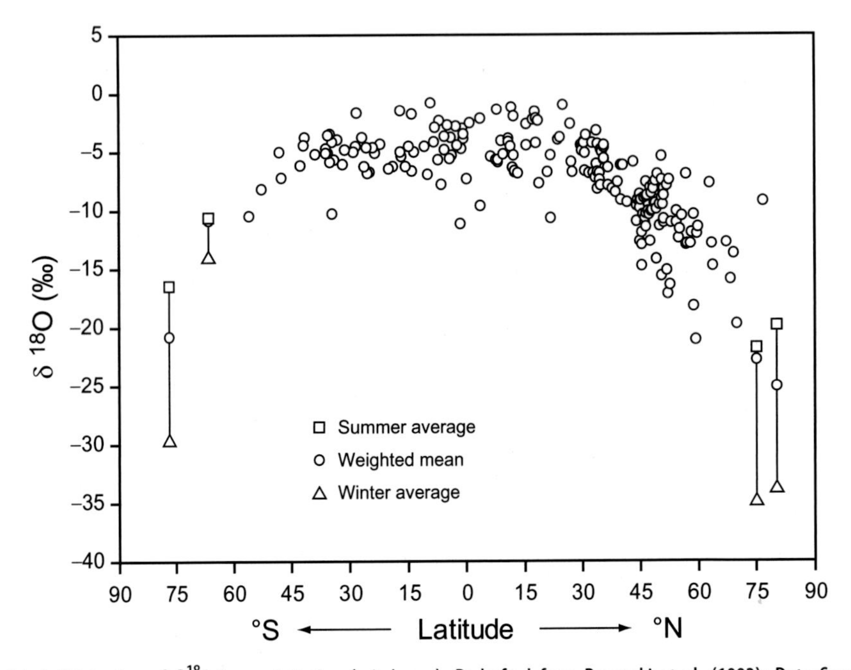

Figure 6.18 The global distribution of $\delta^{18}O$ in precipitation (rain/snow). Redrafted from Rozanski et al. (1993); Data Source: International Atomic Energy Association (IAEA); WMO Global Network of Isotopes in Precipitation (GNIP) Database, 2019, https://nucleus.iaea.org/wiser.

its source, seawater. As one travels toward higher latitudes and the clouds that originated in low latitudes lose more water, the isotope ratios become much more negative. At sub-zero temperatures like those on the Antarctic or Greenland ice sheets, the cloud and the snow falling from it can both have extremely low $\delta^{18}O$ values of less than –30‰.

Discussion Box 6.3

Now that we've learned about Rayleigh fractionation, let's turn our attention back to the change in $\delta^{18}O$ of O_2 in the deep ocean presented earlier in Fig. 6.12 (Lavine et al., 2009).

- Based on the values shown there, which isotope does respiration discriminate against? What would the data look like if there was no fractionation during respiration?
- Two curves are shown in Fig. 6.12 representing the expected trajectory of data for Rayleigh fractionation. Imagine a water parcel at the ocean surface with 100% O_2 saturation. How does Rayleigh fractionation conceptually apply as this water mass is subducted into the ocean interior? What is the difference between the two Rayleigh curves? How well do they fit the data, in your opinion?
- Another curve on Fig. 6.12 shows the expected values for conservative mixing between two endmembers, one at 10% saturation with $\delta^{18}O = 45.5‰$ and the other at 100% saturation with $\delta^{18}O = 0.75‰$. Why isn't this mixing line linear? Imagine a 50:50 mixture of each of these two water sources. What proportion of the O_2 in such a mixture originates from each source? How does that explain the non-linearity of the mixing line?
- Consider how fractionation during denitrification depends on whether the reaction takes place in the sediments versus the water column discussed above in Section 6.4.4. If O_2 were only consumed by respiration in sediments, how would you expect the data in Fig. 6.12 to look different?
- Based on these concepts, how would you explain the data shown in Fig. 6.12? How could you test your hypothesis?

6.6 Anthropogenic Influences

The examples of applying stable isotopes to problems in chemical oceanography presented here represent only a minute sampling of the many different stable isotope studies now being applied to inorganic and organic marine samples. A massive amount of new information is currently being produced by the international GEOTRACES program, a global hydrographic initiative focused on trace metals and isotope tracers, which is underway at the time of writing this book.

Isotope ratios can be valuable tracers for determining the extent and sources of marine anthropogenic contamination. We present two classic examples: changes in the isotope ratios of carbon and lead since the beginning of the heightened industrial period of the twentieth century. As we discussed earlier in the chapter, marine DIC has a $\delta^{13}C$ value near 0‰, making the atmospheric CO_2 value about –7‰. Since organic matter is about 23‰

lighter (less ^{13}C) than marine DIC, the addition of fossil fuel CO_2 (from combustion of coal, oil, and gas that are geologic remnants of plants and animals) to the natural carbon reservoir is steadily making it lighter. This is seen clearly in the $\delta^{13}C$ of CO_2 measured by both direct atmospheric measurements starting in the late 1950s and from measurements of the $\delta^{13}C$ of CO_2 in air bubbles in Antarctic ice cores for earlier periods (Fig. 6.19). Preindustrial (~1900) levels of about 290 μatm fCO_2 with an isotope ratio of ~ –6.7‰ have evolved to fCO_2 values greater than 410 μatm with an isotope ratio of ~ –8.5‰. As atmospheric fCO_2 increases due to fossil fuel combustion, the $\delta^{13}C$ of atmospheric CO_2

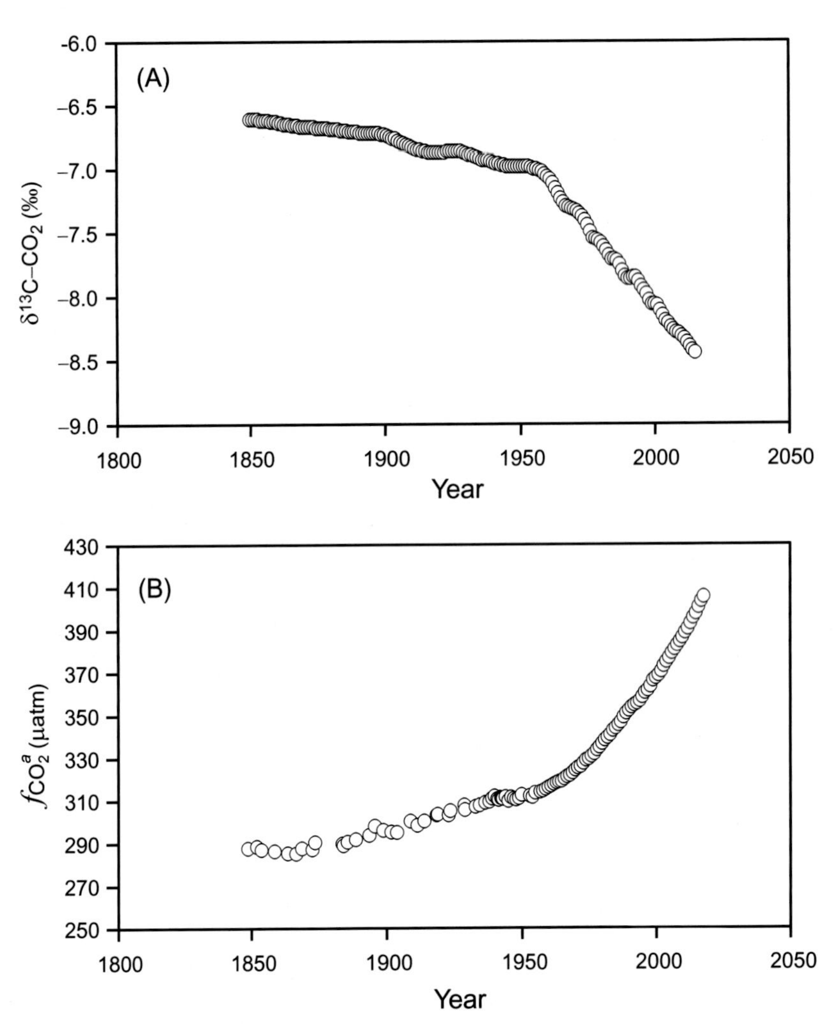

Figure 6.19 Trends in the (A) $\delta^{13}C$ of atmospheric CO_2 (‰) and (B) the fCO_2 of the atmosphere (μatm) since the 1850s. $\delta^{13}C$ values in (A) are the global, annual means of atmospheric measurements from the period 1850–2015 including values from flasks, firn, and ice cores. (Compiled from measurements of many researchers by Graven et al., 2017.). The atmospheric fCO_2 values in (B) are the average of flask measurements from Hawaii and the South Pole merged with published ice core data (R. Keeling and S. Piper, https://scrippsCO2.ucsd.edu).

decreases. The combination of atmospheric CO_2 and $\delta^{13}C$-CO_2 measurements are used to estimate the importance of marine vs. terrestrial sinks for anthropogenic CO_2. Seawater observations with time and depth in the ocean also show that the $\delta^{13}C$ of DIC in the upper ocean is decreasing as anthropogenically produced CO_2 penetrates this sphere (Quay et al., 2003). Determining the change in $\delta^{13}C$ of DIC over time is one of the methods used to demonstrate that the ocean has absorbed approximately one-third of anthropogenically produced carbon (see Chapter 8).

The second example of anthropogenic isotope tracers is that of lead. Humans have increased the flux of reactive Pb into the environment by more than a factor of 10. Lead concentrations in the surface ocean have increased from background values of about 10 pmol kg^{-1} to values in the early 1980s of over 100 pmol kg^{-1} (Fig. 1.8D). The first uncontaminated measurements in seawater were made in the late 1970s, and ocean profiles have been determined in the Atlantic Ocean about every 5 years since then (Boyle et al., 2014). Anthropogenic lead contamination occurred primarily because of the addition of tetraethyl lead to gasoline as an antiknock compound. This practice was gradually banned around the world starting in the 1970s. At present, depth profiles of lead concentration in the Atlantic Ocean have peak values in the ocean thermocline that are remnants of high surface water concentrations 40 years ago. Concentrations below about 3000 m are very low (~10 pmol kg^{-1}) and probably uncontaminated by lead pollution.

The isotope ratio of the anthropogenic lead that contaminates the upper ocean indicates the origin of the lead deposits used to manufacture the tetraethyl lead added to gasoline. Lead deposits used by the United States came from geologic formations with a $^{206}Pb/^{207}Pb$ ratio > 1.17 while deposits used by European countries had a $^{206}Pb/^{207}Pb$ ratio < 1.15. Because the United States used several times more lead than the Europeans during the early years of contamination, the ratio measured in Atlantic Ocean waters was dominated by isotope ratios greater than 1.17 before 1990, but this ratio has declined in more recent years because the United States decreased its use of the lead additive more rapidly.

6A.1 Appendix 6A.1 Relating the Stable Isotope Terms K, α, ε, and δ

The following is a derivation of the relationships between K, α, ε, and δ, for the oxygen isotope exchange reaction between CO_3^{2-} and H_2O that is the basis of the ^{18}O paleotemperature method. Consider the reaction

$$CO_3^{2-} + H_2{}^{18}O \longleftrightarrow C^{18}OO_2^{2-} + H_2O, \quad (6A.1.1)$$

where the superscript for ^{16}O has been dropped for simplicity. K is related to α as

$$K_{1-2} = \frac{[C^{18}OO_2^{2-}][H_2O]}{[CO_3^{2-}][H_2{}^{18}O]} = \frac{[C^{18}OO_2^{2-}]}{[CO_3^{2-}]} \times \left(\frac{[H_2{}^{18}O]}{[H_2O]}\right)^{-1} = r_1 r_2^{-1} = \frac{r_1}{r_2} = \alpha_{1-2}, \quad (6A.1.2)$$

where r is the $^{18}O/^{16}O$ ratio and the subscripts 1 and 2 refer to CO_3^{2-} and H_2O, respectively. The subscript, 1–2, represents the reaction direction as written. Then α_{1-2} is related to ε by

$$\varepsilon_{1-2} = (\alpha_{1-2} - 1) \times 1000. \quad (6A.1.3)$$

Thus, α_{1-2} can be to related to the difference in the product and reactant δ values by generating a relative difference expression in r:

$$\alpha_{1-2} - 1 = \frac{r_1}{r_2} - 1 = \frac{r_1 - r_2}{r_2}. \quad (6A1.4)$$

By definition:

$$\delta^{18}O_i \equiv \left(\frac{r_i}{r_{std}} - 1\right) \times 1000, \quad (6A.1.5)$$

which can be rearranged to give

$$r_i = \left(\frac{\delta^{18}O_i}{1000} + 1\right) r_{std}. \quad (6A.1.6)$$

Substitution of Eq. 6A1.6 into Eq. 6A1.4 (with r_{std} canceling out) results in

$$\frac{r_1 - r_2}{r_2} = \frac{\left(\frac{\delta^{18}O_1}{1000} + 1\right) - \left(\frac{\delta^{18}O_2}{1000} + 1\right)}{\left(\frac{\delta^{18}O_2}{1000} + 1\right)} \quad (6A1.7)$$

and

$$\alpha_{1-2} - 1 = \frac{\delta^{18}O_1 - \delta^{18}O_2}{\delta^{18}O_2 + 1000}, \quad (6A.1.8)$$

$$\varepsilon = (\alpha_{1-2} - 1) \times 1000 \cong \delta^{18}O_1 - \delta^{18}O_2. \quad (6A.1.9)$$

6A.2 Appendix 6A.2 **Derivation of the Rayleigh Fractionation Equation**

Changes in the concentration of light, $[L]$, and heavy, $[H]$, isotopes with respect to time as the result of a first-order reaction are proportional to the rate constant, k, and the concentrations of the isotopes:

$$d[L]/dt = k_L[L], \tag{6.39}$$

$$d[H]/dt = k_H[H]. \tag{6.40}$$

Dividing Eq. 6.40 by Eq. 6.39 gives

$$\frac{d[H]}{d[L]} = \frac{k_H}{k_L}\frac{[H]}{[L]} = \alpha\frac{[H]}{[L]} \tag{6.41}$$

where the ratio of the rate constants is the isotope fractionation factor, α:

$$\alpha = \frac{k_H}{k_L}. \tag{6A.2.1}$$

Rearranging Eq. 6.41 yields

$$\frac{d[H]}{H} = \alpha\frac{d[L]}{[L]}. \tag{6A.2.2}$$

Integrating from the initial values ($t = 0$, $[H] = [H^0]$, $[L] = [L^0]$) to the values $[H]$ and $[L]$ at time, t,

$$\int_{H^0}^{H}\frac{d[H]}{[H]} = \alpha\int_{L^0}^{L}\frac{d[L]}{L} \tag{6A.2.3}$$

gives

$$\ln\left(\frac{H}{H^0}\right) = \alpha\ \ln\left(\frac{[L]}{[L^0]}\right), \tag{6A.2.4}$$

which can be rewritten as

$$\left(\frac{H}{H^0}\right) = \left(\frac{[L]}{[L^0]}\right)^{\alpha}. \tag{6A.2.5}$$

Dividing both sides by L/L^0 givest

$$\left(\frac{[H]/[L]}{[H^0]/[L^0]}\right) = \left(\frac{[L]}{[L^0]}\right)^{\alpha-1} = f^{\alpha-1}, \tag{6A.2.6}$$

where f is the fraction of the light isotope remaining at time t. If you substitute r for the concentration ratios, then you have the Rayleigh distillation equation:

$$\frac{r_t}{r_{t=0}} = f^{\alpha-1}. \tag{6.42}$$

Now, this equation can be transformed to δ notation by rearranging the definition of the delta notation (Eq. 6.1):

$$r_t = \left(\frac{\delta_t}{1000} + 1\right) r_{std};\ r_{t=0} = \left(\frac{\delta_{t=0}}{1000} + 1\right) r_{std}, \tag{6A.2.7}$$

and substituting these values into Eq. 6.42 results in

$$\delta_t = (\delta_{t=0} + 1000) f^{(\alpha-1)} - 1000. \tag{6.43}$$

References

Adkins, J. F., and D. P. Schrag (2001) Pore fluid constraints on deep ocean temperature and salinity during the Last Glacial Maximum, *Geophysical Research Letters*, **28**, 771–774, doi:10.1029/2000GL011597.

Altabet, M. A. (2007) Constraints on oceanic N balance/imbalance from sedimentary ^{15}N records, *Biogeosciences*, **4**, 75–86.

Altabet, M. A., and R. Francois (1994) The use of nitrogen isotopic ratio for reconstruction of past changes in surface ocean nutrient utilization, in *Carbon Cycling in the Glacial Ocean: Constraints on the Ocean's Role in Global Change* (eds. R. Zahn, T. F. Pederson, M. A. Kaminski, and L. Labeyrie), pp. 281–306. Berlin: Springer-Verlag.

Bemis, B. E., H. J. Spero, J. Bijma, and D. W. Lea (1998) Reevaluation of the oxygen isotopic composition of planktonic foraminifera: experimental results and revised paleotemperature equations, *Paleoceanography*, **13**, 150–160.

Bourg, I. C., and G. Sposito (2008) Isotopic fractionation of noble gases by diffusion in liquid water: molecular dynamics simulations and hydrologic applications, *Geochimica et Cosmochimica Acta*, **72**, 2237–2247, doi:10.1016/j.gca.2008.02.012.

Boyle, E. A., J. M. Lee, Y. Echegoyen, et al. (2014) Anthropogenic lead emissions in the ocean: the evolving global experiment, *Oceanography*, **27**, 69–75.

Brandes, J. A., and A. H. Devol (2002) A global marine-fixed nitrogen isotopic budget: implications for Holocene nitrogen cycling, *Global Biogeochemical Cycles*, **16**, doi:10.1029/2001GB001856.

Broecker, W. S. (2002) *The Glacial World According to Wally*. Palisades: Eldigio Press.

Carpenter, E. J., H. R. Harvey, B. Fry, D. G. Capone (1997) Biogeochemical tracers of the marine cyanobacterium *Trichodesmium*, *Deep-Sea Research I*, **44**, 27–38, doi:10.1016/S0967-0637(96)00091-X.

Chappell, J., and N. J. Shackleton (1986) Oxygen isotopes and sea level, *Nature*, **324**, 137–140.

Crowley, T. J. (1983) The geologic record of climate change, *Reviews of Geophysics*, **21**, 828–877, doi:10.1029/RG021i004p00828.

DeVries, T., C. Deutsch, F. Primeau, B. Chang, and A. Devol (2012) Global rates of water-column denitrification derived from nitrogen gas measurements, *Nature Geoscience*, **5**, 547–550, doi:10.1038/NGEO1515.

Dansgaard, W. (1964) Stable isotopes in precipitation, *Tellus*, **16**, 436–468.

Eiler, J. M. (2011) Paleoclimate reconstruction using carbonate clumped isotope thermometry, *Quaternary Science Reviews*, **30**, 3575–3588, doi:10.1016/j.quascirev.2011.09.001.

Emerson, S. R., and J. I. Hedges (2008) *Chemical Oceanography and the Marine Carbon Cycle*, 453pp. Cambridge: Cambridge University Press.

Emerson, S., P. Quay, D. M. Karl, C. Winn, L. M. Tupas, and M. Landry (1997), Experimental determination of the organic carbon flux from open-ocean surface waters, *Nature*, **389**, 951–954, doi:10.1038/40111.

Foster, G. L., and J. W. B. Rae (2016) Reconstructing ocean pH with boron isotopes in foraminifera, *Annual Review of Earth and Planetary Science*, **44**, 207–237, doi:10.1146/annurev-earth-060115-012226.

Granger, J., D. M. Sigman, M. F. Lehmann, and P. D. Tortell (2008), Nitrogen and oxygen isotope fractionation during dissimilatory nitrate reduction by denitrifying bacteria, *Limnology and Oceanography*, **53**, 2533–2545, doi:10.4319/lo.2008.53.6.2533.

Grauel, A.-L., T. W. Schmid, B. Hu, et al. (2013) Calibration and application of the 'clumped isotope' thermometer to foraminifera for high-resolution climate reconstructions, *Geochimica et Cosmochimica Acta*, **108**, 125–140, doi:10.1016/j.gca.2012.12.049.

Graven, H., et al. (2017) Compiled records of carbon isotopes in atmospheric CO_2 for historical simulations in CMIP6, *Geoscientific. Model Development.*, **10**, 4405–4417, doi:10.5194/gmd-10-4405-2017.

Grootes, P. M., M. Stuiver, J. W. C. White, S. Johnsen, and J. Jouzel (1993) Comparison of oxygen isotope records from the GISP2 and GRIP Greenland cores, *Nature*, **366**, 552–554, doi:10.1038/366552a0.

Gruber, N., C. D. Keeling, and T. F. Stocker (1998) Carbon-13 constraints on the seasonal inorganic carbon budget at the BATS site in the northwestern Sargasso Sea, *Deep-Sea Research I*, **45**, 673–717, 2533–2545, doi:10.1016/S0967-0637(97)00098-8.

Juranek, L. W., and P. D. Quay (2013) Using triple isotopes of dissolved oxygen to evaluate global marine productivity, *Annual Review of Marine Science*, **5**, 503–524, doi:10.1146/annurev-marine-121211-172430.

Kiddon, J., M. L. Bender, and J. Orchardo (1993) Isotopic fractionation of oxygen by respiring marine organisms, *Global Biogeochemical Cycles*, **7**, 679–694.

Kim, S.-T., and J. R. O'Neil (1997) Equilibrium and non equilibrium oxygen isotope effects in synthetic carbonates, *Geochimica et Cosmochimica Acta*, **61**, 3461–3475.

Knox, M., P. D. Quay, and D. Wilbur (1992) Kinetic isotopic fractionation during air-water gas transfer of O_2, N_2, CH_4, and H_2, *Journal of Geophysical Research – Oceans*, **97**, 20335–20343, doi:10.1029/92JC00949.

Lansdown, J. M., P. D. Quay, and S. L. King (1992) CH_4 production via CO_2 reduction in a temperate bog: a source of ^{13}C-depleted CH_4, *Geochimica et Cosmochimica Acta*, **56**, 3493–3503, doi:10.1016/0016-7037(92)90393-W.

Lavine, N. M., M. L. Bender, and S. C. Doney (2009) The $\delta^{18}O$ of dissolved O_2 as a tracer of mixing and respiration in the mesopelagic ocean, *Global Biogeochemical Cycles*, **23**, GB1006, doi:10:1029/2007GB003162.

Luz, B., and E. Barkan (2000) Assessment of oceanic productivity with the triple-isotope composition of dissolved oxygen, *Science*, **288**, 2028–2031.

Luz, B., and E. Barkan (2005) The isotopic ratios of $^{17}O/^{16}O$ and $^{18}O/^{16}O$ in molecular oxygen and their significance in biogeochemistry, *Geochimica et Cosmochimica Acta*, **69**, 1099-1110.

Luz, B., and E. Barkan (2011) The isotopic composition of atmospheric oxygen, *Global Biogeochemical Cycles*, **25**, GB3001, doi:10.1029/2010GB003883.

O'Leary, M. H. (1981) Carbon isotope fractionation in plants, *Phytochemistry*, **20**, 553–567.

Palevsky, H. I., P. D. Quay, D. E. Lockwook, and D. P. Nicholson (2016) The annual cycle of gross primary production, net community production, and export efficiency across the North Pacific Ocean, *Global Biogeochemical Cycles*, **30**, 361–380, doi:10.1002/2015GB005318.

Penman, D. E., B. Hönisch, R. E. Zeebe, E. Thomas, and J. C. Zachos (2014) Rapid and sustained surface ocean acidification during the Paleocene-Eocene Thermal Maximum, *Paleoceanography and Paleoclimatology*, **29**, 357–369, doi:10.1002/2014PA002621.

Petit, J. R., J. Jouzel, D. Raynaud, et al. (1999) Climate and atmospheric history of the past 420,000 years from the Vostok ice core, Antarctica, *Nature*, **399**, 429–436.

Quay, P. D., R. Sonnerup, T. Westby, J. Stutsman, and A. McNichol (2003) Changes in the $^{13}C/^{12}C$ of dissolved inorganic carbon in the ocean as a tracer of anthropogenic CO_2 uptake, *Global Biogeochemical Cycles*, **17**, 1004, doi:10.1029/2001GB001817.

Quay, P., and J. Stutsman (2003) Surface layer carbon budget for the subtropical N. Pacific: $\delta^{13}C$ constraints at station ALOHA, *Deep-Sea Research I*, **50**, 1045–1061, doi:10.1016/S0967-0637(03)00116-X.

Rayleigh, J. W. S. (1896) Theoretical considerations respecting the separation of gases by diffusion and other similar processes, *Philosophical Magazine*, **42**, 493–498.

Robinson, R. S., and D. M. Sigman (2008) Nitrogen isotopic evidence for a poleward decrease in surface nitrate within the ice age Antarctic, *Quaternary Science Reviews*, **27**, 1076–1090, doi:10.1016/j.quascirev.2008.02.005.

Romanek, C. S., E. L. Grossman, and J. W. Morse (1992) Carbon isotope fractionation in synthetic aragonite and calcite: effects of temperature and precipitation rate, *Geochimica et Cosmochimica Acta*, **56**, 419–430.

Rozanski, K., L. Araguás-Araguás, and R. Gonfiantini (1993) Isotopic patterns in modern global precipitation, in *Climate Change in Continental Isotopic Records* (eds. P. K. Swart, K. C. Lohmann, J. Mckenzie and S. Savin), Geophysical Monograph 78, Washington, DC: American Geophysical Union.

Ruddiman, W. F. (2001) *Earth's Climate: Past and Future*, 465 pp. New York: W. H. Freeman and Co.

Shackleton, N. J., and N. D. Opdyke (1973) Oxygen isotope and palaeomagnetic stratigraphy of equatorial Pacific core V28–238: Oxygen isotope temperatures and ice volume on a 10^5 and 10^6 year scale, *Quaternary Research*, **3**, 39–55.

Siegenthaler, U. (1976) Stable hydrogen and oxygen isotopes in the water cycle, in *Lectures in Isotope Geology* (eds. Jäger, E. and J. C. Hunziker), 264–273. Heidelberg: Springer-Verlag.

Sowers, T., M. Bender, L. Labeyrie, et al. (1993) A 135,000-year Vostok-Specmap Common temporal framework, *Paleoceanography*, **8**, 737–766, doi:10.1029/93PA02328.

Tempest, K. E., and S. R. Emerson (2013) Kinetic isotopic fractionation of argon and neon during air-water gas transfer, *Marine Chemistry*, **153**, 39–47, doi:10.1016/j.marchem.2013.04.002.

Urey, H. C., H. A. Lowenstam, S. Epstein, and C. R. McKinney (1951) Measurement of paleotemperatures and temperatures of the upper Cretaceous of England, Denmark and the southeastern United States, *GSA Bulletin*, **62**, 399–416.

Yang, B., S. R. Emerson, and P. D. Quay (2019) The subtropical ocean's biological carbon pump determined from O_2 and DIC/DI^{13}C tracers, *Geophysical Research Letters*, **46**, 5361–5368, doi:10.1029/2018GL081239.

Zachos, J. C., M. W. Wara, S. Bohaty, et al. (2003) A transient rise in tropical sea surface temperature during the Paleocene-Eocene Thermal Maximum, *Science*, **302**, 1551-1554.

Zhang, J., P. D. Quay, and D. O. Wilbur (1995) Carbon isotope fractionation during gas-water exchange and dissolution of CO_2, *Geochimica et Cosmochimica Acta*, **59**, 107–114, doi:10.1016/0016-7037(95)91550-D.

Problems for Chapter 6

6.1. Measurements of δ^{18}O-$CaCO_3$ in the calcite shells of foraminifera found in a well-dated sediment core yield values of –0.73‰ at the core top but values of 1.24% at 20 000 years ago. Compared to today, sea level 20 000 years ago was estimated to be 125 m lower.

a. Estimate the δ^{18}O-H_2O of the mean global seawater 20 000 years ago. You may assume that sea level change is due entirely to the accumulation of land ice with a δ^{18}O of –40‰ and that the current δ^{18}O of ocean water is 0‰.

b. Estimate the temperature of the waters in which these foraminifera grew.

c. What temperature values would you have calculated for part b if you did not consider the changing δ^{18}O of the water?

6.2. Figure 6.8 shows δ^{11}B in foraminifera over the PETM.

a. Calculate the apparent change in pH during this event based on average δ^{11}B values during the Paleocene and during the Carbon Isotope Excursion (CIE). You may assume $T = 25°C$ and $S_P = 35$.

b. The δ^{11}B of seawater during this long-ago time is poorly known. If the δ^{11}B of seawater at that time were 4‰ lower than today, how would that affect your calculation of ΔpH?

6.3. Add carbon and oxygen isotopes to the two-box model. The mean DIC of the deep box is 2258 μmol kg^{-1} with a δ^{13}C-DIC of 0‰ with respect to the PDB standard. Air–sea gas exchange keeps the average oxygen concentration of the surface ocean close to equilibrium with the atmosphere (assume a temperature of 2°C and a salinity of 35 and that atmospheric O_2 is the standard). Use values from Chapter 3 for necessary constants. You may ignore the river and burial fluxes for this problem.

a. Calculate the DIC concentration and the δ^{13}C-DIC of the surface box for different surface-to-deep particle flux rates of 0.25, 0.50, and 0.75 mol-C m^{-2} y^{-1}.

b. Calculate the O_2 concentration and δ^{18}O-O_2 of the deep box under the same particle flux conditions as in part a.

6.4. This problem explores what controls the global mean δ^{15}N of nitrate in the ocean. For this problem, think of the ocean as a single, well-mixed box where the fluxes are at steady state.

a. Calculate what the global mean δ^{15}N-NO_3^- would be for a range of different proportions of sedimentary vs. water-column denitrification (from all denitrification takes place in the sediments to all denitrification takes place in the water-column).

b. What proportion of sedimentary vs. water-column denitrification is necessary to explain the current global mean δ^{15}N-NO_3^- of +5‰?

c. Test the sensitivity of your answer in part b to the following sources of uncertainty in the calculation: ε for nitrogen fixation may be as small as 0‰ to as large as –4‰, ε for water column denitrification may be as small as –20‰ to as large as –30‰, ε for sedimentary denitrification may be as small as 0‰. Which uncertainty had the largest impact on your result?

6.5. At the end of the summertime productive season, the mixed layer is 30 m deep and contains a nitrate concentration of 2 μmol kg^{-1} with a δ^{15}N-NO_3^- of +15‰. During the winter, the mixed layer deepens to 100 m. The entrained water has a nitrate concentration of 15 μmol kg^{-1} with a δ^{15}N-NO_3^- of +5‰.

a. Calculate the nitrate concentration and the δ^{15}N-NO_3^- at the end of the wintertime mixing period (approximately March 1).

b. Over the course of the following spring and summer, the mixed layer nitrate concentrations and the δ^{15}N-NO_3^- values are measured (listed in the following table). Determine the fractionation factor associated with nitrate assimilation (photosynthesis) using all these data. Express the fractionation factor as both an α and an ε value. Assume that no additional nitrate is added to the mixed layer after March 1.

Date	$[NO_3^-]$ (μmol kg^{-1})	δ^{15}N-NO_3^- (‰)
April 10	9	6.7
May 20	7	7.6
July 1	5	9.2
August 10	3	12.4

6.6. One mole of water vapor with an isotopic composition of $\delta^{18}O$-H_2O = −13.0‰ is placed into a large glass reservoir at 25°C. The glass reservoir is slowly cooled so that the water condenses at equilibrium with the vapor. Once 0.6 moles of water have condensed, the liquid water is removed all at once through a valve at the bottom of the reservoir. The liquid water has an isotopic composition of $\delta^{18}O$-H_2O = −9.3‰.

a. Calculate the $\delta^{18}O$ of the water vapor remaining in the reservoir.
b. Calculate the equilibrium fractionation factor between liquid water and water vapor for $^{18}O/^{16}O$. Express the fractionation factor as both an α and an ε value.
c. Suppose instead that the condensate had been continuously removed as it formed. Assume the same amount and isotopic composition for the total condensate collected. Calculate the fractionation factor between liquid water and water vapor in this case. Express as both an α and an ε value.

7 Radioisotope Tracers

7.1 Radioactive Decay Mechanisms and Equations 275
7.2 Atmospheric Spallation 278
7.2.1 Natural Carbon-14 278
7.2.2 Beryllium-7 285
Discussion Box 7.1 287
7.3 Uranium and Thorium Decay Series 288
7.3.1 Secular Equilibrium 289
7.3.2 The Geochemistry of Decay Series Isotopes in the Ocean 291
7.3.3 Uranium–Thorium Activities and Ocean Particle Dynamics 294
7.3.4 ^{222}Rn-^{226}Ra Disequilibrium in Surface Waters: A Tracer of Air–Sea Gas Exchange 301
7.3.5 Excess ^{210}Pb and ^{234}Th: Tracers of Coastal Sediment Accumulation 303
7.4 Anthropogenic Influences 304
Discussion Box 7.2 308
References 309
Problems for Chapter 7 311

The study of radioisotopes provides the added dimension of time to the chemical perspective of oceanography. Stable elements and isotopes (Chapter 6) are useful tracers of the sources and transformations of marine materials, but they carry no direct information about the rates and dates of their associated processes. However, such temporal distinctions are made possible by many different, naturally occurring radioactive isotopes with their wide range of elemental forms and decay rates. These highly dependable atomic clocks decay by nuclear processes that are insensitive to temperature and pressure and that allow them to be detected at very low concentrations. A fanciful analogy of how these tracers are used is an experiment in which you attach hundreds of clocks that operate only at pressures less than one atmosphere to helium-filled balloons and release them to float in the atmosphere. After the helium leaks out of the balloons, the clocks would fall to the Earth scattered across the landscape. Because each clock records only the time spent aloft, their position and time would convey information about the direction and speed of local winds. Similarly, fundamental aspects of ocean dynamics, like rates of ocean circulation, air–sea interaction, particles sinking from the ocean surface, and sedimentation at the sea floor, have been determined by measuring the distribution of radioactive elements in the sea.

After we introduce the mechanisms of radioactive decay and the equations used to describe that process, we present examples of how radioactive tracer distributions have been used to determine rates of ocean processes. This discussion is arranged by the sources of radioisotopes to seawater and sediments, beginning with radioisotopes produced via interactions in the Earth's atmosphere with high-energy cosmic rays from space, proceeding to elements produced from long-lived radioisotopes that have existed in the crust since the time Earth formed, and finally ending with environmental contamination by anthropogenically produced isotopes. The most familiar of the cosmic ray–produced elements is carbon-14, which plays a prominent role in environmental radiochemistry because it travels in the carbon cycle and has a half-life that is useful for elucidating many natural processes. The long-lived radioisotopes of uranium and thorium (^{238}U, ^{235}U, and ^{232}Th) are the sources of three separate decay series, each of which involves about a dozen radioactive daughters with remarkably diverse decay rates. Both the parent and daughter isotopes bear unique chemistries that trace environmental reactions or fluxes that would be otherwise difficult to measure. Finally, nuclear weapons testing in the 1950s and 1960s contaminated the atmosphere with spallation products until a treaty among nations banned the practice of atmospheric testing.

7.1 Radioactive Decay Mechanisms and Equations

Radioactivity is characterized by the emission of energy (as electromagnetic radiation or as a particle) from the nucleus of an atom, usually with an associated elemental conversion. These nuclear reactions are independent of temperature and chemical surroundings. There are four basic types of radioactive decay (Table 7.1), of which alpha (α) and beta minus (β^-) decay are the most common in nature. Alpha decay is the only type of decay that causes a net mass change in the parent nuclide by loss of two protons and two neutrons. Because two essentially weightless orbiting electrons are also lost when the equivalent of a helium nucleus is emitted, the parent nuclide transmutes into a daughter element two positions to the left on the periodic table. For example, ^{238}U decays by α emission to ^{234}Th, skipping over protactinium, Pa. The β^- decay involves emission of a nearly weightless negatively charged particle (like an electron) from the nucleus, and thus causes little change in atomic mass. The electron is created from a neutron, which is converted to a proton of essentially equal mass in the process ($n^0 \rightarrow e^- + p^+$). This

Table 7.1 Different types of radioactive decay processes

Type	Emitted particle	Δ protons	Δ neutrons	Comments
α	He^{2+} (helium nucleus)	–2	–2	loss of four atomic mass units
β^-	e^- (electron)	+1	–1	no mass loss
K-capture	None	–1	+1	no mass loss, x-ray emission
β^+	e^+ (positron)	–1	+1	no mass loss, x-ray emission

transition moves the element one step to the right on the periodic table. For example, ^{14}C decays by β^- decay to ^{14}N.

The two other decay processes in Table 7.1 are less common in nature. In K-capture, an orbiting electron (usually in the K or L inner shell) combines with a proton in the nucleus to form a neutron. This relatively rare nuclear transformation process $(e^- + p^+ \rightarrow n^0)$ is just the opposite of that for β^- decay, meaning that the nucleus formed also has the same mass but in this case is displaced one element to the left on the periodic table. For example, ^{40}K decays by K-capture to ^{40}Ar, which means that a nonvolatile alkali metal, potassium, has been the main source on Earth of argon, the third most abundant gas in the atmosphere. Although K-capture emits no nuclear particle, the attending cascade of electrons into lower orbitals leads to x-ray emission at a characteristic energy that can be measured by appropriate detectors. The last decay process (also rare) involves emission of a positron (β^+), a positively charged electron. This nuclear process $(p^+ \rightarrow n^0 + \beta^+)$ has the same net effect as K-capture and is also characterized by x-ray emission.

Conversion of a radioactive parent element to a stable daughter element is the simplest form of nuclear decay and is a perfect example of a first-order irreversible reaction. Although it is impossible to determine when a specific radioactive nucleus will convert, decay rates become predictable for large populations of radionuclides. The general equation describing decay of a parent isotope to a stable daughter is

$$\frac{dN}{dt} = -\lambda N, \tag{7.1}$$

where N is the total number of radioactive atoms present in the system (atoms), dN/dt is the nuclear decay rate (or nuclear activity, atoms per time, e.g., atoms d^{-1}), and λ is the first-order decay constant in units of inverse time (e.g., d^{-1}). Integration of Eq. 7.1 from an initial condition at $t = 0$ to a later time t results in the classic first-order decay equation:

$$N = N_0 e^{-\lambda t}. \tag{7.2}$$

This equation allows the number of radioisotope atoms, N, remaining at a given time, t, to be calculated from the number that was present initially, N_0, and the decay constant for that particular nuclide. An example of this relationship for the decay of carbon-14, which is discussed in the next section, is presented in Fig. 7.1. The total number of radioactive atoms in a sample is often difficult to directly measure, although such analyses are now possible by mass spectrometry. Frequently the activity (A, disintegrations per time) is measured rather than the number of atoms. Since activity is the rate of decay, it is defined as

$$A = -\frac{dN}{dt} = \lambda N. \tag{7.3}$$

Thus, the number (or concentration) of radioactive isotopes, N, is related to a measurable corresponding activity, via the known decay constant, λ. This relationship also allows the fundamental decay equation to be rewritten in terms of more readily measured activities by simply multiplying the concentrations on both sides of Eq. 7.2 by the decay constant

$$A = A_0 e^{-\lambda t}. \tag{7.4}$$

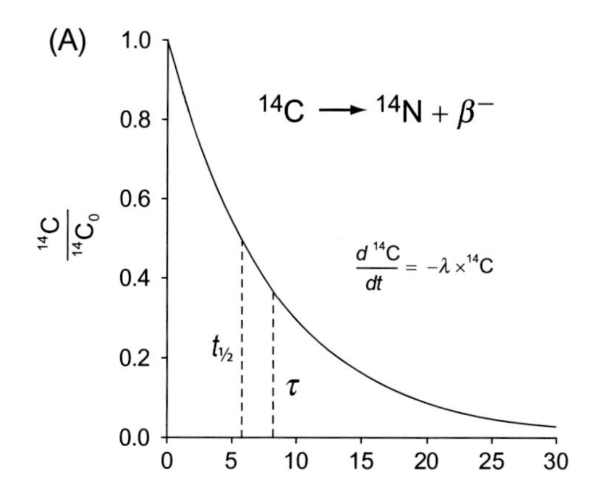

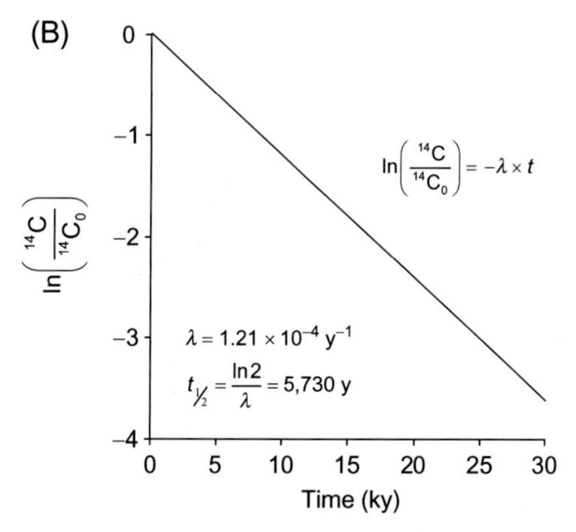

Figure 7.1 The solution to the first-order radioactive decay equation (Eq. 7.2) for ^{14}C in linear (A) and natural log (B) forms. Dashed vertical lines in (A) indicate the relative activity ratio at the half-life, $t_{1/2}$, and the mean life, τ.

The traditional unit by which activity was measured is the *curie*, which is equal to 3.70×10^{10} disintegrations per second (dps) or *becquerels*, which is the present international unit. (One becquerel = 1 dps = 2.703×10^{-11} curie.) The curie is the activity exhibited by 1.00 g of pure ^{226}Ra and derives its name from Marie Curie, who won the Noble Prize twice for her work on radioactivity and radium. By convention, the relative rate at which a radionuclide decays is expressed in terms of its half-life, $t_{½}$, which is the amount of time necessary for exactly half of the radionuclide to have decayed away. The half-life is defined as the time when $A = A_0/2$ and is related to the decay constant by

$$\ln\left(\frac{A}{A_0}\right) = \ln\left(\frac{1}{2}\right) = -0.693 = -\lambda t_{½}; \tag{7.5}$$

thus,

$$t_{½} = \frac{0.693}{\lambda}. \tag{7.6}$$

A related expression called the mean life, τ, is defined as the average time that a radioisotope exists before decay. Conceptually, the mean life is somewhat similar to the idea of residence time but for the non-steady-state decay of radioisotopes. Mathematically, it is the integral of all lives of the atoms in a particular nuclide divided by the initial quantity, which is equivalent to the integral of the proportion of atoms remaining at each time (the area under the curve in Figure 7.1A):

$$\tau = \int_{t=0}^{t=\infty} \frac{N}{N_0} dt = \frac{1}{N_0} \int_{t=0}^{t=\infty} N_0 e^{-\lambda t} dt = \int_{t=0}^{t=\infty} e^{-\lambda t} dt = \frac{1}{\lambda} = \frac{t_{1/2}}{0.693} \quad (7.7)$$

The mean life, τ, is greater than $t_{1/2}$ (Fig. 7.1) because some nuclides persist far past twice the half-life and "drag out" the mean toward higher values.

7.2 Atmospheric Spallation

Many of the low mass radionuclides used in geochemical studies are produced in the upper atmosphere when high-energy cosmic rays from space shatter the nucleus of the most abundant atmospheric gases – N_2 and O_2. Among the fragments produced during this "spallation" process are neutrons, which are then slowed by subsequent collisions and can penetrate the nucleus of ambient gases. Most of these reactions take place in the upper atmosphere near the poles where the atmosphere is unshielded from cosmic rays by the Earth's magnetic lines of force. Among the isotopes produced during this process are carbon-14 and beryllium-7.

7.2.1 Natural Carbon-14

The radioisotope with the longest history and most impressive utility in the environment is carbon-14. When neutrons produced by cosmic ray spallation reactions penetrate the nucleus of a ^{14}N atom (in N_2), a proton is released (conserving mass and charge), and ^{14}N is converted to ^{14}C (Fig. 7.2):

$$^{14}_{7}N + n^0 \rightarrow {}^{14}_{6}C + p^+ + e^-. \quad (7.8)$$

The carbon-14 finds its way into the most abundant form of carbon in the atmosphere – CO_2 gas. Carbon-14, with eight neutrons and six protons, is an unstable nucleus and converts by β^- decay back to ^{14}N with a half-life of 5,730 y. Once $^{14}CO_2$ mixes into the lower atmosphere and enters biomass, carbonate minerals, or the deep ocean, it is effectively separated from its source and undergoes first-order decay (Fig. 7.1) with no replenishment. The isotope is then amenable to a variety of dating and mass balance calculations.

Measurements of ^{14}C in organic samples are made by combustion to CO_2, which is then purified and converted to a suitable form for analysis. Before the widespread availability of accelerator mass spectrometers (AMS), the β^- emission activity of ^{14}C samples was "counted" directly in CO_2 or after the carbon was concentrated in a liquid (e.g., benzene)

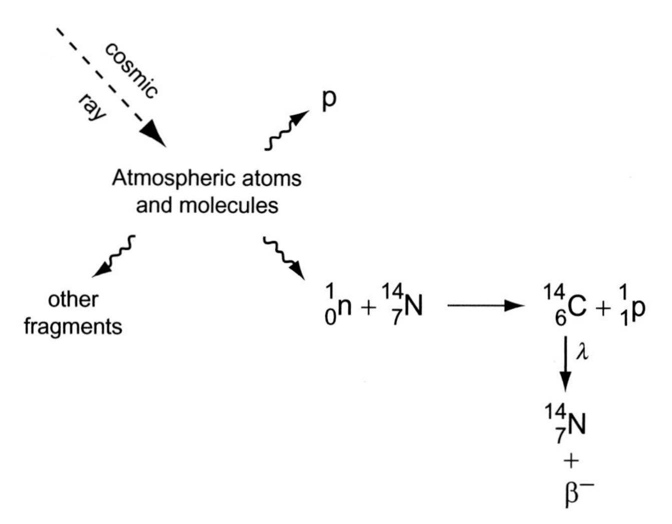

Figure 7.2 Schematic depiction of the formation of ^{14}C in the atmosphere by cosmic ray spallation. High-energy particles from space enter the Earth's upper troposphere and stratosphere and collide with molecules of atmospheric gas. These collisions create neutrons, protons, and other fragments. Neutrons produced in this way react with atmospheric N_2, exchanging a proton for a neutron ((n,p) reaction) forming ^{14}C. Carbon-14 is unstable and decays (indicated by λ) back to nitrogen by emitting a β^- particle.

or solid (e.g., graphite) form. Direct counting of this type requires more than 100 mg of C and is not feasible for small samples. Currently, almost all radiocarbon measurements are made by direct counting of individual ^{14}C atoms in a concentrated graphite "target" using AMS methodology requiring less than 100 μg of C. Whatever the method of analysis, the final abundance results are usually calculated and reported in "delta" ^{14}C format (Stuiver and Polach, 1977):

$$\Delta^{14}C = \delta^{14}C - \left\{(2\delta^{13}C + 50)\left(1 + \frac{\delta^{14}C}{1000}\right)\right\}, \tag{7.9}$$

where $\delta^{14}C$ is defined in an analogous way to stable isotopes only using activities:

$$\delta^{14}C = \left(\frac{^{14}A_{sample}}{0.95^{14}A_{ox}} - 1\right) \times 1000 \tag{7.10}$$

and A_{sample} and A_{ox} are the ^{14}C activities of the sample and standard (oxalic acid) carbon. The term $\left[(2\delta^{13}C + 50)\left(1 + \delta^{14}C/1000\right)\right]$ in Eq. 7.9 corrects the measured $\delta^{14}C$ for isotope fractionation processes that affected the sample differently from the standard, as determined by the $\delta^{13}C$ of the same sample. This term is formulated to be equal to zero for a sample having a $\delta^{13}C$ of –25.0‰, typical of the wood that was the most common early sample type. The $\delta^{13}C$ is multiplied by 2 to account for the fact that isotope fractionation of ^{14}C for a given sample will be twice as large as for ^{13}C, since the magnitudes of carbon isotope effects are proportional to the mass difference from ^{12}C. The $\Delta^{14}C$ terminology, thus, has a built-in correction for isotopic fractionation such that the observed changes are due only to radioactive decay. Eqs. 7.9 are 7.10 are formulated to give a $\Delta^{14}C$ of zero for wood formed in 1850 measured against 95% of the oxalic acid radiocarbon standard. Normalization to a date approximately 150 y ago avoids complications resulting from recent anthropogenic effects on the ^{14}C content of atmospheric CO_2. One of these perturbations is the "Suess Effect," which refers to the net decrease in the ^{14}C content of atmospheric CO_2 since 1850 due to the combustion of fossil fuels, such as coal and petroleum, which are old enough that all their original ^{14}C has decayed away. The other major anthropogenic perturbation arose from nuclear weapons testing that injected large quantities of ^{14}C into the atmosphere (see Section 7.4).

The Carbon-14 Production Rate. The activity of ^{14}C in the atmosphere is not constant through time but instead varies based on changes in its production rate and its cycling through the Earth's carbon reservoirs. The production rate in the atmosphere is influenced primarily by the strength of the geomagnetic field, which shields the Earth from cosmic rays. Among carbon cycling processes, ocean ventilation has the strongest influence on the ^{14}C of atmospheric CO_2, because ventilation alters communication of atmospheric carbon with the dissolved inorganic carbon of the deep ocean – the largest carbon reservoir able to exchange on short timescales. The history of atmospheric ^{14}C activity has been determined empirically by measuring $\Delta^{14}C$ in organic carbon or $CaCO_3$ of substances that have independently known ages. The original method used to determine the variability of carbon-14 production involved comparing the ^{14}C age of tree rings with the age determined independently by counting. This method involves matching tree ring patterns in overlapping wood borings taken from different trees to form a continuous record of ^{14}C production over thousands of years.

Comparison of tree-ring age and ^{14}C age is limited to the timescale of trees with known ages (~5 ky), which is much shorter than the useful timescale of ^{14}C (40–50 ky). Longer records of atmospheric ^{14}C activity have been determined from simultaneous ^{230}Th and ^{14}C dating of corals (Bard et al., 2004) and cave stalagmites (Wang et al., 2001), and by measuring the ^{14}C content of planktonic foraminifera tests in well dated marine sediments. In the last example, the $^{14}C/^{12}C$ and $^{18}O/^{16}O$ ratios of foraminifera tests from varved sediments of the anoxic Cariaco Basin on the continental shelf of Venezuela were determined (Hughen et al., 2006). The calendar age of the varves was determined independently of the ^{14}C content by correlation with $\delta^{18}O$ variations in a Greenland ice core, which had been precisely dated back to 40 ky ago by counting of annual ice layers. The results of all these methods, summarized in Fig. 7.3, indicate offsets of up to 5 ky between ^{14}C age and calendar age over a period 25–44 ky ago and an abrupt shift at a calendar age of ~44 ky in which 7 ^{14}C ky elapsed in only 4 ky of real time. These results have been explained as variations in ^{14}C production rates and changes in the ventilation rate of the deep sea (Hughen et al., 2006).

The Carbon-14 Age of the Ocean. The circulation time of the deep ocean has been determined from ^{14}C measurements in the DIC of seawater. These ages (mean time since water was last at the surface) are illustrated in Fig. 7.4 at a depth of 3500 m in the world's oceans. Deep circulation rates are sufficiently slow that ^{14}C decays measurably along flow paths allowing age to be estimated once water has left contact with the surface. Ages in Fig. 7.4 are not calculated simply from the measured $\Delta^{14}C$ and its half-life for two reasons.

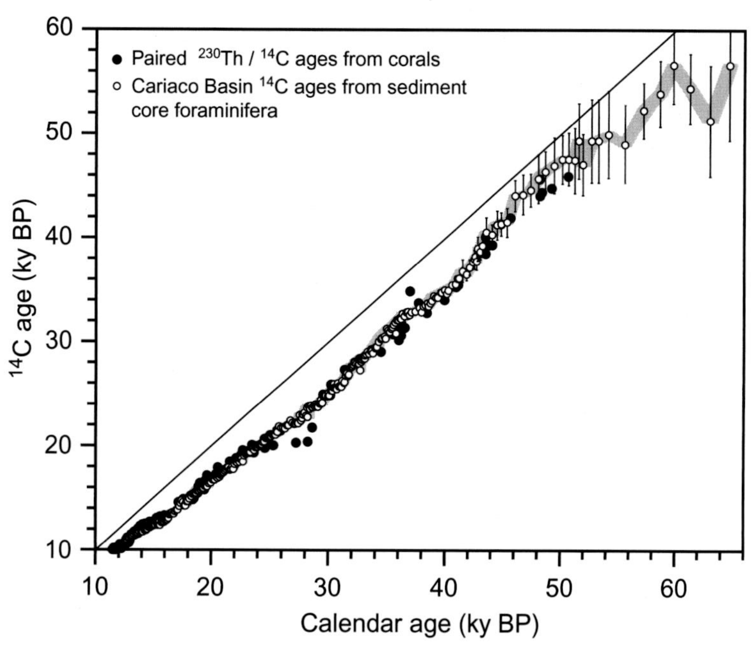

Figure 7.3 ^{14}C age versus calendar age indicating variations in production in the atmosphere and exchange with the ocean over the past 50 ky. ^{14}C ages are from measurements in planktonic foraminifera shells ($CaCO_3$) sieved from more than 400 depths in ocean drilling sediment cores in the Cariaco Basin (open circles) and paired ^{230}Th & ^{14}C dates from corals (solid dots). The calendar age scale was determined by correlation of varves in Cariaco Basin sediments with $\delta^{18}O$ of ice cores combined with calcite $\delta^{18}O$ and ^{230}Th measurements of Hulu Cave speleothems. The solid line is the 1:1 line. Modified from Hughen et al. (2006).

First, values at locations where surface waters are in contact with the atmosphere do not have an apparent $\Delta^{14}C$ age of zero, because surface values contain some older water from upwelling creating a "preformed age" and because air–sea equilibration rates are slow for $^{14}CO_2$. (See Section 8.4.3 of the next chapter.) Second, ^{14}C-DIC ages in the deep sea are strongly influenced by mixing between water masses of different ages, because the exponential nature of radioactive decay gives young waters containing more ^{14}C a disproportionate impact on the apparent mean age. The age estimates shown in Fig. 7.4 from Gebbie and Huybers (2012) account for these effects by considering the distribution of surface $\Delta^{14}C$ and determining the contribution of multiple water masses to any one location.

Younger ages are recorded in the DIC of the North Atlantic Ocean, where convection carries recently ventilated waters to great depths, forming North Atlantic Deep Water (see Section 1.3.3). As waters flow south in the Atlantic Ocean, age increases several hundred years. Ventilation ages in the Southern Ocean are particularly tricky to quantify, because deep waters do not remain at the surface long enough to reequilibrate with the atmosphere, and the deep waters in this region are a mixture of flow from all the different ocean basins. As waters move north in the Indian and Pacific Oceans, age increases dramatically. Waters do not move from the surface to the deep at the northern end of either

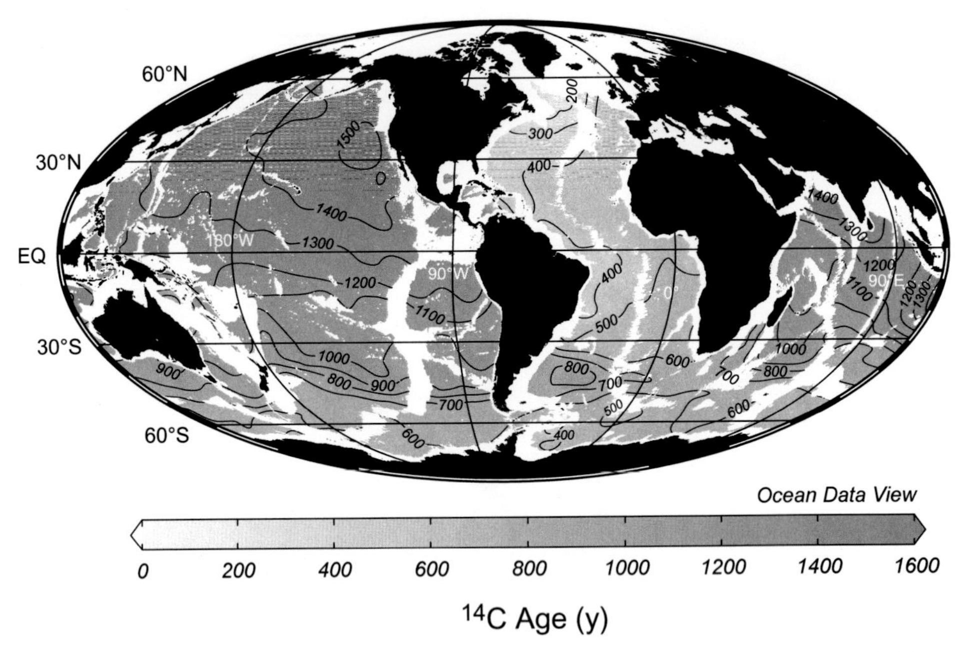

Figure 7.4 The ^{14}C age of DIC in the world's ocean at a depth of 3500 m based on radiocarbon observations compiled by the Global Ocean Data Analysis Project (GLODAP) and a matrix inversion method (from Gebbie and Huybers, 2012). A black and white version of this figure will appear in some formats. For the color version, refer to the plate section.

of these oceans, so water flows mainly northward at the deepest depths with a return flow at mid-depth. (See Fig. 1.16.) The oldest deep-water ages at a depth of 3500 m are found in the North Pacific at about 1500 y.

Next, we take a step back from the model solution of Gebbie and Huybers (2012, Fig. 7.4) and demonstrate how measurements of DIC and ^{14}C-DIC can be used in a simple two-layer model to quantify deep ocean residence time with respect to mixing. The steady-state equation for the deep ocean describing the DIC concentration difference between the surface and deep ocean is (Section 3.2.1 and Fig. 3.6):

$$Q_M\left([\mathrm{DIC}]_D - [\mathrm{DIC}]_S\right) = J_\mathrm{C} - B_\mathrm{C}. \tag{7.11}$$

The left side of Eq. 7.11 describes the net flux of DIC from the deep reservoir to the surface ocean by vertical mixing, which must equal the sinking particle flux of carbon from the surface reservoir, J_C, minus the sediment burial rate of carbon, B_C. The equation for DI^{14}C is exactly identical with the exception of an additional loss term for radioactive decay:

$$Q_M([\mathrm{DI^{14}C}]_D - [\mathrm{DI^{14}C}]_S) = J_{^{14}\mathrm{C}} - B_{^{14}\mathrm{C}} - [\mathrm{DI^{14}C}]_D \lambda_{^{14}\mathrm{C}} V_D. \tag{7.12}$$

The last term on the right side represents the decay of ^{14}C where $\lambda_{^{14}\mathrm{C}}$ is the decay constant ($\lambda_{^{14}\mathrm{C}} = 0.693/5730\ \mathrm{y} = 1.21 \times 10^{-4}\ \mathrm{y}^{-1}$). ^{14}C values presented in units of Δ^{14}C are related to the ratio of ^{14}C and ^{12}C, in DIC (Eqs. 7.9 and 7.10). For convenience, we define the term $r^* = [\mathrm{DI^{14}C}] / [\mathrm{DIC}]$ and rewrite the above equation as

$$Q_M\left([\mathrm{DIC}]_D r_D^* - [\mathrm{DIC}]_S r_S^*\right) = J_C r_S^* - B_C r_S^* - [\mathrm{DIC}]_D r_D^* \lambda_{^{14}\mathrm{C}} V_D. \tag{7.13}$$

We demonstrated in Chapter 3 (Section 3.2.1) that the burial flux on the right side of this equation is much smaller than the particle transport so we will ignore it here. Substituting Eq. 7.13 into Eq. 7.11 to eliminate the biological flux term, J_C, gives

$$Q_M\left(r_S^* - r_D^*\right) = r_D^* \lambda_{^{14}\mathrm{C}} V_D. \tag{7.14}$$

The turnover time or residence time of the deep ocean with respect to mixing is $\tau_M = V_D/Q_M$ ($\mathrm{m^3/m^3y^{-1}} = \mathrm{y}$) so the above equation can be rewritten as

$$\tau = \frac{V_D}{Q_M} = \left(\frac{r_S^*}{r_D^*} - 1\right)\left(\frac{1}{\lambda_{^{14}\mathrm{C}}}\right). \tag{7.15}$$

The relationship between $\Delta^{14}\mathrm{C}$ and r* is approximately

$$\Delta^{14}\mathrm{C} \approx \left(\frac{r^*_{sample}}{r^*_{std}} - 1\right) \times 1000. \tag{7.16}$$

Here we have deleted the corrections for stable isotope fractionation because we are comparing samples with the same source – seawater. Since r^*_{std} cancels, we are left with

$$\frac{r_S^*}{r_D^*} \approx \frac{\left(1 + \dfrac{\Delta^{14}\mathrm{C}_S}{1000}\right)}{\left(1 + \dfrac{\Delta^{14}\mathrm{C}_D}{1000}\right)}. \tag{7.17}$$

The residence time, τ, can now be written in terms of the natural $\Delta^{14}\mathrm{C}$ of the surface and deep water of the ocean.

At present, ocean surface waters are contaminated with bomb produced $^{14}\mathrm{C}$ (see later); however, the few measurements from pre-atmospheric weapons testing and more recent measurements made on corals that grew at that time suggest that surface waters had $\Delta^{14}\mathrm{C}$ values of about –50‰ (see later Fig. 7.18). Since the Antarctic circumpolar water is a mixing region for the whole ocean, we take its value as an estimate of the mean value for deep waters ($\Delta^{14}\mathrm{C} = -160$‰, see later Fig. 7.19). (This value is similar to the mean derived from more rigorous methods of volume averaging the deep-water values, Broecker and Peng, 1982.) Thus, average deep-water $\Delta^{14}\mathrm{C}$ is 110‰ lower than that of the surface, giving a value for r_S^*/r_D^* of 1.13. Now, from Eq. 7.15, the residence time of water in the deep ocean becomes

$$\tau = \frac{V_D}{Q_M} = (1.13 - 1)(8267\ \mathrm{y}) = 1075\ \mathrm{y}. \tag{7.18}$$

The $^{14}\mathrm{C}$ age map of DIC in the ocean's deep waters (Fig. 7.4) indicates the age difference between deep waters in the northern North Atlantic and that in the Northeast Pacific is ~1300 y. This value compares the most recent and most ancient ages of ocean deep water, whereas the box model compares the mean deep-water age of the entire ocean, ~ –160‰ (~1440 y) with that of the surface ocean, ~ –50‰ (420 y) to yield

$1440 - 420 = 1020$ y. In some ways, the largest task in developing the two-layer model is determining representative $\Delta^{14}C$ values for the mean surface and deep ocean. The box-model calculation is a simple demonstration of how $\Delta^{14}C$ values are capable of yielding water mass ages; however, it ignores the complexities of different surface ocean $\Delta^{14}C$ values and the non-linear mixing of $\Delta^{14}C$ values from different water masses.

Ocean Sediment Accumulation Rates. The use of carbon-14 to determine the date that different sediment layers were deposited has facilitated fundamental advances in our understanding of the marine carbon cycle and how it has changed in the past. Burial of organic carbon and $CaCO_3$ in sediments is the most important long-term sink for carbon from the ocean-atmosphere system. Using modern methods of ^{14}C dating it is possible to quantify dates that span the last 30 000 years. This range reaches past the last glacial–interglacial transition (~15 ky BP), which is the most important interval for assessing how the ocean was involved in the most recent naturally forced climate change.

There are many complications involved in interpreting the age of sediment archives, like non-steady accumulation rates and multidimensional transport processes, particularly on continental margins, where accumulation rates are relatively rapid making it possible to observe high-resolution paleoclimate records. Our discussion of using ^{14}C (or any radio-tracer) to determine sediment chronology begins by assuming that to a first approximation sediment accumulation can be interpreted as a one-dimensional process.

Adopting a coordinate system in which the sediment–water interface is at zero depth, $z = 0$, and where depth increases moving downward, the change in property p at a given depth, z^*, below the sediment–water interface is given by

$$\left.\frac{\partial p}{\partial t}\right|_{z=z^*} = \frac{dp}{dt} - s\left.\frac{\partial p}{\partial z}\right|_t . \tag{7.19}$$

In this formulation, the sediment–water boundary and any specific depth within the sediment is moving upward with respect to the center of the Earth as additional sediment accumulates. This equation states that the change in property p with respect to time at a specific depth, z^*, below the sediment–water interface is equal to the total derivative of p with respect to time minus the flux of p transported by sedimentation rate s (cm yr^{-1}) along the depth gradient in p ($\partial p/\partial z$) at time t. The total derivative refers to factors that change p as a function of t in a layer that is stationary with respect to the Earth's center and moving away from the interface ($z = 0$) with velocity equal to the sedimentation rate, s. Equation 7.19 relates the Lagrangian coordinate system that moves with the flow (left-hand side) to the Eularian total derivative that is fixed at a given distance from the centre of the Earth (first term on the right-hand side) via the flux caused by sedimentation (the second term on the right-hand side). A steady-state condition is one in which property p does not change at a given depth below the interface, i.e., the left side of Eq. 7.19 is zero. In this case, the change of p depicted by the total derivative is exactly balanced by the flux of p by sedimentation:

$$s\left.\frac{\partial p}{\partial z}\right|_t = \frac{dp}{dt} \tag{7.20}$$

This is the general equation for the evolution of sediment properties at steady state without compaction or bioturbation. See Berner (1980), Boudreau (1997), and Burdige (2006) for further details. Substituting the concentration of a radioisotope (C, atoms cm^{-3}) for p and the decay rate (λC, atoms cm^{-3} y^{-1}) for the total derivative yields a general equation for determining sediment accumulation rates, s, using radioisotope tracers:

$$s\left(\frac{\partial[C]}{\partial z}\right) = \lambda[C]. \tag{7.21}$$

This equation has the following solution:

$$\ln\left(\frac{[C]}{[C_{t=0}]}\right) = -\left(\frac{\lambda}{s}\right)z, \tag{7.22}$$

where the initial condition ($C_{t=0,}$ atoms cm^{-3}) is the concentration at the sediment–water interface. Notice the similarity between the last two equations and the very first equations of this chapter describing the relationship between radioisotope concentration and time (Eqs. 7.1, 7.2, and 7.4) They are the same except that early equations in the chapter had time, t, in the place of z/s. Thus, the age, t, determined by the fraction of the radioisotope that remains, $[C]/[C_{t=0}]$, is equal to the sediment depth divided by the sedimentation rate, z/s (cm/cm s^{-1}) which is also equal to the left-hand side of Eq. 7.22 divided by the decay constant, λ:

$$\text{age} = -\ln\left(\frac{[C]}{[C_{t=0}]}\right)\frac{1}{\lambda} = \frac{z}{s}. \tag{7.23}$$

Hundreds of marine sediment cores have been dated using the ^{14}C content of $CaCO_3$ shells in the sediments, and a few are presented in Fig. 7.5. Profiles of ^{14}C age versus depth for these carbonate rich sediment cores reveal a surface layer of constant age in the top 5–10 cm followed by a linear decrease in age with depth that reflects sedimentation rates ranging from 1 to 4 cm ky^{-1}. The uniform top section is mixed by bioturbation (benthic animals like worms physically moving sediments) to a depth of 5–10 cm in these deep-sea deposits. This bioturbation depth extends even deeper in shallow continental-margin sediments but is relatively uniform in the deep sea because of the similar size and burrowing habits of benthic fauna there. Bioturbation has ramifications for the utility of any tracer in marine sediments to resolve temporal changes because it acts as a smoothing filter (a running mean) with a timescale equal to the mean age of the bioturbated layer. For sedimentation rates of 1 cm ky^{-1}, bioturbation means that every point represents an average of about 7 ky; while for sedimentation rates of 0.1 cm ky^{-1}, characteristic of $CaCO_3$ poor regions of the North Pacific, it means an average of 70 ky! To investigate millennial-scale climate change, one must sample regions with accumulation rates significantly faster than 1 cm ky^{-1}. This criterion limits suitable locations to either continental margins or deep-sea locations rich in $CaCO_3$ (Fig. 5.8). Applications of radiotracers with different half-lives to ocean sediment dating are described later in this chapter.

7.2.2 Beryllium-7

Another radioisotope produced by cosmic ray spallation that has important applications to oceanographic processes is beryllium-7 with a half-life of 53.3 days. Beryllium-7 is

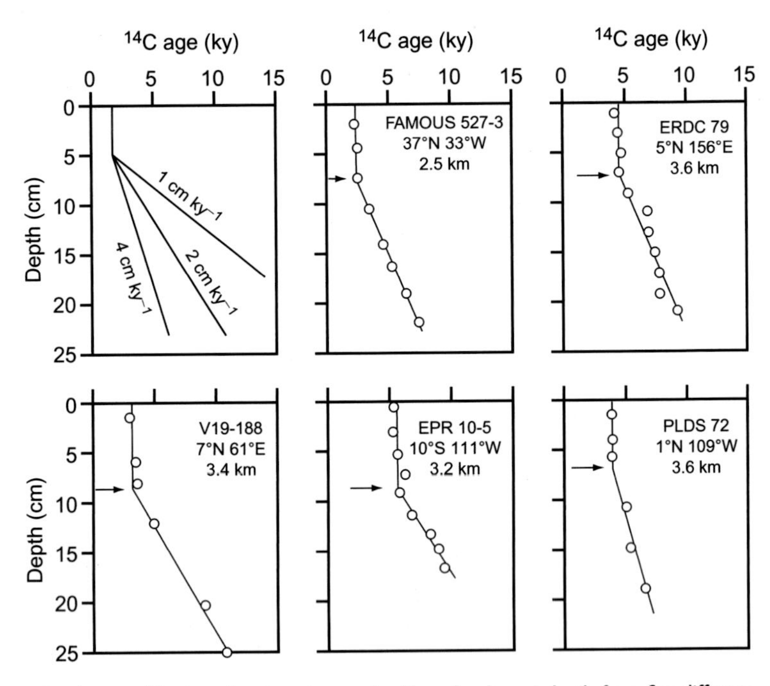

Figure 7.5 The ^{14}C age of carbonate-rich abyssal sediments as a function of sediment depth from five different locations. Expedition names and core numbers are indicated along with location and water column depth. Individual measurements are open circles. Arrows mark the base of the bioturbated layer. The panel in the upper left illustrates the idealized relationship between profile slope and sedimentation rate assuming a 5 cm bioturbated mixed layer at the sediment surface and different sediment accumulation rates. Redrawn from Peng and Broecker (1984).

delivered from the atmosphere to the ocean by rain and stays primarily dissolved in seawater. It has been applied to estimating the rate of upper ocean mixing processes that occur on timescales of several months. The activity of ^{7}Be isotopes in seawater is very low, so it must be concentrated by pumping several hundred liters of water through manganese oxide fibers in order to detect its characteristic gamma radiation during decay. Depth profiles obtained in this way in the NE subtropical Atlantic Ocean (Fig. 7.6) indicate a characteristically uniform value in the surface mixed layer with decreasing values in deeper waters (Kadko and Olson, 1996). The isotope is relatively unreactive in open ocean waters, so that its only loss from the mixed layer is by a combination of decay and transport to greater depths through mixing or advection. These physical processes are frequently intermittent and difficult to characterize by measuring temperature and salinity values without a detailed time series of measurements. The distribution of ^{7}Be with depth (Fig. 7.6) averages the result of these intermittent processes over timescales equal to its mean life allowing one to determine characteristic transport rates integrated over several months.

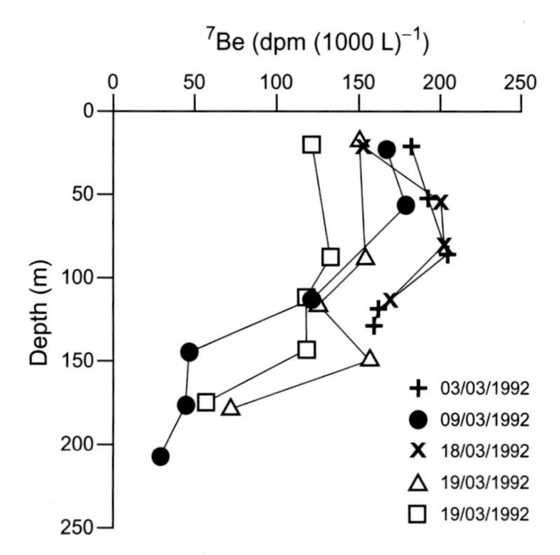

Figure 7.6 The activity of ^{7}Be in the upper ocean near Bermuda. Different symbols indicate data from five separate profiles at the Bermuda Atlantic Time-series site in March 1992. Data are from Kadko and Olson (1996).

Discussion Box 7.1

Let's further consider how ^{7}Be measurements can yield information about upper ocean processes.

- Compare the profiles in Fig. 7.6 collected on March 9 to those collected on March 19. What do these data tell you about how the mixed layer depth changed over these 10 days?
- How would you expect the depth distribution of ^{7}Be to depend on the vertical mixing rate (K_z)? Draw two idealized depth profiles of ^{7}Be, one showing a circumstance with high mixing rate (K_z) and one with low mixing.
- Imagine that a storm suddenly delivered a lot of rain to the surface ocean. How would you expect depth profiles of salinity to look before and after the storm? How would you expect depth profiles of ^{7}Be to look before and after the storm? Draw some idealized example profiles.
- Imagine measuring ^{7}Be in a coastal region. How would you expect a depth profile of ^{7}Be to look in a plume of newly upwelled water versus a nearby location where upwelling is not actively occurring? Draw some idealized example profiles.
- How would you expect the depth distribution of ^{7}Be to evolve from late winter when the mixed layer is at its maximum depth to late summer when the mixed layer has been shallow and the water column beneath has been stratified for some time? Draw some idealized example profiles.

7.3 Uranium and Thorium Decay Series

All elements found in nature with atomic numbers greater than 83 (bismuth) are radioactive. These elements are one of three long-lived radioisotopes: ^{238}U, ^{232}Th, and ^{235}U or their daughters. All three of these parent isotopes have half-lives near or in excess of 10^9 y, which is why appreciable amounts of them still exist on Earth after formation long ago in the Sun by nucleosynthesis. Isotopes in the three decay series are illustrated in Fig. 7.7, which consists of 10 elements and 36 radioactive isotopes that terminate in three stable Pb daughters (^{206}Pb, ^{208}Pb, and ^{207}Pb, respectively). Each row in the chart represents an element, decreasing in atomic number by one from top to bottom. Isotopes in the chart decay by releasing an α or β^- particle, which correspond to transformations on the chart of, respectively, two rows down or one step up and to the right. The 10 elements in Fig. 7.7 represent a huge diversity of chemical characteristics ranging from elements that are relatively soluble in seawater (U, Ra, and Rn, the last of which is a gas), to highly

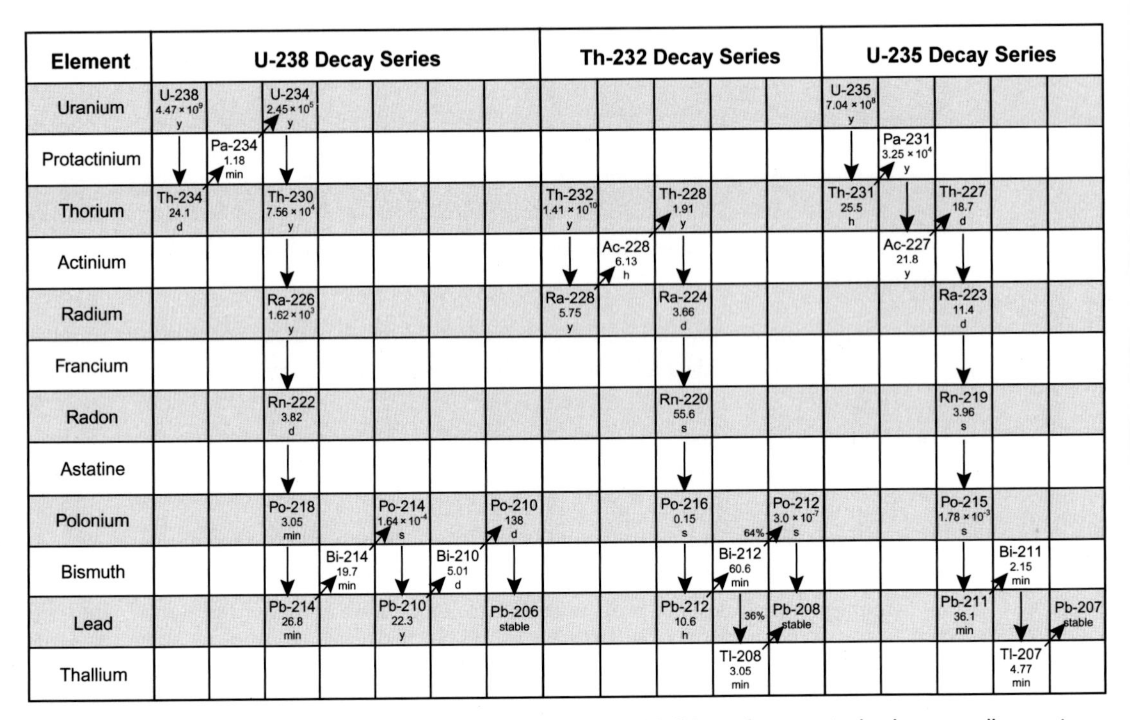

Figure 7.7 A chart of the nuclides showing the decay pathways and half-lives of isotopes in the three naturally occurring U/Th decay chains. The axes are arranged so that the ultimate source nuclide of each decay series is in the upper left corner. Rows indicate atomic number, decreasing toward the bottom. The x-axis makes space for different isotopes of the same element. Arrows that point downward indicate α decay, decreasing the atomic number by 2 and the atomic mass by 4. Arrows that slant upward to the right indicate β^- decay, in which a neutron in the nucleus becomes a proton, causing an increase in atomic number but little change in atomic mass. Modified from Emerson and Hedges (2008).

surface-reactive nuclides (Th, Pa, Po, and Pb) that are readily adsorbed (scavenged) onto particle surfaces. The half-lives of the U and Th series daughters range from fractions of a second (e.g., Po^{212}, Po^{214}, Po^{215}, and Po^{216}) to hundreds of thousands of years (^{234}U, ^{230}Th, and ^{231}Pa). No daughter has a half-life greater than 0.01 percent of its ultimate parent, which has important ramifications for the activity of the daughter isotopes in the decay chain compared to their parent.

7.3.1 Secular Equilibrium

In a closed system in which radioactive parents and daughters remain in contact, the activity of all daughters will eventually, after many daughter half-lives, equal the activity of the parent. This condition of equal radioactivity between a long-lived parent and a coexisting shorter-lived daughter is referred to as "secular equilibrium" and represents one of the most useful concepts in radiochemistry. Secular equilibrium can be demonstrated mathematically by solving for the relationship between the number of radioactive atoms, N, of the daughter, $_{Da}$, and parent, $_{Pa}$, nuclides. At any given time, the change in the amount of the daughter isotope is equal to the rate of parent decay (its source) minus the rate of the daughter decay (its sink):

$$\frac{dN_{Da}}{dt} = \lambda_{Pa} N_{Pa} - \lambda_{Da} N_{Da}. \tag{7.24}$$

Combining this equation with Eq. 7.2, gives

$$\frac{dN_{Da}}{dt} = \lambda_{Pa} N_{Pa,0} e^{-\lambda_{Pa} t} - \lambda_{Da} N_{Da}, \tag{7.25}$$

where $N_{Pa,0}$ is the number of parent atoms at time $t = 0$. The solution of this differential equation for the number of daughter atoms is

$$N_{Da} = \frac{\lambda_{Pa}}{\lambda_{Da} - \lambda_{Pa}} N_{Pa,0} \left(e^{-\lambda_{Pa} t} - e^{-\lambda_{Da} t}\right) + N_{Da,0} e^{-\lambda_{Da} t}. \tag{7.26}$$

This generally applicable equation simplifies considerably when the half-life of the parent is much longer than the half-life of its radioactive daughter. When $t_{½,Pa} \gg t_{½,Da}$, then $\lambda_{Pa} \ll \lambda_{Da}$ and hence $e^{-\lambda_{Pa} t} \gg e^{-\lambda_{Da} t}$. In this circumstance, we can also assume that any daughter isotope initially present at $t = 0$, $N_{Da,0}$, has completely decayed away. These simplifications allow both the final term and the second exponential term in the difference expression on the right-hand side of Eq 7.26 to be dropped, to yield

$$N_{Da} = \frac{\lambda_{Pa}}{\lambda_{Da} - \lambda_{Pa}} N_{Pa,0} \left(e^{-\lambda_{Pa} t}\right) = \frac{\lambda_{Pa}}{\lambda_{Da} - \lambda_{Pa}} N_{Pa} \approx \frac{\lambda_{Pa}}{\lambda_{Da}} N_{Pa}. \tag{7.27}$$

Since the definition of activity is the decay constant times the number of atoms (Eq. 7.3), the equation simplifies to

$$A_{Da} = A_{Pa}. \tag{7.28}$$

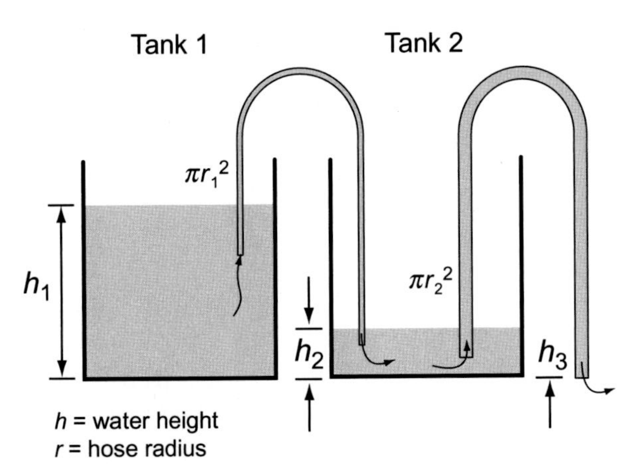

Figure 7.8 An illustration of the two tanks and siphon analogy for secular equilibrium. Water flows from Tank 1 to Tank 2 and from Tank 2 to the ground via siphons. Here h is water height in the tank (cm) and r is the radius of the hoses connecting the tanks. When Tank 2 is at steady state, the flow of water out of Tank 1 equals the flow out of Tank 2. (See text.) Modified from Emerson and Hedges (2008).

A conceptual analogy for secular equilibrium will help clarify these relationships. Imagine two linked water tanks of equal bottom area, each being drained by siphons of different diameter (Fig. 7.8). The water height in Tank 1 is decreasing so slowly that we can assume it is constant in time. The flow of water in the siphons, Q, is equal to the velocity of the water in the tube, v, times the cross-sectional area, $Area_X$, of the tube, $Q = v Area_X (\text{cm s}^{-1}(\text{cm}^2) = \text{cm}^3\ \text{s}^{-1})$. At steady state, the flow of water through the tube entering Tank 2 must equal the flow of water through the tube exiting Tank 2. Bernoulli's equation states that the velocity of water flow is proportional to the square root of the hydrostatic head (the difference in water heights in the tanks), $v_{1\to2} = \{2g(h_1 - h_2)\}^{0.5}$, where g is gravitational acceleration (9.8 m s^{-2}). The flux of water out of the hoses can then be rewritten as $Q_{1\to2} = \{2g(h_1 - h_2)\}^{0.5} Area_X (\text{cm}^3\ \text{s}^{-1})$. In this analogy, the difference in water height ($h_1 - h_2$) between Tank 1 and Tank 2 represents the number of parent radioisotope atoms, and the difference in water height ($h_2 - h_3$) between Tank 2 and the exit represents the number of daughter radioisotope atoms. The cross-sectional area of the siphon tubes, $Area_X$, is analogous to the decay constant. In the radiochemical counterpart, the isotope activity is the product of the decay constant and the number of atoms (Eq. 7.3). Tank 2 is drained by a siphon of greater cross-sectional area (faster decay rate), thus Tank 2 requires a smaller hydrostatic head to generate an output flow that matches the input from Tank 1.

This analogy helps clarify several aspects of secular equilibrium. First, the feedback between water level and flux from the tank is what causes Tank 2 to come to steady state. If the input exceeded the output, the level in Tank 2 would rise until the output grew to the input rate, and vice versa if the output exceeded the input. Radioisotopes work the same way because the activity (dN/dt) depends on the amount of the radioisotope present (N). Second, just like the level in Tank 2 will be much lower than Tank 1, due to its much larger output tube cross-sectional area, the number of atoms of the radioactive daughter supported by the longer-lived parent will be proportionately smaller. Finally, the time needed to reach steady state for Tank 2 depends on the diameter of the outlet siphon. For example, if Tank 2 is empty when the water flow from Tank 1 began, Tank 2 will reach its final volume quickly if the outlet siphon is large and more slowly if it is smaller. For radioisotopes, the time to reach secular equilibrium is determined by the half-life of the daughter not the parent. At first, this may seem counter-intuitive, because the rate at which daughter atoms are produced depends on the half-life or decay rate of the parent. However, from the analogy, the rate at which the volume of water in Tank 2 approaches its steady state depends on the cross-sectional area of the hose that exits Tank 2, not that of the smaller input hose.

7.3.2 The Geochemistry of Decay Series Isotopes in the Ocean

Given the wide range of isotopes (and corresponding half-lives) for many of the individual elements (e.g., Th, Ra, Po, Pb) in the U/Th decay series, some guidelines are useful for understanding which radioisotope is best suited to measure the rate of a particular natural process. First, the chemical behavior of the element in question must undergo the process to be investigated (e.g., mixing in dissolved form, gas exchange, or particle adsorption). Second, some portion of the radioisotope of interest must not be continuously supported by a local source such as an "upstream" parent within the U/Th decay series. Rate determinations are only possible when the chain of secular equilibrium is physically broken between a parent and its daughter. Finally, the rate of the process to be investigated and the half-life of the isotope used to investigate the process should be similar in magnitude. Thus, rapidly decaying isotopes (e.g., ^{234}Th and ^{222}Rn) are used to measure the rate of fast processes such as upper ocean particle fluxes or air–sea gas exchange, whereas radioisotopes with long half-lives (e.g., ^{230}Th or ^{231}Pa) are used to measure the rate of slow processes such as the accumulation rates of deep-sea sediments. An isotope that decays much faster than the characteristic rate of a targeted process will essentially be gone before the process fully expresses itself, whereas an isotope that decays much more slowly will not be measurably changed.

Key nuclear and physical transformations that ^{238}U and its longer-lived daughters undergo in the ocean and atmosphere are illustrated in Fig. 7.9. Mean activity concentrations dissolved in the ocean are presented in Table 7.2. The schematic presentation in Fig. 7.9 is a simplification that excludes some physical sources and sinks, does not specify the chemical forms of the elements, and ignores daughters with half-lives less than a day. Like most seawater elements, ^{238}U contained in continental rocks is released by weathering and carried by rivers to the ocean where it occurs in a highly soluble dissolved form. Uranium is strongly complexed by CO_3^- ions in seawater, relatively inert to particle adsorption, and not readily used by marine biota, so it behaves conservatively in the ocean. When dissolved ^{238}U decays in the ocean it initiates a fascinating series of reactions (Fig. 7.9) that begins with conversions to ^{234}Th $\left(t_{1/2} = 24.1\ \text{d}\right)$ and then to ^{234}U $\left(t_{1/2} = 2.45 \times 10^5\text{y}\right)$. The bulk of both isotopes remain in solution, but ^{230}Th, the following daughter isotope $\left(t_{1/2} = 7.57 \times 10^4\ \text{y}\right)$, is nearly completely removed from the water column (Table 7.2) following its generation by α decay from dissolved ^{234}U. The different behavior of the two isotopes of thorium is caused by their very different half-lives compared to the timescale of particle absorption and is described in the next section.

Table 7.2 Mean activity concentrations of selected radioisotopes in the ^{238}U, ^{235}U, and ^{232}Th decay series in the ocean. Modified from Broecker and Peng (1982) and Yu et al. (1996).

Isotope	Half-life (y)	Warm Surface Water	N. Atlantic Bottom Water	N. Pacific Bottom Water
			dpm (100 kg)$^{-1}$	
^{238}U(Parent)	4.47×10^9	240	240	240
^{234}Th	0.066	230	240	240
^{234}U	245 000	275	275	275
^{230}Th	75 700	~0	~0	0.15
^{226}Ra	1 620	7	13	34
^{222}Rn	0.010	5	>13	>34
^{210}Pb	22.3	20	8	16
^{210}Po	0.38	10	8	16
^{235}U(Parent)	0.704×10^9	10.8	10.8	10.8
^{231}Pa	32 800	~0	~0	0.05
^{232}Th(Parent)	14.1×10^9	~0	~0	~0
^{228}Ra	5.75	3	0.4	0.4
^{228}Th	1.91	0.4	0.3	0.3

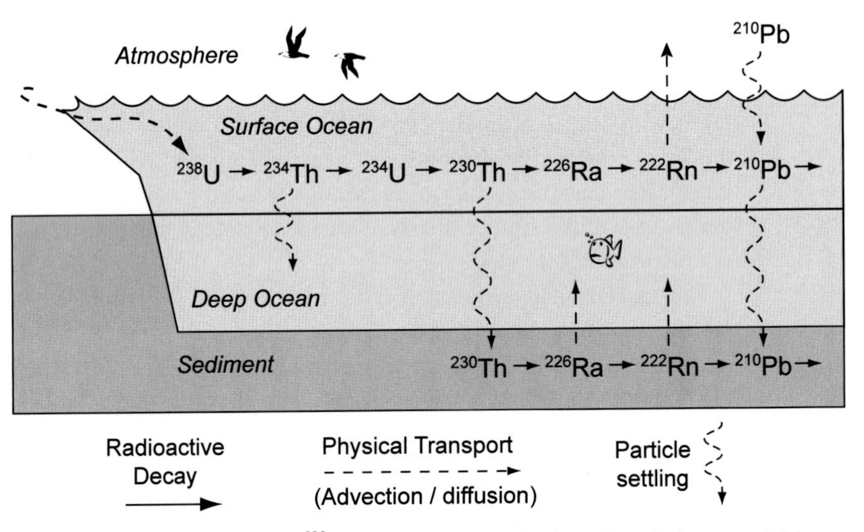

Figure 7.9 A schematic diagram of the pathways of ^{238}U and its longer-lived daughters through the ocean. Solid arrows indicate radioactive decay and dashed arrows indicate physical transport. For simplicity, physical transport of water between the surface and deep ocean is not shown. Modified from Broecker (1974).

Because ^{230}Th scavenged from the water is concentrated on the surface of particles accumulating on the ocean floor, rather than within them, any soluble daughters it forms are in a position to escape to ocean bottom waters. The first daughter of ^{230}Th, ^{226}Ra $(t_{1/2} = 1620\ y)$, is sparingly surface active, such that about 10 percent of this nuclide escapes from sediments to the overlying water column. The half-life of ^{226}Ra is similar to the residence time of the deep ocean, allowing this isotope to increase in concentration along deep ocean flow paths (Table 7.2). In turn, ^{226}Ra decays to ^{222}Rn $(t_{1/2} = 3.82\ d)$, an inert gas with no affinity for mineral surfaces. Because ^{222}Rn is a gas with a short half-life, it is an ideal tracer for determining rates of gas exchange between the air and water. The next daughter of ^{222}Rn with a significant half-life is ^{210}Pb $(t_{1/2} = 22.3\ y)$, a particle reactive element that is scavenged from the water column into sediments, where it can serve as a tracer of relatively fast sediment accumulation. The ^{238}U decay series finally ends with decay of ^{210}Pb to stable ^{206}Pb, which permanently accumulates in sediments.

A cartoon similar to Fig. 7.9 for the ^{235}U decay series would be much simpler. Particle-active ^{231}Pa $(t_{1/2} = 3.28 \times 10^4\ y)$ forms via the short-lived intermediate, ^{231}Th $(t_{1/2} = 25.5\ h)$, and is completely removed to the sediments. Essentially all further radioactivity in this decay series remains in the sediments, because subsequent daughters are either surface active (e.g., ^{227}Ac, $t_{1/2} = 21.8\,y$) or short-lived (e.g., ^{219}Rn, $t_{1/2} = 3.96\ s$). The major application deriving from this decay series is the use of ^{231}Pa as a tracer of particle flux.

The ^{232}Th decay series is different from the other two in that it begins in the sediments, because thorium is particle active. The only radioactivity to escape to the overlying seawater is ^{228}Ra $(t_{1/2} = 5.75\ y)$, which is partially released into bottom waters. This isotope is also enriched in groundwater on land because radium is more soluble in water than its parent thorium. Due to its relatively short half-life, most of the ^{228}Ra found in ocean surface waters comes from the continental margins. Measurements of this isotope in surface waters of the Atlantic Ocean were used to demonstrate that the flux of submarine groundwater to the oceans was greater than previously supposed (Moore et al., 2008). This finding led to the conclusion that nutrients and metals are supplied to the ocean by groundwater at a similar rate as the flux from rivers. ^{228}Ra decays via ^{228}Ac $(t_{1/2} = 6.13\ h)$ to ^{228}Th $(t_{1/2} = 1.91\ y)$, which has the potential to be used to measure scavenging rates but has not yet received widespread use. The rest of the lower-mass daughters have very short half-lives and limited geochemical application.

The processes illustrated in Fig. 7.9 along with more subtle geochemical and oceanographic behavior cause differences in the mean concentrations between surface and deep waters of the isotopes in the three decay series. For example, although the concentration of each uranium isotope is the same in surface and deep water (uranium is conservative in seawater), the activity of the daughter isotope, ^{234}U, is about 15 percent greater than ^{238}U (Table 7.2). This greater activity of ^{234}U is inconsistent with secular equilibrium and is caused by microenvironments in the source rock for uranium. During decay, the high-energy α particle emitted by ^{238}U shatters a portion of the host crystal, which then weathers more rapidly than the surrounding intact crystals, releasing more ^{234}U than ^{238}U to the environment. Rivers carry ^{234}U-enriched water from the continents to the ocean, where this daughter is sufficiently long lived to maintain its excess activity.

Another major pattern evident in the concentrations listed in Table 7.2 is that the seawater activities of all daughter isotopes except ^{234}U and ^{234}Th are very small compared to those of their ultimate parents. This stems from the particle reactive nature of thorium and protactinium described in the next section, which acts to remove these radioisotopes from the water column to the sediments. Radium-226 is released from ocean sediments because it is more soluble than thorium-230, allowing it, along with its daughter ^{222}Rn, to increase "downstream" along the thermohaline circulation from surface waters to the deep North Atlantic and then to the deep North Pacific. Along the way, ^{226}Ra behaves like a nutrient element in the sea and is associated with the cycles of barite, silica, and $CaCO_3$.

In the following sections, we present applications of decay series isotopes to problems in chemical oceanography, focusing on the ^{238}U decay chain and how the varied chemical behavior of the elements affects the utility of these isotopes as tracers.

7.3.3 Uranium–Thorium Activities and Ocean Particle Dynamics

The difference between the activities of ^{234}Th and ^{230}Th shown Table 7.2 and illustrated in Fig. 7.9 stems from what Karl Turekian once called "the great particle conspiracy" in the ocean (Turekian, 1977). Why should two isotopes of the same element, generated from parent isotopes of the same chemistry and nearly the same concentration in seawater, be so dramatically different in concentration? The reason stems from the competition between the timescale required to reach secular equilibrium and the timescale required to remove an element from the dissolved phase by particle adsorption (Fig. 7.10). Earlier in this chapter, we demonstrated by analogy (Fig. 7.8) that the activity of a radioactive daughter grows into secular equilibrium with its parent's activity on a timescale controlled by the daughter's half-life. It turns out that adsorption of thorium onto particles from solution in seawater is slow enough that the concentration of ^{234}Th has time to grow most of the way to secular equilibrium with ^{238}U before it is adsorbed to particles. In the next section, we will show how the small difference between actual ^{234}Th activities and those predicted from secular equilibrium with ^{238}U allows the flux of particles from the surface ocean to be determined. Because the half-life of ^{230}Th is so much longer than that of ^{234}Th, particle adsorption has plenty of time to remove essentially all of the ^{230}Th from solution before its concentration reaches even partway to secular equilibrium.

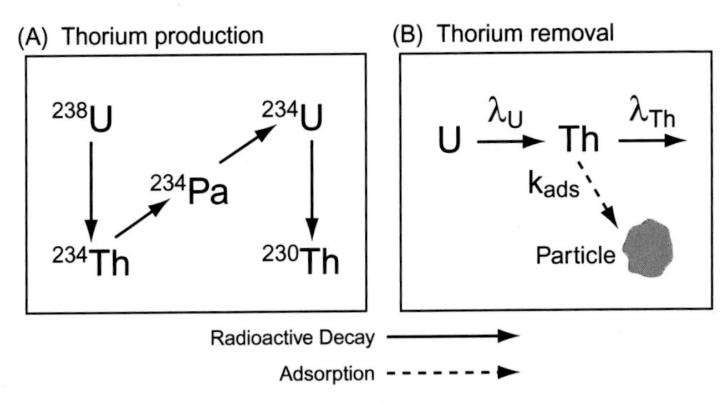

Figure 7.10 A schematic depiction of thorium production and removal from the dissolved phase in the ocean. (A) Both thorium isotopes are produced by decay of uranium; and (B) thorium is removed from the ocean by either radioactive decay, λ, or adsorption to sinking particles via the rate constant k_{ads}. The different half-lives of ^{234}Th and ^{230}Th compared to the rate of adsorption control the proportion of each isotope removed by adsorption.

^{238}U–^{234}Th Disequilibrium: A Tracer for Surface Ocean Particle Flux. The rate at which organic matter sinks from the surface ocean via particles is of great interest to oceanographers because it is a major part of the biological carbon pump and a control over atmospheric CO_2 (Chapters 3 & 8). The best radioisotopic tracer of this process is ^{238}U–^{234}Th because of the differences in the chemical behavior of these two elements and ^{234}Th's short half-life. Uranium is not influenced by chemical reactions, while thorium is extremely particle reactive. The relatively short half-life of ^{234}Th (24.1 d) is nearly ideal for capturing seasonal processes in the ocean euphotic zone. Below the ocean's mixed layer (~100 m) where particles are less abundant, ^{234}Th and ^{238}U are near secular equilibrium.

Particles adsorbing thorium and sinking out of the surface ocean create a deficiency in ^{234}Th activity from that expected at secular equilibrium with ^{238}U. At steady state, the depth-integrated deficiency in the ^{234}Th activity concentration in the euphotic zone (the "missing" ^{234}Th) is equal to its flux from the surface ocean on particles. Combining this ^{234}Th flux with the measured OC:^{234}Th ratio in the particles allows the flux of particulate organic carbon to be calculated. An example of ^{234}Th measurements in the surface waters of the equatorial Pacific (Fig. 7.11) indicates that the ^{234}Th activity is about one third lower than that of secular equilibrium with ^{238}U in the upper ocean at this location.

A ^{234}Th mass balance in the upper ocean equates the change in the concentration of dissolved thorium-234, $[^{234}Th_{dis}]$ (atoms m^{-3}), with time to its production rate by uranium-238 decay, $[^{238}U] \cdot \lambda_{238}$ (atoms m^{-3} d^{-1}), minus the decay rate of dissolved thorium-234, $[^{234}Th_{dis}]\lambda_{234}$, and adsorption of dissolved ^{234}Th onto particles, where k_{ads} is the adsorption rate constant:

$$\frac{d\left[^{234}\mathrm{Th}_{dis}\right]}{dt} = \left[^{238}\mathrm{U}\right]\lambda_{238} - \left[^{234}\mathrm{Th}_{dis}\right]\lambda_{234} - k_{ads}\left[^{234}\mathrm{Th}_{dis}\right]. \qquad (\text{atoms m}^{-3}\text{d}^{-1}) \quad (7.29)$$

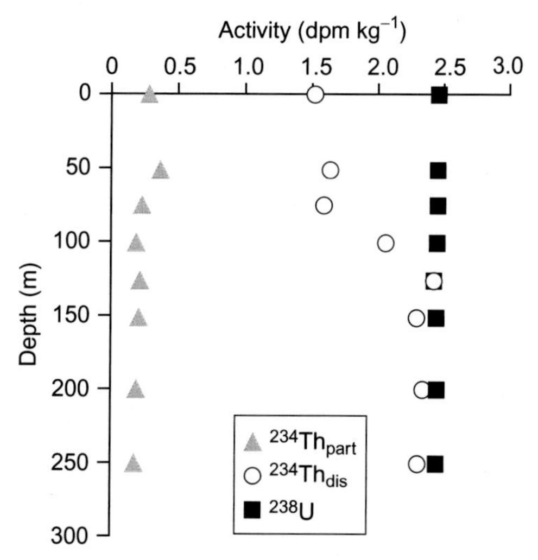

Figure 7.11 An example of the ^{234}Th and ^{238}U activities in the upper ocean at 2°N in the central equatorial Pacific. Subscript "dis" indicates dissolved and "part" indicates particulate. Redrawn from Dunne et al. (1997).

This mass balance is a simplification of the reversible nature of the adsorption mechanism, which would also consider that some ^{234}Th desorbs from particles back to the dissolved phase (Bacon and Anderson, 1982). Multiplying each term of Eq. 7.29 by the decay constant for ^{234}Th changes the concentrations to activity concentrations ($A = [N] \cdot \lambda$, with units of disintegrations per minute per m^{-3}, dpm m^{-3}), which is convenient because it is the activities of radioisotopes that are usually measured:

$$\begin{aligned}\frac{d\left[{}^{234}A_{dis}\right]}{dt} &= \lambda_{234}\left[{}^{238}\mathrm{U}\right]\lambda_{238} - \lambda_{234}\left[{}^{234}\mathrm{Th}_{dis}\right]\lambda_{234} - \lambda_{234}k_{ads}\left[{}^{234}\mathrm{Th}_{dis}\right] \\ &= \lambda_{234}\left({}^{238}A - {}^{234}A_{dis}\right) - k_{ads}{}^{234}A_{dis} \quad \left(\mathrm{dpm\ m^{-3}d^{-1}}\right).\end{aligned} \tag{7.30}$$

A similar mass balance for particulate thorium states that the change in particulate thorium activity in the water, $^{234}A_{part}$, with time is equal to the gain from adsorption of dissolved thorium minus the decay of particulate thorium and the removal of particulate thorium by sinking particles:

$$\frac{d\left[{}^{234}A_{part}\right]}{dt} = k_{ads}{}^{234}A_{dis} - \lambda_{234}{}^{234}A_{part} - \lambda_{234}\,\Psi_{234} \qquad \left(\mathrm{dpm\ m^{-3}d^{-1}}\right). \tag{7.31}$$

where Ψ_{234} is the removal of particulate thorium atoms by sinking particles (atoms m^{-3} d^{-1}) and $\lambda_{234}\Psi$ has units of activity instead of atoms (dpm m^{-3} d^{-1}). Substituting 7.31 into 7.30 to eliminate the adsorption term and assuming steady state gives:

$$0 = \lambda_{234}\left({}^{238}A - {}^{234}A_{dis}\right) - \lambda_{234}{}^{234}A_{part} - \lambda_{234}\,\Psi_{234}. \tag{7.32}$$

Measurements of ^{234}Th in the upper ocean indicate that most ^{234}Th exists in the dissolved phase even though thorium is relatively particle reactive, because dissolved ^{234}Th grows into secular equilibrium faster than it is depleted by adsorption to particles. Combining both particulate and dissolved thorium activities, $^{234}A = {}^{234}A_{dis} + {}^{234}A_{part}$, gives a relationship between the uranium-thorium difference and the particulate thorium flux at any depth in the upper ocean:

$$0 = \lambda_{234}\left({}^{238}A - {}^{234}A\right) - \lambda_{234}\,\Psi_{234}. \tag{7.33}$$

Integrating over the depth of the mixed layer (h) results in a relationship between the particulate thorium activity flux (in units of dpm m^{-2} d^{-1}) at the base of the mixed layer, J_{234} ($z=h$), and the integrated activities of uranium and thorium:

$$\left(\lambda_{234}{}^{234}J\right)_{z=h} = \lambda_{234}\int_{z=0}^{z=h}\Psi_{234}dz = \lambda_{234}\int_{z=0}^{z=h}\left({}^{238}A - {}^{234}A\right)dz. \tag{7.34}$$

There are two serious caveats to using ^{238}U – ^{234}Th dynamics as a tracer for the ocean's biological pump. First, the 24.1-day half-life of ^{234}Th means that any one set of measurements incorporates particle fluxes over about the previous month. Because most locations in the ocean vary strongly over a seasonal cycle, measurements every month are needed to capture the annual particle flux. At the Hawaii Ocean Time-series, monthly measurements revealed that most of the annual flux measured by the thorium-234 method occurred in a few very active months (Benitez-Nelson et al., 2001). Second, the goal is to determine the flux of organic carbon (OC) rather than ^{234}Th, which means that the flux resulting from Eq. 7.34 must be multiplied by the OC : ^{234}Th ratio of the sinking particles. This is complicated by different OC : ^{234}Th ratios on particles of different sizes that sink at different rates. For example, observations indicate that the OC : ^{234}Th ratio in particles captured by sediment traps and by in situ pumping and filtering are sometimes very different.

$^{234}U - ^{230}Th$ Disequilibrium: A Tracer for Whole-Ocean Particle Flux. Because virtually all the ^{230}Th produced in ocean water by ^{234}U decay ends up in particles that accumulate on the ocean floor, the production rate of ^{230}Th by ^{234}U in the water column, $\lambda_{230}{}^{234}A$, must equal the ^{230}Th decay rate in the sediments (Fig. 7.12). "Excess" (subscript "ex" in Fig. 7.12) refers to ^{230}Th activity concentrations that are greater than that produced directly in the sediments (in situ) by decay of ^{234}U in the particulate material. Because uranium is a soluble element in seawater, ^{234}U activity in surface sediments is low compared to ^{230}Th. If both the water column and the sediments below are included in one grand closed system, the parent and daughter radioisotopes will be in secular equilibrium. If all the sediments came from directly above, the inventory of excess ^{230}Th preserved in

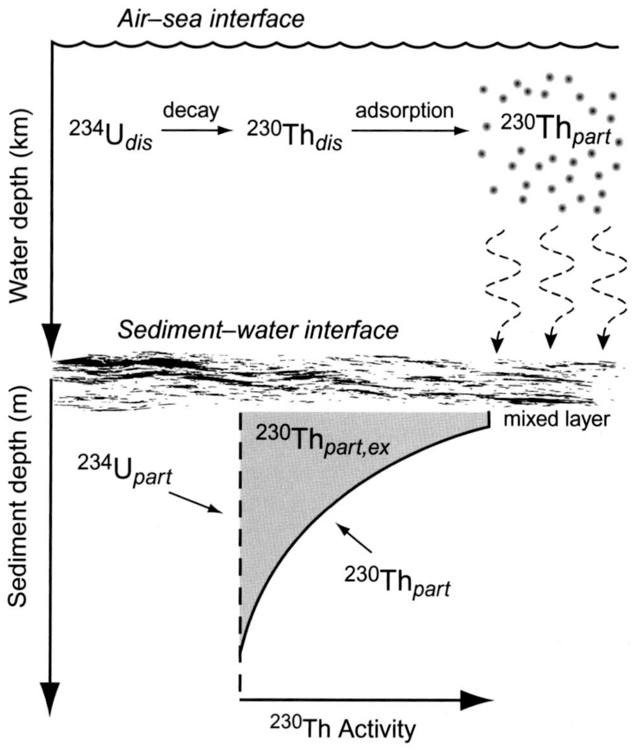

Figure 7.12 A schematic illustration of the relationship between dissolved ^{234}U and its daughter ^{230}Th in the ocean and sediments. Uranium is dissolved in the water ("*dis*"), but thorium is quantitively adsorbed to particles ("*part*") that fall to the sea floor. At secular equilibrium, the integrated activity of excess ^{230}Th in the sediments ("*ex*") should equal the depth integrated activity of ^{234}U in the water column above the sediments. Modified from Emerson and Hedges (2008).

the sediments would be equal to the ^{234}U activity times the water depth. (Since ^{234}U is conservative in the ocean, it has a constant ratio to salinity, and the activity of ^{234}U, ^{234}A, is 2.84 dpm kg^{-1} at $S = 35$.) If the excess ^{230}Th inventory is larger than expected, then the study location is receiving extra sediment from laterally adjacent locations. If the excess ^{230}Th inventory is less than expected, then the region is losing sediments to other locations. This method allows geochemists to correct sediment accumulation rates as a function of time for the influence of changes in horizontal transport.

Sediment redistribution complicates interpretation of the sedimentary record. For example, the mass accumulation rates of $CaCO_3$, opal, and organic carbon sometimes change between sediments of glacial and interglacial age. Such differences could be interpreted either as a change in the rain rate from above, which indicates a change in the production rate in the euphotic zone, or as a change in the direction and strength of bottom currents, which redirected or focused sediments into or out of the area. To understand the ocean's role in climate change, one would like to use the sedimentary record to quantify what was happening in surface waters, so changes due to sediment focusing or erosion need to be removed. This can be done by correcting sediment accumulation rates so that the integrated activity of excess ^{230}Th in the sediments equals the production rate in the overlying water.

The other main application of the isotope ^{230}Th has been to date marine sediments older than the 30–40 ky reach of ^{14}C. With a half-life of ~75 ky, ^{230}Th can be used to date a full 100 ky glacial cycle. Unlike with ^{14}C, the ^{230}Th activity is measured in the bulk sediments, because most of the isotope is found adsorbed to the fine clay minerals instead of the $CaCO_3$ shells where carbon isotopes are found. The excess ^{230}Th (the activity greater than that of ^{234}U) in the sediment decays until it reaches local secular equilibrium with the sediment background ^{234}U activity, at which time it is *supported* by the parent activity (Fig. 7.12). The systematics of the sediment accumulation equations presented earlier for ^{14}C (Eqs. 7.19–7.22) can also be applied to the *excess* (or *unsupported*) ^{230}Th in long cores. One of the original classic examples for a 10-m-long Caribbean core is presented in Fig. 7.13, where the ^{230}Th data yield a mean sedimentation rate between 2 and 3 cm ky^{-1}. Notice the increasing size of the ^{230}Th measurement error bars with depth in Fig. 7.13. These measurements were determined by counting the decay rate of ^{230}Th, which has a much larger error where ^{230}Th activities are low deeper in the sediment. Modern measurements determine the concentration of ^{230}Th using mass spectrometry, which is more accurate than counting radioactive decays and can be applied to much smaller samples.

^{231}Pa/^{230}Th Ratio in Sediments: A Tracer of NADW Flow Rate. The two radioactive decay series that begin with soluble uranium isotopes (^{238}U and ^{235}U) produce relatively long-lived daughter isotopes that are particle-reactive (^{230}Th and ^{231}Pa, Fig. 7.7). These isotopes have half-lives long enough for adsorption to particles to remove them from seawater before they have time to grow into secular equilibrium with their parents (Table 7.2). As we discussed in the last section, the excess radioisotope found in the sediments for particle reactive species is determined by the activity of the parent in the water column. The production rate of the daughter is equal to the decay constant times the water column activity, λA, as demonstrated for ^{234}U by the first term on the right side of

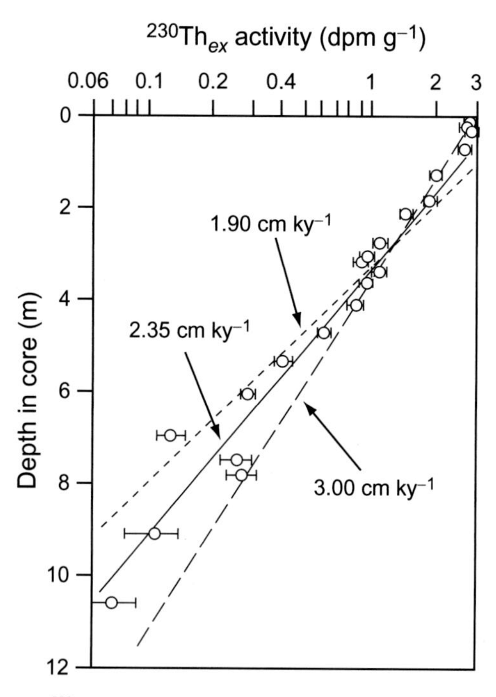

Figure 7.13 The relationship between excess ^{230}Th activity as a function of depth from the top 10 meters of a sediment core from the Caribbean Sea. The x-axis uses a logarithmic scale. Lines are drawn for different sedimentation rates for a one-dimensional model that assumes constant sediment accumulation (Eq. 7.22). This is an early application of the ^{230}Th dating method, and measurements were made by radioactive decay counting. Data from Ku (1976), and Broecker and van Donk (1970). Redrafted from the latter source.

Eq. 7.30. Since uranium is a conservative element in seawater, the production ratio ^{231}Pa/^{230}Th is, $\lambda_{231}\,{}^{235}A/(\lambda_{230}\,{}^{234}A)$, where "$A$"s are measured water column activities (21.32×10^{-6} y^{-1} × 108 dpm m^{-3}/(9.22×10^{-6} y^{-1} × 2750 dpm m^{-3}) = 0.093 using values from Yu et al. (1996)). If all the reactive daughter isotope were adsorbed by particles and transported to the sediments, the protactinium / thorium decay ratio, $\lambda_{231}\,{}^{231}A/(\lambda_{230}\,{}^{230}A)$, deposited to the sediments would be 0.093.

However, thorium and protactinium do not have exactly the same chemistry in seawater. Both are particle-reactive, but Th adsorbs to particles more strongly such that its residence time with respect to particle adsorption is 20–40 years, while that for Pa is 100–200 years (Yu et al., 1996). The removal time for Pa is very close to the time it takes modern North Atlantic Deep Water to flow from its source near Greenland to the Antarctic, where it joins the Circumpolar Deep Water (Broecker, 1979). As a result, roughly half of the ^{231}Pa produced in the water column in the Atlantic Ocean reaches sediments in the Atlantic, while the other half escapes in dissolved form to the Southern Ocean and beyond, where it is eventually adsorbed and deposited. Because dissolved ^{230}Th has an adsorption removal time that is much shorter than the NADW transit time, most all of the ^{230}Th produced in the water column is adsorbed and shunted to the underlying Atlantic Ocean sediments. Modern North Atlantic sediments thus contain a protactinium / thorium decay ratio,

$\lambda_{231}\ ^{231}A\ /\ (\lambda_{230}\ ^{230}A)$, that is significantly lower than the production ratio. Observations of this activity ratio in surface North Atlantic sediments indicate that the ratio is about 0.06 north of 45°S (Yu et al., 1996). The $^{231}A/^{230}A$ ratio deposited in Atlantic Ocean sediments depends on the rate of flow of NADW – the faster the flow, the more dissolved Pa escapes the basin before it is removed to the sediments. When NADW flow is very slow, so that deep water remains in the basin for many ^{231}Pa residence times, the $^{231}A/^{230}A$ ratio in the sediments should approach the 0.093 production ratio from the uranium parents in the water column (Marchal et al., 2000).

Early work on Cd/Ca and other tracers in Atlantic foraminifera as a proxy for nutrient concentrations suggested that NADW flow may have been less during glacial periods than it is today (e.g., Boyle and Keigwin, 1987). However, the timing of these changes was unclear until analyses were made of the $^{231}A/^{230}A$ in sediments in a fast sedimentation rate core from the Bermuda Rise in the northeast subtropical Atlantic Ocean (McManus et al., 2004). The ^{231}Pa/^{230}Th activity ratio, $^{231}A/^{230}A$, in the sediments is plotted versus time in Fig. 7.14 along with the δ^{18}O of contemporaneous planktonic foraminifera, which record some combination of ice volume and temperature (Chapter 6). The ^{14}C measurements and δ^{18}O stratigraphy identify major glacial–interglacial events. The activity ratio during the

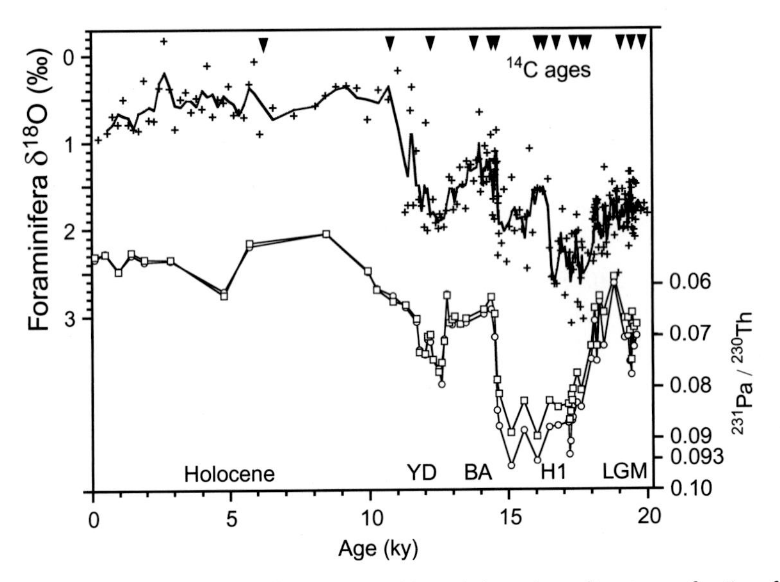

Figure 7.14 The change in the excess protactinium-231/thorium-230 activity ratio in marine sediments as a function of age in a fast sedimentation rate core from the Bermuda Rise in the subtropical Atlantic (open symbols and lower curve). A $^{231}A/^{230}A$ activity ratio of 0.093 is equal to the production ratio of their dissolved parent isotopes, ^{235}U and ^{234}U, in the overlying water. Values are corrected for decay since deposition of the sediments. Crosses (+) and upper curve indicate the δ^{18}O of planktonic foraminifera tests. Sediment age was determined by ^{14}C dates of $CaCO_3$ of planktonic foraminifera indicated by the arrowheads near the upper x-axis. Glacial–interglacial climate intervals indicated near the lower x-axis are identified by the shape of the δ^{18}O curve: Holocene warm interval, Younger Dryas (YD) cold reversal, Bølling-Allerød (BA) warm interval, Heinrich iceberg discharge Event 1 (H1), and Last Glacial Maximum (LGM). Redrawn from McManus et al. (2004).

Holocene period of the last 10 ky is relatively constant and much less than the uranium source, indicating a period of stable and rapid NADW flow similar to that of today. In contrast, values approach the source ratio of 0.093 during the time period labeled H1 (Heinrich event 1), which is characterized by ice-rafted debris in the sediments, indicating the presence of icebergs in surface waters. This period is believed to coincide with a large discharge of meltwater to the surface Atlantic that decreased the formation of NADW. The very rapid changes in $^{231}A/^{230}A$ indicate that the flow of NADW varied dramatically during this glacial–interglacial transition.

7.3.4 ^{222}Rn-^{226}Ra Disequilibrium in Surface Waters: A Tracer of Air–Sea Gas Exchange

The radioisotope pair ^{226}Ra and ^{222}Rn has been applied extensively to determining the rate of air–sea gas exchange. Seawater has a rather high and easily measurable activity of ^{226}Ra with a half-life of 1620 years (^{226}A = 5–15 dpm (100 kg)$^{-1}$, Table 7.2), which decays to the noble gas ^{222}Rn with a half-life of only 3.82 days (Fig. 7.7). Because ^{222}Rn has a much shorter half-life than ^{226}Ra, the activities of the two isotopes would be in secular equilibrium in a closed system, and indeed ^{222}A equals ^{226}A within the ocean's interior, away from the air–water interface. At the surface, however, radon escapes to the atmosphere, because

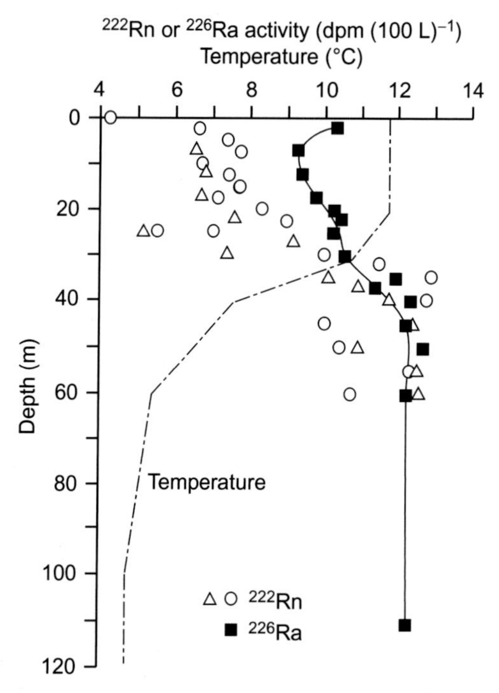

Figure 7.15 Upper ocean profiles of the activity of ^{222}Rn and ^{226}Ra, along with temperature versus depth in the subarctic Pacific Ocean at Ocean Station Papa. Open symbols are ^{222}Rn data, with different symbols indicating profiles measured on different days. Filled squares represent ^{226}Ra data from a single profile. Redrawn from Emerson et al. (1991).

its concentration in the marine atmosphere far from land is very low. Exchange to the atmosphere creates a deficit in ^{222}A relative to ^{226}A in the surface ocean, which is proportional to the rate of gas exchange across the air–water interface.

Hundreds of radon profiles have been measured in the surface ocean, and an example is presented in Fig. 7.15. These data can be used to determine the rate of gas exchange at the atmosphere–ocean interface using a mass balance for ^{222}Rn. The change with time in ^{222}Rn concentration (mol m^{-3}) integrated from the surface to depth z^* at which secular equilibrium is achieved, usually 200 m or less, is equal to the sum of three terms (mol m^{-2} d^{-1}): production by the ^{226}Ra concentration, loss due to radioactive decay, and loss by flux to the atmosphere:

$$\int_{z=0}^{z=z^*} \frac{d\left[{}^{222}\mathrm{Rn}\right]}{dt} dz = \int_{z=0}^{z=z^*} \lambda_{226}\left[{}^{226}\mathrm{Ra}\right] dz - \int_{z=0}^{z=z^*} \lambda_{222}\left[{}^{222}\mathrm{Rn}\right] dz + F_{A-W} \quad \left(\mathrm{mol\,m^{-2}d^{-1}}\right). \tag{7.35}$$

The air–sea flux (Section 2.2 and Appendix 2A.3) is the product of a gas exchange mass transfer coefficient, k_{Rn} (m d^{-1}), and the concentration difference between the measured radon values in surface waters and the value in chemical equilibrium with the partial pressure of ^{222}Rn in the atmosphere, p^{222}Rn:

$$F_{A-W} = -k_{S,\mathrm{Rn}}\left\{\left[{}^{222}\mathrm{Rn}\right]_{ml} - p^{222}\mathrm{Rn}_{atm} K_{H,\mathrm{Rn}}\right\}, \tag{7.36}$$

where $[^{222}\mathrm{Rn}]_{ml}$ is the radon concentration in the ocean mixed layer and $K_{H,\mathrm{Rn}}$ is the Henry's law coefficient for radon. Here we follow the convention in Chapter 2 that the air–sea flux to the ocean is positive. For ^{222}Rn, this equation can be simplified, since the atmospheric partial pressure of radon is negligible several tens of kilometers away from land:

$$F_{A-W} = -k_{S,\mathrm{Rn}}\left[{}^{222}\mathrm{Rn}\right]_{ml} \qquad \left(\mathrm{mol\ m^{-2}d^{-1}}\right). \tag{7.37}$$

Combining Eqs. 7.35 and 7.37 and multiplying by λ_{Rn} to change all concentrations to activity concentrations (dpm m^{-3}) gives

$$k_{S,\mathrm{Rn}}{}^{222}A = \lambda_{222}\int_{z=0}^{z=z^*}\left({}^{226}A - {}^{222}A\right) dz - \int_{z=0}^{z=z^*} \frac{d\left[{}^{222}A\right]}{dt} dz \qquad \left(\mathrm{dpm\ m^{-2}d^{-1}}\right). \tag{7.38}$$

Activity profiles of radon and radium, can be used along with Eq. 7.38 to determine the gas exchange mass transfer coefficient for radon, $k_{S,\mathrm{Rn}}$.

The activity of radon-222 in the surface ocean can change dramatically over short timescales due to its rather short half-life and the inherent variability of surface ocean turbulence. For this reason, the last, non-steady-state term in Eq. 7.38 can be quite large. Evaluating the change in ^{222}Rn activity with time has been done both by measuring a series of radon profiles at a single location and by averaging many individual profiles in different ocean basins. A summary of the radon-determined mass transfer coefficients from time-series experiments and basin-scale averaging is presented in Fig. 7.16 as a function of wind speed during the experiments. The data are plotted along with the regression line (labeled H11) determined from purposeful tracer release experiments (see Fig. 2A.3.3 in Chapter 2). Although there is considerable scatter, the mass transfer coefficient derived from ^{222}Rn

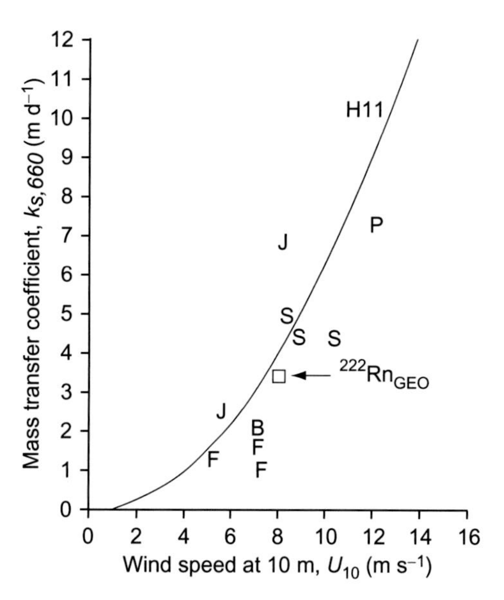

Figure 7.16 The gas exchange mass transfer coefficient, $k_{S,660}$, normalized to a Schmidt number of 660 (m d^{-1}) plotted against the wind speed at 10 meters above the ocean surface, U_{10} (m s^{-1}). (See Chapter 2.) Letters indicate different time-series experiments with repeated ^{222}Rn measurements in the upper ocean, allowing the non-steady-state term to be evaluated: P and S are from the subarctic Pacific (Peng et al., 1974 and Emerson et al., 1991, respectively); B is from the subtropical Atlantic (Broecker and Peng, 1971); F and J are from the North Atlantic (Kromer and Roether, 1983). The box labeled $^{222}Rn_{GEO}$ represents the average of over 100 ^{222}Rn profiles made globally during the GEOSECS program (Peng et al., 1979). The line labeled H11 is the relationship derived from purposeful tracer release experiments : $k_{S,660}$ (cm hr^{-1}) $= 0.262\ U_{10}^2$ or $k_{S,660}$ (m d^{-1}) $= 0.0629\ U_{10}^2$ (Ho et al., 2011, Fig. 2A.3.3 in Chapter 2).

measurements increases with increasing wind speed, and generally follows the best-fit line derived from tracer release experiments. The scatter in the shorter-term radon measurements almost certainly represents real environmental variability in the gas exchange rate due to factors other than wind speed.

7.3.5 Excess ^{210}Pb and ^{234}Th: Tracers of Coastal Sediment Accumulation

The only radioactive daughter of ^{222}Rn with a significantly long half-life is ^{210}Pb ($t_{1/2}$ = 22.3 y), which is particle-reactive like thorium and protactinium and thus ends up in sediments after being produced by radon decay in the water or atmosphere. In deep ocean sediments, which have sedimentation rates of 0.1–1.0 cm per thousand years (Fig. 7.5), any ^{210}Pb observed below the very surface is the result of mixing by burrowing benthic animals – bioturbation. In coastal and inland waters, which have much faster sedimentation rates, ^{210}Pb can be used to judge both the depth of bioturbation and the sediment accumulation rate. For example, sediment profiles of ^{210}Pb and ^{234}Th from the Puget Sound (Fig. 7.17, Carpenter et al., 1984) indicate the overlapping effects of both

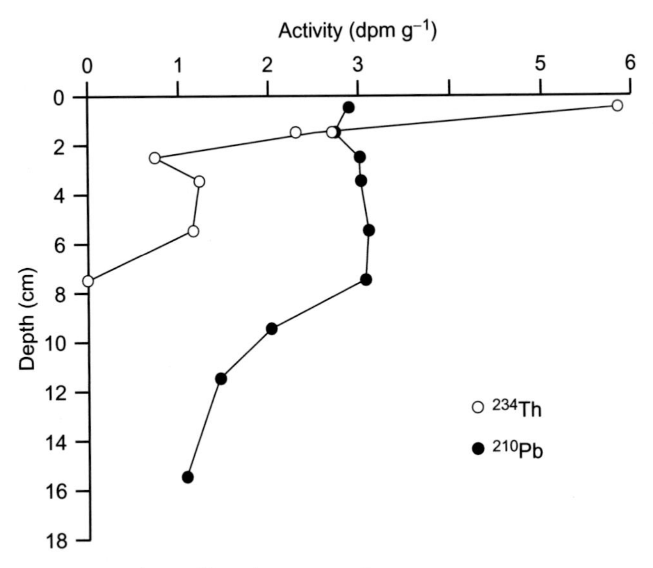

Figure 7.17 Activities of ^{210}Pb $(t_{1/2} = 22.3\text{ yr})$ and ^{234}Th $(t_{1/2} = 24.1\text{ d})$ in sediments of Puget Sound in the northwestern United States. Data are from Carpenter et al. (1984).

processes. Lead-210 is mixed uniformly to a depth of about 8 cm, demonstrating active bioturbation to this depth. The decrease in ^{210}Pb activity below this depth is due primarily to radioactive decay and sediment accumulation, with a resulting sedimentation rate of ~0.3 cm y^{-1} – up to 1000 times the rate in $CaCO_3$-rich deep ocean sediments. The activity of ^{234}Th, which has a much shorter half-life (24.1 d), decreases from high values at the surface to undetectable levels within the ^{210}Pb-determined bioturbated layer, making it possible to evaluate variability in bioturbation rates unresolvable by the longer timescale ^{210}Pb profile. Organisms mix the near-surface sediment at higher frequency than they mix slightly deeper sediments, so even short half-life isotopes are detected at shallow depths below the surface, while ^{210}Pb persists longer and is mixed more deeply by slower bioturbation. The different profiles for ^{14}C in deep ocean sediments (Fig. 7.5) and those in Fig. 7.17 demonstrate how different half-lives make it possible to use radioisotopes to resolve mechanisms with very different timescales.

7.4 Anthropogenic Influences

Man-made radionuclides have been released into the environment by nuclear weapons testing, nuclear power plants, and nuclear fuel reprocessing plants. In all these nuclear releases, two of the most-abundant, long-lived fission products have been ^{137}Cs and ^{90}Sr. Here, we use the documented flux of ^{137}Cs to estimate the total relative escape of radioactivity to the environment. (The units are petabecquerals, PBq $=10^{15}$ Bq, where 1 Bq is one disintegration per second.) By far the largest and most global contamination

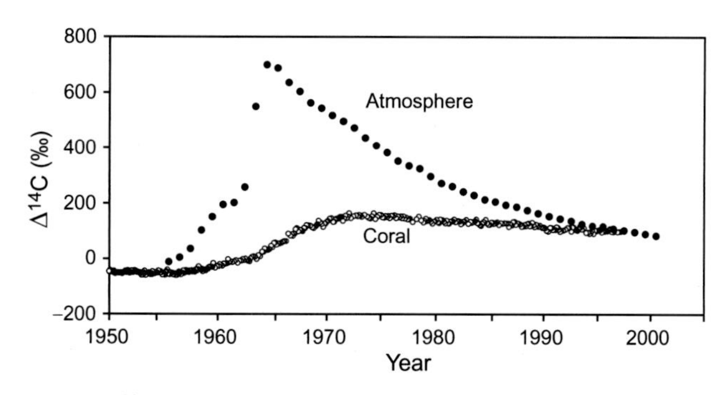

Figure 7.18 Radiocarbon measurements ($\Delta^{14}C$, ‰) in the atmosphere and the surface ocean as a function of time from 1950 to ~2000. Atmospheric data are globally averaged annual values from Hua and Barbetti (2004). Surface ocean $\Delta^{14}C$ values were determined by measurements of ^{14}C activity in growth bands of corals (Guilderson et al., 2000). Modified from Emerson and Hedges (2008).

was caused by atmospheric nuclear weapons testing primarily in the 1950s and 1960s (~600 PBq ^{137}Cs, Aarkrog, 2003). Much smaller intentional releases (~40 PBq ^{137}Cs) occurred in the 1970s and 1980s by nuclear fuel reprocessing facilities at Cap de la Hague, France, and Sellafeld, United Kingdom. Since that time there have been two large nuclear power plant accidents: at Chernobyl, Ukraine, in 1986 (which released about 100 PBq ^{137}Cs to the atmosphere) and after an earthquake and tsunami devastated Fukushima, Japan, in 2011 (which released about 10–40 PBq ^{137}Cs, Buesseler et al., 2012; Steinhauser et al., 2014). The amount of radionuclides that ended up in the ocean from these planned and accidental releases varied depending on the location of the source.

Here we focus on the earliest and most extensive injection of radionuclides into the marine system: atmospheric thermonuclear weapons testing mostly by the United States and Russia with much smaller contributions by the United Kingdom, France, and China. We describe the distribution of atmospheric spallation products, ^{14}C and ^{3}H, resulting from the atmospheric bomb tests, rather than the fission products mentioned in the previous paragraph because these two isotopes have been the most widely measured since the 1950s. The bomb tests nearly doubled the level of ^{14}C in atmospheric CO_2 (Fig. 7.18). The spike in the mid-1960s resulted from nations increasing their thermonuclear detonations in anticipation of an international moratorium on atmospheric testing. Since that time, the $\Delta^{14}C$ of atmospheric CO_2 has decreased at a much faster rate than can be explained by radioactive decay. Active exchange of carbon between high ^{14}C atmospheric CO_2 and low ^{14}C in other major reservoirs, primarily dissolved inorganic carbon (DIC) in the ocean and organic matter in land plants and soil, explains the rapid decrease.

The corresponding increase in the radiocarbon content of surface ocean DIC is readily measurable, but few measurements were made in the 1950s and 1960s, making ^{14}C recorded in the $CaCO_3$ of corals that grew in mid-latitude surface waters the best record of changes in surface ocean DIC-$\Delta^{14}C$ (Fig. 7.18). After the maximum in the $\Delta^{14}C$ of atmospheric CO_2, surface ocean $\Delta^{14}C$ values at mid-latitudes increased from pre-bomb values of approximately −50‰ to ~+150‰. By the mid-1990s, when the global ocean WOCE surveys took place, bomb ^{14}C ($\Delta^{14}C$ values greater than zero, Fig. 7.19) was

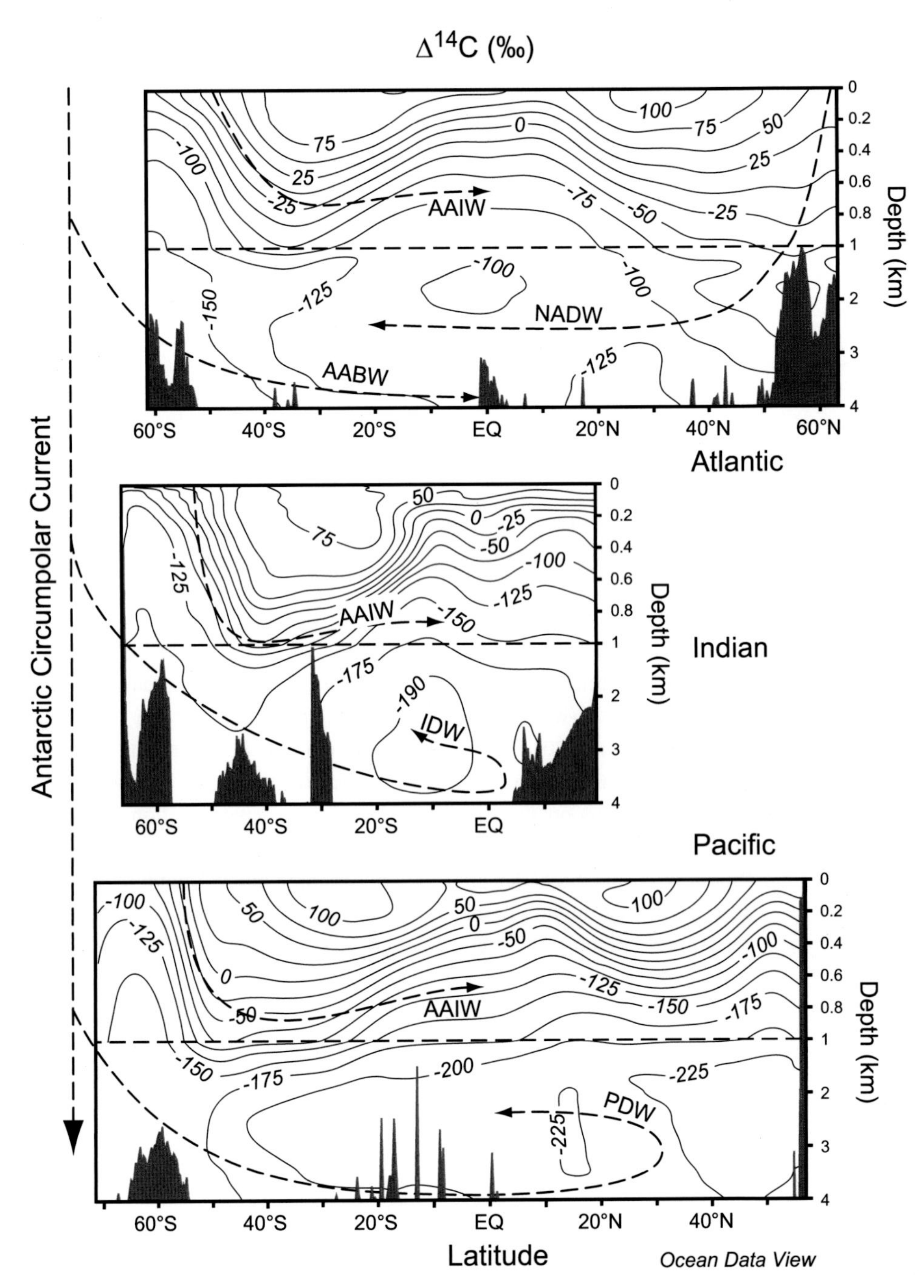

Figure 7.19 Cross sections of total Δ^{14}C-DIC (bomb + natural) in the Atlantic, Indian, and Pacific Oceans showing the penetration of bomb-produced ^{14}C into the thermocline in all oceans and to much greater depths in the North Atlantic. Δ^{14}C values greater than about −50‰ are contaminated with bomb ^{14}C. Note the expanded y-axis scale for the upper 1 km. Data are from the Global Ocean Data Analysis Project version 2 (GLODAP v2; Key et al., 2004) via the Ocean Carbon Data System (OCADS) website: www.ncei.noaa.gov/products/ocean-carbon-data-system/

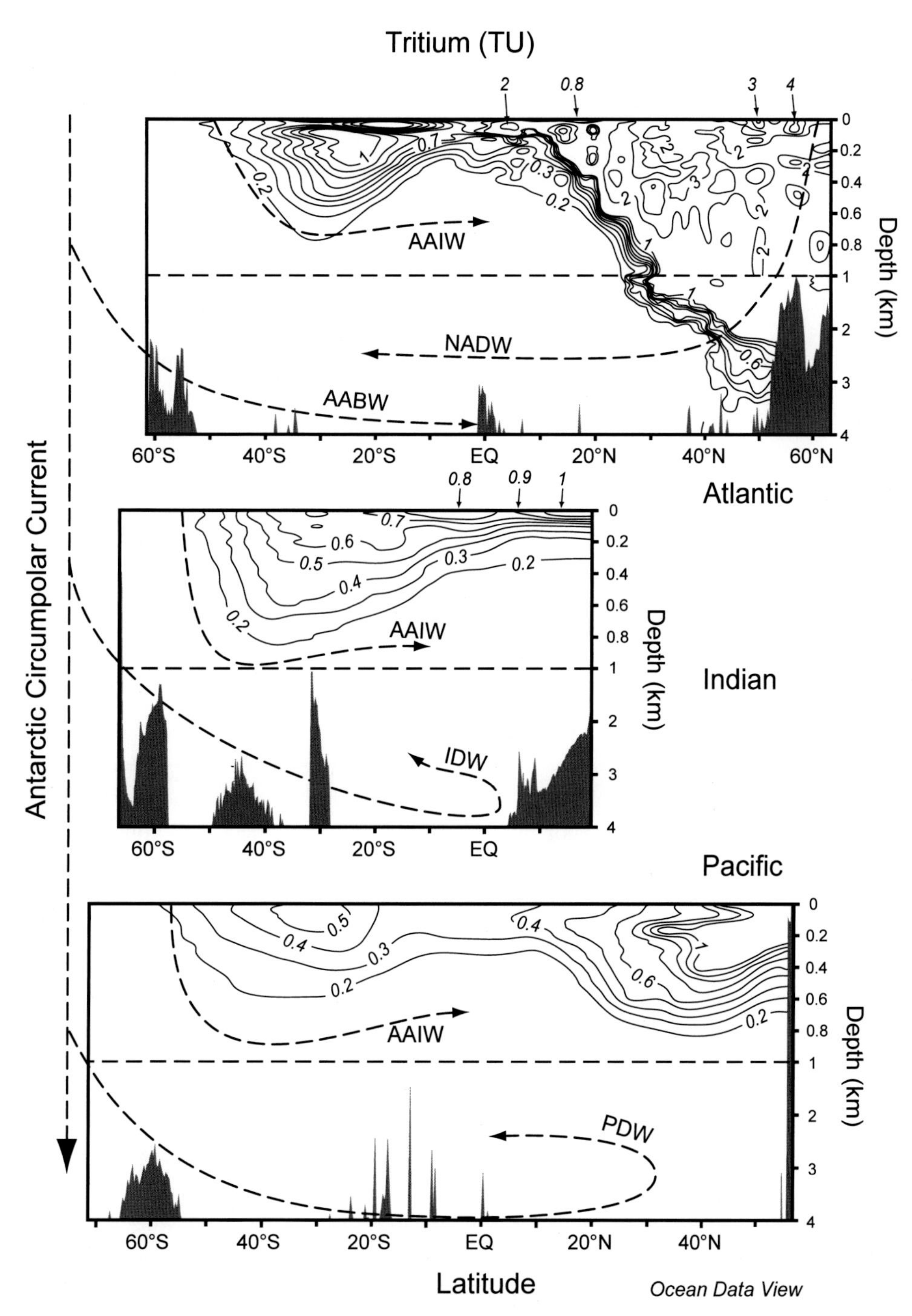

Figure 7.20 A cross section of ^{3}H-H_2O values in the world's ocean. ^{3}H concentrations are given in tritium units, where 1 TU = 1 ^{3}H atom per 10^{18} hydrogen atoms. Note the expanded y-axis scale for the upper 1 km. Data are from OCADS.

concentrated in the intermediate waters associated with the bowls of the subtropical gyres, showing that these water masses were in recent contact with the surface. Lower $\Delta^{14}C$ values near the Equator demonstrate the upwelling of older waters toward the surface. Values in the North Atlantic are elevated most deeply where the thermohaline circulation brings surface waters to abyssal depths (see Section 1.3.3). However, the ^{14}C content of DIC in the vast majority of deep-sea waters is still largely unaffected by bomb ^{14}C. Although injection of bomb ^{14}C into the environment has compromised the age determination methods already discussed in this chapter, it has provided new opportunities to quantify gas exchange rates, as atmospheric CO_2 diffuses into the ocean, as well as mixing rates between surface and deeper ocean waters.

The other atmospheric spallation product created in great abundance, compared to its natural background, by thermonuclear bomb testing is tritium (^3H), a radioactive hydrogen atom with two neutrons and a half-life of 12.3 y. In contrast to ^{14}C, tritium was quickly incorporated into water vapor and removed from the atmosphere by rain, which transferred the ^3H activity from the atmosphere to the ocean and land much more rapidly than for $^{14}CO_2$ (Broecker and Peng, 1982). Tritium is an excellent tracer for the water cycle, because it is physically part of water molecules. Distributions of tritium in the ocean demonstrate similar patterns as bomb ^{14}C, with high concentrations in the subtropical bowls and deep penetration only in the North Atlantic (Fig. 7.20).

Discussion Box 7.2

This is a game for two to four players. The object is to collect the most pairs of an isotope matched with an application for which it is typically used. Six cards are dealt to each player, face down. The remaining cards are placed face down in the center to form the stock. Decide who will begin.

Player 1 asks any other player for an isotope *or* application. If the other player has the card requested, they must give it to Player 1. If they do not, the other player says, "develop a new method" and Player 1 draws a card from the stock. When a match is made between an isotope and application, the player may place that pair on the table, where they cannot be taken by other players. Other players may challenge a pair if they think that isotope is not typically used for that application. The play then passes to the left. When a player has no cards, they take two cards from the stock and then continue. The game ends when no cards are left in the stock, and no more matches are available. At that time, the player with the most pairs on the table wins.

Make the following cards or download them for printing from the Cambridge University Press website (www.cambridge.org/emerson-hamme). *Isotopes:* four cards each for ^{14}C and ^{230}Th; three cards each for ^{234}Th, ^7Be, ^3H, ^{231}Pa, ^{222}Rn, and ^{210}Pb. *Applications:* seven cards for sediment accumulation rates, six cards for upper ocean circulation, five cards for deep ocean circulation, three cards each for surface ocean particle flux and for gas exchange, two cards for sediment focusing.

References

Aarkrog, A. (2003) Input of anthropogenic radionuclides into the World Ocean, *Deep-Sea Research II*, **50**, 2597–2606.

Bacon, M. P., and R. F. Anderson (1982) Distribution of thorium isotopes between dissolved and particulate forms in the deep sea, *Journal of Geophysical Research Oceans*, **87**, 2045–2056.

Bard, E., G. Ménot-Combes, F. Rostek (2004) Present status of radiocarbon calibration and comparison records based on Polynesian corals and Iberian margin sediments, *Radiocarbon*, **46**, 1189–1202.

Benitez-Nelson, C., K. O. Buesseler, D. M. Karl, and J. Andrews (2001) A time-series study of particulate matter export in the North Pacific Subtropical Gyre based on ^{234}Th : ^{238}U disequilibrium, *Deep-Sea Research I*, **48**, 2595–2611.

Berner, R. A. (1980) *Early Diagenesis: A Theoretical Approach*. Princeton: Princeton University Press.

Boudreau, B. P. (1997) *Diagenetic Models and Their Implementation*. Berlin: Springer-Verlag.

Boyle, E. A., and L. Keigwin (1987) North Atlantic thermohaline circulation during the past 20,000 years linked to high-latitude surface temperature, *Nature*, **330**, 35–40.

Broecker, W. S. (1979) A revised estimate of the radiocarbon age of North Atlantic Deep Water, *Journal of Geophysical Research Oceans*, **84**, 3218–3226.

Broecker, W. S. (1974) *Chemical Oceanography*. New York: Harcourt, Brace, Jovanovich.

Broecker, W. S., and T.-H. Peng (1971) The vertical distribution of radon in the Bomex area, *Earth and Planetary Science Letters*, **11**, 99–108.

Broecker, W. S., and T.-H. Peng (1982) *Tracers in the Sea*, 690 pp. Palisades: Eldigio Press.

Broecker, W. S., and J. van Donk (1970) Insolation changes, ice volumes, and the O^{18} record in deep-sea cores, *Reviews of Geophysics and Space Physics*, **8**, 169–198.

Buesseler, K. O., S. R. Jayne, N. S. Fisher, et al. (2012) Fukushima-derived radionuclides in the ocean and biota off Japan, *Proceedings of the National Academy of Sciences of the United States of America*, **109**, 5984–5988, doi:10.1073/pnas.1120794109.

Burdige, D. J. (2006) *Geochemistry of Marine Sediments*. Princeton: Princeton University Press.

Carpenter, R., M. L. Peterson, J. T. Bennett, and B. L. K. Somayajulu (1984) Mixing and cycling of uranium, thorium and ^{210}Pb in Puget Sound sediments, *Geochemica et Cosmochimica Acta*, **48**, 1949–1963.

Dunne, J. P., J. W. Murray, J. Young, L. S. Balistrieri, and J. Bishop (1997) ^{234}Th and particle cycling in the central equatorial Pacific, *Deep Sea Research Part II*, **44**, 2049–2083.

Emerson, S. R., and J. I. Hedges (2008) *Chemical Oceanography and the Marine Carbon Cycle*, 453 pp. Cambridge: Cambridge University Press.

Emerson S., P. Quay, C. Stump, D. Wilbur, and M. Knox (1991) O_2, Ar, N_2, and ^{222}Rn in surface waters of the subarctic Pacific Ocean: net biological O_2 production, *Global Biogeochemical Cycles*, **5**, 49–69.

Gebbie, G., and P. Huybers (2012) The mean age of ocean waters inferred from radiocarbon observations: sensitivity to surface sources and accounting for mixing histories, *Journal of Physical Oceanography*, **42**, 291–305.

Guilderson, T. P., D. P. Schrag, E. Goddard, et al. (2000) Southwest subtropical Pacific surface water radiocarbon in a high-resolution coral record, *Radiocarbon*, **42**, 249–256.

Ho, D. T., R. Wanninkhof, P. Schlosser, et al. (2011) Toward a universal relationship between wind speed and gas exchange: gas transfer velocities measured with $^3He/SF_6$ during the Southern Ocean Gas Exchange Experiment, *Journal of Geophysical Research Oceans*, **116**, C00F04, doi:10.1029/2010JC006854.

Hua, Q., and M. Barbetti (2004) Review of tropospheric bomb ^{14}C for carbon cycle modeling and age calibration purposes, *Radiocarbon*, **46**, 1273–1298.

Hughen, K., J. Southon, S. Lehman, C. Bertrand, and J. Turnbull (2006) Marine-derived ^{14}C calibration and activity record for the past 50,000 years updated from the Cariaco Basin, *Quaternary Science Reviews*, **25**, 3216–3227.

Kadko, D., and D. Olson (1996) Beryllium-7 as a tracer of surface water subduction and mixed-layer history, *Deep-Sea Research I*, **43**, 89–116.

Key, R. M., A. Kozyr, C. L. Sabine, et al. (2004) A global ocean carbon climatology: Results from Global Data Analysis Project (GLODAP), *Global Biogeochemical Cycles*, **18**, GB4031, doi:10.1029/2004GB002247.

Kromer, B., and W. Roether (1983) Field measurements of air-sea gas exchange by the radon deficit method during JASIN 1978 and FGGE 1979, *Meteor Forschungsergebnisse*, **A/B24**, 55–75.

Ku, T-L (1976) The uranium-series methods of age determination, *Annual Review of Earth and Planetary Sciences*, **4**, 347–379.

Marchal, O., R. François, T. F. Stocker, and F. Joos (2000) Ocean thermohaline circulation and sedimentary $^{231}Pa/^{230}Th$ ratio, *Paleoceanography and Paleoclimatology*, **15**, 625–641.

McManus, J. F., R. Francois, J.-M. Gheradi, L. D. Keigwin, and S. Brown-Leger (2004) Collapse and rapid resumption of Atlantic meridional circulation linked to deglacial climate changes, *Nature*, **428**, 834–837.

Moore, W. S., J. L. Sarmiento, and R. M. Key (2008) Submarine groundwater discharge revealed by ^{228}Ra distribution in the upper Atlantic Ocean, *Nature Geoscience*, **1**, 309–311.

Peng, T-H., and W. S. Broecker (1984) The impacts of bioturbation on the age difference between benthic and planktonic foraminifera in deep sea sediments, *Nuclear Instruments and Methods in Physics Research*, **B5**, 346–352.

Peng, T.-H., T. Takahashi, and W. S. Broecker (1974) Surface radon measurements in the North Pacific Ocean Station Papa, *Journal of Geophysical Research Oceans*, **79**, 1772–1780.

Peng, T.-H., W. S. Broecker, G. G. Mathieu, Y.-H. Li, and A. E. Bainbridge (1979) Radon evasion rates in the Atlantic and Pacific oceans as determined during the GEOSECS Program, *Journal of Geophysical Research Oceans*, **84**, 2471–2486.

Steinhauser, G., A. Brandl, and T. E. Johnson (2014) Comparison of the Chernobyl and Fukushima nuclear accidents: a review of the environmental impacts, *Science of the Total Environment*, **470**–471, 800–817.

Stuiver, M., and H. A. Polach (1977) Discussion: reporting of ^{14}C data, *Radiocarbon*, **19**, 355–363.

Turekian, K. K. (1977) The fate of metals in the ocean, *Geochemica et Cosmochimica Acta*, **41**, 1139–1144.

Wang, Y. J., H. Cheng, R. L. Edwards, et al. (2001) A high-resolution absolute-dated late Pleistocene monsoon record from Hulu Cave, China, *Science*, **294**, 2345–2348.

Yu, E.-F., R. Francois, and M. P. Bacon (1996) Similar rates of modern and last-glacial ocean thermohaline circulation inferred from radiochemical data, *Nature*, **379**, 689–694.

Problems for Chapter 7

7.1. Surface dwelling foraminifera found at different depths in a sediment core from a high productivity region in the Southern Ocean have the $\delta^{14}C$ and $\delta^{13}C$ values listed in the following table. Each measurement is performed on multiple foraminifera and is the average of many individual shells.

a. Calculate the $\Delta^{14}C$ values in these foraminifera at each depth.

b. Calculate the age at each depth in the sediment core. You may assume that atmospheric ^{14}C production rates were constant over this period, i.e., a perfect correlation between ^{14}C age and calendar age. Justify any choices you make in your calculation. Make a plot of your calculated ages.

c. Calculate the average sedimentation rate at this location.

d. Determine the depth to which the sediments are mixed by bioturbation.

e. Is the calculated sedimentation rate affected by bioturbation? Why or why not? Is the age at a given depth in the sediments affected by bioturbation? How?

depth (cm)	$\delta^{14}C$ (‰)	$\delta^{13}C$ (‰)
3	−63	1.1
7	−58	1.09
10	−65	1.09
19	−134	1.19
30	−213	1.08
59	−379	1.35
69	−388	1.01
99	−419	1.09
139	−564	1.11

7.2. Assume that the data shown in Fig. 7.11 represent a steady-state situation. Pay careful attention to the units for this problem.

a. Estimate the integrated difference in the upper water column between the ^{234}Th activity expected at secular equilibrium and that actually measured.
b. Calculate the ^{234}Th flux out of the upper ocean.
c. If the measured organic C / ^{234}Th ratio on sinking particles is 4×10^{13} mol C / mol Th, calculate the particulate organic C flux from the surface ocean.

7.3. Excess ^{230}Th determined at different depths in a deep-sea sediment core have the values listed in the following table. For this problem, assume steady state, that the density of the sediment is 1.5 g cm^{-3}, and that the seawater above this site has a constant salinity of 35. The water column at this location is 3200 m deep.

a. Calculate the sedimentation rate for this core. Make a plot of the data and compare your values to Fig. 7.13.
b. Determine the depth-integrated excess ^{230}Th activity in the sediments.
c. Calculate whether this location is receiving sediments from other locations, losing sediments to other locations, or only receiving sediments from directly above.

Depth (cm)	Excess ^{230}Th (dpm g^{-1})
0	2.07
97	1.68
205	1.10
299	1.04
407	0.84
488	0.55
597	0.45
706	0.49
803	0.35
908	0.29
1 001	0.27

7.4. (a) Calculate the mass transfer coefficient for air–sea exchange of Rn from the data in Fig. 7.15. Assume that the ocean at the time these data were collected was at steady state with respect to Rn and that the ^{222}Rn content of the atmosphere is negligible at this location.

(b) Determine the air–sea gas transfer coefficient for CO_2 at 20°C, $k_{S,660}$, from the value determined here, $k_{S,Rn}$. (Consult Chapter 2, Appendix 2A.3, and Appendix E.)

8 The Role of the Ocean in the Global Carbon Cycle

8.1 Carbon Reservoirs and Fluxes 314
Discussion Box 8.1 317
8.2 Natural Ocean Processes Controlling Atmospheric CO_2 317
8.2.1 The Solubility Pump 321
8.2.2 The Biological Pump 321
8.3 Past Changes in Atmospheric CO_2 324
8.3.1 Three-Box Ocean and Atmosphere Model 326
8.3.2 Carbonate Compensation 327
8.4 Anthropogenic Influences 329
8.4.1 Atmospheric CO_2 Observations 330
8.4.2 Anthropogenic CO_2 in the Ocean (the Revelle Factor) 331
8.4.3 The Residence Time of Dissolved Carbon with Respect to Air–Sea Gas Exchange 335
Discussion Box 8.2 336
8.4.4 Measuring Ocean Anthropogenic CO_2 Uptake 338
Discussion Box 8.3 345
8.4.5 The Oxygen Cycle: Carbon's Mirror Image 345
8.4.6 Future Challenges of the Anthropogenic Influence 350
Appendix 8A.1 The Solution to the Three-Box Ocean and Atmosphere Model 351
References 353
Problems for Chapter 8 356

Carbon is the vehicle for the collaboration between life on Earth and the climate of the planet. The Sun supplies the energy for plants to create organic carbon by linking reduced carbon atoms together with O, H, N, and minor amounts of many other elements in a vast variety of organic matter forms (see online Chapter 8 of Emerson and Hedges (2008), www.cambridge.org/emerson-hamme). Reduced organic carbon is thermodynamically unstable, and many organisms derive energy by oxidizing it to CO_2, the most stable form of inorganic carbon in an oxygen-containing world. In our atmosphere, CO_2 causes a strong greenhouse effect warming the planet, and paleoclimate observations indicate past changes in the concentration of atmospheric CO_2 have driven profound climate shifts. This chapter describes the natural processes that control the fluxes of carbon among the Earth's reservoirs and the impact of recent anthropogenic changes, with an emphasis on the role of the ocean. Burning fossil fuels to extract energy for our industrial society has impacted the natural carbon cycle to the point that the addition of anthropogenically produced CO_2 to the atmosphere is of the same magnitude as fundamental natural processes. We must understand the processes controlling the carbon cycle, because predicting future global climate and managing human influence on it will depend on knowing how anthropogenic CO_2 partitions among the atmosphere, land, and ocean.

There are three main natural carbon cycle mechanisms that respond on different timescales to the introduction of excess CO_2 to the atmosphere (Archer, 2005). The first, and shortest, of these mechanisms involves the ocean's circulation and dissolved carbonate chemistry. We shall see in the next section that there is about 50 times more inorganic carbon dissolved in the sea than CO_2 in the atmosphere, which means the ocean controls the fCO_2 of the atmosphere. A pulse of CO_2 injected into the atmosphere will be absorbed into the ocean, equilibrating with the natural dissolved carbonate system, with a timescale of ~300 y – similar to the ocean's thermohaline overturning time. While this is the fastest of the three major removal mechanisms, it is still long from the point of view of a human lifetime. The second mechanism involves the response of the marine $CaCO_3$ cycle. When anthropogenic CO_2 enters the ocean, it shifts the carbonate buffer system to lower the pH and carbonate ion concentration (see Chapter 5). This process causes deep waters bathing sedimentary calcium carbonate to become thermodynamically undersaturated, resulting in the dissolution of $CaCO_3$. The associated increase in ocean alkalinity and pH lowers atmospheric CO_2. The additional time necessary for seawater–sediment interaction makes the timescale for this feedback mechanism several thousand years, longer than simple air–sea transfer and ocean circulation (Archer, 2005). The longest feedback mechanism controlling the CO_2 content of the atmosphere involves the ultimate sources and sinks for marine dissolved inorganic carbon (DIC) – weathering on land, $CaCO_3$ formation and burial, deep-sea hydrothermal processes, and reverse weathering in ocean sediments, all discussed in detail in Chapter 2. The timescale for these processes is similar to or greater than the seawater residence time for DIC – 100 000 years (Chapter 2). While each of these three processes lowers the burden of anthropogenic CO_2 in the atmosphere, the model of Archer (2005) indicates that 10 to 15 percent of the fossil fuel CO_2 will remain in the atmosphere 10 000 years after addition of a pulse of anthropogenic CO_2. In this regard, the fossil fuel CO_2 we are currently adding to the atmosphere will be with us for a very long time, unless mitigation mechanisms, like purposeful removal of CO_2 from the atmosphere, are eventually employed.

8.1 Carbon Reservoirs and Fluxes

Earth system carbon reservoirs include the atmosphere, land biosphere and soils, the ocean, and sedimentary rocks. The amount of carbon in each of these main global reservoirs and the exchange fluxes between them determines how much each of the reservoirs depend on the others. In Table 8.1 and Fig. 8.1, we focus on the most important reservoirs and fluxes involved in the shortest term carbon cycle mechanisms described in the last section. Among the atmosphere, land, and ocean reservoirs, the atmosphere is by far the smallest

Table 8.1 Global carbon reservoirs (Pg-C, petagrams, 1×10^{15} g) and fluxes (Pg-C y^{-1})

Reservoirs (Pg-C):	
Natural	
Atmosphere:[a] CO_2 (286 ppm in 1850)*	607
(407 ppm in 2018)*	865
Oceans:[b] Dissolved Inorganic Carbon (DIC)	38 000
Dissolved Organic Carbon (DOC)	700
Biota	3
Terrestrial: Rocks,[c] carbonate (inorganic C)	6×10^7
shale (organic C)	1.5×10^7
Soils[b]	1 500–2 400
Permafrost[b]	1 700
Biota[b]	450–650
Fossil fuels (reserves)[b]	
Coal	445–540
Oil	175–265
Gas	385–1 135
Anthropogenic:	
(Cumulative changes from 1800 to 2007)[f]	
Fossil fuel burning and cement production (C_{anth})	338
Atmosphere accumulation (A)	215
Ocean uptake (O)	140
Net terrestrial change**	–17
Fluxes (Pg-C y^{-1})	
Natural:	
Atm–ocean gas exchange[b]***	60
Net Primary Production,[d] Ocean	50
Land	60
Ocean biological pump[e]	10
Ocean CO_2 outgassing from riverine carbon[g]	0.45
Sediment burial[c]	0.06

a. Global Carbon Project, Friedlingstein et al. (2019), b. Ciais et al. (2013), c. Berner (2004), d. Field et al. (1998), e. Emerson (2014), f. Gruber et al. (2019), g. Jacobson et al. (2007)
* Pg-C in the atmosphere can be converted using a factor of 2.124 Pg-C per ppm or μatm.
** The net terrestrial change is calculated from the values above (C_{anth}-A-O).
*** This value from Ciais et al. (2013) is similar to the one-way air–sea exchange flux for CO_2, $F_{A-W} = k_{S,CO_2}\,[CO_2]$, where k_{S,CO_2} and $[CO_2]$ are the global mean mass transfer coefficient and CO_2 concentration (~4 m d^{-1}, and ~10 μmol kg^{-1}, respectively).

reservoir of carbon (~600 petagrams, Pg-C, before anthropogenic changes; Pg = 10^{15} g), DIC of the ocean is by far the largest (38 000 Pg-C), and the exchangeable reservoirs in land plants and soils are somewhere in between (500–2000 Pg-C).

Exchange fluxes among the atmosphere, ocean, and land reservoirs are mainly driven by photosynthesis and respiration, which are nearly in balance except for the small fraction of

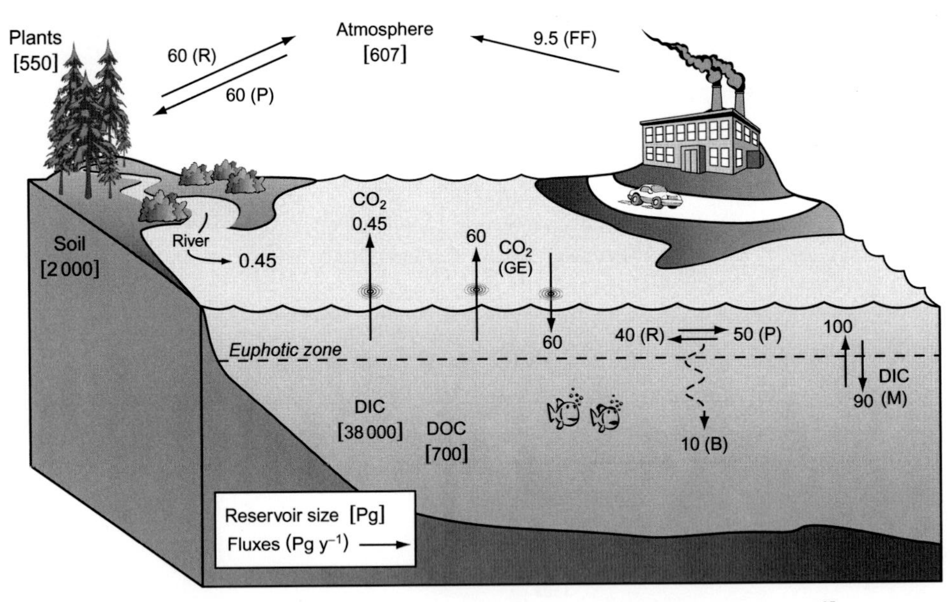

Figure 8.1 The global carbon cycle. Values in brackets are preindustrial reservoir sizes (petagrams, Pg-C = 10^{15} grams of C). Values next to the arrows are fluxes (Pg-C y^{-1}), which are preindustrial except for fossil fuel and cement production (FF – average value in 2009–2018). Symbols: P, photosynthesis rate; R, respiration rate; B, biological pump (export) flux; River, carbon delivery to the atmosphere via river fluxes to the ocean; GE, one-way gas exchange flux; M, mixing between the surface and deeper ocean. Both Plants and Soil are part of the active terrestrial biosphere. DIC: dissolved inorganic carbon; DOC: dissolved organic carbon. Modified from Emerson and Hedges (2008).

organic carbon that is buried. In the case of atmosphere–land exchange, CO_2 is taken up directly from the atmosphere via photosynthesis by plants and released to the atmosphere via respiration both above ground and in soils. In the ocean, light penetration to only about 100 meters creates a physical separation between photosynthesis and respiration. Net primary production of organic carbon in the marine euphotic zone is about 20 percent greater than respiration in the same waters, with the difference equal to the transport of organic carbon to the ocean interior by sinking particulate material and mixing of dissolved organic carbon (DOC). This is the marine "biological carbon pump" discussed in Chapter 3. While the marine biological carbon pump is only about 20 percent of net primary production in the ocean, it represents a large flux of carbon (~10 Pg-C y^{-1}) away from contact with the sunlit surface ocean and atmosphere into the deep ocean DIC reservoir, where it is sequestered for hundreds of years until mixed back to the surface by the ocean's overturning circulation.

The last major flux considered in Fig. 8.1 is CO_2 transfer between the ocean and atmosphere via air–sea gas exchange. The values presented in Table 8.1 and Fig. 8.1 are the one-way gas exchange fluxes, separately accounting for CO_2 molecules moving out of the ocean and those moving in.

A small but important carbon flux from the land through the ocean to the atmosphere is the transport of both organic and inorganic carbon via rivers to the ocean. Riverine organic carbon is oxidized to CO_2 and the inorganic carbon is precipitated as $CaCO_3$ also yielding CO_2 (Jacobson et al., 2007). This flux is estimated to have created a natural mean global fCO_2 supersaturation of about 3 μatm, driving an ocean–atm flux of ~0.45 Pg-C y^{-1}. While this value is relatively small, it becomes important when estimating the flux of anthropogenic CO_2 to the ocean based on surface ocean fCO_2 measurements, because the mean anthropogenically produced atmospheric fCO_2 excess over that in the ocean is only 8–10 μatm (see later).

The values of fluxes and reservoirs depicted in Fig. 8.1 are preindustrial estimates except for the anthropogenic fossil fuel and cement production flux to the atmosphere (FF), which was 9.5 Pg-C y^{-1} on average in 2009–2018. This anthropogenic flux perturbing our planet's carbon cycle is currently almost exactly the same as estimates for the marine biological carbon pump, indicating that it is by all measures a significant influence on the global carbon cycle. The amount of fossil fuel that has been burned between 1800 and 2007 (~338 Pg-C, Table 8.1) is of the same magnitude as known economically feasible reserves of oil, but smaller than those of gas and coal. Estimated fossil fuel reserves are uncertain, and the values given here do not include more difficult to extract reserves or estimates of undiscovered resources. The fossil fuel values indicated in Table 8.1 are of the same magnitude as the natural atmospheric carbon content, but still far smaller than the ocean's DIC reservoir. In the following sections, we demonstrate the effects ocean circulation and biological processes have on the fCO_2 of the atmosphere in a simple quantitative way and review what is known about the fate of carbon anthropogenically introduced into the atmosphere.

Discussion Box 8.1

Consider the relationship among the reservoirs and fluxes depicted in Fig. 8.1 by calculating residence times for carbon.

- Recall the definition of residence time from Section 2.1.2. What information do you need to calculate residence times and where can you find it in Fig. 8.1?
- Which inputs or outputs are most important to consider in calculating residence times in the different reservoirs and which can you ignore?
- Calculate the residence time of carbon in the atmosphere. Do the same for the land biosphere and for the ocean as a whole.
- Based on your calculations, which reservoir is easiest to perturb or change quickly?
- Which reservoir if perturbed slightly (e.g., by increasing the total amount by 5 percent) would have the greatest impact on the others?

8.2 Natural Ocean Processes Controlling Atmospheric CO_2

Air–sea gas exchange is rapid enough that atmospheric fCO_2 is forced to be within a few percent of average surface ocean fCO_2. Mechanisms that determine surface ocean fCO_2 are carbonate chemistry and the carbon pumps that create a vertical gradient in dissolved inorganic carbon (DIC) and alkalinity. These processes move carbon from the upper ~100 m of the ocean down to the ocean's interior, decreasing the amount of carbon in the surface ocean and thus in the atmosphere. This carbon transfer has been divided into the "solubility" and "biological" carbon pumps (Volk and Hoffert, 1985). The solubility pump arises because dense colder water can hold more DIC at thermodynamic equilibrium with atmospheric fCO_2. Since the deep ocean is much colder than most of the surface ocean, this process increases the DIC content of deeper waters relative to the surface. The strength of the solubility pump is measured mostly by the gradient in temperature between the surface and deep ocean. The biological pumps arise from the transport of organic carbon and $CaCO_3$ in the form of particles and dissolved organic matter (DOM) from the euphotic zone to deeper waters. The strength of the biological pump is determined by the vertical gradient of nutrients between the surface and deep ocean. The stoichiometric relationships between nutrients and carbon during biological processes determine the biologically produced vertical gradient of DIC in the ocean.

The importance of ocean temperature, marine biology, and ocean dynamics to atmospheric fCO_2 concentration is demonstrated here using the same two-layer ocean model introduced in Chapter 3 but with the addition of an atmospheric box in equilibrium with the surface ocean (Fig. 8.2). In this calculation, we focus on the ocean's natural, steady-state mechanisms that are uninfluenced by human activities. While this model suffers from approximations, it is a jewel in demonstrating the internal ocean mechanisms that are implemented in more complicated models with more boxes and in global circulation models. After introduction of the model here, we demonstrate in the next section (8.3)

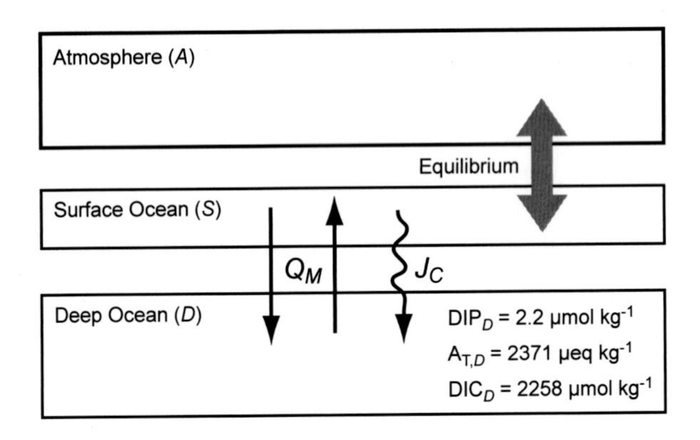

Dynamics: $V_D \frac{d[C]_D}{dt} = Q_M([C]_S - [C]_D) + J_C$

Stoichiometry: $\Delta P : \Delta N : \Delta DIC : \Delta A_T : \Delta Ca$
1 : 16 : 136 : 44 : 30

Carbonate Equilibrium (Chapter 5)

Figure 8.2 Schematic of the two-layer ocean and atmosphere model (see Chapter 3). Equations indicate circulation and biological export dynamics. Q_M is the mixing rate between the surface and deep ocean ($m^3\ y^{-1}$), and deltas indicate stoichiometry between nutrients and carbon species in the particulate matter flux, J_C (mol y^{-1}). Carbonate equilibrium among fCO_2, A_T, and DIC is presented in Chapter 5. Modified from Emerson and Hedges (2008).

how a conceptually small increase in the complexity of the model – adding an additional high-latitude box – dramatically improves the model's accuracy in representing the mechanisms controlling the ocean's carbon cycle.

The two-layer ocean and atmosphere model assumes the atmosphere, the surface ocean, and the deep ocean each have homogeneous concentrations. The well-mixed approximation for the atmosphere is not bad; measurements of atmospheric fCO_2 indicate that differences from place to place are only a few percent. Currently, the north–south hemispheric gradient is only 1–2 percent, caused primarily by greater fossil fuel usage in the north. However, the well-mixed approximation for the surface and deep ocean is extreme, because of the vast differences in temperature and nutrient concentrations across the surface ocean and the strong gradients with depth below the mixed layer. The averages used undoubtedly compromise accuracy of the results.

The model assumes that the atmosphere and surface ocean are in chemical equilibrium with respect to carbon – the fCO_2 of the atmosphere equals the fCO_2 of the surface ocean. While this is not the case in many ocean areas, global surveys indicate that the mean atmospheric fCO_2 is currently only 8–10 μatm higher than the average surface ocean fCO_2.

Since we assume that the atmosphere and ocean are in chemical equilibrium and that the input from rivers and burial in the sediments are small compared to the other fluxes (as in Chapter 3), the entire dynamics of the model is reduced to the rate of surface–deep mixing of inorganic carbon and nutrients and the transport of biological materials to the deep ocean by sinking particles and DOC fluxes (Fig. 8.2). One can see that for a steady state to be achieved, the flux of inorganic nutrients and carbon to the surface ocean must equal the transport of organic nutrients and carbon to the deep layer. The surface–deep mixing rate is quantified from the mean residence time of water in the deep ocean, ~1000 y as determined by natural ^{14}C measurements (see Section 7.2.1).

The chemical currency of the ocean model consists of three dissolved compounds – DIC and alkalinity (A_T) for carbon, and phosphate for nutrients. DIC and alkalinity are used to represent the carbon system because they are total carbon and charge quantities that are independent of temperature and pressure (i.e., they do not change concentration because of pressure differences during transport between deep and surface waters in the way their constituents, HCO_3^-, CO_3^{2-}, and CO_2, do). Biological transformations change DIC and A_T relative to nutrients in clearly defined ways (see Sections 3.1.2 and 5.3.2). Dissolved inorganic phosphate (DIP) represents nutrients in the model. We could also have chosen nitrate (NO_3^-); however, we avoid the complications of nitrogen fixation and denitrification by choosing phosphate, and biological uptake and release of both nutrients are related by a constant ratio (Fig. 3.2). Micronutrients like iron are not considered in this simplified model. For the deep concentrations of alkalinity, DIC, and DIP, we assign the mean values for the world's ocean: 2371 μeq kg^{-1}, 2258 μmol kg^{-1}, and 2.2 μmol kg^{-1}, respectively (Toggweiler and Sarmiento, 1985).

Model geometry, dynamics, and biogeochemistry are summarized in Fig. 8.2. We will use the model dynamics together with the deep ocean values to determine the concentrations of A_T and DIC in the surface ocean layer, and then we calculate the fCO_2 in thermodynamic equilibrium with these values at surface water temperature and salinity. (See Chapter 5 for a discussion of the carbonate equilibrium calculations.) Mass balance equations can be written for both deep and surface layers, but they are not independent since input to one layer is the output from the other. The change in concentration of a dissolved constituent, $[C]$, in these layers is

$$0 = V_D \frac{d[C]_D}{dt} = Q_M([C]_S - [C]_D) + J_C$$
$$0 = V_S \frac{d[C]_S}{dt} = Q_M([C]_D - [C]_S) - J_C, \tag{8.1}$$

where Q_M is the surface–deep mixing rate ($m^3\ y^{-1}$), $[C]_S$ and $[C]_D$ are the surface and deep concentrations, respectively ($mol\ m^{-3}$), J_C is the biological flux to the deep layer ($mol\ y^{-1}$), and V_S and V_D are the volumes of the surface and deep boxes, respectively (m^3). Because this is a steady-state calculation, we assume the left side of Eq. 8.1 is equal to zero. To a good first approximation, the only concentration that varies among our different calculations is $[C]_S$, because the volume of the deep ocean is so much greater than the surface that the deep concentration cannot be appreciably perturbed from the global average values. One can write three equations of this type for DIP, DIC, and A_T. These equations are related through the biological flux terms, J_C, and the stoichiometric ratios in biological material ("Redfield Ratios").

Calculations of atmospheric fCO_2 arising from various changes in the solubility and biological pumps are presented in Table 8.2. The *standard case* (Std) is determined using a weighted mean surface temperature of 20°C. This is a crude estimate as equatorial mean

Table 8.2 The effect of changes in temperature, organic matter export, ocean circulation, and carbonate compensation on the fugacity of CO_2 in the atmosphere, fCO_2^a, determined by the two-layer ocean and atmosphere model (Fig. 8.2). The DIC_S and $A_{T,S}$ are determined from Equations 8.6 and 8.7 using mean deep ocean DIP, DIC, and A_T of 2.2, 2258, and 2371 $\mu mol\ kg^{-1}$, respectively, and a C:P ratio of 106 in organic matter. The fCO_2 and $[CO_3^{2-}]$ are determined from carbonate equilibria (Chapter 5) using an ocean salinity of 35.

	Temp °C	DIP_S $\mu mol\ kg^{-1}$	τ_{mix} y	$R_{OC:Ca}$	DIC_S $\mu mol\ kg^{-1}$	$A_{T,S}$ $\mu eq\ kg^{-1}$	$[CO_3^{2-}]$ $\mu mol\ kg^{-1}$	fCO_2 μatm
Case								
(a) Std	20	0.5	1 000	3.5	2 026	2 295	190	378
(b) T effect	15	0.5	1 000	3.5	2 026	2 295	188	307
	25	0.5	1 000	3.5	2 026	2 295	192	463
(c) Bio. pump								
OM Flux	20	2.2	1 000	3.5	2 258	2 371	98	997
	20	0.0	1 000	3.5	1 958	2 273	219	295
Circulation	20	1.35	500	3.5	2 142	2 333	142	595
	20	0.0	1 300	3.5	1 958	2 273	219	295
OC:$CaCO_3$	20	0.5	1 000	10	2 060	2 362	214	340
	20	0.5	1 000	1.5	1 958	2 158	143	489
(d) CO_3^{2-} Comp*	20	0.5	1 000	3.5	2 076	2 395	226	324

* Add A_T & DIC in 2:1 ratio (100 & 50 $\mu mol\ kg^{-1}$) to surface and deep layers

values are 27°C, subtropical annual mean values are 23°C, and high-latitude annual mean values are <10°C. The surface water phosphate value also represents a weighted mean between near-zero values in subtropical surface waters and higher values at the Equator and high latitudes, where factors other than macronutrient limitation limit surface productivity rates. The standard case DIP_S value of 0.5 μmol kg^{-1} is chosen to generate an fCO_2^a value that is near that measured in the atmosphere today (Table 8.2a). Perhaps the most serious weakness of this simple model is that it does not account for geographic variability in surface nutrient concentrations. We will address this weakness in Section 8.3.1.

8.2.1 The Solubility Pump

Changes in fCO_2 brought about by solubility are caused by the temperature dependence of the carbonate system equilibrium constants. Because there is no contact between the atmosphere and deep ocean in this simple model, the temperature dependence of fCO_2 in equilibrium with surface water A_T and DIC is the only important factor. This model is a dramatic simplification of the real ocean, where water upwelled at the Equator is heated and degasses to the atmosphere as it flows north and south, while at high latitudes the same water is cooled to the point where it reabsorbs some of the CO_2 it lost to the atmosphere before it downwells to the ocean interior. Although the temperature response of CO_2 depends on all three carbonate equilibrium constants, K_H, K'_1, and K'_2, and the borate equilibrium constant, K'_B, the combined temperature dependence of CO_2 solubility trends in the same direction as that for unreactive gases; colder water can hold more dissolved carbon. As the surface ocean warms, CO_2 is outgassed, causing atmospheric fCO_2 to increase. In the two-layer model, atmospheric fCO_2 changes by about 15 μatm per degree change in surface water temperature (Table 8.2b). This result suggests that global warming has a positive feedback; higher temperatures will expel more CO_2 from the ocean, leading to further warming. Conversely, decreases in global temperature during the last ice age would have drawn down atmospheric fCO_2, transferring it to the ocean. This is in the direction observed for fCO_2 changes during glacial periods (see later), but the change in our simple model is too large, about twice that of more realistic global models (Sigman and Boyle, 2000).

8.2.2 The Biological Pump

In order to determine the sensitivity of atmospheric fCO_2 to the flux of organic material from the ocean surface, J_C in Fig. 8.2, one must first establish relationships among the chemical constituents of organic matter. While present studies suggest that changes in P, N, and C during organic matter production and respiration are geographically variable in the ocean, this is a developing field of research at the time of writing this book, and we follow the traditional values described in Chapter 3 here:

$$\Delta P : \Delta N : \Delta C_{OM} = 1:16:106. \tag{8.2}$$

For every mole of DIP added to the water by organic matter (OM) respiration, 16 moles of DIN and 106 moles of C (as DIC) are added. The oxidation of NH_3 to NO_3^- generates a proton and thus decreases alkalinity by one equivalent for each of the 16 moles of NH_3 oxidized per mol of DIP added during organic matter respiration. (See Section 5.3.2.) Thus, for every unit change in DIP by organic matter production or respiration, we can also predict the change in DIC and in alkalinity with a stoichiometry of

$$\Delta P : \Delta N : \Delta DIC : \Delta A_T = 1 : 16 : 106 : -16. \tag{8.3}$$

Calcium carbonate is the only other constituent of biologically produced matter that significantly impacts the DIC and alkalinity of seawater. Opal production and dissolution by diatoms change the concentration of dissolved silicic acid, H_4SiO_4, but such changes have only a very small effect on alkalinity because of the relatively low concentration and pK of H_4SiO_4 in seawater (Chapter 5). Observations of DIC and A_T changes in the upper 1000 m of the Atlantic demonstrate that respiration of organic carbon and dissolution of $CaCO_3$ occurs in a ratio of about 10 in the upper water column, while data following deep ocean circulation pathways demonstrate that at greater depths these processes occur in a ratio between 1 and 2 (Fig. 5.6). For the two-layer ocean, a composite value of 3.5 for the ocean below 100 meters is used (Broecker and Peng, 1982). Thus, 1 mole of $CaCO_3$ is produced or dissolves for every 3.5 moles of organic carbon produced or respired. The DIC change attributed to $CaCO_3$ for one mole of phosphorus change is thus $106/3.5 = 30$. Since production or dissolution of $CaCO_3$ creates a 2:1 change in A_T : DIC, the composite stoichiometric change for the modern ocean in our model becomes

$$\Delta P : \Delta N : \Delta DIC : \Delta A_T = 1 : 16 : 136 : 44. \tag{8.4}$$

To determine the surface ocean concentration of DIC and A_T as a function of the surface–deep gradient of DIP, one can write three versions of steady-state Eq. 8.1 for DIP, DIC, and A_T. These equations are related through the biological flux terms J_C, and the stoichiometric ratios, r, given in equations like Eq. 8.4:

$$J_{DIC} = J_P r_{DIC:P};\ J_{A_T} = J_P r_{A_T:P}. \tag{8.5}$$

By combining the phosphorus equation with those for A_T and DIC, the ratio Q_M/J_C can be eliminated to demonstrate the steady-state, surface–deep water differences for DIP, DIC, and A_T:

$$DIC_D - DIC_S = (DIP_D - DIP_S) r_{DIC:P}, \tag{8.6}$$

$$A_{T,D} - A_{T,S} = (DIP_D - DIP_S) r_{A_T:P}. \tag{8.7}$$

Assuming the mean deep-water concentrations are constant, because the deep layer is so much larger than the surface layer, one can solve for the surface water values of DIC and A_T as a function of the surface to deep differences in DIP. Once this is done, the carbonate equilibrium equations are used to determine fCO_2.

The response of atmospheric fCO_2 to the biological pump in our ocean and atmosphere model is illustrated in three ways (Table 8.2c): by changes in the organic matter flux (J_C), by changes in the mixing rate (Q_M), and by changes in the organic C to $CaCO_3$ ratio of the organic matter flux. The dissolved concentration difference between the surface and deep ocean responds to changes in the biological pump. Because of Eq. 8.1, which can be rearranged to $J_C = Q_M([C]_D - [C]_S)$, any change in the ratio of organic carbon export to surface–deep mixing, J_C/Q_M, must be accompanied by a change in the surface to deep concentration differences. In response to an increase in organic matter flux with no change in mixing or to a decrease in mixing with no change in organic matter flux, the gradient in concentrations between the deep and surface ocean must increase.

We first demonstrate the impact of a change in organic matter flux (J_C) using two extreme scenarios in which the DIP concentration in surface waters is either equal to the deep value or equal to zero (Table 8.2c "OM Flux"). The former case is the "Strangelove Ocean" in which all organisms are dead and there is no biological pump at all. This is the chemical equilibrium ocean. Something like this may have happened at the time of the last great mass extinction in geologic history. Sixty million years ago at the Cretaceous/Tertiary boundary it is believed that a meteor collided with the Earth and 80 percent of existing species became extinct. Without a biological carbon pump, the model predicts that the atmosphere would have an fCO_2 of nearly 1000 μatm. The actual value is not accurate because of the simple model architecture, but the trend toward increasing fCO_2 is clear. In the other extreme, where the organic carbon export increases until the nutrients in surface waters are entirely depleted, the maximum effect of the biological pump is realized. In our two-layer ocean atmosphere model, this causes an ~80 μatm decrease from the standard case. This large effect has inspired marine scientists and paleoclimatologists to focus on a change in the efficiency of nutrient utilization in surface waters as one mechanism for explaining past changes in atmospheric fCO_2. (See next section.)

If the organic matter export remains unchanged but the mixing rate (Q_M) is altered, the DIP gradient between surface and deep waters is forced to change. We demonstrate the impact of mixing rate changes with two scenarios that have different residence times of water in the deep ocean of 500 and 1300 y (Table 8.2c "Circulation"). If the residence time becomes smaller (i.e., circulation faster), the gradient between surface and deep concentrations becomes smaller, and atmospheric fCO_2 increases, creeping in the direction of the chemical equilibrium ocean. The opposite is true if circulation decreases. In this case with nutrients supplied to the surface at a slower rate, organic matter export consumes a greater share of the surface nutrients, and surface DIC and atmospheric fCO_2 decrease. Once the circulation decreases to a deep-water residence time of 1300 y, all the surface nutrients in the model are consumed, and the surface to deep gradient has reached its maximum value. When surface nutrients are totally depleted, any further decrease in mixing must be matched exactly by a decrease in J_C to maintain a constant Q_M/J_C ratio. Atmospheric fCO_2 is no longer sensitive to changes in the biological pump when surface nutrients are totally depleted.

Finally, we demonstrate the dependence of fCO_2 on the chemical character of the biological matter produced in the surface ocean (Table 8.2c "OC:$CaCO_3$"). If the organic carbon to $CaCO_3$ ratio of the particulate material, $R_{OC:Ca}$, becomes greater than in the standard case (less $CaCO_3$) while maintaining the same phosphorus to organic carbon ratio, there will be less alkalinity removed from the surface per mole of DIC removed. Since the A_T–DIC in surface waters is about equal to $[CO_3^{2-}]$ (Chapter 5), this difference and the carbonate ion concentration become greater under this scenario. Higher CO_3^{2-} concentrations mean more basic waters and lower fCO_2 (Fig. 5.2). Changing the particulate carbon flux ratio $R_{OC:Ca}$ from 3.5 to 10 in the model decreases the atmospheric fCO_2 by about 40 μatm.

If the particulate organic carbon to carbonate ratio instead decreased (i.e., a greater flux of $CaCO_3$), the alkalinity of surface waters would decrease relative to DIC because the formation and removal of $CaCO_3$ decreases A_T and DIC in a ratio of 2:1. A greater decrease in A_T relative to DIC would make the A_T–DIC difference smaller, decreasing the surface water CO_3^{2-} concentration and increasing fCO_2. In the model, a decrease in the $R_{OC:Ca}$ ratio from 3.5 to 1.5 causes an *increase* in fCO_2 by ~110 μatm.

These model calculations show that export of organic matter and export of $CaCO_3$ have opposite effects on the fCO_2 of the atmosphere. An organic carbon rich flux from the euphotic zone decreases atmospheric CO_2 while a $CaCO_3$ rich flux increases it. Thus, the ratio in the ocean's surface waters of coccolithophorids (the most common $CaCO_3$ secreting phytoplankton) to other non-$CaCO_3$-secreting organisms has important consequences for the fCO_2 of the atmosphere. If surface waters become more hostile to $CaCO_3$ secreting organisms because of increases in anthropogenic CO_2, as some controlled experiments on the effects of ocean acidification suggest, but the organic carbon portion of the biological pump remains the same, the ocean will respond with a negative feedback, lowering the fCO_2 of the atmosphere by slowing down the carbonate related portion of the biological pump. In some sense, the carbonate pump is a "counter pump." While it does act to lower surface ocean DIC, it changes the speciation of the carbonate species in such a way that fCO_2 is increased, rather than decreased.

Results from the two-layer ocean–atmosphere model in Fig. 8.2 and Table 8.2 demonstrate the fascinating ramifications of sea-surface temperatures and the interplay between biological fluxes and ocean circulation for maintaining atmospheric fCO_2. Projections from much more complex coupled atmosphere–ocean models suggest that a doubling of atmospheric fCO_2 due to anthropogenic emissions would result in a 2–4°C mean global temperature increase. Ocean circulation and sea-surface ecology are also sensitive to temperature, in that surface warming increases stratification, creating a barrier to vertical mixing, and organisms are finely adapted to specific temperature and nutrient regimes. How global warming will affect the biological pump and hence feedbacks to atmospheric fCO_2 is very uncertain, even the sign of the change is still unknown.

8.3 Past Changes in Atmospheric CO_2

"*The past is the key to the future*" is a phrase used by paleoclimatologists, meaning that investigating the geologic record of how the Earth system reacted to past changes will reveal clues to present and future changes. Thus, understanding the relationship between past changes in atmospheric fCO_2 and temperature may help in understanding how the Earth will respond to increased burning of fossil fuels. Over the very long timescales of the Cenozoic Period (the last ~60 My), oxygen and carbon isotopes in marine planktonic foraminifera indicate that Earth's climate cooled substantially (Zachos et al., 2008). During the early Eocene Climatic Optimum (~50–54 My ago), temperatures where up to 10°C warmer than today, and atmospheric fCO_2 was much higher – on the order of 1000 μatm (though this value is highly uncertain). The geologic record also contains several shorter "events" when carbon isotopes decreased dramatically indicating a pulse of CO_2 was released to the atmosphere. The most prominent of these is the PETM (Paleocene-Eocene Thermal Maximum) that occurred ~ 55 My ago (see Fig. 6.8 and the accompanying discussion.) Oxygen isotope changes suggest temperatures increased by ~5°C over a period of about 10 000 y at the PETM, and the accompanying carbon isotope changes require that something like 2000 Pg-C was released to the atmosphere. This massive carbon exhalation event lasted less than 200 000 y and resulted in widespread dissolution of deep-sea sedimentary carbonate (Fig. 6.8 and Zachos et al., 2008).

Throughout the Cenozoic Period, the Earth's climate became gradually cooler with time, eventually bringing on polar ice caps and the Pleistocene glacial cycles (~2 My ago till now). The accumulation of continuous polar ice has created a valuable scientific record of atmospheric gases trapped in bubbles within the ice. Climate scientists have devised accurate methods for measuring the contents of the atmospheric gas trapped in the bubbles and determining its age. A classic example is the 800 ky record of temperature (determined from the δD of ice) and atmospheric fCO_2 (determined in the trapped atmospheric bubbles) from Antarctic ice cores (Fig. 8.3; Jouzel et al., 2007; Lüthi et al., 2008). The two records reveal repeating glacial and interglacial periods when temperature and

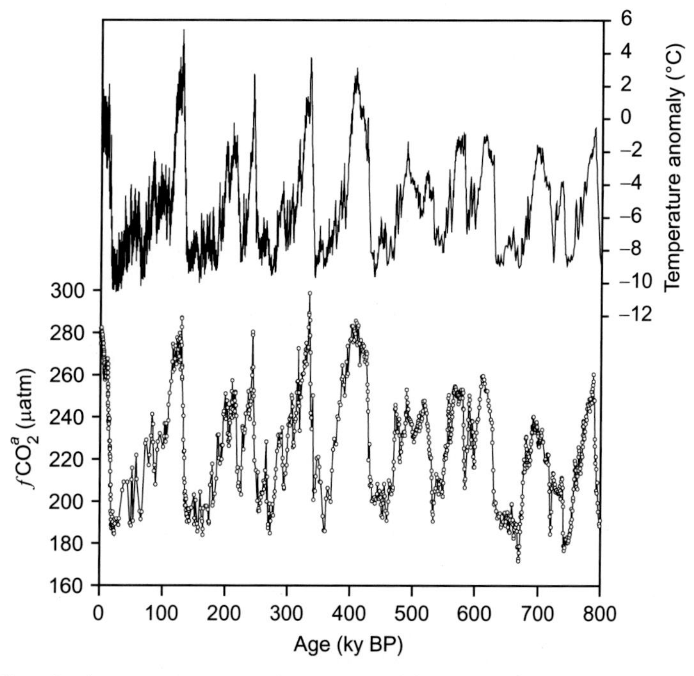

Figure 8.3 Atmospheric fCO_2 and polar temperature anomaly reconstructed from Antarctic ice cores as a function of time before the present (BP in kiloyears). The temperature anomaly is from deuterium data at Dome C relative to the mean temperature of the last millennium. The fCO_2 data are from ice cores at Dome C, Taylor Dome, and Vostok. Data are from Lüthi et al. (2008) and Jouzel et al. (2007).

atmospheric fCO_2 varied together. These records are also correlated with oxygen isotope changes in the calcium carbonate shells of foraminifera in deep sea sediments (Fig. 6.5), and with the Earth's solar insolation changes caused by idiosyncrasies in the rotation of the Earth and its orbit around the Sun. When these profound glacial–interglacial atmospheric fCO_2 cycles were first discovered, environmental scientists were challenged to explain the mechanism linking global temperature and atmospheric fCO_2 changes. It was realized almost immediately that such rapid changes in atmospheric fCO_2 had to be driven by changes in the ocean, because only its massive DIC reservoir could absorb and release enough carbon so quickly (Broecker, 1982).

8.3.1 Three-Box Ocean and Atmosphere Model

It is generally accepted that there is a positive feedback between the glacial–interglacial changes in temperature caused by solar forcing and fCO_2 changes brought about by ocean–atmosphere interaction (Sigman et al., 2010). However, the mechanism and magnitude of the feedback are still debated. An important breakthrough in the search for the mechanism by which the ocean controls atmospheric fCO_2 came with an incremental improvement to the two-layer ocean–atmosphere model described in the last section. In the early 1980s, three papers (Knox and McElroy, 1984; Sarmiento and Toggweiler, 1984; Siegenthaler and Wenk, 1984) were published almost simultaneously that showed the limitation of a model that assumes a single reservoir for the surface ocean. This model violated observations in the high-latitude ocean of elevated surface-ocean nutrient concentrations and vigorous mixing with the ocean interior. By splitting the single surface layer into high-latitude and low-latitude surface boxes, the entire biogeochemical and physical dynamics of ocean–atmosphere interaction in the model were changed to demonstrate atmospheric fCO_2 dependence on interactions among the deep ocean, high-latitude surface ocean, and atmosphere.

An example of the three-box ocean and atmosphere model (Sarmiento and Toggweiler, 1984; Toggweiler and Sarmiento, 1985) is presented in Fig. 8.4. Mass balance equations and a method to solve for atmospheric fCO_2 are presented in Appendix 8A.1. In this model, each surface box has its own biological pump, J_l and J_h. Physical mixing is represented by two different mechanisms: a unidirectional overturning circulation, T, and an additional bidirectional mixing term between the high-latitude and deep boxes simulating deep convection, f_{hd}. The phosphate concentration in the low-latitude box is set to 0.2 $\mu mol\ kg^{-1}$, which approximates observations. The dynamics of high-latitude mixing with the deep reservoir, f_{hd}, and biological pump, J_h, control the phosphate concentration in the high-latitude surface box, which in turn controls surface DIC and A_T and strongly influences atmospheric fCO_2. This dynamic is no different from that in the two-layer ocean and atmosphere model, where surface-to-deep nutrient gradients and surface temperature control atmospheric fCO_2. However, the addition of the high-latitude box provides a means for a realistic connection between the atmosphere and deep ocean, without violating the nutrient concentrations and more stratified nature of the low-latitude ocean.

Model-determined relationships between the biological pump, J_h, high-latitude – deep-mixing rates, f_{hd}, and fCO_2^a, (Fig. 8.5) indicate that fCO_2^a is strongly dependent on the

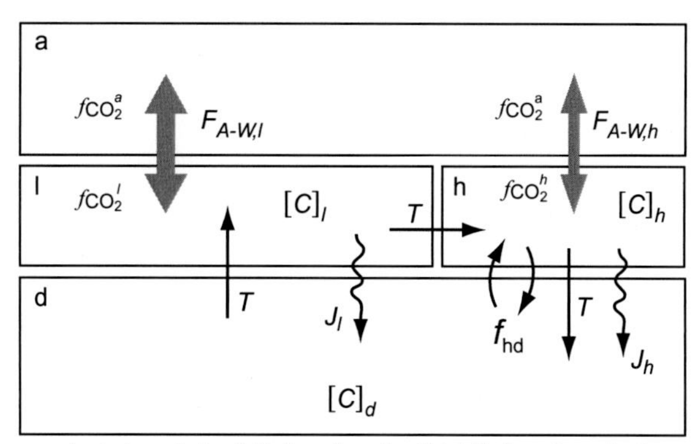

Figure 8.4 Schematic diagram of the three-box ocean and atmosphere model. C = concentration (mol m^{-3}); J = organic matter flux (mol m^{-2} y^{-1}); T = overturning circulation (m^3 y^{-1}); f_{hd} = high-latitude convection (m^3 y^{-1}); $F_{A\text{-}W}$ = air–water gas exchange flux (mol y^{-1}). Modified from Sarmiento and Toggweiler (1984).

high-latitude mixing parameter and somewhat dependent on the biological pump. Both weaker mixing and a stronger biological pump act to decrease the high-latitude concentrations of DIP (Fig. 8.5(B)), which in turn decreases the high-latitude DIC, A_T, and hence atmospheric fCO_2 (Fig. 8.5(A)). Weaker mixing but with a constant biological pump (imagine moving left along a horizontal line across Fig. 8.5) essentially creates a longer residence time for water in the high-latitude surface box, allowing more time for biology to consume surface nutrients. A stronger biological pump but with constant mixing (imagine moving up a vertical line in Fig. 8.5) draws down the available nutrient supply. Just as in the simpler two-layer ocean and atmosphere model, it is the surface-to-deep nutrient gradient, quantified here with the high-latitude phosphate concentration, that controls fCO_2^a. Glacial atmospheric fCO_2 values of 180–200 μatm (Fig. 8.3) are achievable by this model.

Both of the mechanisms that have the potential to lower fCO_2^a, more sluggish mixing between the surface and deep ocean and a stronger flux of organic matter to the ocean interior, will also result in lower oxygen concentrations in the deep ocean. In fact, the necessary conditions for decreasing atmospheric fCO_2 to the level observed in glacial periods creates anoxia in the three-box ocean and atmosphere model (Toggweiler and Sarmiento, 1985). This prediction created doubt that either of these mechanisms for creating the glacial–interglacial fCO_2 cycles were correct, because there is no evidence of anoxic bottom waters in well dated deep-sea sediment cores. However, this inconsistency is less problematic in more complex models that do not treat the entire deep ocean as a single box. More boxes allow more accurate simulation of deep ocean circulation, so that the deepest regions, which today have the highest oxygen concentrations, experience the greater proportion of the oxygen decrease and are not driven completely anoxic (e.g., Toggweiler, 1999).

8.3.2 Carbonate Compensation

If deep ocean waters were more isolated from the atmosphere during glacial periods, such that respiration decreased oxygen concentrations, increased DIC concentrations, and drove

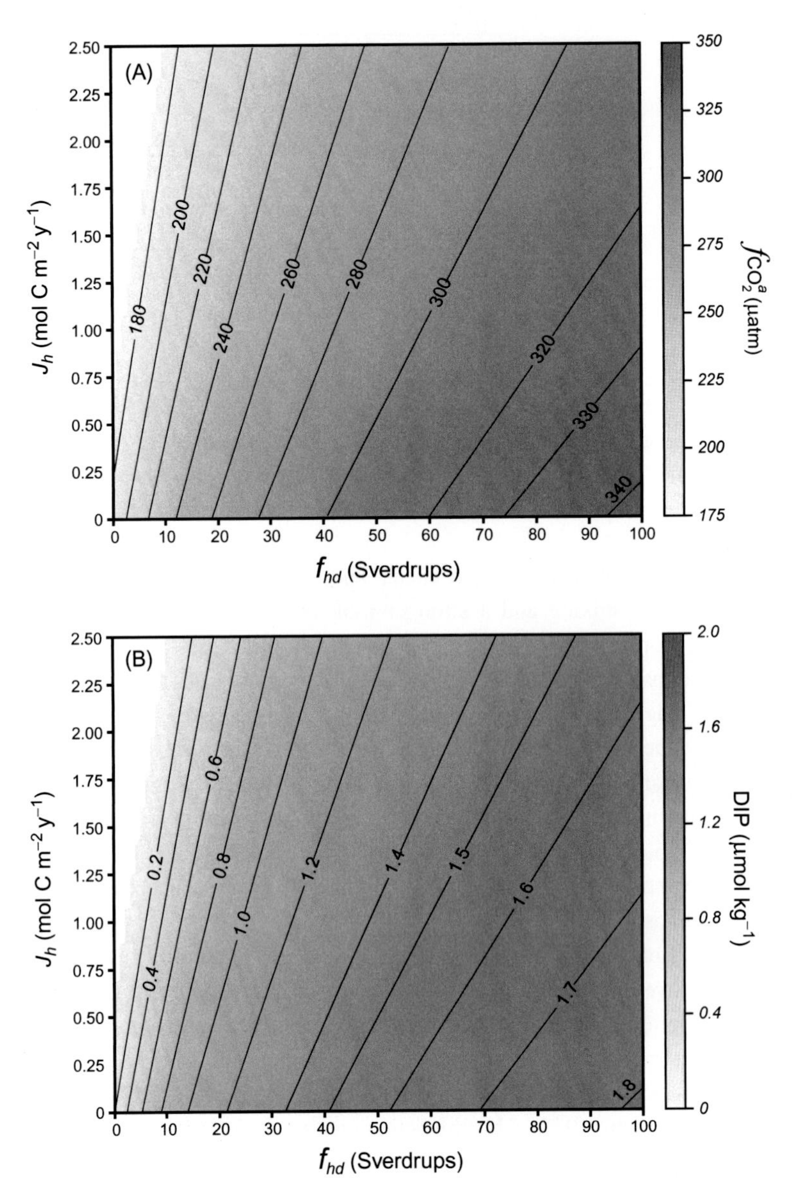

Figure 8.5 Three-box ocean and atmosphere results. (A) Dependence of atmospheric carbon dioxide concentration, fCO_2^a, on the high-latitude convection, f_{hd}, and organic matter flux, J_h. (B) Dependence of the high-latitude phosphate concentration on the same parameters. White areas in the upper left-hand corner indicate physically unrealistic model solutions with negative high-latitude phosphate concentrations.

pH lower, this would decrease the CO_3^{2-} concentration making these waters more corrosive to $CaCO_3$ – the main constituent of Atlantic deep-sea sediments. As illustrated in Figs. 5.10 and 5.11 in Chapter 5, a decrease in the $[CO_3^{2-}]$ of deep waters would expose a greater fraction of the sea floor to waters undersaturated with respect to calcium carbonate. This would cause $CaCO_3$, which was formerly stable, to dissolve and the calcite lysocline (Fig. 5.11) to shoal. When $CaCO_3$ dissolves, A_T and DIC increase in the seawater by a 2:1 ratio, which alters carbonate speciation in a way that decreases fCO_2. As the higher A_T and DIC water mixes to the ocean surface, the ocean becomes able to absorb more atmospheric CO_2. This "carbonate compensation" effect has been suggested as an important driver of glacial–interglacial fCO_2 cycles in most ocean models constructed to explain past atmospheric changes (Archer and Maier-Reimer, 1994; Boyle, 1988; Broecker, 1982; Sigman et al., 2010; Toggweiler, 1999).

The carbonate compensation effect cannot be illustrated as a feedback with a simple two- or three-layer ocean, because the dissolution effect requires bottom topography and depth gradients in the degree of saturation of $CaCO_3$ in the deep ocean. We demonstrate the effect of $CaCO_3$ dissolution on atmospheric fCO_2 by adding A_T and DIC in a 2:1 ratio to the standard case in Table 8.2d. An arbitrary addition of 100 μeq kg^{-1} of A_T and 50 μmol kg^{-1} of DIC decreases the atmospheric fCO_2 by ~55 μatm. The atmospheric fCO_2 decreases because $CaCO_3$ dissolution caused the A_T–DIC difference in surface waters to increase. The extent of the carbonate compensation that occurred in the past has been constrained by observations in deep ocean sediments. Observations support the effect (e.g., Hodell and Venz, 2003), but indicate that the lysocline during glacial ages was probably not more than 1 km shallower (Sigman and Boyle, 2000).

8.4 Anthropogenic Influences

For over a century, human activities have been increasing atmospheric CO_2 levels, but not all CO_2 emitted to the atmosphere remains there. Two seminal efforts initiated and advanced our understanding of the fate of fossil fuel CO_2 and the role of the ocean: monitoring of the fCO_2 of the atmosphere begun in 1958 by Charles David Keeling (Keeling, 1960) and studies of the CO_2 buffering capacity of the ocean (Revelle and Suess, 1957). These investigations demonstrated that only a portion of the CO_2 introduced to the atmosphere was accumulating there and that the ocean was an important sink for the rest. Subsequent studies quantified CO_2 adsorption by the ocean using a series of box and multilayer models, which were forerunners for the introduction of carbonate chemistry into present-day general circulation models (GCMs).

Even though the magnitude of the biological pump (~10 Pg y^{-1}) is similar to the anthropogenic CO_2 flux to the atmosphere (~9.5 Pg y^{-1}), there is not a clear cause-and-effect relationship between these fluxes. The rate of biological organic carbon export from the surface ocean is largely controlled by availability of nutrients and light. Since CO_2 is already abundant in marine waters, there is little direct effect of increased anthropogenic CO_2 on the rate of biological processes. In the rest of this chapter, we will focus on the most important and well-understood processes controlling anthropogenic CO_2 uptake by the ocean: thermodynamic equilibrium and physical mixing.

8.4.1 Atmospheric CO_2 Observations

The amount of anthropogenic CO_2 in the atmosphere is very well known because it has been directly measured since 1958, and ice core studies demonstrate how atmospheric CO_2 evolved between the preindustrial period and 1958 (Fig. 8.6). The amount of fossil fuel combusted in this period is also known from industrial and governmental records. From these numbers, the atmospheric CO_2 increase over the period 1860–2007 is estimated to be only about 65 percent of total emissions (Table 8.1). The only possible locations for the missing anthropogenic CO_2 are the large global carbon reservoirs that turn over on time periods of decades to centuries, which are DIC in the ocean and organic matter in the terrestrial biosphere (Table 8.1, Fig. 8.1). While the size of the terrestrial carbon reservoir has changed over time, it turns out that the ocean has been the only significant long-term sink other than the atmosphere for anthropogenic CO_2 (Khatiwala et al., 2013, our Table 8.1).

Deforestation releases carbon to the atmosphere while "greening" of the biosphere results in net storage of carbon in the terrestrial reservoir. "Greening" refers to the anthropogenic enhancement of photosynthesis by abandoning farmland to forest regrowth and by fertilization of forests by increased CO_2 concentrations and nutrient loading. Unlike marine plants, terrestrial plants can be carbon limited. The changes to the terrestrial reservoir are challenging to measure directly because this reservoir is extremely heterogeneous. Imagine how you would go about measuring the biomass in just a single forest! Because the ocean's reservoir is more homogeneous, the role of terrestrial carbon in sequestering anthropogenic CO_2 is frequently determined as the difference between total

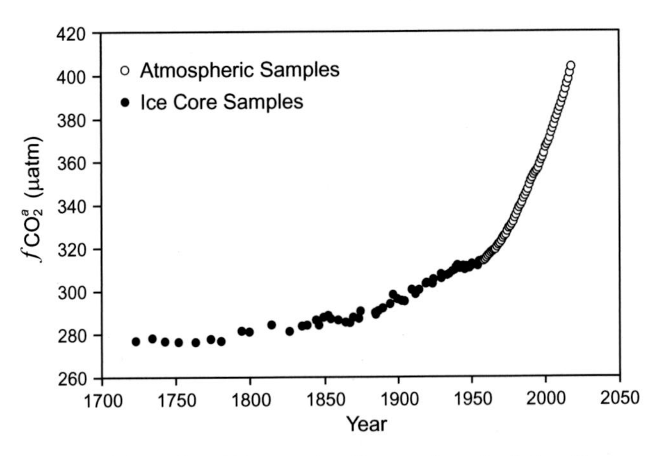

Figure 8.6 The atmospheric fCO_2 record from the 1700s to today. Data are from the Scripps CO_2 program website (scrippsco2.ucsd.edu/data). Before 1958, data are from ice core measurements (Etheridge et al., 1996; MacFarling Meure et al., 2006). After 1958, data are yearly averages of direct measurements at Mauna Loa, Hawaii, and the South Pole.

anthropogenic carbon emissions and the sum of the observed amounts in the atmosphere and ocean. We begin our discussion of the ocean's role in removal of anthropogenic CO_2 added to the atmosphere by demonstrating the theoretical chemical equilibrium response of the marine carbonate system to an atmospheric perturbation; we then discuss the methods that have been used to measure CO_2 uptake by the oceans.

8.4.2 Anthropogenic CO_2 in the Ocean (the Revelle Factor)

The first step to understanding the role of the ocean in adsorbing anthropogenic CO_2 is to determine the response of the carbonate equilibrium in the surface waters to the addition of CO_2 to the atmosphere. Since the ocean on average mixes to a depth of about 100 m annually, this part of the ocean's dissolved inorganic carbon reservoir should be nearly in chemical equilibrium with the atmosphere on timescales of one year. The response of the surface ocean carbonate system to changes in fCO_2 involves the carbonate buffer system and a number that has come to be known as the "Revelle factor."

To demonstrate the response of marine surface water DIC to an increase in atmospheric fCO_2, consider the huge 1 m × 1 m × 10, 100 m box shown in Fig. 8.7. The box contains 10,000 m^3 of air, V_{atm}, and 100 m^3 of seawater, V_{ocean}, and is maintained at a temperature of 20°C and a pressure of 1 atm at the air–water interface. The seawater in the box has a total alkalinity, A_T, of 2300 μeq kg^{-1} and a DIC concentration of 2000 μmol kg^{-1}. The atmosphere and air are at chemical equilibrium. As we learned in Chapter 5, all species

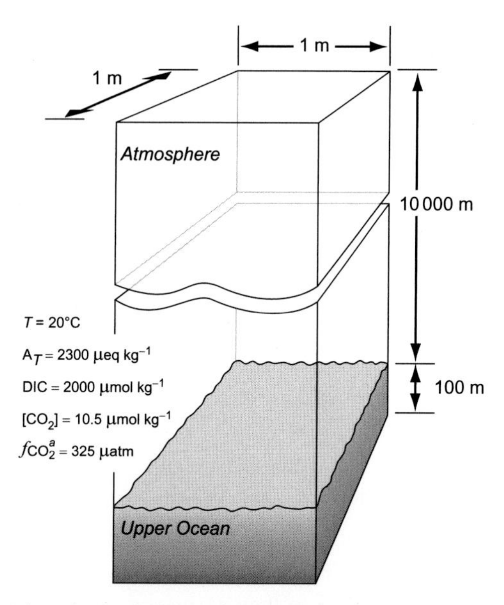

Figure 8.7 Sketch of a tall box with 10 km of air and 100 m of seawater. Air and water carbon concentrations are indicated. This hypothetical air and seawater arrangement is used to illustrate the effect of carbonate equilibrium reactions in controlling the air and sea distribution of a perturbation in atmospheric CO_2. Modified from Emerson and Hedges (2008).

of the carbonate system at chemical equilibrium can be computed if two quantities are known; the equilibrium CO_2 concentration in the water in this case is 10.5 μmol kg^{-1}. If the atmosphere and ocean are in solubility equilibrium, Henry's law can be used to calculate that the fCO_2 of the atmosphere is 325 μatm. The total inventory of carbon in the atmosphere and water in the box in moles is

$$\Sigma C = fCO_2^a \cdot M_{atm} + DIC \cdot V_{ocean} \cdot \rho, \tag{8.8}$$

where M_{atm} is the number of moles of gas in the box's atmosphere and ρ is the density of seawater (kg m^{-3}). If we assume that the pressure at the air–water interface in the box is one atmosphere (1 atm = 101 325 pascals, Pa or kg m^{-1} s^{-2}) and that the pressure at the top of the box is 0.27 atm (27 300 Pa, typical at 10 000 m elevation), we can calculate the number of moles in the box's atmosphere by dividing the difference in pressures by the acceleration due to gravity (9.8 m s^{-2}) and the molecular weight of the atmosphere (0.029 kg $mole^{-1}$) and multiplying by the surface area of the box (1 m^2) ($M_{atm} = (101\,325 - 27\,358)/(9.8 \times 0.029) = 2.6 \times 10^5$ mol atm). The total amounts of carbon in the atmosphere and seawater are

$$\Sigma C_{atm} = \left(0.325 \times 10^{-3} \text{ mol-C/mol atm}\right)\left(2.6 \times 10^5 \text{ mol atm}\right) = 85 \text{ mol-C}, \tag{8.9}$$

$$\Sigma C_{ocean} = \left(2.0 \times 10^{-3} \text{ mol-Ckg}^{-1}\right)\left(100 \text{ m}^3\right)\left(1025 \text{ kg m}^{-3}\right) = 205 \text{ mol-C} \tag{8.10}$$

and the inventory of the CO_2 gas in the water is

$$\Sigma CO_{2,ocean} = \left(10.5 \times 10^{-6} \text{ mol-Ckg}^{-1}\right)\left(10^5 \text{ L}\right)\left(1.025 \text{ kg L}^{-1}\right) = 1.08 \text{ mol-C}. \tag{8.11}$$

The amount of carbon in one square meter of seawater 100 m deep is the same order of magnitude as that in 10 km of air above, which is analogous to the situation in the real ocean mixed layer and atmosphere.

Now, imagine that we inject some CO_2 into the atmosphere of the box and ask how the added carbon distributes itself between the atmosphere and ocean once the system reestablishes a new chemical equilibrium. If CO_2 behaved like an inert gas (no chemical reactions) then the added carbon would be distributed following Henry's law and the fraction of the added carbon that would enter the seawater, $f_{SW\text{-}unreactive}$, would be the same as the CO_2 gas distribution in the original equilibrium:

$$f_{SW-unreactive} = \frac{\Sigma CO_{2,ocean}}{\Sigma C_{atm} + \Sigma CO_{2,ocean}} = \frac{1.08}{85 + 1.08} = 0.01. \tag{8.12}$$

Only about 1 percent of the CO_2 added would enter the surface ocean in such a case.

If, on the other hand, carbon were distributed evenly among the carbon species in the water and atmosphere and these species maintained the same proportions as before the addition (i.e., the added carbon divided itself among HCO_3^-, CO_3^{2-}, dissolved CO_2, and CO_2 in the atmosphere so they each increased by the same percent) then the fraction that would enter the seawater would be

$$f_{SW-distributed} = \frac{\Sigma C_{ocean}}{\Sigma C_{atm} + \Sigma C_{ocean}} = \frac{205}{85 + 205} = 0.71. \tag{8.13}$$

A much greater fraction of the added carbon (71 percent!) enters the seawater in this scenario.

In fact, neither of these estimates is correct because the fate of the CO_2 that enters the water is determined by chemical equilibrium among the carbonate species that make up DIC. Most of the CO_2 that enters the water reacts to become HCO_3^-, releasing H^+, reducing the pH of the seawater, and further altering the distribution of species. The equilibrium change in DIC for a given change in fCO_2 can be determined using the carbonate equilibrium program CO2SYS (Chapter 5) at a given temperature and A_T. This fractional value was first presented by Revelle and Suess (1957) in their classic paper on the fate of fossil fuel CO_2, and later called the Revelle factor, R_{Rev}, by Broecker and Peng (1982):

$$R_{Rev} = \frac{\Delta fCO_2 / fCO_2}{\Delta DIC / DIC}. \tag{8.14}$$

Once we know how the change in fCO_2 affects DIC through its chemical equilibrium, we can determine how much of the CO_2 added to the atmosphere enters the ocean. The change in carbon content of the atmosphere and ocean carbon reservoirs in the box in Fig. 8.7 is given by

$$\Delta\Sigma C = \Delta fCO_2^a \cdot M_{atm} + \Delta DIC \cdot V_{ocean} \cdot \rho. \tag{8.15}$$

As before, defining the fraction, $f_{SW\text{-}true}$, of the CO_2 taken up by the seawater in the box as the ratio of the changes in the seawater to those in the whole box (atmosphere plus seawater) gives

$$f_{SW-true} = \frac{\Delta DIC \cdot V_{ocean} \cdot \rho}{\Delta fCO_2^a \cdot M_{atm} + \Delta DIC \cdot V_{ocean} \cdot \rho}. \tag{8.16}$$

The Revelle factor for the conditions of our experiment and as a function of fCO_2 are presented in Fig. 8.8. For $fCO_2 = 325$ μatm, $R_{Rev} = 9.7$. Rearranging and combining Eqs. 8.14 and 8.16 gives

$$\begin{aligned} f_{SW-true} &= \frac{\Delta DIC \cdot V_{ocean} \cdot \rho}{R_{Rev} fCO_2 \left(\Delta DIC / DIC\right) \cdot M_{atm} + \Delta DIC \cdot V_{ocean} \cdot \rho} \\ &= \frac{V_{ocean} \cdot \rho}{R_{Rev} \left(fCO_2 / DIC\right) \cdot M_{atm} + V_{ocean} \cdot \rho} = 0.20. \end{aligned} \tag{8.17}$$

About 20 percent of the carbon added to the system is ultimately sequestered in the upper 100 m of the experimental ocean, a value between our first two scenarios. More CO_2 finds its way into the ocean than in the unreactive gas scenario because of the reactions of the carbonate system. The majority of the CO_2 entering the ocean from the atmosphere is consumed by reaction with carbonate ion:

$$CO_2 + CO_3^{2-} + H_2O \leftrightarrow 2HCO_3^-,$$

thereby making way for more CO_2 to enter via gas exchange than for an unreactive gas. A Revelle factor near 10 indicates that the fractional change in fCO_2 at equilibrium is

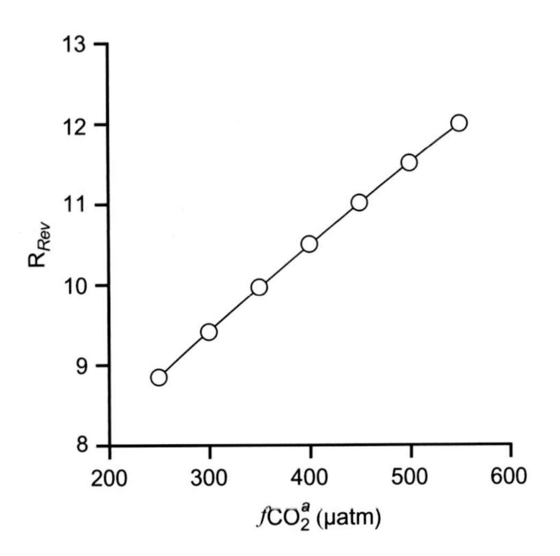

Figure 8.8 The change in the Revelle factor, R_{Rev}, as a function of fCO_2 in equilibrium with a seawater solution with a total alkalinity of 2300 μeq kg^{-1} at 20°C.

about 10 times that of the DIC. Since DIC is about 200 times greater than $[CO_2]$, this represents a carbon uptake by the water that is almost 20 times greater than if there were no chemical reactions.

Notice that in Fig. 8.8 there is a 25 percent increase in R_{Rev} as the fCO_2 increases from preanthropogenic levels of 280 μatm to 550 μatm. A higher R_{Rev} indicates an ocean that is less efficient at absorbing CO_2 from the atmosphere. One way to think about the reason for the decreased efficiency is that CO_2 uptake decreases pH and the concentration of CO_3^{2-} so that the carbonate consumption reaction above is driven less strongly to the right, resulting in more of the CO_2 that enters the ocean remaining as CO_2 rather than shifting to other carbon species.

Now that we have established our calculations for the confined box, we can expand our box to include the whole atmosphere and ocean (Appendix A). The ocean is not currently at full equilibrium with the atmosphere, mainly because vertical mixing limits the depth of the water that is in contact with the atmosphere. The equilibrium calculation will allow us to estimate the fraction of anthropogenic CO_2 taken up by the ocean as a function of the depth, h, of the layer into which anthropogenic CO_2 has equilibrated. Following exactly the reasoning for the confined box, the fraction of anthropogenic CO_2 in the ocean is

$$f_{SW-true}(h) = \frac{\Sigma C_{ocean}\left(\Delta DIC_{anth}/DIC\right)}{\Sigma C_{atm}\left(\Delta fCO^a_{2,anth}/fCO^a_2\right) + \Sigma C_{ocean}\left(\Delta DIC_{anth}/DIC\right)}, \tag{8.18}$$

where ΣC_{ocean} is the total amount of carbon in the ocean from the surface to depth h:

$$\Sigma C_{ocean} = [DIC]\rho A_{ocean}h = \left(2.0 \times 10^{-3}\ \text{mol kg}^{-1}\right)\left(1025\ \text{kg m}^{-3}\right)\left(3.62 \times 10^{14}\text{m}^2\right)h, \tag{8.19}$$

and ΣC_{atm} is the atmospheric inventory of CO_2

$$\Sigma C_{atm} = fCO_2^a \cdot M_{atm} = \left(325 \times 10^{-6}\,\text{atm}\right)\left(1.77 \times 10^{20}\,\text{mol atm}^{-1}\right) = 5.8 \times 10^{16}. \quad (8.20)$$

Using the definition of the Revelle factor and rearranging:

$$f_{SW-true}(h) = \frac{\Sigma C_{ocean}}{\Sigma C_{atm} R_{Rev} + \Sigma C_{ocean}} = \frac{7.4 \times 10^{14} h}{5.8 \times 10^{16} R_{Rev} + 7.4 \times 10^{14} h}. \quad (8.21)$$

With a mean depth of the ocean mixed layer of 100 m, the surface ocean will take up approximately 10 percent of the CO_2 added to the atmosphere, which is that calculated in the box analogy when adjusted to include the upper atmosphere and the greater area of the atmosphere compared to the ocean. This 10 percent uptake by the ocean might be viewed as a lower limit on the amount of anthropogenic CO_2 that penetrates the ocean, since it is calculated for a situation where anthropogenic CO_2 does not penetrate deeper than the mean mixed layer depth.

In fact, ocean transient tracers, like bomb-produced ^{14}C and ^{3}H, indicate that anthropogenic gases do penetrate through the mixed layer well into the thermocline (Figs. 7.19 & 7.20). In order for the ocean to take up 40 percent of the fossil fuel CO_2 (roughly what is missing from the atmosphere), the upper ~500 m of the ocean would have to be in equilibrium with the atmosphere. Performing the calculation with the ocean's full depth yields a value of >80 percent uptake of anthropogenic CO_2 by the ocean. Thus, the limiting factor for the uptake of fossil fuel CO_2 by the ocean is the rate of mixing of anthropogenically contaminated surface water into the thermocline.

8.4.3 The Residence Time of Dissolved Carbon with Respect to Air–Sea Gas Exchange

In the previous section, we assumed atmospheric fCO_2 was in equilibrium with DIC in the ocean mixed layer; however, this cannot be exactly true. The CO_2 content of the atmosphere is rising rapidly, creating a gradient in fCO_2 across the air–water interface and driving a net flux into the ocean. Seasonal temperature changes and biological processes create localized and temporary air–sea gradients, but it is the global average that is of interest here. How quickly can air–sea gas exchange bring the upper ocean back toward equilibrium with the atmosphere? The concept of the Revelle factor is also useful here.

As discussed in Appendix 2A.2, we can quantify the residence time of a substance in two ways: (1) the time necessary for an initial concentration to reach 63 percent of the way to its final equilibrium concentration or (2) more simply as its reservoir size divided by the flux in or out. If we assume for the moment that CO_2 gas is unreactive, the flux across the air–water interface will cause its mixed layer concentration to evolve toward equilibrium as

$$h\frac{dCO_2}{dt} = -k_{S,CO_2}\left([CO_2] - [CO_{2,eq}]\right), \quad (8.22)$$

where h is the mixed layer depth, k_{S,CO_2} is the gas exchange mass transfer coefficient (units length/time), and $[CO_{2,eq}]$ is the concentration of the gas at equilibrium for the temperature

Discussion Box 8.2

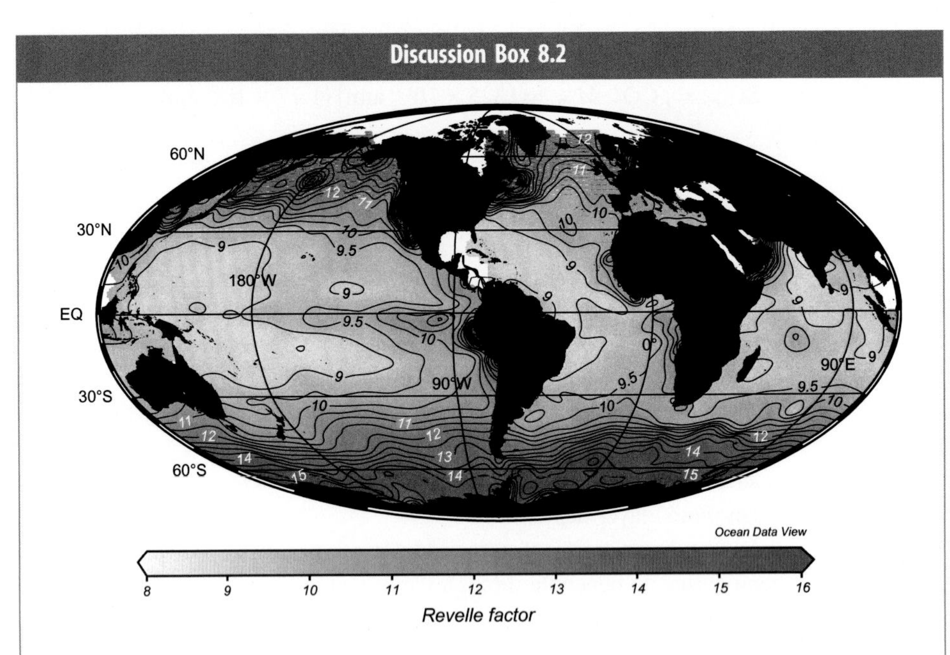

This figure shows a map of the Revelle factor at the ocean surface based on data from GLODAP (Global Ocean Data Analysis Project) v2.2020.

- Which regions of the ocean have the greatest capacity to absorb anthropogenic carbon?
- What do regions with a high Revelle factor have in common? Based on this, what are the likely causes of the high Revelle factor there?
- How does the Revelle factor relate to pH? Based on this map, where would you expect surface ocean pH to be high or low?
- Present-day Revelle factors are more than 1 unit higher than in the preindustrial ocean. Why is this?
- What does the increasing Revelle factor of the ocean imply about its future ability to continue to absorb anthropogenic carbon from the atmosphere?

and salinity of the water. Note that we could have written $[CO_2]$ as fCO_2 using the relationship $[CO_2] = K_{H,CO_2} fCO_2$ as is done in most of this chapter. Solving this differential equation yields:

$$[CO_2] - [CO_{2,eq}] = ([CO_2]_0 - [CO_{2,eq}]) \exp\left(-t \Big/ \left(\frac{h}{k_{S,CO_2}}\right)\right). \quad (8.23)$$

where $[CO_2]_0$ is the initial concentration of CO_2 in the mixed layer, and the residence time is defined as the time when the difference $([CO_2] - [CO_{2,eq}])$ has decreased to $1/e$ of its initial value, $\tau_{unreactive} = \frac{h}{k_{S,CO_2}}$. The same residence time is achieved by dividing the total amount of gas in the mixed layer by the one-way gas exchange flux:

$$\tau_{unreactive} = \frac{h[CO_2]}{k_{S,CO_2}[CO_2]} = \frac{h}{k_{S,CO_2}}. \quad (8.24)$$

We use the one-way flux because the full reservoir of gases in the mixed layer is constantly exchanging with the atmosphere, not just the fraction that is over- or undersaturated. Even in a situation where the air and ocean are exactly in saturation equilibrium, gases in each reservoir are exchanging and have a finite residence time with respect to gas exchange. Using a mean gas exchange mass transfer coefficient of 4 m d^{-1} and an ocean mixed layer depth of 100 m results in a residence time of 25 d (about one month) if CO_2 were an unreactive gas.

The true reservoir size for carbon dioxide, however, is larger than the CO_2 concentration because of the carbonate system equilibrium reactions. The flux of carbon across the air–sea interface due to the CO_2 gradient is the same as in the unreactive gas case, but the total change in carbon in the ocean necessary to reach equilibrium is larger. This situation can be written as

$$h\frac{d\mathrm{DIC}}{dt} = -k_{S,\mathrm{CO_2}}\left([\mathrm{CO_2}] - [\mathrm{CO}_{2,eq}]\right). \tag{8.25}$$

Here we use the Revelle factor (Eq. 8.14 with $[CO_2]$ substituted for fCO_2) to relate the change in DIC to that of CO_2:

$$h\frac{d\mathrm{DIC}}{dt} = \frac{h}{\mathrm{R}_{Rev}}\frac{\mathrm{DIC}}{[\mathrm{CO_2}]}\frac{d[\mathrm{CO_2}]}{dt} = -k_{S,\mathrm{CO_2}}\left([\mathrm{CO_2}] - [\mathrm{CO}_{2,eq}]\right). \tag{8.26}$$

The solution is analogous to Eq. 8.23:

$$[\mathrm{CO_2}] - [\mathrm{CO}_{2,eq}] = \left([\mathrm{CO_2}]_0 - [\mathrm{CO}_{2,eq}]\right)\exp\left(-t\Big/\left(\frac{h}{k_{S,\mathrm{CO_2}}}\frac{1}{\mathrm{R}_{Rev}}\frac{\mathrm{DIC}}{[\mathrm{CO_2}]}\right)\right) \tag{8.27}$$

with a residence time of

$$\tau_{\mathrm{CO_2}} = \frac{h}{k_{S,\mathrm{CO_2}}}\left(\frac{\mathrm{DIC}}{\mathrm{R}_{Rev}[\mathrm{CO_2}]}\right). \tag{8.28}$$

Using values of DIC, CO_2, R_{Rev}, and $h/k_{S,\mathrm{CO_2}}$ of 2000 and 10.5 μmol kg^{-1}, 9.67, and 25 d, respectively, results in a residence time of about 500 days. The residence time for CO_2 is actually more than one year for a 100 m deep mixed layer! This indicates that the surface ocean anthropogenic increase will lag that in the atmosphere, but not by much compared to the decadal to centuries-long anthropogenic perturbation.

Calculating the residence time for isotopes of carbon, ^{13}C and ^{14}C, is different again because isotopes must exchange with the entire carbon reservoir. Using the notation in which r is the carbon isotope ratio, $^{13}C/^{12}C$ or $^{14}C/^{12}C$, we can write the mass balance for carbon isotopes as

$$h\frac{d(r\cdot\mathrm{DIC})}{dt} = -k_{S,\mathrm{CO_2}}\left(r[\mathrm{CO_2}] - r_{eq}[\mathrm{CO}_{2,eq}]\right), \tag{8.29}$$

where r_{eq} is the isotope ratio in equilibrium with the atmosphere. If the total carbon is already in equilibrium but the isotopes are not, $[\mathrm{CO_2}] = [\mathrm{CO}_{2,eq}]$, and both $[CO_2]$ and DIC are constants. Eq. 8.29 then simplifies to

$$h \cdot \mathrm{DIC}\frac{dr}{dt} = -k_{S,\mathrm{CO_2}}[\mathrm{CO_2}](r - r_{eq}) \tag{8.30}$$

with a solution of

$$r - r_{eq} = (r_0 - r_{eq}) \exp\left(-t \Big/ \left(\frac{h}{k_{S,\mathrm{CO_2}}} \frac{\mathrm{DIC}}{[\mathrm{CO_2}]}\right)\right). \tag{8.31}$$

The residence time in this case is

$$\tau_{isotope} = \frac{h[\mathrm{DIC}]}{k_{S,\mathrm{CO_2}}[\mathrm{CO_2}]} = 25\,\frac{2000}{10.5} = 4762 \text{ d.} \tag{8.32}$$

The residence time for the upper ocean with respect to changing its isotopic values is about 10 times longer than that for CO_2 and almost 200 times that for an unreactive gas: $\tau_{isotope} \sim 13$ y!

8.4.4 Measuring Ocean Anthropogenic CO_2 Uptake

The amount of anthropogenic CO_2 already absorbed by the ocean has been determined primarily by four main measurement strategies. As with all geochemical fluxes where no standard exists, accuracy can only be assessed by checking whether different data-based methods agree. Global circulation models are useful for elucidating mechanisms, but they offer little in the effort to quantify anthropogenic CO_2 uptake without constraints by data.

The most widely used method for determining the ocean's uptake of anthropogenic CO_2 is to separate the fractions of measured DIC that come from natural and anthropogenic processes. As described below, this is accomplished by calculating the expected oceanic distributions of natural DIC from carbon, oxygen, and alkalinity data with the assumption of constant metabolic ("Redfield") ratios. The difference between the calculated natural value and that measured is the anthropogenic perturbation. The challenge for this method is that the anthropogenic perturbation is a small fraction of the total measured DIC. A second method involves calculating the flux of CO_2 across the ocean–atmosphere interface from global observations of the fCO_2 in surface waters. This calculation involves compiling millions of global measurements of the CO_2 fugacity and wind speed over the past couple of decades and normalizing them to a single time period. The result of this calculation is a flux rather than a total inventory, which provides valuable information about locations and mechanisms of anthropogenic CO_2 uptake. The third method uses isotopic measurements. Fossil fuel CO_2 contains less ^{13}C compared with ^{12}C (see Chapter 6) than carbon found naturally in the ocean and atmosphere. Measurements of the isotopic composition of carbon in seawater and the atmosphere can be used to trace the fraction of the DIC that has been anthropogenically introduced (Quay et al., 1992, 2003). The fourth method involves using simultaneous measurements of atmospheric fCO_2 and O_2 to infer the fate of anthropogenic CO_2 – whether it stays in the atmosphere or is taken up by the ocean or terrestrial biosphere. The first two of the oceanic methods are described in the following discussion while the atmospheric CO_2 and O_2 mass balance procedure is outlined in the next section on the oxygen cycle.

The Anthropogenic Component of Measured DIC. Methods for distinguishing the fraction of measured DIC that originates from natural versus anthropogenic processes were first applied by Brewer (1978) and Chen and Millero (1979) in their simplest form, and then improved by Gruber et al. (1996), and applied to the whole ocean by Sabine et al. (2004) using data normalized to the year 1994. Most recently, Gruber et al. (2019) used an extension of this method to determine the change in anthropogenic CO_2 between the inventory determined by Sabine et al. in 1994 and that in 2007. Here, we describe just the fundamental concepts of the method along with the main results.

The interior ocean is ventilated with atmospheric gases by the transport and mixing of surface waters along layers of constant density (isopycnal surfaces) into the upper thermocline (recall Fig. 4.5). The DIC measured at any depth, x, along one of these surfaces is equal to the natural (N) value that the water would have had preindustrially plus that added from the atmosphere by anthropogenic processes (*anth*) when the water was in contact with the surface:

$$\mathrm{DIC}_{meas,x} = \mathrm{DIC}_{N,x} + \mathrm{DIC}_{anth,x}. \tag{8.33}$$

We would expect the anthropogenic amount to be the highest near the surface outcrop, since it has been recently in contact with a higher fCO_2 atmosphere, and to decrease moving toward deeper, older parts of the isopycnal. The anthropogenic component is determined by subtracting the calculated natural value ($\mathrm{DIC}_{N,x}$) from the measured value ($\mathrm{DIC}_{meas,x}$). The natural DIC concentration along the isopycnal comes from three sources: (1) DIC that the water would have had when last at the surface during preindustrial times ($\mathrm{DIC}_{N,S}$), (2) DIC added to the water by organic matter respiration, and (3) DIC added to the water by calcium carbonate dissolution. The last two can be calculated from AOU and total alkalinity ($\mathrm{A_T}$) changes:

$$\mathrm{DIC}_{N,x} = \mathrm{DIC}_{N,S} + \frac{106}{153}\mathrm{AOU}_x + \frac{1}{2}\left(\mathrm{A}_{\mathrm{T},x} - \mathrm{A}_{\mathrm{T},S} + \frac{16}{153}\mathrm{AOU}_x\right), \tag{8.34}$$

where the subscript S denotes the value at the surface, 106/153 is the ratio of DIC released to oxygen consumed during organic matter respiration that we adopt in this text, and 16/153 is the ratio of nitrate released to oxygen consumed during organic matter respiration. The value in parentheses in the last term of Eq. 8.34 is the alkalinity change due to calcium carbonate dissolution. As we learned in Chapter 5, calcium carbonate dissolution increases alkalinity while organic matter respiration slightly reduces alkalinity due to the release of a hydrogen ion during nitrification. The adjustment by $16/153\times\mathrm{AOU}$ converts the measured alkalinity change between the surface outcrop ($\mathrm{A}_{\mathrm{T},S}$) and the location of the measurement ($\mathrm{A}_{\mathrm{T},x}$) to the alkalinity change expected from calcium carbonate dissolution alone.

To use this equation, one has to have a scheme for evaluating the surface ocean alkalinity, $\mathrm{A}_{\mathrm{T},S}$, and the surface ocean DIC before anthropogenic contamination, $\mathrm{DIC}_{N,S}$. We call these preformed values, similar to the concept of preformed nutrients being the nutrient concentration a water mass last had at the surface (Section 4.1.2). This is the most conceptually challenging aspect of the method because preindustrial surface values are uncertain. We present the basic idea for reconstructing preformed values here, and refer the reader to Gruber et al. (1996) for more details. Traditionally, preformed alkalinity values were estimated from present surface values at the appropriate density or by the salinity at depth x times the present surface A_T:salinity relationship. Preformed DIC could then be estimated as the value of DIC at chemical equilibrium with $fCO_2 = 280$ μatm (the preindustrial value) and the preformed alkalinity at the temperature and salinity of depth x. The problem with such a determination is that we know that surface fCO_2 in different regions of today's ocean is not uniformly at equilibrium, and this was undoubtedly true in the past. Gruber et al. (1996) developed ways to address the preformed value uncertainties. One of these methods is to find a location along the isopycnal surface that is far enough from the ocean surface to be uncontaminated by anthropogenic CO_2. At this location, preformed DIC can be determined using Eq. 8.34 since $DIC_{meas,x} = DIC_{N,x}$. This value of $DIC_{N,S}$ can then be used to determine DIC_{anth} on shallower regions of the isopycnal. Other methods are more involved and require estimating the water mass age. Comparison of the improved methods with the early versions of Brewer (1978) and Chen and Millero (1979) indicates that the early procedures captured the general trends, but the improved methods of determining preformed A_T and DIC have clearly improved accuracy.

Sabine et al. (2004) applied these methods to global survey data determined during the 1980s and early 1990s from the World Ocean Circulation Experiment (WOCE) and Joint Global Ocean Flux Study (JGOFS). Their data analysis revealed that 118 ± 18 Pg-C of anthropogenic CO_2 had entered the ocean from the beginning of the industrial age until 1994, about 30 percent of the total anthropogenic emissions before 1994. Most recently Gruber et al. (2019) built on this analysis using data in the Global Ocean Data Analysis Project version 2 (GLODAP v2) to provide a value from 1994 to 2007. These results (Table 8.3, Ocean DIC column) indicate that a similar fraction (~30 percent) of the total anthropogenic emissions during this 13-year period was taken up by the ocean.

The global vertical distribution of anthropogenic carbon (summing the pre-1994 and the 1994–2007 estimates) has the characteristics of any unreactive tracer that originates in the atmosphere (Fig. 8.9). The highest values are found in the surface layer, which is in contact with the present-day atmosphere. Values here are what would be expected based on the rapid increase of atmospheric CO_2 and the Revelle factor. Concentrations decrease rapidly with depth, reaching half the surface value by 375 m and one-tenth of the surface value at 1000 m. The distribution of anthropogenic carbon is controlled by the circulation of intermediate and deep waters, so that the timescale of ocean uptake of anthropogenic carbon is set by that of ocean circulation. The bowls centered on 40°N and 40°S are the location of the subtropical gyres underlain by colder subarctic waters. Water mass ages are fairly young everywhere on these isopycnals, so that significant uptake and storage have occurred throughout to depths between 500 and 1000 m, particularly in the Antarctic Intermediate Water from its outcrop region in the Southern Ocean. The deepest penetration is found in the North Atlantic to depths below 2500 m. This is the beginning of the North Atlantic branch of the thermohaline circulation. Most of the rest of the deep ocean shows no anthropogenic carbon, because these waters have not been at the surface since before the industrial age.

Table 8.3 Global anthropogenic carbon flux balance over three separate time intervals. The flux balance equates emissions to the atmosphere, E, with three separate sinks, S:

$$E_{ff} + E_{luc} = S_{atm} + S_{oce} + S_{lr} \qquad (\text{Pg-C y}^{-1})$$

where subscripts are as follows: $_{ff}$ = fossil fuel and cement production; $_{luc}$ = land use change; $_{atm}$ = the amount that remains in the atmosphere; $_{oce}$ = the amount that enters the ocean; and $_{lr}$ = the amount taken up by land regrowth. The ocean sink, S_{oce}, is determined by three separate methods described in the footnotes and main text. Values in parentheses indicate what percent of E_{ff} is in the atmosphere, ocean, and land carbon reservoirs.

Method	Ocean DIC	Oce. Surf. fCO_2	Atm. O_2/N_2 & fCO_2
Period	1994–2007	1998–2011	1991–2011
Flux (Pg-C y^{-1})			
E_{ff}^{*}	7.1 ± 0.4	7.8 ± 0.4	7.4 ± 0.4
S_{atm}^{*}	4.0 ± 0.2 (56)	4.1 ± 0.2 (52)	3.7 ± 0.2 (50)
S_{oce}	2.2 ± 0.4^{a} (31)	2.0 ± 0.6^{b} (26)	2.5 ± 0.6^{c} (34)
S_{lr}-E_{luc}^{**}	0.9 ± 0.6 (13)	1.7 ± 0.7 (22)	1.2 ± 0.7 (16)
E_{luc}^{*}	1.4 ± 0.7	1.4 ± 0.7	1.4 ± 0.7
S_{lr}^{***}	2.3 ± 0.9	3.1 ± 1.0	2.6 ± 1.0

[a] Gruber et al. (2019). MLR and C^{*} methods yield an ocean uptake of 2.6 Pg-C y^{-1}. A loss of natural CO_2 by the ocean of 0.4 Pg-C y^{-1} has been subtracted.
[b] Landschützer et al. (2014). The measured global fCO_2 flux is 1.4 Pg-C y^{-1}. A natural outgassing of 0.45 Pg-C y^{-1} is assumed along with an estimated Arctic flux of 0.1 Pg-C y^{-1}.
[c] Keeling and Manning (2014). From the measured annual atmospheric O_2/N_2 and fCO_2 change due to anthropogenic activities.
* From the Global Carbon Budget (Friedlingstein et al., 2019)
** $(S_{lr} - E_{luc}) = E_{ff} - S_{atm} - S_{oce}$
*** $S_{lr} = (S_{lr} - E_{luc}) + E_{luc}$

The flux of anthropogenic CO_2 to the ocean necessary to accommodate the 13-year buildup quantified by Gruber et al. (2019) is about 2.2 Pg-C y^{-1}. We compare this value with results of the other two methods described later in the row titled S_{oce} in Table 8.3. Results from all three methods agree within the estimated uncertainties, indicating that the ocean sink is known to within these error estimates.

The Air–Sea CO_2 Flux. The second method of determining the uptake of anthropogenic CO_2 by the ocean is accomplished by calculating the flux across the air–water interface using observations of surface ocean fCO_2. As described in Chapter 2, the air–sea gas exchange flux can be determined from the fCO_2 difference between the atmosphere and ocean times the gas exchange mass transfer coefficient, k_{S,CO_2} (m y^{-1}), and the Henry's law coefficient, K_{H,CO_2} (mol kg^{-1} atm^{-1}):

$$F_{CO_2} = k_{s,CO_2} K_{H,CO_2} \left(fCO_2^{a} - fCO_2^{SW} \right). \qquad (8.35)$$

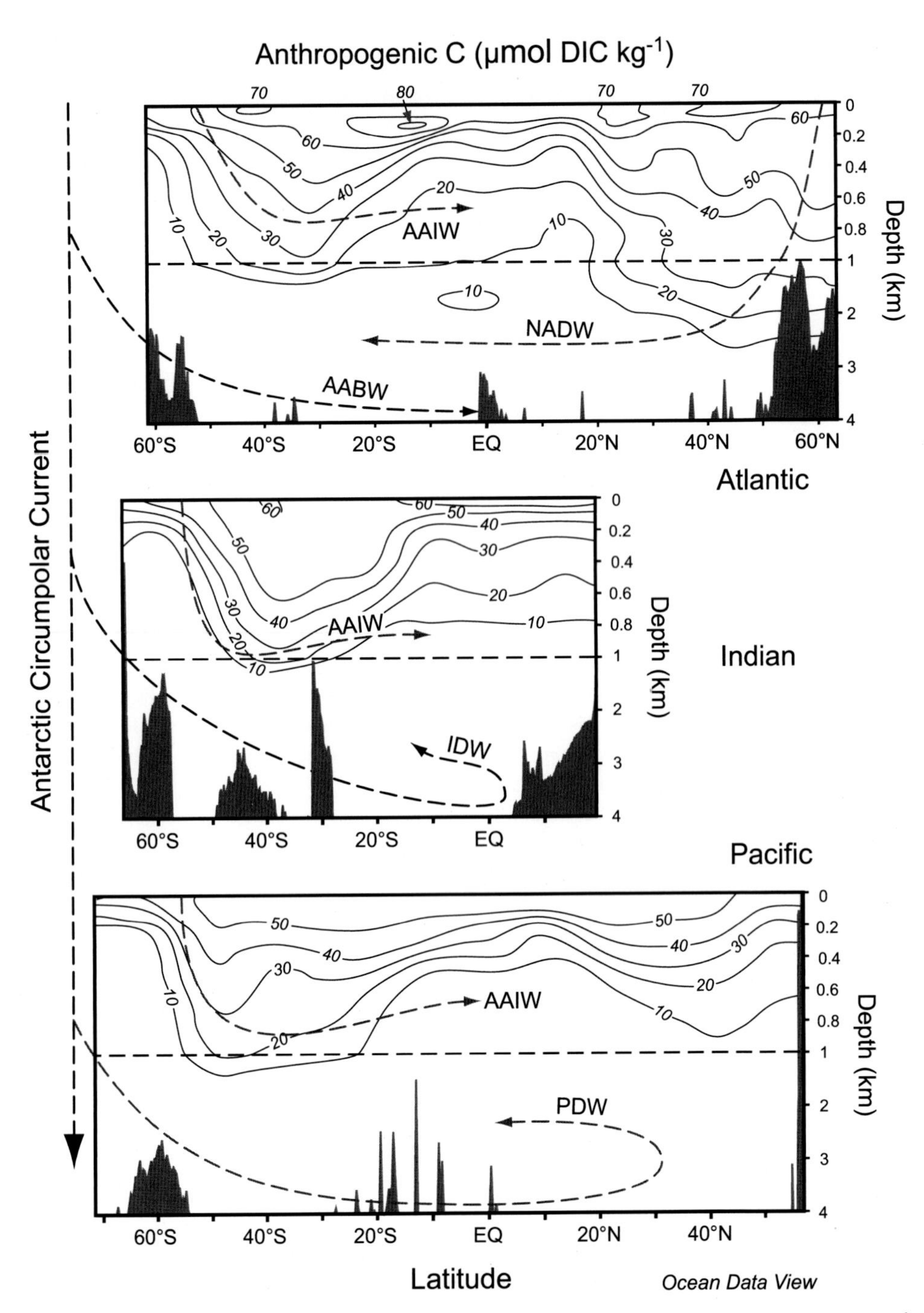

Figure 8.9 Vertical sections of the concentration of anthropogenic carbon in the ocean as of 2007. The contours are μmol kg^{-1} anthropogenic carbon in the measured DIC. These estimates are determined by a combination of multilinear regression (MLR) and DIC_{anth} techniques. Values are the sum of data from the preindustrial period to 1994 (Sabine et al., 2004, as gridded by Key et al., 2004) and the period 1994 to 2007 (Gruber et al., 2019). A black and white version of this figure will appear in some formats. For the color version, refer to the plate section.

The mass transfer coefficient is evaluated using a wind speed relationship, such as Wanninkhof (1992, Fig. 2A.3.3), and high-resolution wind speeds from satellite observations or weather models, such as the Cross-Calibrated Multi-Platform (CCMP) database.

This effort and analysis was initiated by Taro Takahashi et al. (1999), who spearheaded efforts to collect surface ocean fCO_2 measurements from research and cargo vessels, and regularly updated through 2009 (e.g., Takahashi et al. 2009). The most recent interpretation at the time of writing this book is by Landschützer et al. (2014), who determined the mean annual, seasonal, and interannual air–sea CO_2 flux in the years 1998 to 2011 using 10 million surface ocean fCO_2 observations in the Surface Ocean CO_2 Atlas version 2 (SOCAT v2) database (Bakker et al., 2014). Because the observations are sparse, not collected at all places at all times, Landschützer et al. (2014) used a neural network approach to extrapolate the SOCAT data in space and time to create monthly maps at a spatial resolution of 1° latitude × 1° longitude (Fig. 8.10).

The advantage of the air–sea CO_2 flux approach is that it yields detailed information about the geographic distribution of ocean uptake and release of CO_2, both natural and anthropogenic. The disadvantages are that fCO_2 surface ocean measurements have not been made in all locations, seasons, and years; that the wind speed-mass transfer coefficient relationships have significant error; and that different wind speed products give variable results. The map of mean air–sea CO_2 flux during the 1998–2011 period (Fig. 8.10B) indicates that the most important regions of flux from the ocean to the atmosphere are at the Equator, especially the eastern equatorial Pacific and the northwestern Indian Ocean. These are regions of upwelling (see Section 1.3.2) of deeper waters that are enriched in CO_2. The subtropical oceans are regions of very low net flux because surface fCO_2 is nearly in equilibrium with the atmosphere in these regions. Carbon dioxide strongly invades the ocean poleward of the subtropics in both the North and South Pacific and in the North Atlantic. Cooling of the Kuroshio and Gulf Stream currents in the Northern Hemisphere as they move north plays an important role in carbon dioxide uptake at the subtropical-subarctic boundary. Carbon dioxide becomes undersaturated as these grand ocean currents move northward and across the oceans. Carbon dioxide undersaturation caused by cooling combines with CO_2 undersaturation caused by high biological productivity to create strong fCO_2 drawdown in these regions. Regional differences in surface ocean fCO_2 are as large as 100 μatm (Fig. 8.10A), whereas the mean degree of undersaturation required to maintain the measured global anthropogenic flux of CO_2 to the ocean is only 8–10 μatm – a few percent of the mean surface value.

The air–sea CO_2 flux method yields an ocean uptake of 1.4 Pg-C y^{-1} (Table 8.3, "Oce. Surf. fCO_2" column). However, this value is not the same as the anthropogenic CO_2 air–sea flux, because the ocean was a natural source of CO_2 to the atmosphere in the preindustrial era. Rivers supply both inorganic and organic carbon to the ocean, which through creation of $CaCO_3$ and respiration of organic carbon results in a natural net flux of CO_2 to the atmosphere of about 0.45 Pg-C y^{-1}. Thus, the ocean uptake of CO_2 determined from the fCO_2 surface ocean data must be corrected for this riverine-carbon-derived outgassing to arrive at the anthropogenic CO_2 uptake. Adding the riverine value plus an estimated flux to the Arctic ocean of 0.1 Pg-C y^{-1} yields a net invasion of anthropogenic CO_2 of 2.0 Pg-C y^{-1} (Landschützer et al., 2014). The air–sea flux and DIC_{anth} inventory methods yield similar estimates of the flux of anthropogenic CO_2 to the ocean. We shall see that both are similar to the value resulting from atmospheric measurements discussed in the next section (Table 8.3).

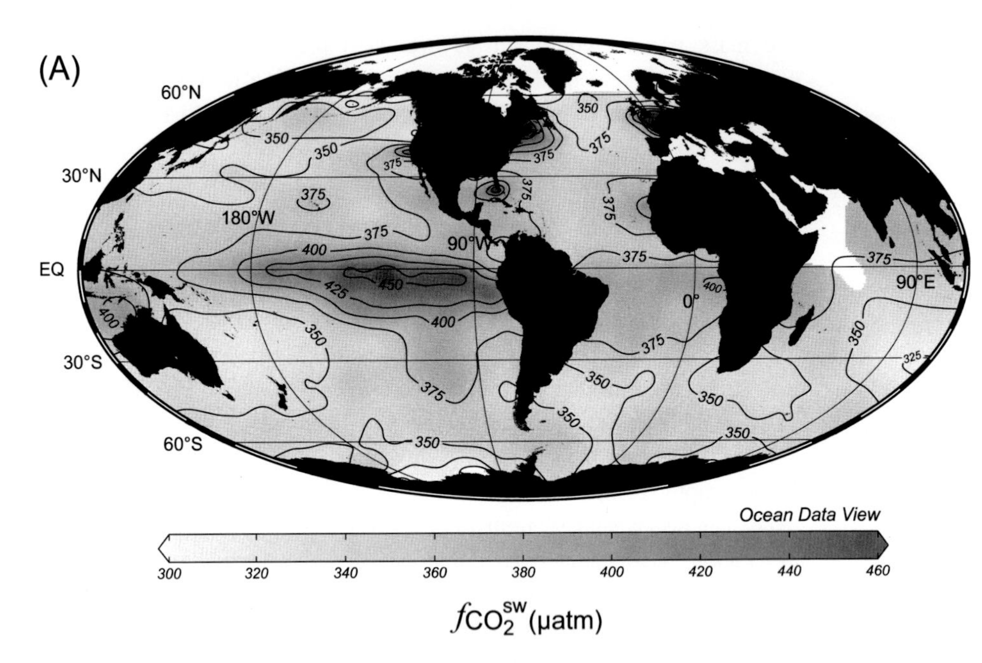

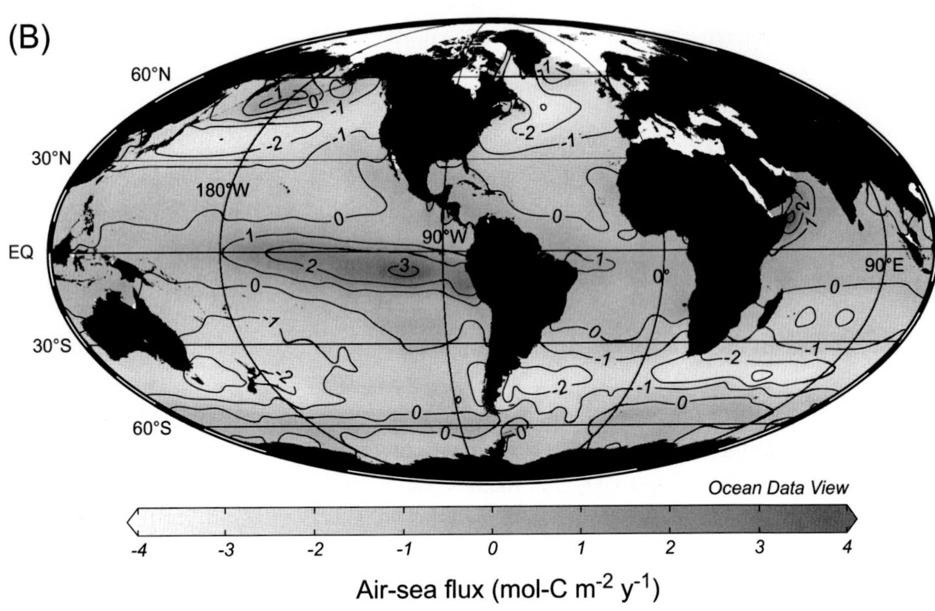

Figure 8.10 The global distribution of surface ocean fCO_2 (µatm) (A) and the air–sea CO_2 flux (mol-C m^{-2} y^{-1}, where a positive flux is to the atmosphere) (B). Values in (A) are the spatial average of all available surface ocean fCO_2 data measured in the 2000–2010 decade and archived in the Surface Ocean CO_2 Atlas, SOCAT, www.socat.info/. Fluxes in (B) are determined from the SOCAT database available in 2020 (Landschützer, personal communication) using the neural network method described in Landschützer et al. (2014). A black and white version of this figure will appear in some formats. For the color version, refer to the plate section.

Discussion Box 8.3

The air–sea CO_2 flux estimate in Fig. 8.10B is derived from surface ocean fCO_2 data like those in Fig. 8.10A, but the patterns show distinct differences.

- In which regions do you see the largest differences? What do these regions have in common?
- What might explain the particularly high values in some small, confined areas?
- Which ocean regions are likely to have the highest data density (number of observations in one location)? Which the lowest? How will such variations affect attempts to estimate a global flux?
- Neural network extrapolation schemes work by detecting regionally specific, non-linear relationships between surface ocean fCO_2 and other factors such as sea surface temperature, salinity, and mixed layer depth. For each of these factors individually, how would you expect surface ocean fCO_2 to be related to them?
- Why would such an extrapolation scheme likely yield more accurate results than simple spatial averaging and interpolation?
- The SOCAT database contains observations from 1957 to the present. How is the changing atmospheric fCO_2 (e.g., in Fig. 8.6) likely affecting surface ocean fCO_2 values? What challenges does that create in estimating an air–sea CO_2 flux over the measurement period?

8.4.5 The Oxygen Cycle: Carbon's Mirror Image

Because carbon and oxygen are the main elements oxidized and reduced during photosynthesis and respiration (with an approximately four-electron transfer between the oxidized and reduced forms, see Section 3.1.1), there must be a near 1:1 $\Delta C:\Delta O_2$ ratio to maintain electron balance. The exact ratio depends on the oxidation state of carbon and the other elements in organic matter (Table 3.1). Because of this relationship, O_2 fluxes can be used as a tracer of organic carbon fluxes. In previous chapters and here, we discuss three timescales on which oxygen is used as a tracer for carbon fluxes. The first and shortest timescale application (annual) uses the oxygen flux out of the upper ocean to quantify annual net community production (ANCP) and the biological pump (Sections 3.2.3 and 3.2.4). The second application involves using the atmospheric mass balance of O_2 and of CO_2 to quantify the fate of anthropogenically produced carbon dioxide (decadal). The third application employs O_2/N_2 measurements in atmospheric gas bubbles trapped in ice cores to reveal changes in organic matter burial and terrestrial weathering on a million-year timescale. We discuss the last two applications in the following sections of this chapter.

Atmospheric O_2/N_2 Ratios as a Tracer of Anthropogenically Produced CO_2. The combustion of fossil fuels both produces CO_2 and consumes atmospheric O_2. While the CO_2 increase (today ~2.4 μatm y^{-1} or ~0.6% y^{-1}) has been readily measurable with great accuracy for at least 60 years, the accompanying oxygen decrease is only a tiny fraction of the total O_2 in the atmosphere (0.20946 atm or 209 460 μatm). The observed decrease in atmospheric O_2 is today ~5.4 μatm y^{-1} or ~0.0026% y^{-1}, a much more analytically challenging measurement. Ralph Keeling (Keeling, 1988) developed the first method to measure O_2/N_2 ratios accurately enough to determine annual and seasonal changes in O_2 in the atmosphere using interferometry. Later, Michael Bender accomplished this measurement with a ratio mass spectrometer (Bender et al., 1994), and now there are several other lower-cost and more portable methods.

Simultaneous determination of both the fCO_2 increase and fO_2 decrease make it possible to identify the difference between the ocean and terrestrial uptake of anthropogenic CO_2 (Keeling et al., 1996). At first, these two tracers may seem redundant because there is an exact stoichiometry between CO_2 release and O_2 consumption during fossil fuel combustion and during photosynthesis/respiration. The reason they are not redundant is that the sinks for CO_2 and O_2 are different. Both the land and ocean are potential sinks (or sources) for atmospheric CO_2 in response to anthropogenic perturbation; however, the ocean does not significantly exchange oxygen in response to the anthropogenic decrease in atmospheric O_2. Because O_2 is a very insoluble gas, about 99 percent of all O_2 in the ocean–atmosphere system is in the atmosphere (Chapter 2). Thus, when atmospheric O_2 decreases due to fossil fuel burning, there is only a tiny release of O_2 from the ocean to make up the deficit in the atmosphere. Any measurable change in the atmospheric O_2 concentration other than that due to fossil fuel combustion must be attributed to exchange with the terrestrial biosphere.

Calculation of the importance of the atmosphere, ocean, and terrestrial biosphere sinks for anthropogenic carbon was elegantly presented in graphical form once there were enough data to determine a trend (Keeling et al., 1996), and later this method was updated to the version shown here in Fig. 8.11 (Keeling and Manning, 2014). The decrease in O_2/N_2 ratio in the atmosphere is plotted versus the increase in atmospheric CO_2 and compared with expected changes and trends. The O_2/N_2 ratio is presented in units similar to those used for isotope ratios (see Chapter 6), except this time the relative ratio difference between sample and standard is multiplied by 10^6 rather than 1000:

$$\delta(O_2/N_2)\ (\text{per meg}) = \left((O_2/N_2)_{\text{sample}}/(O_2/N_2)_{\text{std}} - 1\right) \times 10^6.$$

A negative $\delta(O_2/N_2)$ indicates a decrease in O_2/N_2 relative to the standard, which is a constant. Here, units of per meg can be converted to ppm by multiplying by 0.2094, the dry mole fraction of O_2 in the atmosphere. The ratio change is assumed to represent a change in O_2, since N_2 is largely inert. Mean annual measurements between 1991 and 2011 are presented by the solid circles in Fig. 8.11. The solid arrow reaching from the earliest measurements in 1991 down to the lower right-hand side of the figure is the trend expected based on the known amount of fossil fuel burning during the 20-year period and its $\Delta O_2{:}\Delta CO_2$ combustion ratio. Different fuels have ratios that vary between 1.2 and 2.0, but the ratio based on the mixture of fuels is relatively well known (1.38). The observed atmospheric trend does not follow what would be expected if the CO_2 produced and O_2

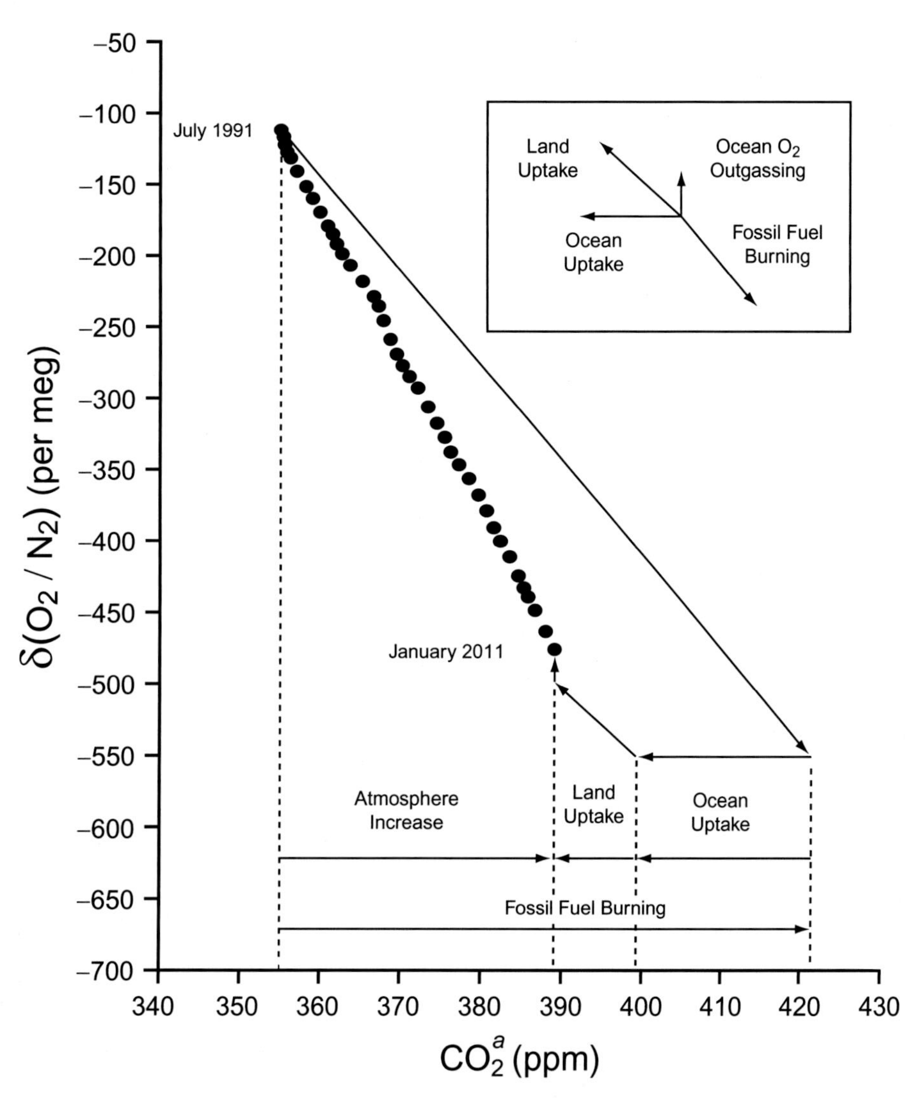

Figure 8.11 The mean change in atmospheric O_2/N_2 ratio and CO_2 from 1991 to 2011. Annually and globally averaged measurements of O_2/N_2 ratio (per meg) and CO_2 (ppm) are indicated by the filled circles. The long downward arrow is the trend expected for fossil fuel burning without exchange with the ocean or land. Arrows in the box in the upper right indicate trends expected for the different mechanisms influencing the measured atmospheric values. Redrawn from Keeling and Manning (2014).

consumed during fossil fuel combustion were confined to the atmosphere! The importance of the ocean versus terrestrial sinks for carbon is controlled by the ratio between CO_2 and O_2 for each of these two sinks. The vector diagram in the upper right of the figure indicates the slopes of the possible pathways that can be used to travel from the expected O_2/N_2 ratio and fCO_2 values in 2011 to the actual observations. Ocean uptake of CO_2 is a horizontal vector, because there is little change in the atmospheric O_2/N_2 ratio caused by atmosphere–ocean exchange in response to fossil fuel burning. The terrestrial reservoir exchange, on the other hand, follows a 1.1:1 ΔO_2:ΔCO_2 trend based on the photosynthesis/

respiration ratio for terrestrial biotic matter. Only one combination of these two vectors with these slopes can span the distance from the expected O_2/N_2 ratio and fCO_2 values in 2011 to the observations, and this defines the fraction of CO_2 that is taken up by the ocean and by the land (shown at the base of the figure).

An important detail about the changes in atmospheric fCO_2 and fO_2 that we haven't considered yet is that the ocean has been warming in response to fossil fuel CO_2 increases in the atmosphere. As the ocean warms, it also becomes more stratified. Both warming and stratification cause atmospheric O_2 to increase. Warm seawater holds less gas at thermodynamic equilibrium, and decreasing ventilation of the ocean interior exposes less low O_2 water at the surface. These effects are indicated as the "Ocean O_2 Outgassing" vertical arrow. This arrow is vertical, because the non-steady-state outgassing effect for CO_2 is already accounted for by the horizontal arrow indicating ocean uptake. The magnitude of the vertical outgassing arrow has been determined independently based on the relationship between ocean heat flux and oxygen flux and the observed warming of the ocean (Keeling and Manning, 2014).

The result of the atmosphere O_2/N_2 and CO_2 analysis is presented in the "Atm. O_2/N_2 & fCO_2" column of Table 8.3. About 50 percent of the anthropogenic CO_2 released into the atmosphere remains there (3.7 Pg-C y^{-1}) and ~35 percent (2.5 Pg-C y^{-1}) has been taken up by the ocean. The ocean uptake value is, within estimated errors, the same as results from the DIC_{anth} and air–sea exchange methods (Table 8.3). The remainder of the anthropogenic CO_2 introduced to the atmosphere must be taken up by the terrestrial biosphere (1.2 Pg-C y^{-1}). This net sink exists in spite of the fact that terrestrial models suggest a release of CO_2 to the atmosphere of 1.4 Pg-C y^{-1} from land use change (E_{luc} in Table 8.3). If the land use change estimate is correct, there must be an uptake of 2.6 Pg-C y^{-1} by fertilization of the terrestrial biosphere by CO_2 and nutrients and by lengthening of the growing season due to global warming.

One of the more fascinating aspects of the global carbon cycle and its response to climate change is that the role of the terrestrial biosphere has changed over the last 40 years. Analysis of atmospheric changes in CO_2, O_2:N_2 ratios, and the $\delta^{13}C$ of the CO_2 indicates that sources and sinks from the terrestrial biosphere were about equal during the 1980s (Battle et al., 2000); i.e., any source to the atmosphere due to land use change was matched by uptake due to enhanced CO_2 and nutrient fertilization. Since then, fertilization, sometimes known as "greening," has dominated the terrestrial signal. The rapidity of this change has been a surprise to carbon cycle scientists and indicates that predictions of the short-term future are challenging.

Atmospheric O_2/N_2 Ratio Change on Glacial–Interglacial Timescales. The final timescale on which atmospheric oxygen can be used to study the carbon cycle is determined by the age of the atmosphere stored in polar ice. Just as how changes in the fCO_2 of the atmosphere have been determined by measuring the CO_2 content of atmospheric bubbles trapped in ice cores (Fig. 8.3), so have changes in the atmospheric O_2/N_2 ratio also been determined. Results from four different ice cores with ages that span the last 800 ky (Fig. 8.12, Stolper et al., 2016) indicate that the atmospheric O_2/N_2 ratio has declined by about 0.7 percent over this timespan. After using simultaneous Ar/N_2 measurements to address known artifacts in the preservation of O_2/N_2 ratios in ice cores, like fractionation during the process of bubble

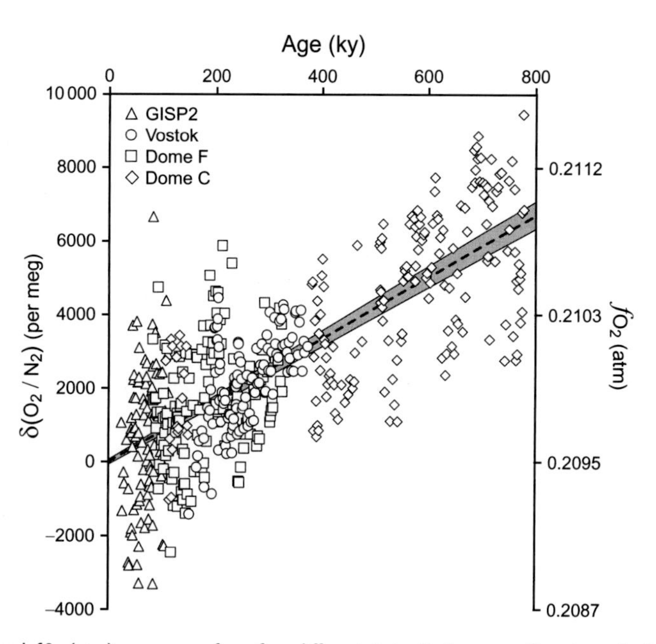

Figure 8.12 O_2/N_2(per meg) and fO_2 (atm) versus age from four different Antarctic ice cores. The atmospheric oxygen content decreased by about 7000 per meg (~0.7 %) over the last 800 000 years. The gray band indicates 95% confidence intervals. Modified from Stolper et al. (2016).

close-off, Stolper et al. (2016) were able to conclude that the observations are consistent only with a real decrease in the atmospheric ratio. Because the residence time for O_2 in the atmosphere is about 2 million years while that for N_2 is on the order of 1 billion years, the O_2/N_2 ratio change is almost certainly due to an oxygen decrease.

The fO_2 of the atmosphere represents a balance between sources, mainly burial of organic carbon and pyrite, and sinks, mainly weathering (oxidization) of organic carbon and pyrite in terrestrial deposits. A 0.7 percent decrease in fO_2 over 800 ky is consistent with a sink that is ~2 percent greater than the source over this period of time. Since oxygen and carbon are linked via photosynthesis and respiration (or weathering and organic carbon burial over longer timescales), a decrease in oxygen ought to be accompanied by an increase in DIC in the ocean and in fCO_2 in the atmosphere. Stolper et al. (2016) suggest that an atmospheric fCO_2 increase of 140 μatm would be necessary to explain a 0.7 percent fO_2 decrease over the 800 ky period. However, the atmospheric fCO_2 record (Fig. 8.3) indicates 100 ky cycles but no long-term, million-year trend of this magnitude. The classic relationship between changing O_2 and C does not appear to apply on timescales that are longer than hundreds to thousands of years. Stolper et al. (2016) conclude that the data for atmospheric fO_2 and fCO_2 can be explained only if there is a relatively rapid feedback to the carbon cycle by CO_2-dependent silicate weathering (Berner, 2004), which would essentially override the relationship between organic carbon and O_2 on million-year timescales. Silicate weathering should affect the C cycle, but have little impact on atmospheric O_2. This explanation is consistent with the timescales for feedbacks: for carbon the dominant response time is that associated with the 100 000-year residence time for HCO_3^- in the ocean, and for oxygen the response time is more like the 2 million year residence time of O_2 in the atmosphere.

8.4.6 Future Challenges of the Anthropogenic Influence

The ocean plays a dominant role in the Earth's carbon cycle by harboring the largest carbon reservoir exchanging with the atmosphere on timescales of less than 1000 years. Marine physical and biological processes have influenced atmospheric fCO_2 changes on glacial–interglacial timescales, and they represent the most important processes controlling the sink for atmospheric CO_2 introduced by fossil fuel burning. The natural coupling between the cycles of oxygen and carbon has provided environmental scientists with tracers that reveal some of the secrets of the carbon cycle. However, even though anthropogenic processes are currently perturbing the natural cycle by easily measurable amounts, it is difficult to be certain about the nature and magnitude of marine carbon cycle feedbacks to climate change. The only way to improve our ability to predict, with some confidence, the impact of humanity's influence on the global carbon cycle is to better understand how it works. This can be accomplished by a combination of monitoring chemical tracers of environmental processes and by constructing more informed models of the physical and biological processes that control feedbacks to change. The ocean is a very large place that is dramatically undersampled by traditional oceanographic methods. The way forward to better quantifying the mechanisms and fluxes in the marine environment is to improve monitoring methods to the point where it is possible to measure changes on a global scale.

8A.1 Appendix 8A.1 The Solution to the Three-Box Ocean and Atmosphere Model

A diagram of the three-box ocean and atmosphere model is presented in Fig. 8.4 and its solution in Fig. 8.5. Symbols used in those figures and in the equations below are described in the figure captions and in Table 8A.1.1. The solution is achieved by solving multiple simultaneous linear equations. Toggweiler and Sarmiento (1985) solve the equations by substitution and an iterative procedure. Here we use a matrix solution for simultaneous equations, suitable for software like MATLAB or Python. In addition to the mass balance equations for DIP, A_T, and DIC (Eqs. 8A.1.1–8A.1.9), we use a linear approximation for the relationship among A_T, DIC, and fCO_2 (Eqs. 8A.1.10 and 8A.1.11) using the CO2SYS carbonate equilibrium program (see Chapter 5) for the temperature and salinity conditions of the model. Here, we show the equations and their rearrangement into matrix form.

Steady-State Mass Balance Equations
For DIP (abbreviated P in mol m^{-3})

$$\text{Box } h: \ 0 = T\,(\mathrm{P}_l - \mathrm{P}_h) + f_{hd}(\mathrm{P}_d - \mathrm{P}_h) - J_h S_h \qquad \left(\text{fluxes in mol-P y}^{-1}\right) \tag{8A.1.1}$$

$$\text{Box } l: \ 0 = T\,(\mathrm{P}_d - \mathrm{P}_l) - J_l S_l. \tag{8A.1.2}$$

For alkalinity (abbreviated A in eq m^{-3})

$$\text{Box } h: \ 0 = T\,(\mathrm{A}_l - \mathrm{A}_h) + f_{hd}(\mathrm{A}_d - \mathrm{A}_h) - J_h S_h r_{\mathrm{A:P}} \qquad \left(\text{fluxes in eq y}^{-1}\right) \tag{8A.1.3}$$

$$\text{Box } l: \ 0 = T\,(\mathrm{A}_d - \mathrm{A}_l) - J_l S_l r_{\mathrm{A:P}} \tag{8A.1.4}$$

$$\Sigma\mathrm{A} = V_l \mathrm{A}_l + V_d \mathrm{A}_d + V_h \mathrm{A}_h. \tag{8A.1.5}$$

For DIC (abbreviated C in mol m^{-3})

$$\begin{aligned}\text{Box } h: \ 0 = {} & T(\mathrm{C}_l - \mathrm{C}_h) + f_{hd}(\mathrm{C}_d - \mathrm{C}_h) \\ & - J_h S_h r_{\mathrm{C:P}} + k K_{H,h} S_h \left(f\mathrm{CO}_2^a - f\mathrm{CO}_2^h\right) \qquad \left(\text{fluxes in mol-C y}^{-1}\right)\end{aligned} \tag{8A.1.6}$$

$$\text{Box } l: \ 0 = T\,(\mathrm{C}_d - \mathrm{C}_l) - J_l S_l r_{\mathrm{C:P}} + k\,K_{H,l} S_l \left(f\mathrm{CO}_2^a - f\mathrm{CO}_2^l\right) \tag{8A.1.7}$$

$$\text{Box } a: \ 0 = k K_{H,h} S_h \left(f\mathrm{CO}_2^a - f\mathrm{CO}_2^h\right) + k K_{H,l} S_l \left(f\mathrm{CO}_2^a - f\mathrm{CO}_2^l\right) \tag{8A.1.8}$$

$$\Sigma\mathrm{C} = V_l \mathrm{C}_l + V_d \mathrm{C}_d + V_h \mathrm{C}_h + M_a \cdot f\mathrm{CO}_2^a. \tag{8A.1.9}$$

Linear approximation relating A, C, and $f\mathrm{CO}_2^{sw}$

$$\text{Box } h: \ (\mathrm{A}_h - \mathrm{C}_h) = \beta_h f\mathrm{CO}_2^h + \gamma_h \tag{8A.1.10}$$

$$\text{Box } l: \ (\mathrm{A}_l - \mathrm{C}_l) = \beta_l f\mathrm{CO}_2^l + \gamma_l. \tag{8A.1.11}$$

Table 8A.1.1 Symbols used in the solution of the three-box ocean and atmosphere model (Figs. 8.4 & 8.5 and Eqs. 8A.1.1–8A.1.11). Most values are taken from Toggweiler and Sarmiento (1985). Temperature and salinity values are those used to determine K_H and fCO_2

Symbol and value	Description and units
$V_o = 1.292 \times 10^{18}$	Ocean volume (m^3)
$S_o = 3.49 \times 10^{14}$	Ocean surface area (m^2)
$S_l = S_o \times 0.85$	Surface area of low-latitude surface box (m^2)
$S_h = S_o \times 0.15$	Surface area of high-latitude surface box (m^2)
$V_l = S_l \times 100$	Volume of low-latitude surface box (m^3)
$V_h = S_h \times 250$	Volume of high-latitude surface box (m^3)
$V_d = V_o - V_h - V_l$	Volume of deep box (m^3)
$M_a = 1.773 \times 10^{20}$	Molar volume of the atmosphere (moles atm^{-1})
$Temp_l = 21.5$	Temperature of low-latitude surface box (°C)
$Temp_h = 2.5$	Temperature of high-latitude surface box (°C)
Sal = 34.7	Salinity of surface ocean
$K_{H,l} = 31.9$	Henry's law coefficient for low-latitude surface CO_2 (mol m^{-3} atm^{-1})
$K_{H,h} = 58.8$	Henry's law coefficient for high-latitude surface CO_2 (mol m^{-3} atm^{-1})
$P_d = 2.15 \times 10^{-3}$	Phosphate concentration of the deep box (mol m^{-3})
$\Sigma C = 3.02 \times 10^{18}$	Total C in the ocean (DIC) and atmosphere (fCO_2) (mol)
$\Sigma A = 3.14 \times 10^{18}$	Total alkalinity in the ocean (eq)
$T = 20$ Sv	Thermohaline circulation (10^6 m^3 s^{-1})
$f_{hd} = 50$ Sv	High-latitude convective mixing (10^6 m^3 s^{-1})
$r_{A:P} = 44$	Ratio of alkalinity to phosphorus in biological flux term
$r_{C:P} = 136$	Ratio of carbon to phosphorus in biological flux term
$k = 3$	Gas exchange mass transfer coefficient (m d^{-1})
$\beta_l = -784.16$	The linear relationship: $(A_l - C_l) = \beta_l fCO_2^l + \gamma_l$ (mol m^{-3} atm^{-1})
$\gamma_l = 0.56759$	The linear relationship: $(A_l - C_l) = \beta_l fCO_2^l + \gamma_l$ (mol m^{-3})
$\beta_h = -620.04$	The linear relationship: $(A_h - C_h) = \beta_h fCO_2^h + \gamma_h$ (mol m^{-3} atm^{-1})
$\gamma_h = 0.37159$	The linear relationship: $(A_h - C_h) = \beta_h fCO_2^h + \gamma_h$ (mol m^{-3})

Rearranging Mass Balance Equations into Matrices

For P (2 × 2) specify f_{hd}, P_l, P_h, unknowns: J_h and J_l

$$\begin{bmatrix} S_h & 0 \\ 0 & S_l \end{bmatrix} \times \begin{bmatrix} J_h \\ J_l \end{bmatrix} = \begin{bmatrix} T(P_l - P_h) + f_{hd}(P_d - P_h) \\ T(P_d - P_l) \end{bmatrix}.$$

For A (3 × 3) unknowns: A_l, A_h, A_d

$$\begin{bmatrix} T & (-T - f_{hd}) & f_{hd} \\ -T & 0 & T \\ V_l & V_h & V_d \end{bmatrix} \times \begin{bmatrix} A_l \\ A_h \\ A_d \end{bmatrix} = \begin{bmatrix} J_h S_h r_{A:P} \\ J_l S_l r_{A:P} \\ \Sigma A \end{bmatrix}.$$

For C (6 × 6) unknowns: C_l, C_h, C_d, $fCO_2^l, fCO_2^h, fCO_2^a$.
Multiplying the last two equations by 10^{15} removes scaling problems where some rows contain much larger numbers than others, creating inaccurate results.

$$\begin{bmatrix} T & (-T-f_{hd}) & f_{hd} & 0 & -kK_{H,h}S_h & kK_{H,h}S_h \\ -T & 0 & T & -kK_{H,l}S_l & 0 & kK_{H,l}S_l \\ 0 & 0 & 0 & kK_{H,l}S_l & kK_{H,h}S_h & -k(K_{H,h}S_h + K_{H,l}S_l) \\ V_l & V_h & V_d & 0 & 0 & M_a \\ 10^{15} & 0 & 0 & \beta_l \times 10^{15} & 0 & 0 \\ 0 & 10^{15} & 0 & 0 & \beta_h \times 10^{15} & 0 \end{bmatrix} \times \begin{bmatrix} C_l \\ C_h \\ C_d \\ fCO_2^l \\ fCO_2^h \\ fCO_2^a \end{bmatrix}$$

$$= \begin{bmatrix} J_h S_h r_{C:P} \\ J_l S_l r_{C:P} \\ 0 \\ \Sigma C \\ (A_l - \gamma_l) \times 10^{15} \\ (A_h - \gamma_h) \times 10^{15} \end{bmatrix}.$$

References

Archer, D. (2005) Fate of fossil fuel CO_2 in geologic time, *Journal of Geophysical Research-Oceans*, **110**, C09S05, doi:10.1029/2004JC002625.

Archer, D., and E. Maier-Reimer (1994) Effect of deep-sea sedimentary calcite preservation on atmospheric CO_2 concentration, *Nature*, **367**, 260–263.

Bakker, D. C. E., B. Pfeil, K. Smith, et al. (2014) An update to the Surface Ocean CO_2 Atlas (SOCAT version 2), *Earth System Science Data*, **6**, 69–90.

Battle, M., M. L. Bender, P. P. Tans, et al. (2000) Global carbon sinks and their variability inferred from atmospheric O_2 and $\delta^{13}C$, *Science*, **287**, 2467–2470.

Bender, M. L., P. P. Tans, J. T. Ellis, J. Orchardo, and K. Habfast (1994) A high-precision isotope ratio mass spectrometry method for measuring the O_2/N_2 ratio of air, *Geochimica et Cosmochimica Acta*, **58**, 4751–4758.

Berner, R. A. (2004) *The Phanerozoic Carbon Cycle: CO_2 and O_2*, 150 pp. Oxford: Oxford University Press.

Boyle, E. A. (1988) Vertical oceanic nutrient fractionation and glacial/interglacial CO_2 cycles, *Nature*, **331**, 55–56.

Brewer, P. G. (1978) Direct observation of the oceanic CO_2 increase, *Geophysical Research Letters*, **5**, 997–1000.

Broecker, W. S. (1982) Ocean chemistry during glacial time, *Geochimica et Cosmochimica Acta*, **46**, 1689–1706.

Broecker, W. S., and T.-H. Peng (1982) *Tracers in the Sea*, 690 pp. Palisades: Eldigio Press.

Ciais, P., C. Sabine, et al. (2013) Chapter 6: Carbon and other biogeochemical cycles, in *Climate Change 2013, The Physical Science Basis* (eds. T. Stocker, D. Qin, G.-K. Platner et al.), Cambridge: Cambridge University Press.

Chen, G.-T., and F. J. Millero (1979) Gradual increase of oceanic carbon dioxide, *Nature*, **277**, 205–206.

Emerson, S. (2014) Net community production and biological carbon flux in the ocean, *Global Biogeochemical Cycles*, **28**, doi:10.1002/2013GB004680.

Emerson, S. R., and J. I. Hedges (2008) *Chemical Oceanography and the Marine Carbon Cycle*, 453 pp. Cambridge: Cambridge University Press.

Etheridge, D. M., L. P. Steele, R. L. Langenfelds, et al. (1996) Natural and anthropogenic changes in atmospheric CO_2 over the last 1000 years from air in Antarctic ice and firn, *Journal of Geophysical Research – Atmospheres*, **101**, 4115–4128.

Field, C. B., M. J. Behrenfeld, J. T. Randerson, and P. Falkowski (1998) Primary production of the biosphere: integrating terrestrial and oceanic components, *Science*, **281**, 237–240.

Friedlingstein, P., M. W. Jones, M. O'Sullivan, et al. (2019) Global Carbon Budget 2019, *Earth System Science Data*, **11**, 1783–1838.

Gruber N., J. L. Sarmiento, and T. F. Stocker (1996) An improved method for detecting anthropogenic CO_2 in the oceans, *Global Biogeochemical Cycles*, **10**, 809–837.

Gruber, N., D. Clement, B. R. Carter, et al., (2019) The oceanic sink for anthropogenic CO_2 from 1994 to 2007, *Science*, **363**, 1193–1199.

Hodell, D. A., K. A. Venz, C. D. Charles, and U. S. Ninnemann (2003) Pleistocene vertical carbon isotope and carbonate gradients in the South Atlantic section of the Southern Ocean, *Geochemistry, Geophysics, Geosystems*, **4**, 1–19, doi:10.1029/2002GC000367.

Jacobson, A. R., S. E. Mikaloff-Fletcher, N. Gruber, J. L. Sarmiento, and M. Gloor (2007) A joint atmosphere-ocean inversion for surface fluxes of carbon dioxide: 1. Methods and global-scale fluxes, *Global Biogeochemical Cycles*, **21**, GB1019, doi:10.1029/2005GB002556.

Jouzel, J., V. Masson-Delmotte, O. Cattani, et al. (2007) Orbital and millennial Antarctic climate variability over the past 800,000 years, *Science*, **317**, 793–796.

Keeling, C. D. (1960) The concentration and isotopic abundances of carbon dioxide in the atmosphere, *Tellus*, **12**, 200–203.

Keeling, R. F. (1988) Measuring correlations between atmospheric oxygen and carbon dioxide mole fractions: a preliminary study in urban air, *Journal of Atmospheric Chemistry*, **7**, 153–176.

Keeling, R. F., and A. C. Manning (2014) Studies of recent changes in atmospheric O_2 content, in *Treatise on Geochemistry*, Vol. 5, *The Atmosphere* (eds. R. Keeling and L. Russell), pp. 385–404. New York: Elsevier.

Keeling, R. F., S. C. Piper, and M. Heimann (1996) Global and hemispheric CO_2 sinks deduced from changes in atmospheric O_2 concentration, *Nature*, **381**, 218–221.

Key, R. M., A. Kozyr, C. L. Sabine, et al. (2004) A global ocean carbon climatology: results from Global Data Analysis Project (GLODAP), *Global Biogeochemical Cycles*, **18**, GB4031, doi:10.1029/2004GB002247.

Khatiwala, S., T. Tanhua, S. Mikaloff-Fletcher, et al., (2013) Global ocean storage of anthropogenic carbon, *Biogeosciences*, **10**, 2169–2191, doi:10.5194/bg-10-2169-2013.

Knox, F., and M. B. McElroy (1984) Changes in atmospheric CO_2: influence of the marine biota at high latitude, *Journal of Geophysical Research – Atmospheres*, **89**, 4629–4637.

Landschützer, P., N. Gruber, D. C. E. Bakker, and U. Schuster (2014) Recent variability of the global ocean carbon sink, *Global Biogeochemical Cycles*, **28**, 927–949.

Lüthi, D., M. Le Floch, B. Bereiter, et al. (2008) High-resolution carbon dioxide concentration record 650,000–800,000 years before present, *Nature*, **453**, 379–382.

MacFarling Meure, C., D. Etheridge, C. Trudinger, et al. (2006) Law Dome CO_2, CH_4 and N_2O ice core records extended to 2000 years BP, *Geophysical Research Letters*, **33**, L14810, doi:10.1029/2006GL026152.

Quay, P. D., B. Tilbrook, and C. S. Wong (1992) Oceanic uptake of fossil fuel CO_2: carbon-13 evidence, *Science*, **256**, 74–79.

Quay, P. Q., R. Sonnerup, T. Westby, J. Stutsman, and A. McNichol (2003) Changes in the $^{13}C/^{12}C$ of dissolved inorganic carbon in the ocean as a tracer of anthropogenic CO_2 uptake, *Global Biogeochemical Cycles*, **17**, 1004, doi:10.1029/2001GB001817.

Revelle, R., and H. E. Suess (1957) Carbon dioxide exchange between atmosphere and ocean and the question of an increase of atmospheric CO_2 during past decades, *Tellus*, **9**, 18–27.

Sabine, C. L., R. A. Feely, N. Gruber, et al. (2004) The oceanic sink for anthropogenic CO_2, *Science*, **305**, 367–371.

Sarmiento, J. L., and J. R. Toggweiler (1984) A new model for the role of the oceans in determining atmospheric pCO_2, *Nature*, **308**, 621–624.

Siegenthaler, U., and T. Wenk (1984) Rapid atmospheric CO_2 variations and ocean circulation, *Nature*, **308**, 624–626.

Sigman, D. M., and E. A. Boyle (2000) Glacial/interglacial variations in atmospheric carbon dioxide, *Nature*, **407**, 859–869.

Sigman, D. M., M. P. Hain, and G. H. Haug (2010) The polar ocean and glacial cycles in atmospheric CO_2 concentration, *Nature*, **446**, 47–55, doi:1038/nature09149.

Stolper, D. A., M. L. Bender, G. B. Dreyfus, Y. Yan, and J. A. Higgins (2016) A Pleistocene ice core record of atmospheric O_2 concentrations, *Science*, **353**, 1427–1430.

Takahashi, T., R. H. Wanninkhof, R. A. Feely, et al. (1999) Net sea-air CO_2 flux over the global oceans: an improved estimate based on the sea-air pCO_2 difference, in *Proceedings Second International Symposium on CO_2 in the Oceans* (ed. Y. Nojiri), pp. 9–14, Tsukuba: National Institute for Environmental Studies.

Takahashi, T., S. C. Sutherland, R. Wanninkhof, et al. (2009) Climatological mean and decadal change in surface pCO_2, and net air-sea flux over the global oceans, *Deep-Sea Research II*, **56**, 554–577.

Toggweiler, J. R. (1999) Variations of atmospheric CO_2 by ventilation of the ocean's deepest water, *Paleoceanography*, **14**, 571–588.

Toggweiler, J. R., and J. L. Sarmiento (1985) Glacial to interglacial changes in atmospheric carbon dioxide: the critical role of ocean surface water in high latitudes, in *The Carbon Cycle and Atmospheric CO_2: Natural Variations Archean to Present*, Geophysical Monograph Series, Vol. 32 (eds. E. Sundquist and W. S. Broecker), pp. 163–184. Washington, DC: American Geophysical Union.

Volk, T., and M. I. Hoffert (1985) Ocean carbon pumps: analysis of relative strengths and efficiencies in ocean-driven atmospheric CO_2 changes, in *The Carbon Cycle and Atmospheric CO_2: Natural Variations Archean to Present*, Geophysical Monograph Series, Vol. 32 (eds. E. Sundquist and W. S. Broecker), pp. 99–110. Washington, DC: American Geophysical Union.

Wanninkhof, R. (1992) Relationship between wind speed and gas exchange over the ocean, *Journal of Geophysical Research – Oceans*, **97**, 7373–7382.
Zachos, J. C., G. R. Dickens, and R. E. Zeebe (2008) An early Cenozoic perspective on greenhouse warming and carbon-cycle dynamics, *Nature*, **451**, 279–283, doi:10.1038/nature06588.

Problems for Chapter 8

8.1. This problem explores the relationships among different parameters in the two-box model, carbonate system, and atmospheric fCO_2. Use the values for the model's Std case in Table 8.2 as the starting point for your calculations. Plot your results and comment on them.

a. Keeping all other parameters the same, vary the temperature of the surface box in increments from 10 to 30°C and calculate the resulting atmospheric fCO_2 and surface ocean pH.

b. Keeping all other parameters the same, including the stoichiometry of the particulate flux, vary the phosphate concentration of the surface box in increments from 0 to 2.2 μmol kg^{-1} and calculate the resulting atmospheric fCO_2 and surface ocean pH.

c. Keeping all other parameters the same, vary the $CaCO_3$ to organic carbon stoichiometry of the particulate flux from 0 (no $CaCO_3$ produced) to 1 (same amount of $CaCO_3$ produced as organic carbon) and calculate the resulting atmospheric fCO_2, surface ocean pH, and surface ocean carbonate ion concentration.

8.2. Using the three-box ocean and atmosphere model, as laid out in Fig. 8.4 and Appendix 8A.1, solve for the high-latitude particle flux and the atmospheric fCO_2 level at different high-latitude phosphate concentrations from 0 to 1.6 μmol kg^{-1}. Assume a low-latitude phosphate concentration of 0.2 μmol kg^{-1}. Explain what causes the trends you observe.

8.3. This problem explores how the Revelle factor depends on other ocean parameters. Use the conditions in Fig. 8.7 as a starting point for your calculations. Plot your results and comment on them.

a. Keeping all other parameters the same, vary the temperature of the surface ocean and calculate the resulting Revelle factor. What is the percent variation in R_{Rev} over the environmental range of ocean temperatures?

b. Using the data in Fig. 8.9 as a basis for how much DIC has increased at the ocean surface, calculate how the Revelle factor has been affected already by the ocean's uptake of anthropogenic carbon. What is the percent variation in R_{Rev} over the anthropogenic change in ocean DIC? What is the relationship between R_{Rev} and carbonate ion concentration as DIC increases?

8.4. This problem demonstrates the origin of the Revelle factor using the carbonate equilibrium equations. Use the seawater conditions in Fig. 8.7 as your starting point.

a. For a given increase in DIC due to the addition of anthropogenic CO_2 to seawater, determine the approximate change in HCO_3^- and in CO_3^{2-} concentrations using the carbonate equilibrium equations below. Then calculate the resulting change in CO_2 concentration. Remember that there is no change in alkalinity during addition of CO_2 to seawater, and assume here that there is no borate so it is the carbonate alkalinity that does not change. This is an approximate calculation because of these assumptions.

(1) $A_C = [HCO_3^-] + 2[CO_3^{2-}]$

(2) $DIC = [HCO_3^-] + [CO_3^{2-}] + [CO_2] \sim [HCO_3^-] + [CO_3^{2-}]$

(3) $K_1' = [HCO_3^-][H^+]/[CO_2] = 1.28 \times 10^{-6}$

(4) $K_2' = [CO_3^{2-}][H^+]/[HCO_3^-] = 8.98 \times 10^{-10}$

b. How does your calculated relationship between the change in $[CO_2]$ and DIC compare to that predicted by the Revelle factor ($\Delta[CO_2]/\Delta DIC$ in Eq. 8.14 but with $[CO_2]$ substituted for fCO_2)?

8.5. The following table displays oxygen, DIC, and alkalinity measurements along an isopycnal in the South Pacific (ordered from surface to deep)

O_2 (μmol kg^{-1})	DIC (μmol kg^{-1})	Alkalinity (μeq kg^{-1})
263	2 089	2 292
243	2 095	2 298
223	2 101	2 304
203	2 103	2 310
183	2 106	2 315
163	2 119	2 321
113	2 159	2 336

The isopycnal has a temperature of 12°C and a salinity of 35. The top value is the surface where the isopycnal outcrops. The surface oxygen concentration is in equilibrium with the atmosphere. The last value listed is for an area of the isopycnal deep enough that it contains no anthropogenic carbon. You can assume that the alkalinity of the ocean has not been affected by anthropogenic carbon.

Calculate the anthropogenic carbon content at each depth.

8.6. Measurements over a 10-y period show atmospheric CO_2 concentrations increased by 17.6 ppm while atmospheric O_2 concentrations decreased by 192 per meg. Over the same period, fossil fuel combustion released 72.4 Pg-C into the atmosphere. Determine how much carbon has been absorbed by the ocean and how much by the terrestrial biosphere over this 10-y period. What percentage of anthropogenic CO_2 ended up in the atmosphere, ocean, and terrestrial biosphere?

Appendix A Critical Quantities for the Ocean–Atmosphere System

Atmosphere Inventory	1.773×10^{20} mol (all dry gases)
	5.132×10^{18} kg (all dry gases)
Air Density 15°C and Mean Sea Level	1.225 kg m^{-3}
Earth Surface Area	5.10×10^{14} m^2
Ocean Surface Area[1]	3.62×10^{14} m^2 (71% of Earth's area)
Ocean Mean Depth[1]	3682 m
Ocean Volume[1]	1.33×10^{18} m^3
Ocean Mass	1.38×10^{21} kg
Proportion of Ocean in Each Basin[2]	
Pacific	52.9%
Atlantic	25.0%
Indian	21.1%
Arctic	1.0%
Mean Ocean Mixed Layer Depth[3]	67 m
River Flow Rate[4]	3.73×10^{13} m^3 y^{-1}
Ocean Mean Temperature[5]	3.53°C
Ocean Mean Salinity[5]	34.72 (PSS-78)
Freezing Point of Seawater $S_P = 35$, P = 0 dbar	–1.922°C
Heat Capacity of Seawater	3 994 J kg^{-1} K^{-1}
$T_{Celsius} = 20$°C, $S_P = 35$, P = 0 dbar	
Typical Threshold for Hypoxia[6]	60 μmol-O_2 kg^{-1}

[1] Charette and Smith (2010) [2] Pilson (2013) [3] Kara et al. (2003) [4] Dai and Trenberth (2002)
[5] Sarmiento and Gruber (2006) [6] Hofmann et al. (2011)

Table A.1 Water densities (in kg m^{-3}) at several temperatures and salinities at 0 dbar sea pressure (surface ocean). See the Thermodynamic Equation of Seawater – 2010 (TEOS-10, www.teos-10.org) for equations to calculate density under other conditions.

	$T_{Celsius} = 0$°C	$T_{Celsius} = 10$°C	$T_{Celsius} = 20$°C	$T_{Celsius} = 30$°C
$S_P = 0$	999.8	999.7	998.2	995.7
$S_P = 35$, $S_A = 35.166$	1 028.1	1 027.0	1 024.8	1 021.7

A practical salinity of $S_P = 35$ is equivalent to an absolute salinity on the TEOS-10 scale of $S_A =$ 35.166 in the surface North Atlantic, $S_A = 35.169$ in the deep North Atlantic, and $S_A = 35.191$ in the deep North Pacific based on representative locations at 50°N, 45°W and 50°N, 145°W.

Appendix B Fundamental Constants and Unit Conversions

Avogadro's constant: $N_A = 6.02214076 \times 10^{23}$ mol^{-1}
Drag Coefficient at the Air–Water Interface (4 m s^{-1} < wind speed < 11 m s^{-1}): $C_D = 0.0012$[1]
Faraday Constant: $F = 96485$ C mol^{-1}
Gravitational Acceleration on Earth (average): $g = 9.81$ m s^{-2}
Ideal Gas Constant: $R = 8.314463$ J K^{-1} mol^{-1} = 8.314463 m^3 Pa K^{-1} mol^{-1} = 0.08205737 L atm K^{-1} mol^{-1}
Molar Volume of an Ideal Gas at STP = 22.4140 L mol^{-1}
Rotation Rate of Earth: $\Omega = 7.2921 \times 10^{-5}$ rad s^{-1}
Standard Atmospheric Pressure = 1 atm = 101 325 Pa
Stephan–Boltzmann constant: $\sigma = 5.6704 \times 10^{-8}$ W m^{-2} K^{-4}

[1] Large and Pond (1981)

Table B.1 International System of Units (SI) base units[a]

Quantity	Unit	Basis
length	meter (m)	defined by the speed of light being exactly 299 792 458 m s^{-1}
time	second (s)	defined by the unperturbed ground-state hyperfine transition frequency of the cesium-133 atom being exactly 9 192 631 770 s^{-1}
amount of substance	mole (mol)	defined as exactly $6.02214076 \times 10^{23}$ elementary entities (molecules, atoms, etc.)
temperature	kelvin (K)	defined by the Boltzmann constant being exactly 1.380649×10^{-23} kg m^2 s^{-2} K^{-1}
mass	kilogram (kg)	defined by the Plank constant being exactly $6.62607015 \times 10^{-34}$ kg m^2 s^{-1}
electric current	ampere (A)	defined by the elementary charge (of an electron or proton) being exactly $1.602176634 \times 10^{-19}$ A s
luminous intensity	candela (cd)	defined by the luminous efficacy of monochromatic radiation of frequency 540×10^{12} Hz being exactly 683 cd sr kg^{-1} m^{-2} s^3 (where sr is steradians)

[a] US National Institute of Standards and Technology

Table B.2 Common unit conversions and equivalences. SI base units are given in bold italics.

Temperature	Degrees Celsius (°C) + 273.15 = ***K***
Gas Concentration[a]	Convert mL L^{-1} to mmol L^{-1} using the molar volume of a gas 22.4140 L mol^{-1}
Pressure	1 atm
	= 101 325 Pa (pascal) = 101 325 N m^{-2} = 101 325 $\boldsymbol{kg\ m^{-1}\ s^{-2}}$ = 1013.25 hPa
	= 1013.25 mbar = 1.01325 bar
	= 760 Torr = 760 mm-Hg (mm of mercury)
	= 14.6959 psi (pounds per square inch)
Distance	1 ***m***
	= 10^{10} Å (angstrom)
	= 5.400 × 10^{-4} nautical mile (1 nautical mile = 1.852 km)
Speed	1 $\boldsymbol{m\ s^{-1}}$ = 1.944 Knot
Volume	1 $\boldsymbol{m^3}$ = 10^3 L = 10^6 mL
Mass	1 ***kg*** = 10^{-3} metric ton
Radioactivity	1 Bq (becquerel) = 1 decays s^{-1}
	= 2.7027 × 10^{-11} Ci (curie)
	= 10^{-6} Rd (rutherford)
Frequency	1 Hz (hertz) = 1 $\boldsymbol{s^{-1}}$
Force	1 N (newton) = 1 $\boldsymbol{kg\ m\ s^{-2}}$
Energy	1 J (joule) = 1 N m = 1 $\boldsymbol{kg\ m^2\ s^{-2}}$
	= 0.2390 cal (calorie)
	= 6.242 × 10^{18} eV (electron volt)
Power	1 W (watt) = 1 J s^{-1} = 1 $\boldsymbol{kg\ m^2\ s^{-3}}$
Quantity of Electricity	1 C (coulomb) = 1 $\boldsymbol{A\ s}$
Electrical potential	1 V (volt) = 1 J C^{-1} = 1 W A^{-1} = 1 $\boldsymbol{kg\ m^2\ s^{-3} A^{-1}}$
Electrical Resistance	1 Ω (ohm) = 1 V A^{-1} = 1 $\boldsymbol{kg\ m^2\ s^{-3} A^{-2}}$
Electrical Conductivity	1 S m^{-1} (Siemens per meter) = 1 $\boldsymbol{kg^{-1}\ m^{-3}\ s^3\ A^2}$

[a] mL L^{-1} is a commonly used unit for dissolved gases, especially for O_2, and is equivalent to cc-STP L^{-1}. These units indicate the volume of pure gas at standard temperature and pressure (STP = 0°C and 1 atm). A nearly accurate conversion can be made between volume of gas and moles of gas (in the numerator) using the molar volume of an ideal gas, 22.414 L mol^{-1}. More accurate conversions should be made using the molar volumes for specific gases in Table D.2. The potential density of the water is used to convert the volume of solution in the denominator to mass of solution.

Appendix C Vapor Pressure of Water

(1) The vapor pressure of pure water over a temperature range of 273 to 647 K (atm, see Appendix B for conversion factors to other units) (from Wagner and Pruß, 2002):

$$\ln\left(\frac{p^a_{H_2O}}{217.75\text{ atm}}\right) = \frac{647.096\text{ K}}{T}\left(\begin{array}{c} -7.85951783\vartheta + 1.84408259\vartheta^{1.5} - 11.7866497\vartheta^3 \\ +22.6807411\vartheta^{3.5} - 15.9618719\vartheta^4 + 1.80122502\vartheta^{7.5} \end{array}\right), \tag{C.1}$$

where $\vartheta = 1 - \frac{T}{647.096\text{ K}}$ and T is in units of K.

(2) The vapor pressure of seawater is related to that of pure water by

$$p^a_{H_2O}(seawater) = p^a_{H_2O}(pure\ water) \cdot exp\left(-0.018\phi\frac{31.998S_P}{1000 - 1.005S_P}\right), \tag{C.2}$$

where ϕ (the osmotic coefficient at 25°C) over a salinity range of 16 to 40 is equal to (from Millero, 1974):

$$\phi = 0.90799 - 0.08992\left(\frac{15.999S_P}{1000 - 1.005S_P}\right) + 0.18458\left(\frac{15.999S_P}{1000 - 1.005S_P}\right)^2 - 0.07395\left(\frac{15.999S_P}{1000 - 1.005S_P}\right)^3 - 0.00221\left(\frac{15.999S_P}{1000 - 1.005S_P}\right)^4. \tag{C.3}$$

The small temperature dependence of ϕ is negligible here. S_P is practical salinity on the PSS-78 scale.

Table C.1 Water vapor pressures (in atm, see Appendix B for conversion factors to other units) at several temperatures and salinities. See www.cambridge.org/emerson-hamme for MATLAB and Python toolboxes containing a function to calculate this quantity.

	$T_{Celsius} = 0°C$	$T_{Celsius} = 10°C$	$T_{Celsius} = 20°C$	$T_{Celsius} = 30°C$
$S_P = 0$	0.006032	0.012120	0.023086	0.041913
$S_P = 35$	0.005919	0.011894	0.022654	0.041129

D

Appendix D Atmospheric Mole Fractions, Molar Volumes, Saturation Concentrations, and Henry's Law Constants for Gases

D.1 Atmospheric Mole Fractions and Molar Volumes

Table D.1 Mole fractions of gases in the atmosphere. Values are given for a dry atmosphere for all gases except water vapor. Values without a footnote are from Glueckauf, 1951.

Gas	X_C (mol_{gas} mol_{atm}^{-1})
N_2	7.8084×10^{-1}
O_2	2.0946×10^{-1}
H_2O [a]	up to 4×10^{-2}
Ar	9.34×10^{-3}
CO_2 [b]	4.10×10^{-4}
Ne	1.818×10^{-5}
He	5.24×10^{-6}
CH_4 [b]	1.87×10^{-6}
Kr	1.14×10^{-6}
N_2O [b]	3.32×10^{-7}
Xe	8.7×10^{-8}
CFC-12 [b]	5.01×10^{-10}
CFC-11 [b]	2.26×10^{-10}
SF_6 [b]	9.96×10^{-12}
DMS [c]	up to 3×10^{-10}

[a] See Appendix C
[b] 2019 averages from the NOAA Annual Greenhouse Gas Index data
[c] Woodhouse et al. (2010)

Table D.2 Molar volumes of selected gases (L mol^{-1} at STP) calculated from the virial coefficient data in Dymond and Smith (1980) except for CO_2, which is calculated from Weiss (1974), and N_2O, which is calculated from Weiss and Price (1980).

N_2	22.4040
O_2	22.3922
CO_2	22.2644
N_2O	22.2532
He	22.4258
Ne	22.4241
Ar	22.3924
Kr	22.3518
Xe	22.2582

D.2 Saturation Concentrations for Gases

All equations in this part, except that for DMS, give the concentration of a dissolved gas to be expected at saturation equilibrium with the atmosphere at 1 atm pressure including saturated water vapor. The order of the equations matches that of the gas's atmospheric mole fractions (Table D.1). The Henry's law constant for CO_2 is given in Appendix G.

The relationship between saturation concentrations and the apparent Henry's law constant for gas C at one atmosphere is

$$K'_{H,C} = \frac{[C]_{eq}}{(1 - p^a_{\mathrm{H_2O}})X_C}, \tag{D.1}$$

where $K'_{H,C}$ is the Henry's law constant, $[C]_{eq}$ is the saturation concentration, $p^a_{\mathrm{H_2O}}$ is the vapor pressure of water in units of atm, and X_C is the mole fraction of gas C in the dry atmosphere. This equation assumes that fugacity and partial pressure are equal, i.e., that gases behave ideally. (See Eqs. 2.12–2.16 in Chapter 2.)

(1) The saturation concentration for N_2 in seawater (μmol kg^{-1}) over a temperature range of 1 to 35°C and a salinity range of 0 to 40 (from Hamme and Emerson, 2004):

$$\begin{aligned}\ln [\mathrm{N_2}]_{eq} &= 6.42931 + 2.92704\, T_S + 4.32531\, {T_S}^2 + 4.69149\, {T_S}^3 \\ &+ S_P\left(-7.44129 \times 10^{-3} - 8.02566 \times 10^{-3}\, T_S - 1.46775 \times 10^{-2}\, {T_S}^2\right),\end{aligned} \tag{D.2}$$

where S_P is practical salinity on the PSS-78 scale and T_S is a scaled temperature calculated as

$$T_S = \ln\left(\frac{298.15 - T_{Celsius}}{273.15 + T_{Celsius}}\right), \tag{D.3}$$

where $T_{Celsius}$ is the water temperature in degrees Celsius.

(2) The saturation concentration for O_2 in seawater (μmol kg^{-1}) over a temperature range of freezing to 40°C and a salinity range of 0 to 42 (from Garcia and Gordon, 1992, 1993):

$$\begin{aligned}\ln [O_2]_{eq} = & \; 5.80871 + 3.20291\, T_S + 4.17887\, {T_S}^2 \\ & + 5.10006\, {T_S}^3 - 9.86643 \times 10^{-2}\, {T_S}^4 + 3.80369\, {T_S}^5 \\ & + S_P(-7.01577 \times 10^{-3} - 7.70028 \times 10^{-3}\, T_S \\ & - 1.13864 \times 10^{-2}\, {T_S}^2 - 9.51519 \times 10^{-3}\, {T_S}^3) \\ & - 2.75915 \times 10^{-7} {S_P}^2,\end{aligned} \tag{D.4}$$

where S_P is practical salinity on the PSS-78 scale and T_S is a scaled temperature calculated as in Eq. D.3.

(3) The saturation concentration for Ar in seawater (μmol kg^{-1}) over a temperature range of 1 to 35°C and a salinity range of 0 to 40 (from Hamme and Emerson, 2004):

$$\begin{aligned}\ln [Ar]_{eq} = & \; 2.79150 + 3.17609\, T_S + 4.13116\, {T_S}^2 + 4.90379\, {T_S}^3 \\ & + S_P\left(-6.96233 \times 10^{-3} - 7.66670 \times 10^{-3}\, T_S - 1.16888 \times 10^{-2}\, {T_S}^2\right),\end{aligned} \tag{D.5}$$

where S_P is practical salinity on the PSS-78 scale and T_S is a scaled temperature calculated as in Eq. D.3. See also Jenkins et al. (2019) for an alternate relationship for Ar solubility.

(4) The saturation concentration for Ne in seawater (nmol kg^{-1}) over a temperature range of 1 to 35°C and a salinity range of 0 to 40 (from Hamme and Emerson, 2004):

$$\begin{aligned}\ln [Ne]_{eq} = & \; 2.18156 + 1.29108\, T_S + 2.12504\, {T_S}^2 \\ & + S_P\left(-5.94737 \times 10^{-3} - 5.13896 \times 10^{-3}\, T_S\right),\end{aligned} \tag{D.6}$$

where S_P is practical salinity on the PSS-78 scale and T_S is a scaled temperature calculated as in Eq. D.3. See also Jenkins et al. (2019) for an alternate relationship for Ne solubility.

(5) The saturation concentration for He in seawater (mol kg^{-1}) over a temperature range of freezing to 35°C and a salinity range of 0 to 40 (from Jenkins et al., 2019):

$$\begin{aligned}\ln [He]_{eq} = & -178.1424 + 217.5991\left(\frac{100}{T}\right) + 140.7506 \ln\left(\frac{T}{100}\right) - 23.01954\left(\frac{T}{100}\right) \\ & + S_P\left(-0.038129 + 0.019190\left(\frac{T}{100}\right) - 0.0026898\left(\frac{T}{100}\right)^2\right) \\ & - 2.55 \times 10^{-6} {S_P}^2,\end{aligned} \tag{D.7}$$

where S_P is practical salinity on the PSS-78 scale and T is in units of K.

(6) The saturation concentration for CH_4 in seawater (nmol kg^{-1}) over a temperature range of –2 to 30°C and a salinity range of 0 to 40 (from Wiesenburg and Guinasso, 1979):

$$\begin{aligned}\ln [CH_4]_{eq} = & \ln X_{CH_4} - 417.5053 + 599.8626\left(\frac{100}{T}\right) + 380.3636 \ln\left(\frac{T}{100}\right) - 62.0764\left(\frac{T}{100}\right) \\ & + S_P\left(-0.064236 + 0.034980\left(\frac{T}{100}\right) - 0.0052732\left(\frac{T}{100}\right)^2\right),\end{aligned} \tag{D.8}$$

where X_{CH_4} is the mole fraction of methane in the dry atmosphere, S_P is practical salinity on the PSS-78 scale, and T is in units of K.

(7) The saturation concentration for Kr in seawater (mL-STP kg^{-1}) over a temperature range of 0 to 40°C and a salinity range of 0 to 40 (from Weiss and Kyser, 1978):

$$\ln[\mathrm{Kr}]_{eq} = -112.6840 + 153.5817\left(\frac{100}{T}\right) + 74.4690\ln\left(\frac{T}{100}\right) - 10.0189\left(\frac{T}{100}\right) + S_P\left(-0.011213 - 0.001844\left(\frac{T}{100}\right) + 0.0011201\left(\frac{T}{100}\right)^2\right), \tag{D.9}$$

where S_P is practical salinity on the PSS-78 scale and T is in units of K. Saturation concentrations calculated from this relationship should be divided by 0.0223518 to convert to units of µmol kg^{-1}. See also Jenkins et al. (2019) for an alternate relationship for Kr solubility.

(8) The saturation concentration for N_2O in seawater (mol kg^{-1}) over a temperature range of 0 to 40°C and a salinity range of 0 to 40 (from Weiss and Price, 1980):

$$\ln[\mathrm{N_2O}]_{eq} = \ln X_{N_2O} - 168.2459 + 226.0894\left(\frac{100}{T}\right) + 93.2817\ln\left(\frac{T}{100}\right) - 1.48693\left(\frac{T}{100}\right)^2 + S_P\left(-0.060361 + 0.033765\left(\frac{T}{100}\right) - 0.0051862\left(\frac{T}{100}\right)^2\right), \tag{D.10}$$

where X_{N_2O} is the mole fraction of N_2O in the dry atmosphere, S_P is practical salinity on the PSS-78 scale, and T is in units of K.

(9) The saturation concentration for Xe in seawater (mol kg^{-1}) over a temperature range of freezing to 35°C and a salinity range of 0 to 40 (from Jenkins et al., 2019):

$$\ln[\mathrm{Xe}]_{eq} = -224.5100 + 292.8234\left(\frac{100}{T}\right) + 157.6127\ln\left(\frac{T}{100}\right) - 22.66895\left(\frac{T}{100}\right) + S_P\left(-0.084915 + 0.047996\left(\frac{T}{100}\right) - 0.0073595\left(\frac{T}{100}\right)^2\right) + 6.69\times10^{-6}S_P^{\,2}, \tag{D.11}$$

where S_P is practical salinity on the PSS-78 scale and T is in units of K.

(10) The saturation concentration for CFC-12 in seawater (mol kg^{-1}) over a temperature range of 0 to 40°C and a salinity range of 0 to 40 (from Warner and Weiss, 1985):

$$\ln[\text{CFC-12}]_{eq} = \ln X_{\text{CFC-12}} - 220.2120 + 301.8695\left(\frac{100}{T}\right) + 114.8533\ln\left(\frac{T}{100}\right) - 1.39165\left(\frac{T}{100}\right)^2 + S_P\left(-0.147718 + 0.093175\left(\frac{T}{100}\right) - 0.0157340\left(\frac{T}{100}\right)^2\right), \tag{D.12}$$

where $X_{\text{CFC-12}}$ is the mole fraction of CFC-12 in the dry atmosphere, S_P is practical salinity on the PSS-78 scale, and T is in units of K.

(11) The saturation concentration for CFC-11 in seawater (mol kg^{-1}) over a temperature range of 0 to 40°C and a salinity range of 0 to 40 (from Warner and Weiss, 1985):

$$\ln[\text{CFC-11}]_{eq} = \ln X_{\text{CFC-11}} - 232.0411 + 322.5546\left(\frac{100}{T}\right) + 120.4956\ln\left(\frac{T}{100}\right) - 1.39165\left(\frac{T}{100}\right)^2 + S_P\left(-0.146531 + 0.093621\left(\frac{T}{100}\right) - 0.0160693\left(\frac{T}{100}\right)^2\right), \tag{D.13}$$

where $X_{\text{CFC-11}}$ is the mole fraction of CFC-11 in the dry atmosphere, S_P is practical salinity on the PSS-78 scale, and T is in units of K.

(12) The saturation concentration for SF_6 in seawater (mol kg^{-1}) over a temperature range of –1 to 40°C and a salinity range of 0 to 40 (from Bullister et al., 2002):

$$\ln[SF_6]_{eq} = \ln X_{SF_6} - 82.1639 + 120.152\left(\frac{100}{T}\right) + 30.6372\ln\left(\frac{T}{100}\right) + S_P\left(0.0293201 - 0.0351974\left(\frac{T}{100}\right) + 0.00740056\left(\frac{T}{100}\right)^2\right), \tag{D.14}$$

where X_{SF_6} is the mole fraction of SF_6 in the dry atmosphere, S_P is practical salinity on the PSS-78 scale, and T is in units of K.

(13) The Henry's law constant for DMS (mol atm^{-1} L^{-1}) over a temperature range of 0–32°C and at a salinity of 0 or 34.5 (from Dacey et al., 1984):

$$\ln K_{H,\text{DMS}} = \frac{3463}{T} - 12.20 \text{ in distilled water,} \tag{D.15}$$

$$\ln K_{H,\text{DMS}} = \frac{3547}{T} - 12.64 \text{ in seawater,}$$

where T is in units of K. Henry's law constants calculated from this relationship should be divided by density to obtain units of mol atm^{-1} kg^{-1}.

Table D.3 Saturation concentrations of gases in seawater at a variety of temperatures and salinities. Atmospheric dry mole fractions used in calculations are those from Table D.1. See www.cambridge.org/emerson-hamme for MATLAB and Python toolboxes containing functions to calculate these quantities.

		$T_{Celsius} = 0°C$	$T_{Celsius} = 10°C$	$T_{Celsius} = 20°C$	$T_{Celsius} = 30°C$
N_2 (μmol kg^{-1})	$S_P = 0$	830.45	653.21	537.44	457.90
	$S_P = 35$	622.03	500.89	419.77	362.47
O_2 (μmol kg^{-1})	$S_P = 0$	457.01	352.86	284.65	237.25
	$S_P = 35$	347.90	274.61	225.54	190.74
Ar (μmol kg^{-1})	$S_P = 0$	22.301	17.260	13.948	11.646
	$S_P = 35$	17.019	13.462	11.075	9.3751
Ne (nmol kg^{-1})	$S_P = 0$	10.084	9.0685	8.3279	7.8080
	$S_P = 35$	8.0608	7.3412	6.8271	6.4822
He (nmol kg^{-1})	$S_P = 0$	2.2238	2.1056	2.0264	1.9680
	$S_P = 35$	1.8108	1.7401	1.6964	1.6658
CH_4 (nmol kg^{-1})	$S_P = 0$	4.7654	3.5753	2.8324	2.3213
	$S_P = 35$	3.5976	2.7530	2.2163	1.8390
Kr (nmol kg^{-1})	$S_P = 0$	5.5531	4.0727	3.1168	2.4711
	$S_P = 35$	4.2127	3.1375	2.4401	1.9676
N_2O (nmol kg^{-1})	$S_P = 0$	19.501	13.119	9.3087	6.8838
	$S_P = 35$	15.356	10.510	7.5593	5.6460
Xe (nmol kg^{-1})	$S_P = 0$	0.88048	0.59859	0.43280	0.32872
	$S_P = 35$	0.65414	0.45584	0.33609	0.25897
CFC-12 (pmol kg^{-1})	$S_P = 0$	4.6927	2.7140	1.7263	1.1875
	$S_P = 35$	3.2378	1.9100	1.2255	0.84114
CFC-11 (pmol kg^{-1})	$S_P = 0$	8.6998	4.7169	2.8443	1.8733
	$S_P = 35$	5.9823	3.2919	1.9921	1.3020
SF_6 (fmol kg^{-1})	$S_P = 0$	6.1460	3.9113	2.6637	1.9258
	$S_P = 35$	4.0944	2.6607	1.8599	1.3874

Table D.4 Henry's law constants (10^3 mol kg^{-1} atm^{-1}) at a variety of temperatures and salinities. These are apparent constants calculated from the saturation concentrations in Table D.3 divided by the dry mole fraction from Table D.1 and by $(1 - p^a_{H_2O})$. These Henry's law constants have been multiplied by 10^3 before tabulating. For example, K'_{H,N_2} $(0°C, S_P = 0) = 1.0700 \times 10^{-3}$ mol kg^{-1} atm^{-1}. See www.cambridge.org/emerson-hamme for MATLAB and Python toolboxes containing functions to calculate these quantities.

		$T_{Celsius} = 0°C$	$T_{Celsius} = 10°C$	$T_{Celsius} = 20°C$	$T_{Celsius} = 30°C$
N_2	$S_P = 0$	1.0700	0.84681	0.70455	0.61208
	$S_P = 35$	0.80136	0.64919	0.55005	0.48412
O_2	$S_P = 0$	2.1951	1.7053	1.3911	1.1822
	$S_P = 35$	1.6708	1.3268	1.1017	0.94969
Ar	$S_P = 0$	2.4022	1.8706	1.5287	1.3015
	$S_P = 35$	1.8330	1.4587	1.2132	1.0468
Ne	$S_P = 0$	0.55803	0.50494	0.46890	0.44827
	$S_P = 35$	0.44603	0.40867	0.38423	0.37185
He	$S_P = 0$	0.42697	0.40676	0.39586	0.39199
	$S_P = 35$	0.34764	0.33608	0.33125	0.33153
CH_4	$S_P = 0$	2.5638	1.9354	1.5505	1.2957
	$S_P = 35$	1.9353	1.4899	1.2127	1.0256
Kr	$S_P = 0$	4.9007	3.6164	2.7987	2.2625
	$S_P = 35$	3.7174	2.7853	2.1901	1.8000
N_2O	$S_P = 0$	59.094	40.000	28.701	21.642
	$S_P = 35$	46.530	32.038	23.297	17.736
Xe	$S_P = 0$	10.182	6.9648	5.0923	3.9437
	$S_P = 35$	7.5637	5.3026	3.9527	3.1043
CFC-12	$S_P = 0$	9.4235	5.4837	3.5271	2.4740
	$S_P = 35$	6.5011	3.8582	2.5029	1.7509
CFC-11	$S_P = 0$	38.728	21.127	12.883	8.6518
	$S_P = 35$	26.628	14.741	9.0188	6.0080
SF_6	$S_P = 0$	0.62081	0.39752	0.27376	0.20181
	$S_P = 35$	0.41353	0.27036	0.19106	0.14527
DMS	$S_P = 0$	1613.1	1031.0	680.36	461.97
	$S_P = 35$	1374.1	869.65	568.44	382.50

E

Appendix E Viscosity, Diffusion Coefficients, and Schmidt Numbers

E.1 Viscosity

Viscosity is the resistance of a fluid to deformation; higher values indicate greater resistance to flow. For example, whipped cream has a greater viscosity than the cream did when it was poured from the carton. The general term is dynamic viscosity, but boundary layer turbulence or air–sea exchange most often uses kinematic viscosity.

E.1.1 Dynamic Viscosity

Dynamic viscosity, μ, has units of centipois (millipascal-second, mPa s, or g m^{-1} s^{-1}) and has been measured as a function of temperature and salinity in water and seawater. Viscosity of water, μ_w, increases as temperature decreases because water molecules move more slowly when cold, and the fluid contains more ice-like structures (Section 1.2.1, Fig. 1.4). Seawater viscosity, μ_{sw}, is greater than pure water viscosity, μ_w, because ions order polar water molecules making them more resistant to flow (Section 1.2.2, Fig. 1.5).

The dynamic viscosity of distilled water (Pa s) over a temperature range of 0 to 180°C (from Sharqawy et al., 2010):

$$\mu_w = 4.2844 \times 10^{-5} + \left(0.157(T_{Celsius} + 64.993)^2 - 91.296\right)^{-1}, \qquad \text{(E.1)}$$

where $T_{Celsius}$ is in units of °C. The dynamic viscosity of seawater (Pa s) over a temperature range of 0 to 180°C and a salinity range of 0 to 150 (from Sharqawy et al., 2010):

$$\mu_{sw} = \mu_w\left(1 + A\frac{S_P}{1000} + B\left(\frac{S_P}{1000}\right)^2\right), \qquad \text{(E.2)}$$

$$A = 1.541 + 1.998 \times 10^{-2} T_{Celsius} - 9.52 \times 10^{-5} T_{Celsius}{}^2,$$

$$B = 7.974 - 7.561 \times 10^{-2} T_{Celsius} + 4.724 \times 10^{-4} T_{Celsius}{}^2,$$

where S_P is practical salinity on the PSS-78 scale.

Table E.1 Dynamic Viscosity, μ, (10^3 kg m^{-1} s^{-1}) at a variety of temperatures and salinities. These dynamic viscosities have been multiplied by 10^3 before tabulating. For example, μ (0°C, $S_P = 0$) = 1.791×10^{-3} kg m^{-1} s^{-1}. See www.cambridge.org/emerson-hamme for MATLAB and Python toolboxes containing a function to calculate this quantity.

	$T_{Celsius} = 0°C$	$T_{Celsius} = 10°C$	$T_{Celsius} = 20°C$	$T_{Celsius} = 30°C$
$S_P = 0$	1.791	1.306	1.002	0.797
$S_P = 35$	1.906	1.397	1.077	0.861

Table E.2 Kinematic viscosity, ν, (10^6 m^2 s^{-1}) at a variety of temperatures and salinities. These kinematic viscosities have been multiplied by 10^6 before tabulating. For example, ν (0°C, $S_P = 0$) = 1.792×10^{-6} m^2 s^{-1}. See www.cambridge.org/emerson-hamme for MATLAB and Python toolboxes containing a function to calculate this quantity.

	$T_{Celsius} = 0°C$	$T_{Celsius} = 10°C$	$T_{Celsius} = 20°C$	$T_{Celsius} = 30°C$
$S_P = 0$	1.792	1.306	1.004	0.801
$S_P = 35$	1.853	1.360	1.051	0.842

E.1.2 Kinematic Viscosity

Kinematic viscosity, ν, is the ratio of dynamic viscosity, μ, and density, ρ, and has units of m^2 s^{-1} (kg m^{-1} s^{-1}/(kg m^{-3}) = m^2 s^{-1}). It is sometimes described as the molecular diffusion of momentum and represents the smallest random movements of the flow.

$$\nu = \frac{\mu}{\rho} \tag{E.3}$$

E.2 Molecular Diffusion

Transport of ions and gases at boundaries like the air–water and water–solid interfaces is frequently limited by the random walk of molecular diffusion. Diffusive fluxes and concentration gradients are related via the molecular diffusion coefficient, which has been measured for individual ions and gases in laboratory experiments.

E.2.1 Ion Molecular Diffusion Coefficients

Ion molecular diffusion coefficients in water have been determined experimentally in laboratory systems that ensure the solution medium is totally free of turbulent motion, such as a bed of glass beads or agar, or in narrow capillary tubes (Li and Gregory, 1974; Poisson and Papaud, 1983). The temperature dependence of diffusive transport of ion $_j$

is described by the Nernst equation that relates the molecular diffusion coefficient, D_j ($m^2\ s^{-1}$), to the ion equivalent conductance, λ_j,

$$D_j = \frac{RT\lambda_j}{|Z|F^2}, \tag{E.4}$$

where T is temperature (K), R is the gas constant (8.3145 J K^{-1} mol^{-1} or kg m^2 s^{-2} K^{-1} mol^{-1}), λ_j is the equivalent conductance (m^2 S mol^{-1} or kg^{-1} s^3 A^2 mol^{-1}), F is the Faraday constant (96 485 A s mol^{-1}), and $|Z|$ is the absolute value of the ion's charge.

Salinity dependence measurements of ion molecular diffusion coefficients indicate values at $S_P = 35$ that are at most 5–10 percent slower than in freshwater (Li and Gregory, 1974; Poisson and Papaud, 1983). This change is closely related to the change in viscosity, where the freshwater–seawater ratio of molecular diffusion coefficients is inversely proportional to the viscosity ratio ($D_{sw}/D_w \propto \mu_w/\mu_{sw}$).

Generally, ions with low charge to radius ratios (low ionic potential, like K^+) diffuse faster than small, highly charged ions (like Mg^{2+}) because higher charge creates a thicker hydration layer of water molecules that must diffuse along with the ion (Li and Gregory, 1974).

It has been demonstrated that the diffusion coefficient of an ion in water depends on the mobility of ions that diffuse along with it as well as its own mobility to maintain charge balance (Ben-Yaakov, 1972; Vinograd and McBain, 1941). Although the effect has important consequences in freshwater systems, it is less critical in seawater because it is a strong electrolyte. Essentially, the small gradient in charge potential created by diffusion of trace concentrations of cations and/or anions of different mobility is neutralized by other ions in great abundance (primarily Na^+ and Cl^- in seawater) so that the trace ions diffuse at a rate consistent with their own individual ion diffusion coefficients.

Table E.3 Molecular diffusion coefficients (10^5 cm^2 s^{-1}) and equivalent conductance (10^4 m^2 S mol^{-1}) of some of the major ions in seawater at a variety of temperatures. Values for λ_j are from the *Handbook of Chemistry and Physics* (1992/1993) and have been multiplied by 10^4 before tabulating. For example, λ_{H^+} (0°C, $S_P = 0$) $= 349.65 \times 10^{-4}$ m^2 S mol^{-1}. Diffusion coefficient values in the table have been reduced by 4.9 percent to account for the salinity effect and have been multiplied by 10^5 before tabulating. For example, D_{H^+} (0°C, $S_P = 0$) $= 8.11 \times 10^{-5}$ cm^2 s^{-1}. See www.cambridge.org/emerson-hamme for MATLAB and Python toolboxes containing functions to calculate these quantities.

		$T_{Celsius} =$ 0°C	$T_{Celsius} =$ 10°C	$T_{Celsius} =$ 20°C	$T_{Celsius} =$ 30°C
ion	λ_j (10^4 m^2 S mol^{-1})	D_j (10^5 cm^2 s^{-1})			
H^+	349.65	8.12	8.42	8.71	9.01
K^+	73.48	1.71	1.77	1.83	1.89
Na^+	50.08	1.16	1.21	1.25	1.29
Ca^{2+}	59.47	0.69	0.72	0.74	0.77
Mg^{2+}	53.0	0.62	0.64	0.66	0.68
OH^-	198	4.60	4.77	4.93	5.10
Cl^-	76.31	1.77	1.84	1.90	1.97
SO_4^{2-}	80.0	0.93	0.96	1.00	1.03
HCO_3^-	44.5	1.03	1.07	1.11	1.15

E.2.2 Gas Diffusion Coefficients

Molecular diffusion coefficients for gases of significant abundance (N_2, O_2, Ar, CO_2, CH_4, and the noble gases, He, Ne, Ar, Kr, and Xe) have been determined by laboratory experiments (e.g., Ferrell and Himmelblau, 1967; Jahne et al., 1987). The temperature dependence of the diffusion coefficients follows the Eyring equation (Eyring, 1936):

$$D = Ae^{-E_a/RT}, \tag{E.5}$$

where E_a is the activation energy for diffusion in water (J mol^{-1}), R is the gas constant (8.3145 J K^{-1} mol^{-1}), T is the temperature (K), and A is the preexponential factor (cm^2 s^{-1}). Values for A and E_a for different gases in water are presented in Table E.4.

Table E.4 Pre-exponential term, A (10^5 cm^2 s^{-1}), and activation energy, E_a (10^{-3} J mol^{-1}), for calculating gas diffusion coefficients (cm^2 s^{-1}) from the Eyring equation (Eq. E.5). Values are from Ferrell and Himmelblau (1967) for N_2 and O_2; Jahne et al. (1987) for CO_2, Ne, He, CH_4, Kr, Xe, and Rn; Zheng et al. (1998) for CFC-12 and CFC-11; King and Saltzman (1995) for SF6; and Saltzman et al. (1993) for DMS. Values for Ar are extrapolated from the other noble gas values of Jahne et al. (1987). Pre-exponential terms have been multiplied by 10^5 and the activation energies by 10^{-3} before tabulating. For example, $A_{N_2} = 3412 \times 10^{-5}$ cm^2 s^{-1} and $E_{a,N_2} = 18.50 \times 10^3$ J mol^{-1}.

	N_2	O_2	Ar	CO_2	Ne	He	CH_4	Kr	Xe
A	3 412	4 286	2 227	5 019	1 608	818	3 047	6 393	9 007
E_a	18.50	18.70	16.68	19.51	14.84	11.70	18.36	20.20	21.61

	CFC-12	CFC-11	SF_6	DMS	Rn
A	3 600	1 500	2 900	2 000	15 877
E_a	20.1	18.1	19.3	18.1	23.26

Table E.5 Molecular diffusion coefficients (10^5 cm^2 s^{-1}) for gases in seawater at a variety of temperatures. Values in the table have been reduced by 4.9 percent to account for the salinity effect and have been multiplied by 10^5 before tabulating. For example, $D_{N_2}(0°C, S_P = 35) = 0.94 \times 10^{-5}$ cm^2 s^{-1}. See www.cambridge.org/emerson-hamme for MATLAB and Python toolboxes containing functions to calculate these quantities.

	$T_{Celsius} = 0°C$	$T_{Celsius} = 10°C$	$T_{Celsius} = 20°C$	$T_{Celsius} = 30°C$
N_2	0.94	1.26	1.64	2.11
O_2	1.08	1.45	1.90	2.45
Ar	1.37	1.78	2.26	2.83
CO_2	0.89	1.20	1.60	2.08
Ne	2.22	2.80	3.47	4.24
He	4.51	5.41	6.41	7.50
CH_4	0.89	1.19	1.55	1.99
Kr	0.83	1.14	1.53	2.01
Xe	0.63	0.88	1.21	1.62
CFC-12	0.49	0.67	0.90	1.18
CFC-11	0.49	0.65	0.85	1.09
SF_6	0.56	0.76	1.00	1.30
DMS	0.66	0.87	1.13	1.45
Rn	0.54	0.77	1.08	1.48

The salinity dependence of the molecular diffusion coefficients of the gases H_2 and He has been determined by laboratory measurements in both freshwater and seawater (Jahne et al., 1987). They found a decrease of about 5 percent at a salinity of 35. This is close to the decrease in the molecular diffusion coefficient of ions in seawater relative to freshwater (see above), and is also similar to the increase in viscosity with salinity in Table E.1. It appears that much of the salinity dependent decrease in the molecular diffusion coefficient of both ions and gases is caused by the changes in the viscosity of the solution.

E.3 The Schmidt Number

The transfer of gas, C, between the ocean and atmosphere, $F_{S,C}$, is proportional to the concentration gradient between the value in the water, $[C]$, and that in equilibrium with the atmosphere, $[C_{eq}]$, with the mass transfer coefficient, $k_{S,C}$ (m d^{-1}), being the proportionality constant (see Chapter 2, Appendix 2A.3):

$$F_{S,C} = -k_{S,C}\{[C] - [C_{eq}]\}. \tag{2A.3.2}$$

Experimental measurements of the mass transfer coefficient, $k_{S,C}$, indicate that it is proportional to the individual gas molecular diffusion coefficient, D_C, and inversely proportional to the kinematic viscosity (or the molecular diffusion of momentum), v, of the fluid to the power n.

$$k_{S,C}\, \alpha\, (D_C/v)^n. \tag{2A.3.3}$$

In the presence of waves, n has been shown experimentally to be ~0.5. (See Appendix 2A.3.) The ratio of the kinematic viscosity and molecular diffusion coefficient is called the Schmidt number, Sc, and is unitless:

$$Sc = (v/D_C). \tag{2A.3.4}$$

Thus, the mass transfer coefficient is inversely proportional to the Schmidt number.

$$k_{S,C}\, \alpha\, (Sc)^{-0.5}. \tag{E.6}$$

This relationship is important for comparing air–sea exchange of different gases in the same fluid or the same gas at different temperatures. Because of the definitions and trends of the molecular diffusion coefficient and kinematic viscosity, the temperature dependence of the Schmidt number is greater than that for the diffusion coefficient – D increases with temperature while the kinematic viscosity decreases (Table E.6).

While the temperature dependence of the Schmidt number is easily envisioned, the salinity dependence is not as clear given available measurements. From Eq. 2A.3.4, the ratio of Schmidt numbers in seawater and freshwater is inversely proportional to the v/D ratio:

$$\frac{Sc_{sw}}{Sc_w} = \frac{v_{sw}/D_{sw}}{v_w/D_w} = \frac{v_{sw}}{v_w}\frac{D_w}{D_{sw}}. \tag{E.7}$$

Since measurements of the salinity dependence of gas diffusion coefficients indicate that they are inversely proportional to the viscosity (see above), the salinity dependence of the Schmidt number is equal to the square of the ratio of the kinematic viscosities:

Table E.6 Schmidt numbers (unitless) in seawater at a variety of temperatures. Values are determined as the ratio in Eq. 2A.3.4. See www.cambridge.org/emerson-hamme for MATLAB and Python toolboxes containing functions to calculate these quantities.

	$T_{Celsius} = 0°C$	$T_{Celsius} = 10°C$	$T_{Celsius} = 20°C$	$T_{Celsius} = 30°C$
N_2	1 969	1 083	640	400
O_2	1 711	939	553	344
Ar	1 353	766	465	297
CO_2	2 088	1 131	659	406
Ne	834	486	303	198
He	411	252	164	112
CH_4	2 073	1 143	677	423
Kr	2 221	1 190	686	419
Xe	2 933	1 538	869	520
CFC-12	3774	2026	1170	714
CFC-11	3755	2079	1236	775
SF_6	3294	1791	1046	646
DMS	2816	1560	927	582
Rn	3441	1759	970	568

$$\frac{Sc_{sw}}{Sc_w} = \left(\frac{\mu_{sw}}{\mu_w}\right)^2 \left(\frac{\rho_w}{\rho_{sw}}\right). \tag{E.8}$$

The squared viscosity ratio at 20°C is about 1.16 and the density ratio is 0.975, resulting in a 13 percent increase in the Schmidt number from freshwater to seawater. On the other hand, one could interpret the experimental data to indicate that the salinity dependence is due to the viscosity difference between freshwater and seawater with little salinity dependence on the molecular diffusion coefficient. If this is the case, the ratio of the Schmidt numbers is equal to the ratio of the kinematic viscosities alone, which is about 8 percent rather than 13 percent.

The values in Table E.6 assume the first of these interpretations and are consistent with the assumptions used currently in the literature (e.g., Wanninkhof, 1992). We will leave it there – that the salinity effect increases the Schmidt number by 8 to 13 percent and that it is currently not possible to distinguish the roles of the molecular diffusion coefficient and the kinematic viscosity.

F

Appendix F Equilibrium Constants of the Carbonate and Borate Buffer Systems

Constants in this appendix are based on the "total" pH scale, pH_T (Dickson, 1984, 1993). Values are first presented at 1 atm pressure and then equations are given for calculating the pressure effect on the equilibrium constants. A complete description of seawater carbonate chemistry requires the dissociation constants for HSO_4^-, HF, H_3PO_4, $H_2PO_4^-$, HPO_4^-, and H_4SiO_4, for which the reader is referred to Dickson et al. (2007).

F.1 Values at 0 dbar Sea Pressure (Surface Ocean)

(1) The Henry's law constant for CO_2 in seawater (mol kg^{-1} atm^{-1}) over a temperature range of 0 to 40°C and a salinity range of 0 to 40 (from Weiss, 1974); see Eq. 5.21:

$$\ln K_{H,CO_2} = -60.2409 + \frac{9345.17}{T} + 23.3585 \ln\left(\frac{T}{100}\right) + S_P\left[0.023517 - 0.00023656T + 0.0047036\left(\frac{T}{100}\right)^2\right], \quad \text{(F.1)}$$

where T is in units of K, and S_P is practical salinity on the PSS-78 scale. Unlike the values in Table D.2, this constant does not assume that fugacity and partial pressure are equal, i.e., it takes into account the non-ideal nature of CO_2.

(2) The first dissociation constant for carbonic acid in seawater (mol kg^{-1}) over a temperature range of 0 to 40°C, a salinity range of 15 to 45, and at fCO_2 values up to 500 μatm (from Lueker et al., 2000); see Eq. 5.18:

$$pK_1' = \frac{3633.86}{T} - 61.2172 + 9.67770 \ln(T) - 0.011555S_P + 0.0001152S_P^2, \quad \text{(F.2)}$$

where T is in units of K, S_P is practical salinity on the PSS-78 scale, and $pK_1' = -\log_{10}K_1'$.

(3) The second dissociation constant for carbonic acid in seawater (mol kg^{-1}) over a temperature range of 0 to 40°C, a salinity range of 15 to 45, and at fCO_2 values up to 500 μatm (from Lueker et al., 2000); see Eq. 5.16:

$$pK_2' = \frac{471.78}{T} + 25.9290 - 3.16967 \ln(T) - 0.01781S_P + 0.0001122S_P^2, \quad \text{(F.3)}$$

where T is in units of K, S_P is practical salinity on the PSS-78 scale, and $pK_2' = -\log_{10}K_2'$.

(4) The concentration of total boron in seawater (mol kg^{-1}) (from Uppström, 1974); see Eq. 5.27:

$$B_T = [B(OH)_3] + [B(OH)_4^-] = 4.157 \times 10^{-4} \times \left(\frac{S_P}{35}\right), \tag{F.4}$$

where S_P is practical salinity on the PSS-78 scale.

(5) The dissociation constant for boric acid in seawater (mol kg^{-1}) over a temperature range of 0 to 46°C and a salinity range of 5 to 45 (from Dickson, 1990); see Eq. 5.26:

$$\begin{aligned}\ln K_B = {} & \frac{-8966.90 - 2890.53\, S_P^{0.5} - 77.942\, S_P + 1.728\, S_P^{1.5} - 0.0996\, S_P^2}{T} \\ & + 148.0248 + 137.1942\, S_P^{0.5} + 1.62142\, S_P \\ & - \left(24.4344 + 25.085\, S_P^{0.5} + 0.2474\, S_P\right) \ln(T) + 0.053105\, S_P^{0.5} T\end{aligned} \tag{F.5}$$

where T is in units of K and S_P is practical salinity on the PSS-78 scale.

(6) The dissociation constant of water (mol^2 kg^{-2}) over a temperature range of 0 to 45°C and a salinity range of 0 to 45 (from Millero, 1995; as reported in Dickson et al., 2007):

$$\begin{aligned}\ln K_W = {} & 148.9652 - \frac{13847.26}{T} - 23.6521 \ln(T) \\ & + \left(\frac{118.67}{T} - 5.977 + 1.0495 \ln(T)\right) S_P^{0.5} - 0.01615\, S_P,\end{aligned} \tag{F.6}$$

where T is in units of K and S_P is practical salinity on the PSS-78 scale.

Table F.1 Henry's law constant for CO_2 in seawater (10^3 mol kg^{-1} atm^{-1}) at several temperatures and salinities. These Henry's law constants have been multiplied by 10^3 before tabulating. For example, $K'_{H,CO_2}(0°C, S_P = 0) = 77.58 \times 10^{-3}$ mol kg^{-1}atm^{-1}. See www.cambridge.org/emerson-hamme for MATLAB and Python toolboxes containing a function to calculate this quantity.

	$T_{Celsius} = 0°C$	$T_{Celsius} = 10°C$	$T_{Celsius} = 20°C$	$T_{Celsius} = 30°C$
$S_P = 0$	77.58	53.67	39.16	29.95
$S_P = 35$	62.87	43.88	32.41	25.17

Table F.2 The first dissociation constant for carbonic acid in seawater (mol kg^{-1}) at several temperatures and salinities. See www.cambridge.org/emerson-hamme for MATLAB and Python toolboxes containing a function to calculate this quantity.

	$T_{Celsius} = 0°C$	$T_{Celsius} = 10°C$	$T_{Celsius} = 20°C$	$T_{Celsius} = 30°C$
$S_P = 0$	4.184×10^{-7}	5.539×10^{-7}	7.004×10^{-7}	8.504×10^{-7}
$S_P = 35$	7.671×10^{-7}	1.016×10^{-6}	1.284×10^{-6}	1.559×10^{-6}

Table F.3 The second dissociation constant for carbonic acid in seawater (mol kg^{-1}) at several temperatures and salinities. See www.cambridge.org/emerson-hamme for MATLAB and Python toolboxes containing a function to calculate this quantity.

	$T_{Celsius} = 0°C$	$T_{Celsius} = 10°C$	$T_{Celsius} = 20°C$	$T_{Celsius} = 30°C$
$S_P = 0$	1.336×10^{-10}	1.998×10^{-10}	2.935×10^{-10}	4.236×10^{-10}
$S_P = 35$	4.089×10^{-10}	6.118×10^{-10}	8.984×10^{-10}	1.297×10^{-9}

Table F.4 The dissociation constant for boric acid in seawater (mol kg^{-1}) at several temperatures and salinities. See www.cambridge.org/emerson-hamme for MATLAB and Python toolboxes containing a function to calculate this quantity.

	$T_{Celsius} = 0°C$	$T_{Celsius} = 10°C$	$T_{Celsius} = 20°C$	$T_{Celsius} = 30°C$
$S_P = 0$	3.144×10^{-10}	4.163×10^{-10}	5.251×10^{-10}	6.346×10^{-10}
$S_P = 35$	1.222×10^{-9}	1.663×10^{-9}	2.207×10^{-9}	2.884×10^{-9}

Table F.5 The dissociation constant for water (mol^2 kg^{-2}) at several temperatures and salinities. See www.cambridge.org/emerson-hamme for MATLAB and Python toolboxes containing a function to calculate this quantity.

	$T_{Celsius} = 0°C$	$T_{Celsius} = 10°C$	$T_{Celsius} = 20°C$	$T_{Celsius} = 30°C$
$S_P = 0$	1.128×10^{-15}	2.888×10^{-15}	6.738×10^{-15}	1.448×10^{-14}
$S_P = 35$	4.940×10^{-15}	1.444×10^{-14}	3.840×10^{-14}	9.388×10^{-14}

F.2 Adjusting for Non-zero Sea Pressure

The effect on equilibrium constants of pressure beneath the surface can be calculated from the changes in molal volume (ΔV) and compressibility (Δκ) for any given reaction via

$$\ln \frac{K_P}{K_0} = -\left(\frac{\Delta V}{RT}\right)P + \left(\frac{0.5\Delta\kappa}{RT}\right)P^2, \tag{F.7}$$

where K_P and K_0 are the equilibrium constants for the reaction of interest at sea pressure P and at 0 (surface ocean), respectively, P is pressure in units of bars, R is the gas constant and equal to 83.1446 cm^3 bar^{-1} mol^{-1} K^{-1}, and T is in units of K. The molal volume (cm^3 mol) and compressibility has been fit to equations of the form:

$$\Delta V = a_0 + a_1 T_{Celsius} + a_2 T_{Celsius}^2 \tag{F.8}$$

$$\Delta\kappa = b_0 + b_1 T_{Celsius} \tag{F.9}$$

where $T_{Celsius}$ is now temperature in degrees Celsius, and $S_P = 35$. Values for the coefficients a and b are presented in Table F.6, while calculated differences in K' at two different pressures are presented in Table F.7.

Table F.6 Constants for use in calculating the pressure correction on dissociation constants for carbonic acid (first and second), boric acid, and water (from Millero, 1995 with corrections from CO2SYS documentation, Lewis and Wallace, 1998)

Constant	a_0	a_1	a_2	b_0	b_1
K'_1	–25.50	0.1271	0.0	-3.08×10^{-3}	8.77×10^{-5}
K'_2	–15.82	–0.0219	0.0	1.13×10^{-3}	-1.475×10^{-4}
K'_B	–29.48	0.1622	-2.608×10^{-3}	-2.84×10^{-3}	0.0
K'_W	–25.60	0.2324	-3.6246×10^{-3}	-5.13×10^{-3}	7.94×10^{-5}

Table F.7 Comparison of dissociation constants at 0 and 300 bar (~2960 m depth). K' values for each dissociation are shown at $T_{Celsius} = 2°C$, $S_P = 35$, and at $P = 0$ and 300 bar, respectively. The final ratio is that of the K' value at 300 bar divided by that at 0 bar (both at $T_{Celsius} = 2°C$, $S_P = 35$).

	K' at $P = 0$ bar	K' at $P = 300$ bar	$K'_{(P=300)}/K'_{(P=0)}$
First dissociation of carbonic acid (K'_1)	8.146×10^{-7}	1.128×10^{-6}	1.385
Second dissociation of carbonic acid (K'_2)	4.439×10^{-10}	5.475×10^{-10}	1.233
Dissociation of boric acid (K'_B)	1.303×10^{-9}	1.899×10^{-9}	1.458
Dissociation of water (K'_W)	6.171×10^{-15}	8.498×10^{-15}	1.377

G

Appendix G Apparent Solubility Products of Calcite and Aragonite

(1) The apparent solubility product $(\mathrm{p}K'_{SP})$ for calcite (mol^2 kg^{-2}) over a temperature range of 5 to 40°C and a salinity range of 5 to 44 (from Mucci, 1983); see Eq. 5.41:

$$\begin{aligned}\mathrm{p}K'_{SP,calcite} &= 171.9065 + 0.077993\,T - \frac{2839.319}{T} - 71.595\log_{10}(T)\\ &+ \left(0.77712 - 0.0028426\,T - \frac{178.34}{T}\right)S_P^{0.5} + 0.07711\,S_P\\ &- 0.0041249\,S_P^{1.5}\end{aligned} \tag{G.1}$$

where T is in units of K, S_P is practical salinity on the PSS-78 scale, and $\mathrm{p}K'_{SP,calcite} = -\log_{10}K'_{SP,calcite}$.

(2) The solubility product for aragonite (mol^2 kg^{-2}) over a temperature range of 5 to 40°C and a salinity range of 5 to 44 (from Mucci, 1983); see Eq. 5.41:

$$\begin{aligned}\mathrm{p}K'_{SP,aragonite} &= 171.945 + 0.077993\,T - \frac{2903.293}{T} - 71.595\log_{10}(T)\\ &+ \left(0.068393 - 0.0017276\,T - \frac{88.135}{T}\right)S_P^{0.5} + 0.10018\,S_P\\ &- 0.0059415\,S_P^{1.5}\end{aligned} \tag{G.2}$$

where T is in units of K, S_P is practical salinity on the PSS-78 scale, and $\mathrm{p}K'_{SP,aragonite} = -\log_{10}K'_{SP,aragonite}$.

(3) Adjusting the apparent solubility products for non-zero pressures can be done following the same procedure laid out in Appendix F.2 for dissociation constants. The values in Table G.3 should be used to calculate ΔV and $\Delta\kappa$, which can then be used to determine the ratio of K'_{SP} at some pressure to that at the sea surface.

Table G.1 The apparent solubility product for calcite (mol^2 kg^{-2}) at several temperatures and pressures ($S_P = 35$ for all values). See www.cambridge.org/emerson-hamme for MATLAB and Python toolboxes containing a function to calculate this quantity.

	$T_{Celsius} = 0°C$	$T_{Celsius} = 10°C$	$T_{Celsius} = 20°C$	$T_{Celsius} = 30°C$
$P = 0$ dbar	4.291×10^{-7}	4.317×10^{-7}	4.300×10^{-7}	4.231×10^{-7}
$P = 3\,000$ dbar	7.982×10^{-7}	7.396×10^{-7}	6.822×10^{-7}	6.248×10^{-7}

Table G.2 The apparent solubility product for aragonite ($mol^2\ kg^{-2}$) at several temperatures and pressures ($S_P = 35$ for all values). See www.cambridge.org/emerson-hamme for MATLAB and Python toolboxes containing a function to calculate this quantity.

	$T_{Celsius} = 0°C$	$T_{Celsius} = 10°C$	$T_{Celsius} = 20°C$	$T_{Celsius} = 30°C$
$P = 0$ dbar	6.828×10^{-7}	6.788×10^{-7}	6.616×10^{-7}	6.317×10^{-7}
$P = 3\,000$ dbar	1.224×10^{-6}	1.122×10^{-6}	1.014×10^{-6}	9.022×10^{-7}

Table G.3 Constants for use in calculating the pressure correction on calcite and aragonite solubility products (from Millero, 1979)

Constant	a_0	a_1	b_0	b_1
$pK'_{SP,calcite}$	−48.76	0.5304	-1.176×10^{-2}	3.692×10^{-4}
$pK'_{SP,aragonite}$	−45.96	0.5304	-1.176×10^{-2}	3.692×10^{-4}

Appendix References

Ben-Yaakov, S. (1972) Diffusion of sea water ions-I. Diffusion of sea water into a dilute solution, *Geochimica et Cosmochimica Acta*, **36**, 1395–1406.

Bullister, J. L., D. P. Wisegarver, and F. A. Menzia (2002) The solubility of sulfur hexafluoride in water and seawater, *Deep-Sea Research I*, **49**, 175–187.

Charette, M. A., and W. H. F. Smith (2010) The volume of Earth's ocean, *Oceanography*, **23**, 112–114, doi: 10.5670/oceanog.2010.51.

CRC Handbook of Chemistry and Physics (1992/93) (ed. D. E. Lide), V. 73. Cleveland: CRC Press.

Dacey, J. W. H., S. G. Wakeham, and B. L. Howes (1984) Henry's law constants for dimethylsulfide in freshwater and seawater, *Geophysical Research Letters*, **11**, 991–994.

Dai, A., and K. E. Trenberth (2002) Estimates of freshwater discharge from continents: latitudinal and seasonal variations, *Journal of Hydrometeorology*, **3**, 660–687.

Dickson, A. G. (1984) pH scales and proton-transfer reactions in saline media such as sea water, *Geochimica et Cosmochimica Acta*, **48**, 2299–2308, doi:10.1016/0016-7037(84)90225-4.

Dickson, A. G. (1990) Thermodynamics of the dissociation of boric acid in synthetic seawater from 273.15 to 318.15 K, *Deep-Sea Research A*, **37**, 755–766.

Dickson, A. G. (1993) The measurement of sea water pH, *Marine Chemistry*, **44**, 131–142.

Dickson, A. G., C. L. Sabine, and J. R. Christian (eds.) (2007) *Guide to Best Practices for Ocean CO_2 Measurements*, PICES Special Publication 3, 191 pp. Sidney: North Pacific Marine Science Organization.

Dymond, J. H., and E. B. Smith (1980) *The Virial Coefficients of Pure Gases and Mixtures: A Critical Compilation*, 518 pp. Oxford: Oxford University Press.

Eyring, H. (1936) Viscosity, plasticity, and diffusion as examples of absolute reaction rates, *Journal of Chemical Physics*, **4**, 283–291.

Ferrell, R. T., and D. M. Himmelblau (1967) Diffusion coefficients of nitrogen and oxygen in water, *Journal of Chemical and Engineering Data*, **12**, 111–115.

Garcia, H. E., and L. I. Gordon (1992) Oxygen solubility in seawater: better fitting equations, *Limnology and Oceanography*, **37**, 1307–1312, doi:10.4319/lo.1992.37.6.1307.

Garcia, H. E., and L. I. Gordon (1993) Erratum: oxygen solubility in seawater: better fitting equations, *Limnology and Oceanography*, **38**, 656.

Glueckauf, E. (1951) The composition of atmospheric air, in *Compendium of Meteorology*, pp. 3–10. Boston: American Meteorological Society.

Hamme, R. C., and S. R. Emerson (2004) The solubility of neon, nitrogen and argon in distilled water and seawater, *Deep Sea Research I*, **51**, 1517–1528, doi:10.1016/j.dsr.2004.06.009.

Hofmann, A. F., E. T. Peltzer, P. M. Walz, P. G. Brewer (2011) Hypoxia by degrees: establishing definitions for a changing ocean, *Deep-Sea Research I*, **58**, 1212–1226, doi:10.1016/j.dsr.2011.09.004.

Jähne, B., G. Heinz, and W. Dietrich (1987) Measurements of the diffusion coefficients of sparingly soluble gases in water, *Journal of Geophysical Research – Oceans*, **92**, 10767–10776.

Jenkins, W. J., D. E. Lott III, and K. L. Cahill (2019) A determination of atmospheric helium, neon, argon, krypton, and xenon solubility concentrations in water and seawater, *Marine Chemistry*, **211**, 94–107.

Kara, A. B., P. A. Rochford, and H. E. Hurlburt (2003) Mixed layer depth variability over the global ocean, *Journal of Geophysical Research – Oceans*, **108**, 3079, doi:10.1029/2000JC000736.

King, D. B., and E. S. Saltzman (1995) Measurement of the diffusion coefficient of sulfur hexafluoride in water, *Journal of Geophysical Research – Oceans*, **100**, 7083–7088.

Large, W. G., and S. Pond (1981) Open ocean momentum flux measurements in moderate to strong winds, *Journal of Physical Oceanography*, **11**, 324–336.

Lewis, E. R., and D. W. R. Wallace (1998) Program developed for CO2 system calculations, ORNL/CDIAC – 105. Oak Ridge: Carbon Dioxide Information Analysis Center, Oak Ridge National Laboratory, U.S. Department of Energy, Oak Ridge, Tenn.

Li, Y.-H., and S. Gregory (1974) Diffusion of ions in sea water and in deep-sea sediments, *Geochimica et Cosmochimica Acta*, **38**, 703–714.

Lueker, T. J., A. G. Dickson, and C. D. Keeling (2000) Ocean pCO_2 calculated from dissolved inorganic carbon, alkalinity, and equations for K_1 and K_2: validation based on laboratory measurements of CO_2 in gas and seawater at equilibrium, *Marine Chemistry*, **70**, 105–119.

Millero, F. J. (1974) Seawater as a multicomponent electrolyte solution, in *The Sea*, Vol. 5 (ed. E. D. Goldberg). New York: John Wiley and Sons.

Millero, F. J. (1979) The thermodynamics of the carbonate system in seawater, *Geochimica et Cosmochimica Acta*, **43**, 1651–1661.

Millero, F. J. (1995) Thermodynamics of the carbon dioxide system in the oceans, *Geochimica et Cosmochimica Acta*, **59**, 661–677.

Mucci, A. (1983) The solubility of calcite and aragonite in seawater at various salinities, temperatures, and one atmosphere total pressure, *American Journal of Science*, **283**, 780–799.

Pilson, M. E. Q. (2013) *An Introduction to the Chemistry of the Sea*, 515 pp. Cambridge: Cambridge University Press.

Poisson, A., and A. Papaud (1983) Diffusion coefficients of major ions in seawater, *Marine Chemistry*, **13**, 265–280.

Saltzman, E. S., D. B. King, K. Holmen, and C. Leck (1993) Experimental determination of the diffusion coefficient of dimethylsulfide in water, *Journal of Geophysical Research – Oceans*, **98**, 16481–16486.

Sarmiento, J. L., and N. Gruber (2006) *Ocean Biogeochemical Dynamics*, 503 pp. Princeton: Princeton University Press.

Sharqawy, M. H., J. H. Lienhard, and S. M. Zubair (2010) The thermophysical properties of seawater: a review of existing correlations and data, *Desalination and Water Treatment*, **16**, 354–380.

Uppström, L. R. (1974) Boron/chlorinity ratio of deep-sea water from the Pacific Ocean, *Deep-Sea Research*, **21**, 161–162.

Vinograd, J. R., and J. W. McBain (1941) Diffusion of electrolytes and of the ions in their mixtures, *Journal of the American Chemical Society*, **63**, 2008–2015.

Wagner, W., and A. Pruß (2002) The IAPWS formulation 1995 for the thermodynamic properties of ordinary water substance for general and scientific use, *Journal of Physical and Chemical Reference Data*, **31**, 387–535.

Wanninkhof, R. (1992) Relationship between wind speed and gas exchange over the ocean, *Journal of Geophysical Research – Oceans*, **97**, 7373–7382.

Warner, M. J., and R. F. Weiss (1985) Solubilities of chlorofluorocarbons 11 and 12 in water and seawater, *Deep-Sea Research A*, **32**, 1485–1497.

Weiss, R. F. (1974) Carbon dioxide in water and seawater: the solubility of a non-ideal gas, *Marine Chemistry*, **2**, 203–215.

Weiss, R. F., and T. K. Kyser (1978) Solubility of krypton in water and seawater, *Journal of Chemical and Engineering Data*, **23**, 69–72.

Weiss, R. F., and B. A. Price (1980) Nitrous oxide solubility in water and seawater, *Marine Chemistry*, **8**, 347–359.

Wiesenburg, D. A., and N. L. Guinasso, Jr. (1979) Equilibrium solubilities of methane, carbon monoxide, and hydrogen in water and sea water, *Journal of Chemical and Engineering Data*, **24**, 356–360.

Woodhouse, M. T., K. S. Carslaw, G. W. Mann, et al. (2010) Low sensitivity of cloud condensation nuclei to changes in the sea-air flux of dimethyl-sulphide, *Atmospheric Chemistry and Physics*, **10**, 7545–7559, doi:10.5194/acp-10-7545-2010.

Zheng, M., W. J. De Bruyn, and E. S. Saltzman (1998) Measurements of the diffusion coefficients of CFC-11 and CFC-12 in pure water and seawater, *Journal of Geophysical Research – Oceans*, **103**, 1375–1379.

Index

acidification, ocean, 214–216
acids *see also* alkalinity
  alkalinity of seawater, 187–191
  chemical equilibrium constant, 178–180
  hydrogen ion exchange, 180–182
  in seawater, 182–186
age, of ocean water, 159–160, 280–284
air–sea chemical equilibrium, 69–73
air–sea gas exchange, 68, 301–303, 335
alkalinity, 10, 55, 191, 319
  changes within oceans, 195–203
  control processes, 194–203
  seawater, 187–191
annual net community production (ANCP), 39, 114, 129–132, 345, *see also* biological pump
Antarctic Bottom Water (AABW), 17, 33, 152, 160
Antarctic Intermediate Water (AAIW), 17, 33, 152
anthropogenic influences
  biological fluxes, 133
  biological respiration, 170
  element classification, 17–22
  global carbon cycle, 329–335
  isotope tracers, 263–265
  marine carbonate chemistry, 214–216
  radioisotope tracers, 304–308
apparent oxygen utilization (AOU), 147–149
authigenesis, 45
authigenic minerals, 45, 77

bacteria, 35
benthic respiration, 162–166, *see also* biological respiration rates:below euphoric zone
beryllium-7, 285–286
biological carbon pump, 38
biological pump, 39, 113–114
  methods of determining ANCP, 128–129
  O_2/Ar and O_2/N_2 tracers, 126–128
  particle fluxes and thorium isotope tracers, 119–121
  relationship to OUR, 160–162
  upper ocean metabolite mass balance, 121–126
  whole-ocean model, 114–118
biological respiration rates, 144, 165
  below euphoric zone, 144–147, 155–156
    benthic respiration, 162–166
    in absence of oxygen, 166–169
    interaction and age, 159–160
    OUR, 156–158
    OUR and biological pump, 160–162
    oxygen and AOU, 147–149
    POC and DOC, 162
boric acid, 185
boron isotopes, 243–249
bubbles, 87–90
Bunsen coefficient, 71

calcium carbonate
  mechanisms of dissolution and burial, 205–206
    kinetics of, 211–214
    thermodynamic equilibrium, 206–210
carbon cycle, global, 313–314
  anthropogenic influence, 330–350
    atmospheric observations, 330–331
    future challenges, 350
    measuring uptake, 338–345
  natural ocean processes, 317–321
    biological pump, 321–324
    residence time in air–sea gas exchange, 335–338
    solubility pump, 321
  past changes in carbon dioxide, 324–326
    carbonate compensation, 327–329
    three-box ocean and atmosphere model, 326–327
  reservoirs and fluxes, 314–317
carbon dioxide (CO_2)
  air–sea flux, 341
  atmospheric, 324–326, 329–331
  in oceans, 331–335
  three-box ocean and atmospheric model, 326–327
carbon export, organic
  ANCP and model predictions, 129
  anthropopgenic influence, 129–132
carbon reservoirs, 350
carbon-14, natural, 278–285
carbonate compensation, 327–329
carbonate compensation depth (CCD), 209, 211
carbonic acid, 183
chemical perspective, 1–3
chlorofluorocarbon (CFC), 157
chlorophyll, 37
circulation. *see* ocean circulation
Circumpolar Deep Water (CDW), 152, 299
clay minerals, 76

climate change, 284–285, 298
  chemical flows, 55
  surface oceans, 132–133
Climate Variability and Predictability (CLIVAR), 2
clumped isotopes, 235, 240–241
CO_2 uptake, 338–343
coastal sediment accumulation, 303–304
conservative elements, 18
conveyor belt circulation, 195
Coriolis force, 26, 28
Cross-Calibrated Multi-Platform (CCMP) winds, 343

dead zones, coastal, 170
decay mechanisms *see* radioisotope tracers
deep-oceans, 32, 63, 118, 282, 293, 304
  global carbon cycle, 326
deforestation, 330, *see also* greening
denitrification, 75, 103, 116, 153–154, 255–257
density of seawater, 10–17
diffusive boundary layer, 81
dissolved inorganic carbon (DIC), 195–203, 319
  anthropogenic component, 339–341
dissolved inorganic nitrogen (DIN), 100
dissolved inorganic phosphorus (DIP), 100, 319
dissolved organic carbon (DOC), 99, 162
dissolved organic matter (DOM), 99
Dole Effect, 252
downwelling, 28–29

ecosystems, 162
Ekman transport, 26, 28
electrostriction, 7, 9
elements, biologically-produced, 113–129
energy production, 3
equilibrium isotope fractionation, 231–241

fluxes, biological, 133
fossil fuels, 22, 133, 214
  global carbon cycle, 314, 324, 329–330, 346, 350
fractionation, isotopic, 231

gas exchange
  air–sea, 80
  flux equations, 80–83
  rates in oceans, 84–87
gas transfer, bubbles, 87–90
gases, 20, 68
  Henry's law solubility, 69–73
  sources and sinks, 74–75
geochemical mass balance, 44–45
  gases, 68–75
  ion mass balance, 45
  Mackenzie and Garrels mass balance, 50–56
  residence times, 47–50
  sinks, 56–68
  source, 45–48

Geochemical Ocean Sections Study (GEOSECS), 2, 178
GEOTRACES, 2, 263
global warming, 5
  deep oceans, 162, 170
  global carbon cycle, 321, 324
  life in surface oceans, 133
GO-SHIP, 2
greening, 330, 348
gross oxygen production (GOP), 250
gross primary production (GPP), 39
gyre circulation, 28–29

Henry's law solubility, 69–73
hydrothermal circulation, 56–65

ice cores, 325
igneous rocks, 76
interglacial times. *see* temperature changes, past
Intergovernmental Oceanographic Commission (IOC), 11
International Association of the Physical Sciences of the Oceans (IAPSO), 11
ion mass balance *see* weathering
ions, seawater, 7–10
isotope tracers, stable, 226–228
  analytical methods and terminology, 229–231
  anthropogenic influences, 263–265
  equilibrium fractionation, 231–243
  kinetic fractionation, 244–245
    biological processes, 246–249
    photosynthesis, 249–251
    respiration, 251–258
  Rayleigh fractionation, 259–263

Joint Global Ocean Flux Study (JGOFS), 2, 178, 340

kinetic isotope fractionation, 244–245
  biological processes, 246–249
  marine nitrogen cycle, 255–258
  photosynthesis, 249–251
  respiration, 251–255

life, chemistry of
  main elements, 106–107
  redox processes, 97–99
  trace elements, 107–113
lysocline, 209, 211

Mackenzie and Garrels mass balance, 50–56
marine carbonate chemistry, 177–178, 218–219
  acids and bases, 191
    alkalinity of seawater, 187–191
    chemical equilibrium constant, 178–180
    hydrogen ion exchange, 180–182
    seawater, 182–186
  anthropogenic influences, 214–216

marine carbonate chemistry (cont.)
calcium carbonate dissolution and burial, 205–206
kinetics of $CaCO_3$ dissolution, 211–214
thermodynamic equilibrium, 206–210
calculating carbonate equilibria and pH, 191–194, 204
processes controlling alkalinity and DIC of seawater
alkalinity and DIC changes in oceans, 195–203
terrestrial weathering and river inflows, 194–195
marine metabolism, 37–39
minerals, authigenic, 45–47, 77–78
minerals, clay, 76
mining, 3, *see also* fossil fuels
model predictions, 129–132

net community production (NCP), 39, 114, *see also* biological pump
net primary production (NPP), 39
nitrogen cycles, 152
nitrogen fixation, 107, 153–154, 255, 258
North Atlantic Deep Water (NADW), 17, 33, 152, 160
flow rates, 298–301
nuclear releases, 304–305, *see also* weapons, nuclear
nutrients, 149–152, 319

ocean acidification, 214–216
ocean biology, 35
marine metabolism, 37–39
types of plankton, 35–37
ocean circulation, 25
interior circulation, 30–34
seasonality, 25
wind-driven circulation, 25–30
Ocean Drilling Project (ODP), 239
ocean particle dynamics, 294–301
ocean processes, natural, 317–321, *see also* biological pump
biological pump, 321–324
solubility pump, 321
Ostwald Solubility Coefficient, 71
oxidation-reduction reactions (redox), 97, 166–169
oxygen cycle, 345–350
oxygen utilization rate (OUR), 156–158
relationship to biological pump, 160–162
oxygen, respiration in the absence of, 166–169

Pacific Decadal Oscillation (PDO), 133
Pacific Deep Water (PDW), 32
Paleocene-Eocene Thermal Maximum (PETM), 243, 325
particle flux
surface ocean, 295
whole-ocean, 297
particulate organic carbon (POC), 99, 162
particulate organic matter (POM), 99
PeeDee Belemnite (PDB) standard, 229
phosphorus (cycles), 102, 152–154
photosynthesis, 35
phytoplankton, 36
plankton, 35–37
production rates
gross primary, 135–136
net primary, 134–135
production, biological *see* surface oceans
pycnocline, 25

radioisotope tracers, 274–275
anthropogenic influences, 304–308
atmospheric spallation, 278, 287
beryllium-7, 285–286
natural carbon-14, 278–285
decay mechanisms and equations, 275–278
uranium and thorium decay series, 288–289
air–sea gas exchange, 303–310
coastal sediment accumulation, 303–304
geochemistry of, 291–294
ocean particle dynamics, 294–301
secular euilibrium, 289–291
Rayleigh distillation model, 252
Rayleigh fractionation, 259–262
releases, intentional nuclear, 305
reservoirs, carbon, 54, 99, 264, 280, 314–317, *see also* Revelle factor
residence times, 47–50, 78–79, 335–338
respiration, 35
Revelle factor, 331–335
reverse weathering, 56, 65–68
river fluxes, 45–48
river inflows, 45, 194–195
rocks, igneous, 76

salinity, 10–17
satellites, 28, 37–38, 343
saturation horizon (SH), 209
Scientific Committee on Ocean Research (SCOR), 11
sea levels, 28
seasonality, 25
Sea-viewing Wide Field-of-view Sensor (SeaWiFS), 133
seawater
acids and bases, 182–186
anthropogenic influences, 22
element classification, 17–22
ions in, 7–10
residence times, 47–50
salinity and density, 10–17
units, 9
water in, 3–7
secular equilibrium, 289
sediment accumulation rates, ocean, 284–285
sediment, coastal, 303–304
sinks, 74–75
hydrothermal circulation, 56–65
reverse weathering, 65–68

solubility pumps, 321
solubility, gas, 72
spallation, atmospheric, 278
  beryllium-7, 285–286
  natural carbon-14, 278–285
Standard Free Energy, 166
Strangelove Ocean, 323
Surface Ocean CO_2 Atlas version 2 (SOCAT v2), 343
surface oceans, 96–97
  biologically-produced elements, 113–114
  main elements in organic matter, 99–106
  organic carbon export, 129–133
  redox processes, 97–99
  trace elements in organic matter, 107–112
Surface Renewal Model, 83
surface waters, disequilibrium in, 303–310
Sverdrup balance (transport), 28, 30

temperature changes, past, 236
temperature, global, 22
temperatures, high *see also* hydrothermal circulation
  sinks, 56–61
thermocline ventilation, 31
Thermodynamic Equation of Seawater 2010 (TEOS-10), 11
thermodynamic equilibrium, 206–210
thermohaline circulation, 32
thorium decay series, 288–289
  air–sea gas exchange, 303–310
  coastal sediment accumulation, 303–304
  geochemistry of, 291–294
  ocean particle dynamics, 294–301
  secular equilibrium, 289–291
three-box ocean and atmosphere model, 326–327
tritium (^3H), 308

upwelling, 28, 30
uranium decay series, 288–289
  air–sea gas exchange, 301–303
  coastal sediment accumulation, 303–304
  geochemistry of, 291–294
  ocean particle dynamics, 294–301
  secular equilibrium, 289–291

ventilation, thermocline, 31
Vienna Standard Mean Ocean Water (VSMOW), 229

water
  in seawater, 3–7
water flows
  hydrothermal systems, 61–63
weapons, nuclear, 3, 304
weathering, 45, *see also* minerals, authigenic
  terrestrial, 194–195
wind-driven circulation, 25–30
World Ocean Circulation Experiment (WOCE), 2, 178, 305, 340

zooplankton, 37

Printed in the United States
by Baker & Taylor Publisher Services